Protein Sequencing Protocols .

METHODS IN MOLECULAR BIOLOGY™

John M. Walker, SERIES EDITOR

METHODS IN MOLECULAR BIOLOGY™

Protein Sequencing Protocols

Second Edition

Edited by

Bryan John Smith

Celltech Chiroscience,
Slough, UK

Humana Press ✳ Totowa, New Jersey

Cover Illustration: Foreground: Diagramatic protein sequence. Illustration provided by Bryan John Smith. Background: PVDF Western Electroblot Transfer of 300 µg human heart proteins. See Fig. 1 on page 46. Illustration provided by Michael J. Dunn.

Cover design by Patricia F. Cleary.

Production Editor: Mark J. Breaugh.

For additional copies, pricing for bulk purchases, and/or information about other Humana titles, contact Humana at the above address or at any of the following numbers: Tel.: 973-256-1699; Fax: 973-256-8341; E-mail: humana@humanapr.com or visit our Website: http://humanapress.com

Printed in the United States of America. 10 9 8 7 6 5 4 3 2 1

Library of Congress Cataloging in Publication Data

Protein sequencing protocols / edited by Bryan John Smith.-- 2nd ed.
    p. cm. -- (Methods in molecular biology ; v. 211)
  Includes bibliographical references and index.
  ISBN 0-89603-975-7 (alk. paper)
  1. Nucleotide sequence--Laboratory manuals. I. Smith, Bryan John. II. Methods in molecular biology (Clifton, N.J.) ; v. 211.

QP625.N89 P76 2002
572.8'633-dc21
2002022740

Preface

Determination of the protein sequence is as important today as it was a half century ago, even though the techniques and purposes have changed over time. Mass spectrometry has continued its recent rapid development to find notable application in the characterization of small amounts of protein, for example, in the field of proteomics. The "traditional" chemical N-terminal sequencing is still of great value in quality assurance of the increasing number of biopharmaceuticals that are to be found in the clinic, checking processing events of recombinant proteins, and so on. It is joined in the armory of methods of protein analysis by such techniques as C-terminal sequencing and amino acid analysis. These methods are continually developing. The first edition of *Protein Sequencing Protocols* was a "snapshot" of methods in use in protein biochemistry laboratories at the time, and this, the second edition, is likewise. Methods have evolved in the intervening period, and the content of this book has similarly changed, the content of some chapters having been superceded and replaced by other approaches. Thus, in this edition, there is inclusion of approaches to validation of methods for quality assurance work, reflecting the current importance of biopharmaceuticals, and also a guide to further analysis of protein sequence information, acknowledging the importance of bioinformatics. Some of the areas are also the subjects of other volumes in the *Methods in Molecular Biology*™ series published by Humana, and the interested reader is directed to such volumes as **59** (*Protein Purification Protocols*), **76** (*Glycoanalysis*), **112** (*2-D Proteome Analysis Protocols*), **143** (*Protein Structure Prediction*), **146** (*Mass Spectrometry of Proteins and Peptides*), and **159** (*Amino Acid Analysis Protocols*).

The style of *Protein Sequencing Protocols* is like that of others in the series, that of a laboratory manual. The aim is to permit ready replication of the methods. It is recognized, however, that various of the techniques require expensive equipment. This can be a barrier to those without such facilities, but who might wish to collaborate with those who do have them. This volume is intended to indicate to those workers what is required by way of sample preparation, and what can be achieved by the techniques, and so be an aid to collaboration. Appendices are provided as handy references to the molecular weights of amino acids and their derivatives, and to consensus sequences.

I would like to thank all the authors for their expert contributions to this volume—their efforts have made the book what it is. I would also like to thank Stef, for her continued support.

Bryan John Smith

Contents

Contents

Contributors

ALASTAIR AITKEN • *Department of Biomedical Sciences, University of Edinburgh, Edinburgh, UK*

NEBOJSA AVDALOVIC • *Dionex Corporation, Sunnyvale, CA*

PEER BORK • *European Molecular Biology Laboratory, Heidelberg, Germany*

FRANCESCO BOSSA • *Dipartimento Scienze Biomediche, Università La Sapienza, Rome, Italy*

ALEX F. CARNE • *Institute of Cancer Research, London, UK*

JUN CHENG • *Dionex Corporation, Sunnyvale, CA*

STEVEN A. COHEN • *Waters Corporation, Milford, MA*

RICHARD G. COOK • *Protein Chemistry Core Lab, Baylor College of Medicine, Houston, TX*

RICHARD R. COPLEY • *EMBL, Heidelberg, Germany*

MARK W. CRANKSHAW • *Department of Molecular Biology and Pharmacology, Washington University School of Medicine, St. Louis, MO*

IAN DAVIDSON • *Department of Molecular & Cell Biology, University of Aberdeen, UK*

NEIL DODSWORTH • *Delta Biotechnology, Nottingham, UK*

BRYAN DUNBAR • *Department of Molecular & Cell Biology, University of Aberdeen, UK*

MICHAEL J. DUNN • *Institute of Psychiatry, Kings College London, London, UK*

DAVID R. DUPONT • *Applied Biosystems, Foster City, CA*

KENNETH S. GRAHAM • *Regeneron Pharmaceuticals, Tarrytown, NY*

GREGORY A. GRANT • *Department of Molecular Biology and Pharmacology, Washington University School of Medicine, St. Louis, MO*

FIONA M. GREER • *M-Scan Ltd, Wokingham, UK*

DAVID J. HARVEY • *Oxford Glycobiology Institute, Oxford University, UK*

HISASHI HIRANO • *Yokohama City University, Yokohama, Japan*

G. BRENT IRVINE • *Division of Biochemistry, School of Biology and Biochemistry, Medical Biology Centre, Queen's University of Belfast, Belfast, UK*

PHILIP J. JACKSON • *Applied Biosystems, Warrington, Cheshire, UK*

PETR JANDIK • *Dionex Corporation, Sunnyvale, CA*

ROZA M. KAMP • *Department of Biotechnology, University of Applied Science and Technology, Berlin, Germany*

MICHELE LEARMONTH • *Department of Biomedical Sciences,
University of Edinburgh, UK*
ADAM LISKA • *MPI of Molecular Cell Biology and Genetics, Dresden, Germany*
HOWARD R. MORRIS • *M-Scan Ltd., Wokingham, UK; Department
of Biological Sciences, Imperial College, London, UK*
JACEK MOZDZANOWSKI • *Analytical Methods Development,
Glaxosmithkline Pharmaceuticals, King of Prussia, PA*
DARRYL J. C. PAPPIN • *Imperial College, University of London, London, UK*
ANDREW J. REASON • *M-Scan Ltd, Wokingham, UK*
ROBERT B. RUSSELL • *European Molecular Biology Laboratory,
Heidelberg, Germany*
ANDREA SCALONI • *Proteomics and Mass Spectrometry Laboratory,
IABBAM, National Research Council, Naples, Italy*
DAVID M. SCHIELTZ • *Torrey Mesa Research Institute, San Diego, CA*
ANDREJ SHEVCHENKO • *European Molecular Biology Laboratory,
Heidelberg, Germany*
ANNA SHEVCHENKO • *MPI of Molecular Cell Biology and Genetics,
Dresden, Germany*
MAURIZIO SIMMACO • *Dipartimento Scienze Biomediche,
Università G. D'Annunzio, Chieti, Italy*
ALAN J. SMITH • *Stanford University Medical Center, Stanford, CA*
BRYAN JOHN SMITH • *Celltech Chiroscience, Slough, UK*
LAUREY STEINKE • *Protein Structure Core Facility,
University of Nebraska Medical Center, Lincoln, NE*
SHAMIL SUNYAEV • *European Molecular Biology Laboratory,
Heidelberg, Germany*
PAUL TEMPST • *Protein Center, Memorial Sloan-Kettering Cancer Center,
New York, NY*
CHRISTOPH W. TURCK • *Max Planck Institute of Psychiatry, Molecular,
Cellular, Clinical Proteomics, Munich, Germany*
JAMES P. TURNER • *Celltech Chiroscience, Slough, UK*
JOHN R. YATES, III • *Torrey Mesa Research Institute, San Diego, CA*
SYLVIA W. YUEN • *Applied Biosystems, Foster City, CA*

1

Strategies for Handling Polypeptides on a Micro-Scale

Bryan John Smith and Paul Tempst

1. Introduction

Samples for sequence analysis frequently are in far from plentiful supply. Preparation of protein without loss, contamination or modification becomes more problematical as the amount of the sample decreases. The most successful approach is likely to include the minimum number of steps, at any of which a problem might arise. The strategy for preparation of a given protein will depend on its own particular properties, but several points of advice apply. These are:

- Minimize sample loss: *see* **Note 1**.
- Minimize contamination of the sample: *see* **Note 2**.
- Minimize artificial modification of the sample: *see* **Note 3**.

When it comes to sample purification, polyacrylamide gel electrophoresis is a common method of choice, since it is suited to sub-µg amounts of sample, entails minimal sample handling, is quick, and has high resolving power. Proteins may be fragmented while in the gel (*see* Chapters 5 and 6), or electroeluted from it using commercially available equipment. Commonly, however, proteins and peptides are transferred onto membranes prior to analysis by various strategies as described in Chapter 4. Capillary electrophoresis (Chapter 8) and high-performance liquid chromatography (HPLC) are alternative separation techniques. Capillary electrophoresis has sufficient sensitivity to be useful for few µg or sub- µg amounts of sample. For maximum sensitivity on HPLC, columns of 1 mm or less inside diameter (id) may be used, but for doing so there are considerations extra to those that apply to use of larger-bore columns. These are discussed below.

From: *Methods in Molecular Biology, vol. 211: Protein Sequencing Protocols, 2nd ed.*
Edited by: B. J. Smith © Humana Press Inc., Totowa, NJ

Although desirable to minimize the amount of handling of a sample, it is frequently necessary to manipulate the sample prior to further purification or analysis, in order to concentrate the sample or to change the buffer, for instance. Some examples of methods for the handling of small samples follow below. They do not form an exhaustive list, but illustrate the type of approach that it may be necessary to adopt.

2. Materials
2.1. Microbore HPLC

1. An HPLC system able to operate at low flow rates (of the order of 30 µL/min) while giving a steady chromatogram baseline, with minimal mixing and dilution of sample peaks in the postcolumn plumbing (notably at the flow cell) and with minimal volume between flow cell and outflow (to minimize time delay, so to ease collection of sample peaks).

   An example design is described by Elicone et al (*1*). These authors used a 140B Solvent Delivery System from Applied Biosystems. The system was equipped with a 75 µL dynamic mixer and a precolumn filter with a 0.5 µm frit (Upchurch Scientific, Oak Harbor, WA) was plumbed between the mixer and a Rheodyne 7125 injector (from Rainin, Ridgefield, NJ) using two pieces (0.007 inch ID, 27 cm long [1 in. = 2.54 cm]) of PEEK tubing. The injector was fitted with a 50 µL loop and connected to the column inlet with PEEK tubing (0.005 inch × 30 cm). The outlet of the column was connected directly to a glass capillary (280 µm OD/75 cm ID × 20 cm; 0.88 µL), which is the leading portion of an U-Z view flow cell (35 nL volume, 8-mm path length; LC Packings, San Francisco, CA), fitted into an Applied Biosystems 783 detector. The trailing portion of the capillary cell was trimmed to a 15 cm length and threaded out of the detector head, resulting in a post flow cell volume of 0.66 µL and a collection delay of 1.3 s (at a flow rate of 30 µL/ min). Alternatively, various HPLC systems suitable for microbore work are available from commercial sources.
2. Clean glassware, syringe, and tubes for collection (polypropylene, such as the 0.5 µL or 1.5 µL Eppendorf type).
3. Solvents: use only HPLC-grade reagents (Fisons or other supplier), including distilled water (commercial HPLC-grade or Milli-Q water). A typical solvent system would be an increasing gradient of acetonitrile in 0.1% (v/v) trifluoroacetic acid (TFA) in water. The TFA acts as an ion-pairing reagent, interacting with positive charges on the polypeptide and generally improving chromatography. If TFA is not added to the acetonitrile stock, the baseline will decrease (owing to decreasing overall content of TFA), which makes identification of sample peaks more difficult. A level baseline can be maintained by adding TFA to the acetonitrile stock, in sufficient concentration (usually about 0.09% v/v) to make its absorbency at 214 or 220 nm equal to that of the other gradient component, 0.1% TFA in water. Check this by spectrophotometry. The absorbency remains stable for days.

4. Microbore HPLC columns of internal diameter 2.1 mm, 1 mm or less, are available from various commercial sources.

2.2. Concentration and Desalting of Sample Solutions

1. HPLC system: not necessarily as described above for microbore HPLC, but capable of delivering a flow rate of a few hundred μL to 1 mL per min. Monitor elution at 220 nm or 214 nm.
2. Clean syringe, tubes, HPLC-grade solvents, and so on as described in **Subheadings 2.1.**, **steps 2** and **3**.
3. Reverse-phase HPLC column, of alkyl chain length C2 or C4. Since analysis and resolution of mixtures of polypeptides is not the aim here, relatively cheap HPLC columns may be used (and reused). The method described employs the 2.1 mm ID × 10 mm C2 guard column. (Brownlee, from Applied Biosystems), available in cartridge format.

2.3. Small Scale Sample Clean-Up Using Reverse-Phase "Micro-tips"

1. Pipet tip: Eppendorf "gel loader" tip (cat. no. 2235165-6, Brinkman, Westbury, NY).
2. Glass fiber, such as the TFA-washed glass fibre disks used in Applied Biosystems automated protein sequencers (Applied Biosystems, cat. no. 499379).
3. Reverse-phase chromatography matrix, such as Poros 50 R2 (PerSeptive Biosystems, Framingham, MA). Make as a slurry in ethanol, 4:1::ethanol:beads (v/v).
4. Wash buffer: formic acid (0.1%, v/v in water). Elution buffer: acetonitrile in 0.1% formic acid, e.g., 30% acetonitrile (v/v).
5. Argon gas supply, at about 10–15 psi pressure, with line suited to attach to the pipet tip.
6. Micro-tubes: small volume, capped, e.g., 0.2 mL (United Scientific Products, San Leandro CA, Cat. no. PCR-02).

3. Methods
3.1. Microbore HPLC (see Notes 4–13)
3.1.1. Establishment of Baseline (see **Notes 4** and **7**)

A flat baseline at high-sensitivity setting (e.g., 15 mAUFs at 214 mm) is required for optimal peak detection. The use of an optimized HPLC and clean and UV absorbency-balanced solvents should generate a level baseline with little noise and peaks of contamination. A small degree of baseline noise originates from the UV detector. Beware that this may get worse as the detector lamp ages. Some baseline fluctuation may arise from the action of pumps and/ or solvent mixer. Slow flow rates seem to accentuate such problems that can go unnoticed at higher flows. Thorough sparging of solvents by helium may reduce these problems. New or recently unused columns require thorough washing before a reliable baseline is obtained. To do this, run several gradients and then run the starting solvent mixture until the baseline settles (this may take an hour or more). Such problems are reduced if the column is used continuously, and to achieve this in between runs, an isocratic mixture of solvents (e.g., 60% acetonitrile) may be run at low flow rate (e.g., 10 µL/min). Check system performance by running standard samples (e.g., a tryptic digest of 5 pmole of cytochrome C).

3.1.2. Identification of Sample Peaks (see **Notes 4, 7,** and **8**)

1. Peaks that do not derive from the sample protein(s), may arise from other sample constituents, such as added buffers or enzymes. To identify these contaminants, run controls lacking sample protein. Once the sample has been injected, run the system isocratically in the starting solvent mixture until the baseline is level and has returned to its pre-inject position. This can take up to 1 h in case of peptide mixtures that have been reacted with UV-absorbing chemicals (4-vinyl pyridine for example) before chromatography.
2. Peaks may be large enough to permit on-line spectroscopy where a diode array is available. Some analysis of amino acid content by second derivative spectroscopy may then be undertaken, identifying tryptophan-containing polypeptides, for instance, as described in Chapter 9.
3. Polypeptides containing tryptophan, tyrosine, or pyridylethylcysteine may be identified by monitoring elution at just three wavelengths (253, 277, 297 nm) in addition to 214 nm. Ratios of peak heights at these wavelengths indicate content of the polypeptides as described in **Note 8**. This approach can be used at the few pmole level.
4. Flow from the HPLC may be split and a small fraction diverted to an on-line electrospray mass spectrograph, so as to generate information on sample mass as well as possible identification of contaminants.

3.1.3. Peak Collection (see **Notes 4, 9–12**)

1. While programmable fraction collectors are available, peak collection is most reliably and flexibly done by hand. This operation is best done with detection of peaks on a flatbed chart recorder in real time. The use of flatbed chart recorder allows notation of collected fractions on the chart recording for future reference. The delay between peak detection and peak emergence at the outflow must be accurately known (*see* **Note 5**).
2. When the beginning of a peak is observed, remove the forming droplet with a paper tissue. Collect the outflow by touching the end of the outflow tubing against the side of the collection tube, so that the liquid flows continuously into the tube and drops are not formed. Typical volumes of collected peaks are 40–60 µL (from a 2.1 mm ID column) and 15–30 µL (from a 1 mm ID column). *See* **Note 9**.
3. Cap tubes to prevent evaporation of solvent. Store collected fractions for a short term on ice, and transfer to freezer (–20°C or –70°C) for long-term storage (*see* **Notes 10** and **11**).

4. Retrieval of sample following storage in polypropylene tubes is improved by acidification of the thawed sample, by addition of neat TFA to a final TFA concentration of 10% (v/v).

3.2. Concentration and Desalting of Sample Solutions (see *Notes 14–24*)

1. Equilibrate the C2 or C4 reverse-phase HPLC column in 1% acetonitrile (or other organic solvent of choice) in 0.1% TFA (v/v) in water, at a flow rate of 0.5 mL/min at ambient temperature.
2. Load the sample on to the column. If the sample is in organic solvent of concentration greater than 1% (v/v), dilute it with water or aqueous buffer (to ensure that the protein binds to the reverse-phase column) but do this just before loading (to minimize losses by adsorption from aqueous solution onto vessel walls). If the sample volume is greater than the HPLC loop size, simply repeat the loading process until the entire sample has been loaded.
3. Wash the column with isocratic 1% (v/v) acetonitrile in 0.1% TFA in water. Monitor elution of salts and/or other hydrophilic species that do not bind to the column. When absorbency at 220 nm has returned to baseline a gradient is applied to as to elute polypeptides from the column. The gradient is a simple, linear increase of acetonitrile content from the original 1% to 95%, flow rate 0.5mL/min, ambient temperature, over 20 min. Collect and store emerging peaks as described above (*see* **Subheading 3.1.2.** and *see* **Note 9**).
4. The column may be washed by isocratic 95% acetonitrile in 0.1% TFA in water, 0.5 mL/min, 5 min before being re-equilibrated to 1% acetonitrile for subsequent use.

3.3. Small Scale Sample Clean-up Using "Micro-tips" (see *Notes 25–28*).

1. Using a pipet tip, core out a small disk from the glass-fiber disk. Push it down the inside of the gel-loader tip (containing 20 µL of ethanol), until it is stuck. Pipet onto this frit 10 µL of reverse-phase matrix slurry (equivalent to about 2 µL of packed beads). Apply argon gas to the top of the tip, to force liquid through the tip and pack the beads. Wash the beads by applying 3 lots of 20 µL of 0.1% formic acid, forcing the liquid through the micro-column with argon, but never allowing the column to run dry. Use a magnifying glass to check this, if necessary. Leave about 5 mm of final wash above the micro-column. The column is ready to use.
2. Apply the sample solution to the micro-column and wash with 3 lots of 20 µL 0.1% formic acid, leaving a minimum of the final wash solution above the micro-column. Pipet 3–4 µL (i.e., about 2 column volumes) of elution buffer into the micro-tip, leaving a bubble of air between the elution buffer and the micro-column in ash buffer. The elution buffer is then forced into the micro-column (but without mixing with the wash buffer, for clearly, this would alter the composition of the buffer and possibly adversely affect elution). Collect the buffer containing the eluted sample. If further elution steps are required, do not let the

micro-column dry out, and proceed as before by leaving a bubble of air between the fresh elution buffer and the preceding buffer. Collect and store eluted fractions as in **Subheading 3.1.2.** and *see* **Notes 9–12**.

4. Notes

1. Small amounts of polypeptide are difficult to monitor and may be easily lost, for instance, by adsorption to vessel walls. Minimize the number of handling maneuvers and transfers to new tubes.

2. Work in clean conditions with the cleanest possible reagents. Consider the possible effects of added components such as amine-containing buffer components such as glycine (which may interfere with Edman sequencing), detergents, protease inhibitors (especially proteinaceous ones such as soybean trypsin inhibitor), agents to assist in extraction procedures (such as lysozyme), and serum components (added to cell culture media).

3. Modification of the polypeptide sample can arise by reaction with reactive peroxide species that occur as trace contaminants in triton and other nonionic detergents *(2)*. The presence of these reactive contaminants is minimized by the use of fresh, specially purified detergent stored under nitrogen (such as is available from commercial sources, such as Pierce). Mixed bed resins, mixtures of strong cation and anion resins (available commercially from sources such as Pharmacia Biotech, BioRad, or BDH) can be used to remove trace ionic impurities from nonionic reagent solutions such as triton X100, urea, or acrylamide. Excess resin is merely mixed with the solution for an hour or so, and then removed by centrifugation or filtration. The supernatant or filtrate is then ready to use. Use while fresh in case contaminants reappear with time. In this way, for example, cyanate ions that might otherwise cause carbamylation of primary amines (and so block the N-terminus to Edman sequencing) may be removed from solutions of urea.

   Polypeptide modification may also occur in conditions of low pH; for instance, N-terminal glutaminyl residues may cyclize to produce the blocked pyroglutamyl residue, glutamine, and asparagine may become deamidated, or the polypeptide chain may be cleaved (as described in Chapter 6). Again, exposure of proteins to formic acid has been reported to result in formylation, detectable by mass spectrometry *(3)*. Problems of this sort are reduced by minimizing exposure of the sample to acid and substitution of formic acid by, say, acetic or trifluoroacetic acid (TFA) for the purposes of treatment with cyanogen bromide (*see* Chapter 6).

4.1. Microbore HPLC

4. When working with µg or sub- µg amounts of sample the problem of contamination is a serious one, not only adding to the background of amino acids and nonamino acid artifact peaks in the final sequence analysis, but also during sample preparation, generating artificial peaks, which may be analyzed mistakenly. To reduce this problem most effectively, for microbore HPLC or other technique, it is necessary to adopt the "semi-clean room" approach, whereby ingress of contaminating protein is minimized. Thus:

a. Dedicate space to the HPLC, sequencer and other associated equipment. As far as possible, set this apart from activities such as peptide synthesis, biochemistry, molecular biology, and microbiology.

b. Dedicate equipment and chemical supplies. This includes equipment such as pipets, freezers, and HPLC solvents.

c. Keep the area and equipment clean. Do not use materials from central glass washing or media preparation facilities. It is not uncommon to find traces of detergent or other residues on glass from central washing facilities, for instance. Remember that "sterile" does not necessarily mean protein-free!

d. Use powderless gloves and clean labcoats. Avoid coughing, sneezing and hair near samples. As with other labs, ban food and drink. Limit unnecessary traffic of other workers, visitors, and so on.

e. Limit the size of samples analyzed, or beware the problem of sample carryover. If a large sample has been chromatographed or otherwise analyzed, check with "blank" samples that no trace of it remains to appear in subsequent analyses.

5. Micro-preparation of peptides destined for chemical sequencing and mass spectrometric analysis often requires high performance reversed-phase LC systems, preferably operated with volatile solvents. Sensitivity of sample detection in HPLC is inversely proportional to the cross-sectional area of the HPLC column used, such that a 1 mm ID column potentially will give 17-fold greater sensitivity than a 4.6 mm ID column. Microbore HPLC tends to highlight shortcomings in an HPLC system, however, so to get optimal performance from a microbore system attention to design and operation is necessary, as indicated in Materials (**Subheading 2.**) and Methods (**Subheading 3.**).

At the slow flow rates used in microbore HPLC, the delay between the detection of a peak and its appearance at the outflow may be significant, and must be known accurately for efficient peak collection. If the volume of the tubing between the UV detector cell and the outflow is known, the time delay (t) may be calculated:

$$t = \frac{\text{tubing volume, } \mu L}{\text{flow rate, } \mu L/min}$$

where t is in minutes. The collection of any peak must be delayed by t minutes after first detection of the peak. The flow rate should be measured at the point of outflow - a nominal flow rate set on a pump controller may be faster than the actual flow rate due to the effect of back pressure in the system (e.g., by the column).

Alternatively, t may be determined empirically as follows:

a. Disconnect the column, replace it with a tubing connector.

b. Set isocratic flow of 0.1% TFA in water at a rate equal to that when the column is in-line and check flow rate by measuring the outflow.

c. Inject 50 μL of a suitable coloured solution, e.g., 0.1% (w/v) Ponceau S solution in 1% acetic acid (v/v).

d. Collect outflow. To see eluted color readily, collect outflow as spots onto filter paper (e.g., Whatman 3MM).

e. Measure the time between first detection of the dye peak, and first appearance of color at the outflow. Repeat this process at the same or different flow rates sufficient to gain an accurate estimate, which may be used to calculate the tubing volume (see equation for t).

The slow flow rate has another consequence too, namely a delay of onset of a gradient. The volume of the system before the column may be significant and a gradient being generated from the solvent reservoirs has to work its way through this volume before reaching the column or UV detector. For instance, a pre-column system volume of 600 µL would generate a 20-min delay if the flow rate were 30 µL/min. If the length of this delay is unknown, it may be measured empirically as follows:

a. Leave the HPLC column connected to the system. Have one solvent (A) as a mixture, 5% (v/v) acetonitrile in 0.1% v/v TFA in water, and another solvent (B) as 95% (v/v) acetonitrile in 0.1% (v/v) TFA in water. (NOTE: solvents not balanced for UV absorption).

b. From one solvent inlet, run solvent mixture A isocratically at, say 30 µL/min, until the baseline is level.

c. Halt solvent flow, replace A with B and resume flow at same flow rate.

d. Measure time from resumption of flow to sudden change of UV absorption. This is the time required for a solvent front to reach the detector, with the column of interest in the system.

Remember to allow for this delay when programming gradients.

6. Reverse-phase columns are commonly used for polypeptide separations. Columns of various chain lengths up to C18 are available commercially in 2.1 or 1 mm ID. As for wider-bore HPLC, the best column for any particular purpose is best determined empirically, though the following may be stated: use larger-pore matrices for larger polypeptides; use shorter-length alkyl chain columns for chromatography of hydrophobic polypeptides. As an example of the latter point, human Tumor Necrosis Factor-α (TNF-α) is soluble in plasma and is biologically active as a homotrimer, but binds so tightly to a C18 reverse-phase column that 99% acetonitrile in 0.1% v/v TFA in water will not remove it. It can be eluted from C2 or C4 columns by increasing gradients of acetonitrile, however.

Gradient systems used in microbore reverse-phase HPLC are also best determined empirically, but commonly would utilize an increasing gradient of acetonitrile (or other organic solvent) in 0.1% (v/v) TFA (or other ion-pairing agent, such as heptafluorobutyric acid) in water. Flow rates would be of the order of 30 µL/min for a 1 mm ID column, or 100 µL/min for a 2.1 mm ID column. Use ambient temperature if possible, to avoid the possibility of baseline fluctuation due to variation in temperature of solvent as it passes from heated column to cooler flow cell.

7. In the various forms of chromatography, elution of polypeptide sample is commonly monitored at 280 nm. However, not only may some polypeptides lack significant absorbency at 280 nm, but also detection is an order of magnitude less sensitive than at 220nm. Absorbency at the lower wavelengths is due to the pep-

tide bond (obviously present in all polypeptides). However, absorbency due to solvent and additives such as TFA and contaminants tends to be higher. This "background" absorbency becomes greater as wavelengths are reduced towards 200 nm and with it the problems of maintaining a stable baseline and detection of contaminants become greater. The trade-off between greater sensitivity and background absorbency is best made empirically with the user's own equipment. Detection at 214 nm or 220 nm is commonly used, with lower wavelengths being more problematical.

8. Sample peaks may be analyzed on-line by spectroscopy. With a diode array and enough sample to generate a reliable spectrum, second derivative spectroscopy may be used as described in Chapter 9. At the few pmole level, monitoring at 253 nm, 277 nm, and 297 nm may indicate peaks that may be of interest by virtue of containing tryptophan, tyrosine or pyridylethylcysteine. A peptide's content of tryptophan, tyrosine, and (pyridylethyl) cysteine may be judged from the ratios of absorbency at 253, 277, and 297 nm. Thus:

    a. Greatest absorbency at 253 nm with minimal absorbency at 297 nm indicates the presence of pyridylethylcysteine.

    b. Greatest absorbency at 277 nm with minimal absorbency at 297 nm indicates the presence of tyrosine.

    c. Greatest absorbency at 277 nm with moderate absorbency at 253 nm and 297 nm indicates the presence of tryptophan.

    If more than one of these three types of residue occur in one peptide, identification is more problematical since the residues' UV spectra overlap. However, comparison with results from model peptides assist analysis, as described by Erdjument-Bromage et al (*4*), whose results are summarized in **Table 1**. The presence of tyrosine is the most difficult to determine, but combinations of tryptophan and pyridylethylcysteine may be identified. As Erdjument-Bromage et al. (*5*) point out, this analysis is only valid when the mobile phase is acidic (e.g., in 0.1% TFA in water and acetonitrile), for UV spectra of tryptophan and tyrosine change markedly with changes in pH. This type of analysis may be performed on 5–10 pmole of peptides.

9. Drops flowing from HPLC have a volume of the order of 25 µL. At the type of flow rate used for microbore HPLC, a drop of this size may take a minute to form and so may contain more than one peak. This is unacceptable. Collection of outflow down the inside wall of the collection tube inhibits droplet formation and allows interruption of the collection (changing to the next fraction) at any time.

10. Once peptides elute from a reverse-phase HPLC column, they are obtained as a dilute solution (1–2 pmoles per 5 µL) in 0.1% TFA/10–30% (v/v) acetonitrile, or similar solvent. At those concentrations and below, many peptides tend to "disappear" from the solutions. The problem of minute peptide losses during preparation, storage, and transfer has either not been fully recognized or has been blamed on unrelated factors, column losses for example. Actually, column effects are minimal (*1*). Instead, it has been shown that losses primarily occur in test tubes and pipet tips (*5*). At concentrations of 2.5–8 pmoles per 25 µL (amounts and volume repre-

Table 1
Reverse-Phase HPLC with Triple Wavelength Detection of Peptides Containing Trp (W), Tyr (Y), or pyridyl ethyl-Cys (pC)[a]

Peptide	Relative Peak Height (in %)			Number of Residents		
	A_{253}	A_{277}	A_{297}	W	Y	pC
pCPSPKTPVNFNNFQ	100	12	2	-	-	1
QNpCDQFEK	100	14	1	-	-	1
GNLWATGHF	45	100	28	1	-	-
ILLQKWE	43	100	26	1	-	-
YEVKMDAEF	33	100	3	-	1	-
TGQAPGFTYTDANK	38	100	2	-	1	-
YSLEPSSPSHWGOLPTP	45	100	21	1	1	-
GITWKEETLMEYLENPK	42	100	24	1	1	-
EDWKKYEKYR	40	100	23	1	2	-
YEDWKKYEKYR	37	100	19	1	3	-
Insulin beta chain / 4VP	100	39	4	-	2	2
Insulin alpha chain / 4VP	100	32	3	-	2	4
DST peptide (25 a.a.)	100	73	23	1	-	1
PepepII (27 a.a.)	100	100	20	1	1	1

[a]Peptides (20 picomoles each, or less) were chromatographed on a Vydac C4 (2.1 × 250 mm) column at a flow of 0.1 mL/min. Peak heights on chromatographs, produced by monitoring at different wavelengths, are expressed in %, relatively to the tallest peak. Total number of W, Y, or pC present in each peptide are listed. Sequences of bovine insulin alpha and beta chains are taken from SWISS and PIR database; PepepII, ISpCWAQIGKEPITFEHINYERVSDR; DST peptide, DLFNAAFVSpCWSELNEDQQDELIR. Insulin was reduced with 2-mercaptoethanol and reacted with 4-vinyl pyridine prior to HPLC. Reprinted with permission from **ref. (4)**.

sentative for a typical microbore LC fraction), about 50% of the peptide is not recovered from storage in 0.1% TFA (from 1 min to 1 wk). When supplemented with 33% TFA, recoveries were 80% on the average. Best transfers, regardless of volume and duration of storage, were obtained in 10% TFA/30% acetonitrile. From those data it follows that, upon storage at −70°C for 24 h or more, up to 45% losses may be incurred for LC collected peptides. Although adding concentrated TFA prior to storage results in best recoveries (> 90%), it might degrade the peptides. Thus, it is best to store HPLC-collected peptides at −70°C and always add neat TFA in a 1 to 8 ratio (TFA: sample) after storage, just before loading on the sequencer disc. Additionally, coating the polypropylene with polyethylenimine may reduce this loss, as indicated by an observed improved retrieval of radiolabeled bradykinin from polypropylene tubes (increased from 30% to 65% yield). Tubes were coated by immersion in 0.5% polyethylenimine in water overnight, room temperature, followed by rinsing in distilled water and thorough drying in a glass-drying oven (Dr J. O'Connell, unpublished observation).

Having collected a sample in a mixture of solvents in which it is soluble, it is unwise to alter this mixture for the sample may then become insoluble. Thus, concentration under vacuum will remove organic solvent before removing the less-volatile water, as changing the solvent mixture. Again, if the sample contacts membranes such as used for concentration, filtration or dialysis it may become irreversibly bound. Complete drying down may also be a problem— redissolving the dried sample may be difficult, requiring glacial acetic acid or formic acid (70% v/v, or greater).

11. Repeated cycles of freezing and thawing may cause fragmentation of polypeptides eventually, this tending to increase adsorption losses. Beware that the temperature inside a (nominally) –20°C freezer may rise to close to 0°C during defrosting or while the door is left open while other samples are being retrieved, such that sample quality may suffer. Storage at –70°C is safer.

12. Another solution to the problem of storage of HPLC fractions, at least for subsequent sequencing by Edman chemistry, is immediate transfer to polyvinylidene difluoride (PVDF) membrane, on which medium (dried) polypeptides are stable for prolonged periods. This may be accomplished by use of the single use Prosorb device from Applied Biosystems. The sample solution is drawn by capillary action through a PVDF membrane, to which polypeptides bind. Addition of polybrene (Biobrene, Applied Biosystems) is recommended for sequencing of PVDF-bound peptides (see the literature that accompanies Biobrene for its method of use). For processing large numbers of samples, PVDF sheets may be used to trap the polypeptides. The membrane is placed in a Hybridot 96-well manifold (BRL), or similar, and the sample solutions are drawn slowly through the membrane. The location of the bound protein spots may be confirmed by staining of the wetted membrane for a few minutes in Ponceau S (Sigma), 0.1% (w/v) in acetic acid (1% v/v in water), followed by destaining in water.

    PVDF requires wetting with organic solvent prior to wetting by water. Dried PVDF membrane may be re-wetted with 20% methanol in water without significant loss of polypeptide sample. Many reverse-phase HPLC fractions (e.g., from a gradient of organic solvent in TFA-water) will likewise wet PVDF directly.

13. Various criteria can be applied to sample peaks in order to decide whether they are suitable for sequencing by Edman Chemistry, i.e., pure and in sufficient quantity. These are:
    a. The peak should not show signs of any shoulders indicative of underlying species.
    b. Spectra collected at multiple points through the peak should be identical-differences indicate multiple species present.
    c. If mass spectrometry is carried out on part of the sample peak, a single mass is a reasonably good indication of purity.

    If a sample peak appears not to be pure by such criteria, collected fractions may be prepared for chromatography on a second, different HPLC system as follows:
    a. Add neat TFA in the ratio 1:8::TFA: sample (v/v), in order to improve recovery (*see* **Subheading 4.1.**, **step 7**, above).

b. Dilute by addition of one volume of water or 0.1% TFA in water (v/v), just before injection. Recoveries after rechromatography are usually of the order of 40–60%.

4.2. Desalting/Concentration

14. The presence of salts and detergents can interfere with analysis by mass spectrometry or protein sequencing by Edman chemistry (if these reagents restrict access of chemicals to the sample, or generate artificial products). Again, if a sample solution is too dilute, analysis may be problematical.

    As an example of the HPLC method for concentration and desalting of sample solutions described in **Subheading 3.2.**, it has been used in preparation of human TNF-α, a hydrophobic protein that can absorb to membranes used for filtration as well as to C18 reverse-phase HPLC columns. TNF-α at as little as 2 ng/mL in 2 L cell culture medium containing 10% (v/v) fetal calf serum (FCS) was prepared at approx 100% yield as follows:

    a. Concentration approx fivefold on a 10 KDa cut-off membrane (using a Filtron miniultra-cassette, with losses of TNF-α being minimized by the presence of other proteins).

    b. Affinity chromatography on a solid-phase-linked, anti-human TNF-α antibody, the TNF-α eluting in 7.5 mL of a buffer of trizma-HCl, 50 mM, pH 7.6, magnesium chloride, 3 M.

    c. Final concentration and desalting by C2 HPLC as described in **Subheading 3.2.**, eluting from the column in 0.5 mL.

15. The concentrating/desalting method described is a basic one for separating hydrophilic and hydrophobic species, the former being salt and the latter being the TNF-α in the example above. The system may be modified in various ways for less hydrophobic polypeptides. Thus, replacement of the C2 HPLC column by a C4 or even C18 column may provide better discrimination between salts and hydrophilic polypeptides. Alternatively, the relatively cheap "guard" column used here may be replaced by an analytical column such that mixtures of polypeptides may be resolved on the column after salts have been removed.

16. Nonionic detergents may not be separated from polypeptide during concentration or desalting on reverse phase columns - Triton X 100 and Tween do not elute with hydrophilic species but do so in the subsequent acetonitrile gradient. A detergent can be removed by dialysis but requires extensive dilution to below the detergent's critical micelle concentration (CMC), followed by prolonged dialysis. n-Octyl-β-glucopyranoside is one of the better detergents in this respect, since it has a relatively high CMC of 20–25 mM. Alternatively, matrices such as Calbiosorb (Calbiochem) may be used to remove detergent chromatographically. Nonionic species may be removed from solutions of proteins by ion-exchange chromatography. One proviso is that the protein should bear charge, i.e., the solution pH should not be equal to the proteins pI. With that condition satisfied the protein may be bound to the ion exchange matrix while non-ionic species may be washed away. Protein may be removed subsequently, by altering pH or salt concentration.

17. The use of an ion exchange pre-column (DEAE-Toyopearl, 4 × 50mm) has been described for removal of SDS and Coomassie brilliant blue R-250 from gel extracts, prior to peptide separation by reverse phase HPLC *(6)*. Hydrophilic interaction chromatography on poly(2-hydroxyethyl-aspartimide)-coated silica (PolyLC Inc.) in n-propanol-formic acid solvent can also remove salts and contaminants that may occur in samples electroeluted form polyacrylamide gels, for example *(7)*.

18. Batch chromatography offers an alternative means of concentration and salt removal (*see* **Note 28**).

19. Sample peaks may be analyzed by on-line spectroscopy during concentration or desalting, as described in **Note 8**.

20. Reverse-phase HPLC may be interfaced with electrospray mass spectrometry, so the method described may, in such a coupled system, be used to desalt samples for analysis.

    To avoid build-up of salty deposits in the mass spectrometer the salt peak may be diverted to waste.

    A similar end may be achieved by using a gel filtration column in-line, ahead of the mass spectrometer, proteins emerging ahead of salts and other small species. Gel filtration dilutes rather than concentrate samples, however.

21. Various commercially available small scale devices offer alternatives to the HPLC method. For example, single use Ultrafree-MC filters (Millipore) are suitable for concentration of samples down to volumes of the order of 50–100 µL: the sample is placed in the device and then centrifuged, driving smaller species through the membrane while retaining larger species. The sample may be repeatedly topped up and centrifuged in order to process larger volumes. Similarly, the sample may be repeatedly concentrated and then diluted with water or alternative buffer for the purposes of buffer exchange. Small-volume dialysis devices are also available for exchange of buffers in samples as small as 10 µL (for instance the Slide-A-Lyzer units from Pierce). Generally these approaches are not suitable for small molecular-weight peptides, but Fierens et al. *(8)* have reported a means (albeit not suited to all circumstances) whereby peptides may be retained by filter membranes with nominal cutoffs greater than the size of the peptides. This involves addition of albumin, to which the peptides may bind, and which does not pass through the filter membrane.

    Beware that buffer exchange and concentration procedures carry with them the danger of sample aggregation and precipitation, and the loss of sample solution that cannot be retrieved from the surfaces and corners of the devices used.

22. Salts may be removed from polypeptide solutions by transfer of the polypeptide to PVDF (*see* **Note 12**). Salts are not retained on PVDF, whereas polypeptides are. Remaining traces of salts or contaminating amino acids may be removed by washing of the membrane in a small volume of methanol, followed by drying in air. Thus samples applied in 100 m*M* Trizma buffer or in 1 *M* NaCl show the same initial and repetitive yields as samples applied in water, with no extra peaks of contamination.

The presence of detergent can interfere with polypeptide binding to PVDF (e.g., 0.1% v/v brij 35 reduces binding by 5- to 10-fold). Dilution of the sample overcomes this problem. Applied Biosystems recommend dilution of Triton X100 to 0.05% or less, and sodium dodecyl sulfate (SDS) to 0.2% to allow efficient binding of protein to ProBlott PVDF membrane. Similarly they recommend dilution of urea or guanidine hydrochloride to 2–3 M. Sample dilution is not a problem in so far as a large volume may be filtered through the PVDF (by repeatedly refilling sample well or ProSorb) but the following should be remembered: large volumes of diluent may introduce significant contamination; dilution of detergent may cause the sample to come out of solution of bind to vessel walls. To minimize the latter, make dilutions immediately before filtration.

23. Various stains are available for detection of proteins on PVDF membranes. This allows location of the protein and may allow approximate quantification. The fluorescent stain Sypro Ruby (Molecular Dynamics) has sensitivity approaching that of silver stains. It does not interfere with subsequent analysis of bound sample. Sypro Ruby may be used for quantification (detection under UV light and scanning in a Bio Rad FluorS scanning densitometer). Methods have also been described for quantification of Ponceau S-stained protein on membranes *(9,10)*. Beware that the handling involved in staining and scanning may introduce contamination.

24. Proteins may be removed from salty solution and concentrated by precipitation. This may be achieved by addition of 1/4 volume of 100% w/v trichloroacetic acid solution (i.e., 100 g TCA in 100 mL solution – beware the highly corrosive nature of this solution: wear protective clothing), thus giving a final TCA concentration of 20%. Stand the mixture on ice for 1 h or so and centrifuge. Discard the supernatant. Remove traces of acid by several washes in acetone and finally dry under vacuum. Other molecules than proteins may co-precipitate, for instance nucleic acids. Sauvé et al. have described a method for concentration of proteins from solutions of 10 ng/mL *(11)* to 100 ng/mL *(12)*. The proteins are extracted in water-saturated phenol, and then extracted from the phenol solution by ether, whence they are isolated by evaporation of the ether. Recoveries were determined to be of the order of 80%, better than about 50% achieved by the TCA precipitation method. Protein in solution with guanidine hydrochloride may be precipitated by addition of sodium deoxycholate and TCA *(13)*. With any precipitation procedure there is a potential problem of rendering the protein insoluble. Heating in SDS-PAGE sample buffer, for subsequent PAGE, overcomes this in most cases.

4.3. Small Scale Sample Clean-up

25. The method described is that of Erdjument-Bromage et al. *(14)*. The method was intended for processing of small samples prior to mass spectrometry, which can be adversely affected by salts, detergents, or other components in the sample. The method is essentially low-pressure reverse-phase chromatography, in which salts do not bind to the column and can therefore be removed. Like other forms of

chromatography, elution of bound material may be achieved by a single step from low to high concentration of organic solvent, or by a succession of smaller incremental steps of increasing solvent concentration. By using incremental steps, bound material may be fractionated. Erdjument-Bromage et al. *(14)* illustrated this with a sample of trypsin digests of 100 fmole of glucose-6-phosphate dehydrogenase in polyacrylamide gel. A two-step elution of the digested peptides from the micro-tip was achieved using 16% and 30% acetonitrile in 0.1% formic acid, and mass spectrometric analysis was subsequently successfully achieved.

The approach may be adapted to other cases. For example, the matrix may be changed to another to achieve a different chromatographic separation. For example, affinity purification of phosphopeptides may be carried out by use of immobilized metal affinity chromatography *(15)*. Gallium (III) ions are immobilised on beads of chelating resin (Poros MC) by washing the beads in a solution of $GaCl_3$. A solution of mixed peptides is loaded onto the column and the phosphopeptides are selectively bound. After washing to remove unbound peptides, phosphopeptides are eluted in a buffer of pH 8.5 in the presence of phosphate, which displace the phosphopeptide. The micro-columns in this case were about 12 µL, and sample volumes were optimally less than 50% the volume of the column, namely about 5 µL. Beware the toxicity of gallium (III) chloride, and its violent reaction with water; wear protective clothing.

26. Versions of micro-tips are now available commercially, such as Zip-Tips from Millipore (similar to the micro-tips described earlier, and operable with a pipet rather than pressurized gas), or Supro-tips from AmiKa Corp. (Columbia, MD) where the matrix is bound to the tip. A variety of matrices is available, allowing desalting and concentration of polypeptides, step-wise fractionation, preparation of phosphopeptides and His-tagged polypeptides, and removal of detergents and other contaminants. Note that these come in fixed sizes, with fixed capacity, whereas the manually prepared version can be adapted and made larger if required.

27. Note that these columns are made small in order to deal with small volume samples (of the order of 10 µL or less and containing 1 µg of protein or less). Beware that the capacity of the small columns may be easily exceeded (for instance by contaminants as well as the desired sample itself), and this may adversely affect the purification.

28. Sample clean-up can be achieved by small-scale batch chromatography without use of micro-tips. This may be useful for larger volumes of sample than are convenient for micro-tips. In essence the approach is: incubate the sample with chromatography medium; centrifuge the mixture to separate supernatant from chromatography matrix; further analyse the supernatant or the chromatography matrix (if the desired molecule is bound). Elute the sample molecule from the chromatography matrix if desired. The choice of chromatography matrix, and conditions for sample binding and elution are dependent on the case in point. If it is necessary to check the pH of a small volume of solution prior to this batch chromatography, this may be done economically by use of the "dip-stick" type of pH indicator strips (e.g., from BDH, Poole, UK). For this, cut the strip into further,

smaller strips of less than 1 mm width. Each of these requires only 1 µL or so of solution to gain a colourimetric reading of pH. Detergents may be removed from samples by use of products such as BioBeads (Bio-Rad) in a similar batch mode.

This approach may be used to remove contaminants from solution. An example of this is the removal of albumin from plasma or cell culture medium, where it may be so abundant as to interfere with analysis of other proteins present. This is achieved by incubation of the sample with Cibacron Blue linked to sepharose or agarose beads. The albumin binds to the Cibacron Blue and can be removed (totally or partly) on the beads. Rengarajan et al. *(16)* have described this approach for dealing with small serum samples. To 10 µL serum mixed with 240 µL phosphate buffer was added 160 mL slurry of Affi-Gel Blue (agarose-bound dye, Bio-Rad, Hercules, CA). This mixture was incubated for 30 min at room temperature prior to centrifugation. The Affi-Gel Blue beads were washed to retrieve supernatant trapped between the beads and the pooled supernatants concentrated prior to further analysis. Blue sepharose (Pharmacia Biotech) works in similar fashion. Beware that proteins other than albumin may bind to the Cibacron Blue dye moiety.

Alternatively, the molecule of interest may be bound to the beads. As an example, Gammabind Plus sepharose (Pharmacia Biotech) can be used to separate molecules containing an immunoglobulin Fc domain such as IgG itself, or proteins genetically fused to an Fc "tag." The sample, at neutral pH, is incubated with beads (with gentle mixing) and then centrifuged. The beads may be directly heated in SDS-PAGE sample buffer prior to electrophoresis. Alternatively, the bound molecules may be eluted for other analyses, by washing the beads in low pH buffer (pH 2.0 or 3.0). Qian et al. *(17)* have described analysis of multi-His-tagged peptides and proteins while they were still bound to an affinity matrix of immobilized metal ion beads. Matrix-assisted laser desorption/ionization mass spectrometry (MALDI) of the samples was possible directly on the polypeptide-loaded beads, and it was also possible to proteolyse the sample on the bead prior to mass spectrometric analysis.

Another example of batch chromatography may be useful for concentration of proteins from dilute solution (100 ng/mL), prior to SDS-PAGE *(12)*. The protein solution is incubated with Strataclean beads (from Stratagene, cat. no. 400714), with shaking at room temperature for 20 min or so. Any protein(s) present bind to the beads and can be pelleted on the beads. They are then released by heating in SDS PAGE sample buffer prior to electrophoresis. High ionic strength (2 *M* ammonium sulphate) and various detergents do not interfere with this process, though 1% deoxycholate did interfere with recovery of albumin and ovalbumin, at least.

Acknowledgments

We thank Dr. J. O'Connell of Celltech R and D. for allowing us to use his example of peptide loss and its remedy given in **Note 10**.

References

1. Elicone, C., Lui, M., Geromanos, S. S., Erdjument-Bromage, H., and Tempst, P. (1994) Microbore reversed-phase high performance liquid chromatographic purification of peptides for combined chemical sequencing/laser-desorption mass spectrometric analysis. *J. Chromatog.* **676**, 121–137.
2. Chang, H. W. and Bock, E. (1980) Pitfalls in the use of commercial nonionic detergents for the solubilisation of integral membrane proteins: sulfhydryl oxidising contaminants and their elimination. *Anal. Biochem.* **104**, 112–117.
3. Beavis, R. C. and Chait, B. T. (1990) Rapid, sensitive analysis of protein mixtures by mass spectrometry. *Proc. Natl. Acad. Sci. USA* **87**, 6873–6877.
4. Erdjument-Bromage, H., Lui, M., Sabatini, D. M., Snyder, S. H., and Tempst, P. (1994) High-sensitivity sequencing of large proteins: partial structure of the rapamycin-FKBP12 target. *Protein Sci.* **3**, 2435–2446.
5. Tempst, P., Geromanos, S., Elicone, C., and Erdjument-Bromage, H. (1994) Improvements in microsequencer performance for low picomole sequence analysis. *Methods: Comp. Methods Enzymol.* **6**, 248–261.
6. Kawasaki, H., Emori, Y., and Suzuki, K. (1990) Production and separation of peptides from proteins stained with Coomassie brilliant blue R-250 after separation by sodium dodecyl sulfate-polyacrylamide gel electrophoresis. *Anal. Biochem.* **191**, 332–336.
7. Jenö, P., Scherer, P. E., Manning-Krieg, U., and Horst, M. (1993) Desalting electroeluted proteins with hydrophilic interaction chromatography. *Anal. Biochem.* **215**, 292–298.
8. Fierens, C., Thienpont, L. M., Stockl, D., and De Leenheer, A. P. (2000) Overcoming practical limitations for the application of ultrafiltration in sample preparation for liquid chromatography/mass spectrometry of small peptides. *Anal. Biochem.* **285**, 168–169.
9. Morcol, T. and Subramanian, A. (1999) A red-dot-blot protein assay technique in the low nanogram range. *Anal. Biochem.* **270**, 75–82.
10. Bannur, S. V., Kulgood, S. V., Metkar, S. S., Mahajan, S. K., and Sainis, J. K. (1999) Protein determination by Ponceau S using digital color image analysis of protein spots on nitrocellulose membranes. *Anal. Biochem.* **267**, 382–389.
11. Sauvé, D. M., Ho, D. T., and Roberge, M. (1995) Concentration of dilute protein for gel electrophoresis. *Anal. Biochem.* **226**, 382–383.
12. Ziegler, J., Vogt, T., Miersch, O., and Strack, D. (1997) Concentration of dilute protein solutions prior to sodium dodecyl sulfate-polyacrylamide gel electrophoresis. *Anal. Biochem.* **250**, 257–260.
13. Arnold, U. and Ulbrich-Hofmann, R. (1999) Quantitative protein precipitation from guanidine hydrpchloride-containing solutions by sodium deoxycholate/trichloroacetic acid. *Anal. Biochem.* **271**, 197–199.
14. Erdjument-Bromage, H., Lui, M., Lacomis, L., Grewal, A., Annan, R. S., McNulty, D. E., et al. (1998) Examination of micro-tip reverse-phase liquid chromatographic extraction of peptide pools for mass spectrometric analysis. *J. Chromatog. A* **826**, 176–181.

15. Posewitz, M. C. and Tempst, P. (1999) Immobilized Gallium(III) affinity chromatography of phosphopeptides. *Anal. Chem.* **71**, 2883–2892.
16. Rengarajan, K., de Smet, B., and Wiggert, B. (1996) Removal of albumin from multiple human serum samples. *BioTechniques* **20**, 30–32.
17. Qian, X., Zhou, W., Khaledi, M. G., and Tomer, K. B. (1999) Direct analysis of the products of sequential cleavages of peptides and proteins affinity-bound to immobilized metal ion beads by matrix-assisted laser desorption/ionization mass spectrometry. *Anal. Biochem.* **274**, 174–180.

2

SDS Polyacrylamide Gel Electrophoresis for N-Terminal Protein Sequencing

Bryan John Smith

1. Introduction

Polyacrylamide gel electrophoresis in the presence of sodium dodecylsulphate (SDS-PAGE) is a very common technique used for analysis of complex mixtures of polypeptides. It has great resolving powers, is rapid, and is suitable for proteins of either acidic or basic pI. The last is because the protein is reacted with SDS, which binds to the protein in the approximate ratio 1.4:1 (SDS:protein, w/w) and imparts a negative charge to the SDS-protein complex. The charged complexes move towards the anode when placed in an electric field, and are separated on the basis of differences in charge and size. SDS-PAGE is commonly used to estimate a protein's molecular weight, but estimates are approximate (being termed "apparent molecular weights") and sometimes prone to marked error. For instance, disproportionately large increases in apparent molecular weight may occur upon covalent phosphorylation of a protein *(1)*, or artificial entrapment of phosphoric acid *(2)*. Most designs of SDS-PAGE employ a "stacking gel." Such a system enables concentration of a sample from a comparatively large volume to a very small zone within the gel. The proteins within this zone are concentrated into very narrow bands, making them not only more easily detected but also better resolved from neighboring bands of other proteins. The principle involved in this protein concentration (or "stacking") is that of isotachophoresis. It is set up by making a stacking gel on top of the "separating gel," which is of a different pH. The sample is applied at the stacking gel and when the electric field is applied the negatively charged complexes and smaller ions move towards the anode. At the pH prevailing in the stacking gel, protein-SDS complexes have mobilities intermediate between the faster Cl⁻ ions (present throughout the electrophore-

From: *Methods in Molecular Biology, vol. 211: Protein Sequencing Protocols, 2nd ed.*
Edited by: B. J. Smith © Humana Press Inc., Totowa, NJ

sis system) and the slower glycinate ions (present in the cathode reservoir buffer). The protein-SDS complexes concentrate in the narrow zone between Cl^- and glycinate ions. When the moving zones reach the separating gel with its different pH their respective mobilities change and glycinate overtakes the protein-SDS complexes, which then move at rates governed by their size and charge in a uniformly buffered electric field. Isotachophoresis is described in more detail in the literature (e.g., **ref.** *3*).

SDS-PAGE requires microgram to submicrogram amounts of each protein sample. That is similar to amounts required for analysis by automated protein sequencing and mass spectrometry. The achievement of interfacing SDS-PAGE with sequencing has brought a notable step forward in sample handling technique: small amounts of a complex mixture may be resolved suitably for sequencing in just a few hours. This is done by transferring or "blotting" proteins which have been resolved by SDS-PAGE to polyvinylidene difluoride (PVDF) or other similar support, as described in Chapter 4. This medium may also be used for other analyses such as characterization by use of specific antibodies (Western blotting), such that specific proteins (on sister blots) may be identified for further analysis by sequencing or by mass spectrometry. It is important to maximize yields of sequencable protein throughout the whole process, however, and conditions for transfer may require optimization to obtain significant amounts of sample bound to the PVDF. Prior to that, however, the conditions for SDS-PAGE need to be such that minimal protein N-terminal blockage occurs by reaction of the free amino group with species in the gel. Usually, these reactive species and the blocking groups that they produce remain unknown but ways to remove them at least partially have been developed empirically. These include electrophoresis of the gel before application of the sample, but this destroys the isotachophoretic stacking system described earlier. For some applications, the accompanying reduction of resolution may be undesirable, but Dunbar and Wilson *(4)* have described a method that minimizes this problem. Their approach to SDS-PAGE for preparation of polypeptide samples for sequencing is described in this chapter.

2. Materials

1. Apparatus for PAGE: Slab gels are used, so as to allow the blotting procedures that follow electrophoresis. There are many commercial suppliers of the units, glass plates, spacers and combs (*see* **Note 1**) that are required for PAGE. Should it be necessary to build apparatus from scratch, refer to the design of Studier *(5)* but, for safety reasons, ensure that access to electrodes or buffers is impossible whilst the apparatus is connected to a live power supply. The direct current power supply required to run gels may also be obtained from commercial sources. Again for safety, check that the power supply has a safety cutout.

2. Stock acrylamide solution: (*see* **Notes 2** and **3**).

    Total acrylamide concentration, %T = 30% w/v. Ratio of crosslinking agent, bis-acrylamide to acrylamide monomer, % C = 2.7% w/w.

    Dissolve 73 g acrylamide and 2 g of *bis*-acrylamide in distilled water (HPLC grade), and make to 250 mL. Filter to remove any particulate matter. Store in brown glass, stable for weeks at 4°C.

    To lessen the problem of protein derivatisation, use the purest reagents available. BioRad and Fluka are sources of suitable acrylamide and *bis*-acrylamide. Beware the irritant and neurotoxic nature of acrylamide monomer, and avoid its contact with skin by wearing gloves, safety glasses, and other protective clothing. Wash thoroughly after any contact. Be especially wary of handling the dry powder acrylamide - use a fume hood and facemask. Ready prepared acrylamide-*bis*-acrylamide solutions are available commercially that obviate the need to handle powders; for instance, "Protogel " (from National Diagnostics) gives satisfactory results.

3. Stock (4X concentrated) Separating Gel buffer Pre-electrophoresis lower reservoir buffer; A: 0.4% (w/v) SDS, 1.5 *M* Tris-HCl, pH 8.8.

    Dissolve 1.0 g SDS and 45.5 g Tris base (tris (hydroxymethyl) amino methane) in about 200 mL distilled water (high-performance liquid chromatography [HPLC] grade), adjust the pH to 8.8 with concentrated HCl, and make the volume to 250 mL with water. Filter and store at 4°C at which it is stable for months. Use Analar grade SDS and tris base (e.g., from Sigma).

4. Stock (4X concentrated) Pre-electrophoresis upper reservoir buffer/Stacking Gel buffer; B: 0.4% (w/v) SDS, 0.5 *M* Tris-HCl, pH 6.8.

    Dissolve 1.0 g SDS and 15.1 g tris base in about 200 mL distilled water (HPLC grade), adjust the pH to 6.8 with HCl, and make to 250 mL with water. Filter and store at 4°C, at which it is stable for months. Use Analar reagents (e.g., from Sigma).

5. Stock ammonium persulphate: (*see* **Note 4**): 10% (w/v) ammonium persulphate in water.

    Dissolve 1.0 g ammonium persulphate (Analar grade) in 10 mL distilled water (HPLC grade). Although apparently stable in the dark at 4°C for weeks, it is probably best practice to renew it every few days.

6. TEMED (N, N, N', N'-tetramethylethylenediamine): use as supplied (e.g., from BioRad, electrophoresis purity grade) (*see* **Note 4**).

7. Water-saturated butanol: In a glass vessel mix some n- or butan-2-ol with a lesser volume of water. Leave to stand. The upper layer is butanol saturated with water.

8. Reservoir buffer (for sample electrophoresis); C: 0.192 *M* glycine, 0.1% (w/v) SDS, 0.025 *M* Tris-HCl, pH 8.3.

    Dissolve 28.8 g glycine, 6.0 g Tris Base, and 2.0 g SDS in distilled water (HPLC grade) and make to 2 L with water. The pH should be about pH 8.3 without adjustment. Store at 4°C. Stable for days.

9. Stock (200X concentrated) Glutathione solution: 10 m*M* reduced glutathione in water.

Dissolve 30.7 mg reduced glutathione (Analar grade, e.g., from Sigma) in 10 mL distilled water (HPLC grade). Store frozen at −20°C or −70°C. Stable for weeks.

10. Stock (1000X concentrated) Sodium thioglycollate solution: 100 mM sodium thioglycollate in water.

    Dissolve 114 mg sodium thioglycollate (Analar grade, e.g., from Sigma) in 10 mL distilled water (HPLC grade). Stable for weeks, frozen to −20°C or −70°C.

11. Stock (2X strength) sample buffer: (*see* **Note 5**): 4.6% (w/v) SDS, 0.124 M Tris-HCl (pH 6.8), 10.0% (v/v) 2-mercaptoethanol, 20.0% (w/v) glycerol, 0.05% (w/v) bromophenol blue

    Dissolve the following in distilled water (HPLC grade) to volume less than 20 mL: 0.92 g SDS; 0.3 g Tris base; 4.0 g glycerol; 2 mL 2-mercaptoethanol; 2 mL bromophenol blue solution (0.1% w/v in water). Adjust pH to 6.8 with HCl, and make volume to 20 mL. Although stable at 4°C for days, over longer periods exposed to air, the reducing power of the 2-mercaptoethanol may wane. Aliquots of stock solution may be stored for longer periods (weeks to months) if frozen to −20°C or −70°C. Use Analar reagents (e.g., from Sigma).

12. Protein staining solution: (*see* **Notes 6–8**):
    Protein stain: Sigma, product number B-8772: Coomassie Brilliant Blue G (C.I. 42655) 0.04% w/v in 3.5% w/v perchloric acid (*see* **Notes 6–8**). Stable for months at room temperature, in the dark. Beware the low pH of this stain. Where protective clothing. Use fresh, undiluted stain, as supplied.

13. Destaining solution: Distilled water.

3. Methods (*see* Notes 1–11)

1. Take the glass plates, spacers, and comb appropriate to the gel apparatus to be used. Thoroughly clean them by washing in soapy water, rinse in distilled water, and then methanol. Allow to air-dry. Assemble plates and spacers as instructed by suppliers in preparation for making the gel.

2. Prepare the separating gel mixture. 30 mL of mixture will suffice for one gel of about 14 × 14 × 0.1 cm, or four gels of 8 × 9 × 0.1 cm. For gel(s) of 15% T, the mixture is made as follows. Mix 15 mL stock acrylamide solution and 7.5 mL of distilled water (HPLC grade); degas on a water vacuum pump; add 7.5 mL of separating gel buffer A, 45 μL stock ammonium persulphate solution and 15 μL of TEMED. Mix gently and use immediately, because polymerisation starts when the TEMED is added (*see* **Notes 2–4**).

3. Carefully pipette the freshly mixed gel mixture between the prepared gel plates, without trapping any air bubbles. Pour to about 1 cm below where the bottom of the well-forming comb will come when it is in position. Carefully overlayer the gel mixture with a few mm-deep layer of water-saturated butanol (to eliminate air, which would inhibit polymerization and to generate a flat top to the gel). Leave the gel until it is set (0.5–1.5 h).

4. Prepare the upper gel mixture. 5 mL will provide a gel layer 1–2 cm deep for one gel 14 cm wide, 0.1 cm thick, or four gels of 9 cm wide, 0.1 cm thick. A 5% T gel

is made as follows: mix 0.83 mL stock acrylamide solution A and 2.92 mL distilled water (HPLC grade); degas on a water vacuum pump; add 1.25 mL of stock separating gel buffer A, 15 µL stock ammonium persulphate, and 5µL TEMED. Mix gently and use immediately.

5. Pour off the butanol from the polymerised separating gel. Rinse the top of the gel with a little water, then a little upper gel mixture (from **step 4**). Fill the gap remaining above the gel with the upper gel mixture from **step 4** Insert the well-forming comb without trapping any air bubbles. Leave to polymerize (0.5–1.5 h).

6. When the gel has finally polymerized store it at 4°C overnight or longer (*see* **Note 9**).

7. When the gel is to be used, remove the comb and the bottom spacer in order to expose the top and bottom edges of the gel. Install in the gel apparatus. Dilute the stock separating gel/pre-electrophoresis lower reservoir buffer A, by mixing one volume of it with three volumes of distilled water (HPLC grade). Pour it into the lower (anode) reservoir of the apparatus. Dilute the stock pre-electrophoresis upper reservoir/stacking gel buffer B by mixing one volume of it with three volumes of distilled water (HPLC grade). Mix in the stock glutathione solution, diluting it 200-fold to a final concentration of 50 µM glutathione. Pour this mixture into the top reservoir.

8. Add a few µL of sample solvent to one well and "pre-electrophorese" the gel by running it at low voltage (25–75v) according to the size of the gel). Continue this pre-electrophoresis until the blue band of bromophenol blue from the sample solvent reaches the boundary between upper and lower gels, and then switch off the power.

9. While pre-electrophoresis is in progress, prepare the sample(s) for electrophoresis as follows: Dissolve the sample in a small volume of water in a small polypropylene vial (e.g., Eppendorf) and mix in an equal volume of sample solvent. The volume of the sample solution should be small enough and the protein concentration great enough to enable sufficient protein to be loaded in a single well on the gel. Heat the sample in the capped vial at 100°C (i.e., in boiling water, or 100°–110°C in a heating block) for 2 min. Allow to cool and briefly centrifuge to bring any condensation to the bottom of the tube and to sediment any solid material present in the sample. The bromophenol blue dye indicates if the sample solution is acidic by turning yellow. If this occurs, add a few µL of NaOH solution, just sufficient to approximately neutralize the solution and turn it blue.

10. When the gel has been pre-electrophoresed (*see* **step 8**) remove the reservoir buffers and replace them with reservoir buffer for sample electrophoresis, C. Add stock sodium thioglycollate to the top (cathode) reservoir buffer, diluting it 1000-fold to a final concentration of 100 µM sodium thioglycollate. Using a microsyringe or pipette, load the prepared samples (*see* **step 9**). Start electrophoresis by applying voltage of, e.g., 150v (or 30–40 mA) for a gel of 8 × 9 × 0.1 cm, and continue until the bromophenol blue dye front (which indicates the position of the smallest, fastest-migrating species present in the sample) reaches the bottom of the gel. Turn off the power and remove the gel from the apparatus.

11. At this stage the gel is ready to be blotted (as described in Chapter 4), or stained. For staining, wash the gel for a few minutes with several changes of water (*see* **Note 6**), then immerse the gel (with gentle shaking) in the Colloidal Coomassie Brilliant Blue G. This time varies with the gel type (e.g., 1.0–1.5 h for a 1–1.5 mm thick SDS polyacrylamide gel slab), but cannot really be overdone. Discard the stain after use, for its efficacy declines with use. At the end of the staining period, decolorize the background by immersion in distilled water, with agitation, and a change of water whenever it becomes colored. Background destaining is fairly rapid, giving a clear background after a few hours (*see* **Notes 6** and **7**). Gels may be likewise stained after blotting to visualize protein remaining there (*see* **Notes 8** and **10**).

4. Notes

1. A factor in improving final sequencing yields is minimization of the size of the band of protein of interest, i.e., minimization of the size of the piece of PVDF which bears the sample band and which is finally put into the sequencer. Thus, use narrow sample wells in the gel, and put as much sample as possible in a single well. In doing this, however, beware that overloading a track with sample may distort electrophoresis and spoil band resolution.

2. The system described is basically the traditional discontinuous SDS-PAGE system of Laemmli *(6)*, set up in the manner described by Dunbar and Wilson (4) in order to generate a stacking buffer system during the pre-electrophoresis. The system described has a separating gel of 15% T, 2.7% C, 0.1% (w/v) SDS, 0.375 *M* Tris-HCl, pH8.8. The upper, stacking gel is 5% T, 2.7% C, 0.1% (w/v) SDS 0.125 *M* Tris-HCl, pH6.8, 5% (v/v) 2-mercaptoethanol, 10% (w/v) glycerol, 0.025% (w/v) bromophenol blue.

   This system may be modified in order to suit the needs of a particular application. Thus, gels of greater or lesser %T (acrylamide content) may be made by increasing or decreasing (respectively) the volume of stock acrylamide solution added to the mixture (*see* **Subheading 3.2.**) and adding proportionately less or more water (respectively). Gradient gels may be made by mixing two different % T mixtures as the gel is poured (e.g., *see* **ref. 7**).

   Another alternative is to vary the SDS content. For instance, Dunbar and Wilson *(4)* use 2% (w/v) SDS. Alternatively a "native" or nondenaturing gel may be made by deleting SDS entirely from all solutions, and also deleting the reducing agent, 2-mercaptoethanol from the sample solvent. In fact, a SDS-free gel may also be used for denaturing SDS-PAGE by inclusion of SDS in sample solvent and reservoir buffers, sufficient SDS deriving from these sources. Resolution of small proteins or peptides, below 5–10 kDa, may be problematical. Schagger and von Jagow *(8)* describe the use of tricine as trailing ion for improved resolution of polypeptides as small as 5 kDa or less.

3. In SDS-PAGE for the purposes of sequencing or mass spectrometry, the aim is to resolve mixtures of proteins while minimizing modification of the N-terminus or of side chains of the sample. Modifications are caused by reactive species in the

gel or sample solvent. As reviewed in **refs.** *(9)* and *(10)*, several modifications have been recognized: the product of cysteinyl reacting with residual acrylamide monomer, cysteinyl-S-β-propionamide; addition of 2-mercaptoethanol, probably to cysteinyl; oxidation of methionyl to methionyl sulphoxide, possibly by persulphate. The procedures described in **Subheading 3.** have been found to minimise problems due to reactive species. Correia et al. *(11)* described other effects, namely cleavage at Asp-Pro bonds during heating in sample buffer, and formation of covalent lysine-dehydroalanine crosslinks.

To minimize sample modification, only the purest available reagents are used. Check that supplies of reagents, especially of acrylamide, are acceptable by making a gel and running standard proteins, blotting them, and sequencing them.

Secondly, the amounts of ammonium persulphate and TEMED used to initiate polymerization are lower than used in some laboratories. Also, it is recommended that prepared gels are stored at 4°C for some time (overnight at least). These two points are meant to minimize the presence of reactive species such as radicals, with the storage of the gel intended to allow complete dissipation of radicals and completion of polymerization.

Thirdly, the purpose of pre-electrophoresis is to run reduced glutathione into the gel so that it runs ahead of the sample and reacts with any reactive species remaining (this strategy, and the use of sodium thioglycollate, being described in **ref.** *12*).

4. As stated in **Note 3**, amounts of ammonium persulphate and TEMED used are lower than those used in some laboratories. Oxygen can inhibit the polymerisation process, so degassing is used to reduce this problem. However, exhaustive degassing (e.g., prolonged degassing on a high vacuum pump) can result in a failure of the gel solution to polymerize evenly and completely. If polymerization fails, repeat the process with less degassing. Additionally, ensure that the ammonium persulphate solution is fresh. Do not increase amounts of persulphate or TEMED, for the reasons discussed in **Note 3**.

5. The sample buffer may be made as much as fourfold concentrated by simple alteration of the volumes given in **Subheading 2., item 11**. The advantage of this is that it is necessary to add less to the sample, so the sample itself is diluted less, and a larger amount of sample may therefore be loaded onto the gel. The sample may be prepared in sample buffer as described, then frozen at −10°C for future use. Dithiothreitol (DTT) has a less unpleasant smell than 2-mercaptoethanol and may be used as the reducing agent at 50 mM in the final sample solution (as suggested by Invitrogen for their "NuPAGE" gels, *see* **Note 11**). Note that if DTT is to be held as a stock solution its oxidation in the presence of trace concentrations of metals such as Fe^{3+} or Ni^{2+} should be inhibited by inclusion of a chelator such as EGTA *(13)*. For nonreduced SDS PAGE, omit the reducing agent. For nondenaturing PAGE, omit both reducing agent and SDS, and do not heat. If the sample is too dilute or contains too much salt (which may cause distortion of migration in the gel), this may be rectified by various concentration and buffer exchange strategies described in Chapter 1.

6. As an alternative to the commercial colloidal Coomassie Brilliant Blue G stain, it may be made from its separate components. Another commercially available alternative is the Gel Code blue stain reagent from Pierce (product number 24590 or 24592). Details of the stain components are not divulged, other than they also include Coomassie (G250), but the stain is used in the same way as described for the Sigma reagent, gives similar results, and costs approximately the same.

   The water wash that precedes staining by Coomassie Brilliant Blue G is intended to wash away at least some SDS from the gel, and so speed up destaining of the background. However, it should be remembered that proteins are not fixed in the gel until in the acidic stain mixture and consequently some loss of small polypeptides may occur in the wash step. Delete the wash step if this is of concern. A fixing step may be used immediately after electrophoresis (e.g., methanol:glacial acetic acid:water::50:7:43, (v/v/v) for 15–30 min, followed by water washing to remove the solvent and acid), though this may counter subsequent attempts to blot or otherwise extract the protein from the gel.

   Destaining of the background may be speeded up by frequent changes of the water, and further by inclusion in this wash of an agent that will absorb free dye. Various such agents are commercially available (e.g., Cozap, from Amika Corp.), but a cheap alternative is a plastic sponge of the sort used to plug flasks used for microbial culture. The agent absorbs the stain and is subsequently discarded. The background can be made clear by these means, and the stained bands remain stained while stored in water for weeks. They may be re-stained if necessary.

   Heavily loaded samples show up during staining with Coomassie Brilliant Blue G, but during destaining of the background the blue staining of the protein bands becomes accentuated. Bands of just a few tens of ng are visible on a 1 mm-thick gel (i.e., the lower limit of detection is less that 10 ng/mm^2). Variability may be experienced from gel to gel, however. For example, duplicate loadings of samples on separate gels, electrophoresed and stained in parallel, have differed in the degree of staining by a factor of 1.5, for unknown reasons. Furthermore, different proteins bind the dye to different extents: horse myoglobin may be stained twice as heavily as is bovine serum albumin (BSA), though this, too, is somewhat variable. While this formulation of Coomassie Brilliant Blue G is a good general protein stain, It is advisable to treat sample proteins on a case by case basis. This Coomassie stain may be used to quantify proteins in gels being quantitative, or nearly so, from about $10–20 \text{ ng/mm}^2$ up to about $1–5 \text{ μg/mm}^2$. The stoichiometry of dye binding is subject to some variation, such that standard curves may be either linear or slightly curved, but even the latter case is acceptable provided standards are run on the same gel as samples. Some reports claim that Coomassie blue staining may create problems in subsequent mass spectrometric analysis by virtue of adduct formation. In this case, alternative stains are available (*see* **Note 7**).

7. There is a variety of alternative stains to the Coomassie method described in **Note 6**. One of note is Sypro Ruby, available commercially from Molecular Probes (Eugene, OR, product number S-12000 or S-12001). The components

of the reagent are not revealed. Its cost is of the order of twice that of the Sigma Brilliant Blue G stain. To stain with this reagent, rinse the gel in water briefly, put it into a clean dish and then cover it with Sypro Ruby gel stain solution. Gently agitate until staining is completed, which may take up to 24 h or longer. Overstaining will not occur during prolonged stained. Do not let the stain dry up on the gel during long staining procedures. Discard the stain after use, for it becomes less efficacious with use. During the staining procedure the gel may be removed from the stain and inspected under UV light to monitor progress. If the staining is insufficient, the gel may be replaced in the stain for further incubation. Destain the background by washing the gel in a few changes of water for 15–30 min.

Generally, the Sypro Ruby method is more sensitive than the Coomassie Brilliant Blue G method, although for best sensitivity prolonged staining (24 h or more) may be required. Protein to protein variation can occur. For instance, horse myoglobin binds about 10-fold less dye than BSA does. Thus, in one experiment, the minimum amount of BSA detectable after Sypro Ruby staining was about 5 ng/mm^2 (about four- to fivefold more sensitive than samples stained in parallel by the Coomassie Brilliant Blue G method), whereas the minimum amount of horse myoglobin detectable was about 50 ng/mm^2 (similar to that detectable by the Coomassie Brilliant Blue G stain). Sypro Ruby requires UV irradiation for detection, but does not entail a fixation step. This is an advantage over traditional silver staining methods, whose sensitivity it approaches.

The silver staining method may be made compatible with subsequent analysis by in-gel proteolysis and mass spectrometry if the gel is fixed in 50% methanol, 5% acetic acid (v/v), but not by glutaraldehyde *(14)*. Gharahdaghi et al. *(15)* recommend an additional step of removal of silver from the protein by washing in fresh reducing solution (15 m*M* potassium ferricyanide, 50 m*M* sodium thiosulfate), prior to in-gel proteolysis and matrix-assisted laser-desorption ionization-time of flight (MALDI-TOF) mass spectrometry of eluted peptides. A mass spectrometry-compatible silver stain is available commercially (SilverQuest from Invitrogen).

Negative stains stain the background while leaving the protein band unstained. Cohen and Chait *(16)* used either copper or imidazole-zinc negative stain prior to protein extraction from the gel for mass spectrometric analysis (*see* **Note 12**).

8. Do not stain the gel before blotting. Blot, and then stain the PVDF with suitable stain to detect bands. A suitable PVDF blot stain is Ponceau S (0.1% w/v in 1% v/v acetic acid in water), the blot being immersed for 1–2 min, then washed in water to destain the background (and with time, bands also). See Chapter 4 for further detail of blot staining. If, for some reason, the gel has been stained by Coomassie stain before it is decided to blot it for further analysis, it can still be blotted in just the same way as used for unstained gels. The stained protein will transfer to PVDF and can be subject to N-terminal sequencing procedures. The yield of sequencable protein is markedly less than if the gel is not stained, but may still be sufficient to provide sequence information.

9. If resolution of bands is not so critical, gels may be prepared with the discontinuous buffer system in place. For this, prepare the separating gel as described in **Subheadings 3.2.** and **3.3.** and prepare the stacking gel as described in **Subheading 3.4.** but substitute buffer B for buffer A. The gel may then be used for nonsequencing purposes, or pre-electrophoresed with 50 μM glutathione in reservoir buffer C, to scavenge reactive species present. Gels with a discontinuous buffer system in place cannot be stored for much longer than a day, for the buffers diffuse into each other. The system described in Methods, with just one buffer (prior to pre-electrophoresis), may be conveniently stored for days, provided that the gel is sealed tightly in a plastic bag or in plastic wrapping such as laboratory sealing film.

10. The overall yield of PAGE, blotting and sequencing is heavily dependent not only on the PAGE stage, but also on the blotting stage. For further detail of the latter refer to Chapter 4.

11. The method described earlier is for preparation of SDS gels from scratch. Ready-made gels are available commercially from various sources, however. Ready-made gels from Daiichi, and Novex (Invitrogen) have proven suitable in our lab. They provide a viable alternative to scratch-made gels; while more expensive on materials, they reduce manpower costs and provide convenience.

    A discontinuous buffer system such as that described earlier (for the stacking gel system) is not stable during long-term storage. Commercial gels from Novex do not have such a system. Novex technical literature explains that at the pH 8.8 of the separating gel, the mobility of glycine is low enough to allow protein-SDS complexes to stack in a low % T gel as well as they would do with a pH 6.8 stacking gel, at least for proteins of about 70 KDa or less. No pre-electrophoresis of Novex gels is necessary, possibly because of the extended period of time between preparation of the gel and its use (*see* **Note 3**). In fact, if the pre-electrophoresis procedure (**Subheading 3**, **steps 7** and **8**) and inclusion of thioglycollate (**Subheading 3.10.**) are carried out in a Novex gel, band resolution is worsened.

    Invitrogen (Carlsbad, CA) market "NuPAGE" gels that may be stored for prolonged periods: a maximum of one year is recommended but older gels may still behave satisfactorily (performance did not deteriorate after 2 yr storage in our lab). The system operates at pH 7.0, using Bis-Tris instead of Tris-chloride buffers. The loading buffer, pH 8.5, uses lithium dodecyl sulphate, which has greater solubility than SDS, and either 2-mercaptoethanol or 0.05 M DTT as reducing agent (*see* **Note 5**). The buffers and other reagents supplied by Invitrogen are compatible with peptide sequencing. Use of an anti-oxidant in the upper buffer reservoir is recommended to inhibit (re)oxidation of protein during electrophoresis and so maintain band sharpness, but beware that its presence is sufficient to cause some reduction of at least some proteins in a nonreducing gel. The gels can successfully resolve proteins of about 5 kDa or less.

12. Proteins may be prepared from polyacrylamide gels by use of equipment available from various companies such as BioRad and AmiKa Corp. This may be useful if procedures such as peptide mapping are to follow. In essence, the protein is electrophoresed out of the piece of gel into solution. It is held in a small

volume of solution by a dialysis membrane of low nominal molecular-weight cut-off (e.g., 10 kDa). Unstained gel may be excised after being located by comparison with stained sister track(s). The yield of protein is in the range of 50–100%, and it is sequencable. Contaminants may emerge with the protein, causing it to look somewhat streaky upon re-electrophoresis, for example. Some sample clean-up may be necessary if these contaminants interfere with subsequent analysis.

Depending on the apparatus used, the final sample volume may be as much as 1 mL. If this is excessively large, the sample may be concentrated, and the buffer may be exchanged for another, as described in Chapter 1. If the sample band is small and it is necessary to minimize the final volume of the gel extract, the gel piece may simply be cut into small cubes and immersed in the minimal volume of a suitable buffer such as 100 mM Tris-HCl, pH 8.5, 0.1% SDS (w/v). Overnight incubation at room temperature, followed by brief centrifugation, may give sufficient protein in the supernatant to allow analysis. Beware that the rate of passive diffusion through and out of the gel decreases as the molecular weight of the sample protein and the degree of polyacrylamide crosslinking (%C) each increases. Cohen and Chait describe passive extraction of protein from polyacrylamide gel prior to mass spectrometric analysis *(16)*. Protein is detected in the gel by negative staining (*see* **Note 7**) and the gel piece cut out, destained and then crushed. Amounts of 25 pmole or more of protein may be extracted by vigorous shaking at ambient temperature for 4 to 8 h in a mixture of water, acid and organic solvent, such as formic acid: water: 2-propanol::1:3:2 (v/v/v). After drying under vacuum the sample may be analysed by mass spectrometry. For samples of less than 25 pmole the crushed gel is shaken with a MALDI matrix solution (e.g., a saturated solution of 4-hydroxy-a-cyano-cinnamic acid in formic acid: water: 2-propanol::1:3:2, v/v/v) for 1–2 h at room temperature, then left exposed to air to allow crystallization of the 4-hydroxy-α-cyano-cinnamic acid. The suspension of crystals may be analyzed directly by MALDI MS.

13. The voltage or current used to run the gels is somewhat arbitrary. Low power will give slow migration of proteins. Higher voltage gives faster migration but generates more heat. The buffers can be cooled by circulating them through a bath of ice, for instance, but migration of bands is then slower. The voltage suggested in **Subheading 3.10.** will run the bromophenol dye to the bottom of the gel of 8 × 9 × 0.1 cm, in about 2 h. Whether the gel is run at constant current or wattage, set an upper limit to the voltage. This prevents the voltage rising to dangerously high levels in the event of a fault in the system causing high resistance. Alternatively, run at constant voltage.

14. Various chemical and enzymatic methods have been described for cleavage of proteins while still in polyacrylamide gel, prior to elution an further analysis by HPLC or mass spectrometry – *see* Chapters 5 and 6.

References

1. Hutchinson, K. A., Dalman, F. C., Hoeck, W., Groner, B., and Pratt, W. B. (1993) Localisation of the ~ 12kDa Mr discrepancy in gel migration of the mouse gluco-

corticoid receptor to the major phosphorylated cyanogen bromide fragment in the transactivating domain. *J. Steroid Biochem. Mol. Biol.* **46**, 681–688.

2. Fountoulakis, M., Vilbois, F., Oesterhelt, G., and Vetter , W. (1995) Phosphoric acid entrapment leads to apparent protein heterogeneity. *Biotechnology* **13**, 383–388.

3. Deyl, Z. (1979) *Electrophoresis: A Survey of Techniques and Applications. Part A.* Elsevier, Amsterdam.

4. Dunbar, B. and Wilson, S. B. (1994) A buffer exchange procedure giving enhanced resolution to polyacrylamide gels prerun for protein sequencing. *Anal. Biochem.* **216**, 227–228.

5. Studier, F. W. (1973) Analysis of bacteriophage T7 early RNAs and proteins on slab gels. *J. Mol. Biol.* **79**, 237–248.

6. Laemmli, U. K. (1970) Cleavage of structural proteins during the assembly of the head of bacteriophage T4. *Nature* **227**, 680–685.

7. Walker, J. M. (1994) Gradient SDS polyacrylamide gel electrophoresis of proteins, in *Methods in Molecular Biology*, vol. 32, *Basic Protein and Peptide Protocols* (Walker, J. M., ed.) Humana Press Inc., Totowa, NJ, pp. 35–38.

8. Schagger, H. and von Jagow, G. (1987) Tricine-sodium dodecyl sulfate-polyacrylamide gel electrophoresis for the separation of proteins in the range from 1 to 100 kDa. *Anal. Biochem.* **166**, 368–379.

9. Patterson, S. D. (1994) From electrophoretically separated protein to identification: strategies for sequence and mass analysis. *Anal. Biochem.* **221**, 1–15.

10. Patterson, S. D. and Aebersold, R. (1995) Mass spectrometric approaches for the identification of gel-separated proteins. *Electrophoresis* **16**, 1791–1814.

11. Correia, J. J., Lipscomb, L. D., and Lobert, S. (1993) Nondisulfide crosslinking and chemical cleavage of tubulin subunits: pH and temperature dependence. *Arch. Biochem. Biophys.* **300**, 105–114

12. Yuen, S., Hunkapiller, M. W., Wilson, K. J., and Yuan, P. M. (1986) *Appl. Biosys. User Bull. No 25.*

13. Getz, E. B., Xiao, M., Chakrabarty, T., Cooke, R., and Selvin, P. R. (1999) A comparison between the sulphydryl reductants tris(2-carboxyethyl)phosphine and dithiothreitol for use in biochemistry. *Anal. Biochem.* **273**, 73–80.

14. Shevchenko, A., Wilm, M., Vorm, O., and Mann, M. (1996) Mass spectrometric sequencing of proteins from silver-stained gels. *Anal. Chem.* **68**, 850–858.

15. Gharahdaghi, F., Weinberg, C. R., Meagher, D. A., Imal, B. S., and Mische, S. M. (1999) Mass spectrometric identification of proteins from silver-stained polyacrylamide gel; a method for the removal of silver ions to enhance sensitivity. *Electrophoresis* **20**, 601–605.

16. Cohen, S. L. and Chait, B. T. (1997) Mass spectrometry of whole proteins eluted from sodium dodecyl sulfate-polyacrylamide gel electrophoresis gels. *Anal. Biochem.* **247**, 257–267.

3

Two-Dimensional Polyacrylamide Gel Electrophoresis for the Separation of Proteins for Chemical Characterization

Michael J. Dunn

1. Introduction

The first complete genome, that of the bacterium *Hemophilus influenzae*, was published in 1995 *(1)*. We now have the complete genomic sequences for more than 80 prokaryotic and eukaryotic organisms, and a major milestone has been reached recently with the completion of the human genome *(2,3)*. A major challenge in the post-genome era will be to elucidate the biological function of the large number of novel gene products that have been revealed by the genome sequencing initiatives, to understand their role in health and disease, and to exploit this information to develop new therapeutic agents. The assignment of protein function will require detailed and direct analysis of the patterns of expression, interaction, localization, and structure of the proteins encoded by genomes; the area now known as "proteomics" *(4)*.

The first requirement for proteome analysis is the separation of the complex mixtures containing as many as several thousand proteins obtained from whole cells, tissues, or organisms. Recently, progress has been made in the development of alternative methods of protein separation for proteomics, such as the use of chip-based technologies *(5,6)*, the direct analysis of protein complexes using mass spectrometry *(7)*, the use of affinity tags *(8,9)*, and large-scale yeast two-hybrid screening *(10)*. However, two-dimensional polyacrylamide gel electrophoresis (2-DE) remains the core technology of choice for separating complex protein mixtures in the majority of proteome projects *(11)*. This is due to its unrivaled power to separate simultaneously thousands of proteins, the subsequent high-sensitivity visualization of the resulting 2-D separations *(12)* that are amenable to quantitative computer analysis to detect differentially

From: *Methods in Molecular Biology, vol. 211: Protein Sequencing Protocols, 2nd ed.*
Edited by: B. J. Smith © Humana Press Inc., Totowa, NJ

regulated proteins *(13)*, and the relative ease with which proteins from 2-D gels can be identified and characterized using highly sensitive microchemical methods *(14)*, particularly those based on mass spectrometry *(15)*.

2. Materials

Prepare all solution from analytical grade reagents (except where otherwise indicated) using deionized double-distilled water.

1. 18 cm Immobiline IPG DryStrip pH 3-10 NL gel strips (Amersham Pharmacia Biotech, Amersham, UK) (*see* **Notes 1** and **2**).
2. IPG Immobiline DryStrip reswelling tray (Amersham Pharmacia Biotech) (*see* **Note 3**).
3. Multiphor II horizontal flatbed electrophoresis unit (Amersham Pharmacia Biotech) (*see* **Note 4**).
4. Immobiline DryStrip kit for Multiphor II (Amersham Pharmacia Biotech) (*see* **Note 5**).
5. Power supply capable of providing an output of 3500 V (*see* **Note 6**).
6. MultiTemp III thermostatic circulator (Amersham Pharmacia Biotech).
7. IEF electrode strips (Amersham Pharmacia Biotech) cut to a length of 110 mm.
8. Urea (GibcoBRL Ultrapure, Life Technologies, Paisley, UK) (*see* **Note 7**).
9. Solution A: 9 M urea (100 mL): Dissolve 54.0 g of urea in 59.5 mL deionized water. Deionize the solution by adding 1 g Amberlite MB-1 monobed resin (Merck, Poole Dorset, UK) and stirring for 1 h. Filter the solution using a sintered glass filter.
10. Solution B: sample lysis buffer: 9 M urea, 2% (w/v) CHAPS, 1% (w/v) dithiothreitol (DTT), 0.8% (w/v) 2-D Pharmalyte pH 3.0–10.0 (Pharmacia, St. Albans, UK) (*see* **Note 8**). Add 2.0 g CHAPS (*see* **Note 9**), 1.0 g DTT and 2.0 mL of Pharmalyte pH 3.0–10.0 to 96.0 mL of solution A.
11. Solution C: 8 M urea solution (40 mL): Dissolve 19.2 g of urea in 25.6 mL deionized water. Deionize the solution by adding 1 g Amberlite MB-1 monobed resin (Merck, Poole Dorset, UK) and stirring for 1 h. Filter the solution using a sintered-glass filter.
12. Solution D: reswelling solution: 8 M urea, 0.5% (w/v) CHAPS, 0.2% (w/v) DTT, 0.2% (w/v) 2-D Pharmalyte 3-10. Add 60 mg DTT, 150 mg CHAPS, and 150 µL Pharmalyte 3-10 to 29.7 mL of solution C.
13. Silicon fluid, Dow Corning 200/10 cs (Merck, Poole, Dorset, UK).
14. Solution E: Electrolyte solution for both anode and cathode: Distilled water.
15. Solution F: equilibration buffer (100 mL): 6 M urea, 30% (w/v) glycerol, 2% (w/v) SDS, 50 mM Tris-HCl buffer, pH 6.8. Add 36 g urea, 30 g glycerol, and 2 g SDS to 3.3 mL 1.5 M Tris-HCl buffer, pH 6.8.
16. Solution G: DTT stock solution: Add 200 mg DTT to 1 mL deionized water. Prepare immediately before use.
17. Solution H: Bromophenol Blue solution: Add 30 mg Bromophenol Blue to 10 mL 1.5 M Tris-HCL buffer, pH 6.8.

3. Methods

1. Sample preparation: Samples of isolated cells can be prepared by suspension in a small volume of lysis buffer, followed by disruption by sonication in an ice bath. Samples of solid tissues should be homogenised while still frozen in a mortar cooled with liquid nitrogen. The resulting powder is then suspended in a small volume of lysis buffer. Plant cells and tissues often require additional treatment *(18)*. The final protein concentration of the samples should be about 10 mg/mL. Protein samples should be used immediately or stored frozen at –80°C.

2. Rehydration of IPG gel strips with the protein sample: Dilute an aliquot of each sample containing an appropriate amount of protein (*see* **Note 10**) with solution D to a total volume of 450 µL (*see* **Note 11**). Pipet each sample into one groove of the reswelling tray. Peel off the protective cover sheets from the IPG strips and insert the IPG strips (gel side down) into the grooves. Avoid trapping air bubbles. Cover the strips with 1 mL of silicon oil, close the lid and allow the strips to rehydrate overnight at room temperature

3. Preparation of IEF apparatus: Ensure that the strip tray, template for strip alignment, and electrodes are clean and dry. Set the thermostatic circulator at 20°C (*see* **Note 12**) and switch on at least 15 min prior to starting the IEF separation. Pipet a few drops of silicon fluid onto the cooling plate and position the strip tray on the plate. The film of silicon fluid, which has excellent thermal conductivity properties and a low viscosity, allows for good contact between the strip tray and the cooling plate. Pipette a few drops of silicon fluid into the tray and insert the IPG strip alignment guide.

4. After rehydration is complete, remove the IPG strips from the reswelling tray, rinse them briefly with deionized water and place them, gel side up, on a sheet of water-saturated filter paper. Wet a second sheet of filter paper with deionised water, blot it slightly to remove excess water and place on the surface of the IPG strips. Blot them gently for a few seconds to remove excess rehydration solution in order to prevent urea crystallization on the surface of the gel during IEF.

5. IPG IEF dimension: Place the IPG gel strips side-by-side in the grooves of the alignment guide of the strip tray, which will take up to 12 strips (*see* **Note 13**). The basic end of the IPG strips must be at the cathodic side of the apparatus. Wet the electrode wicks with about 0.5 mL of the electrode solution (solution E) and remove excess liquid with a tissue. Place the electrode wicks on top of the strips as near to the gel edges as possible. Position the electrodes and press them down onto the electrode wicks. Fill the strip tray with silicon oil to protect the IPG strips from the effects of the atmosphere.

6. IEF running conditions: Run the IPG IEF gels at 0.05 mA per strip, and 5 W limiting. For the higher protein loads used for micro-preparative runs it is recommended to limit the initial voltage to 150 V for 30 min (75 Vh) and then 300 V for 60 min (300 Vh). Continue IEF with maximum settings of 3500 V, 2 mA, and 5 W until constant focusing patterns are obtained. The precise running conditions required depend on the pH gradient, the separation distances used, and the type of sample being analysed (*see* **Note 14**).

7. After completion of IEF, remove the gel strips from the apparatus. Freeze the strips in plastic bags and store them at −80°C if they are not to be used immediately for the second dimension separation.

8. Equilibration of IPG gel strips: Equilibrate IPG gel strips with gentle shaking for 2×15 min in 10 mL equilibration buffer (solution F). Add 500 µL/10 mL DTT stock (solution G) and 30 µL/10 mL Bromophenol Blue stock (solution H) to the first equilibration solution. Add 500 mg iodoacetamide per 10 mL of the second equilibration solution (final concentration iodoacetamide 5% w/v).

9. SDS-PAGE dimension: The second-dimension sodium dodecyl sulfate polyacrylamide gel electrophoresis (SDS-PAGE) separation is carried out using a standard vertical SDS-PAGE system (*see* **Note 15**) of the normal Laemmli type *(22)* (*see* **Note 16**) as described in Chapter 2. The gels can be either of a suitable constant percentage concentration of polyacrylamide or of a linear or nonlinear polyacrylamide concentration gradient. We routinely use 1.5 mm-thick 12% T SDS-PAGE gels (26 cm $\times$ 20 cm). No stacking gel is used.

10. Rinse the equilibrated IPG gel strips with deionized water and blot them on filter paper to remove excess liquid.

11. Apply the IPG gel strips to the SDS-PAGE gels by filling the space in the cassette above the separation gel with upper reservoir buffer and gently slide the strips into place. Good contact between the tops of the SDS gels and the strips must be achieved and air bubbles must be avoided. Cement the strips in place with 1% (w/v) agarose in equilibration buffer.

12. The gels are run in a suitable vertical electrophoresis apparatus. We use the Ettan DALT II vertical system (Amersham Pharmacia Biotech) which allows up to 12 large-format (26 $\times$ 20 cm) second-dimension SDS-PAGE gels to be electrophoresed simultaneously. The gels are run at 5 W/gel at 28°C for 45 min and then at 200 W maximum overnight at 15°C until the Bromophenol Blue tracking dye reaches the bottom of the gels. This takes approx 5 h for a full set of 12 gels.

13. The gels can be subjected to any suitable procedure to detect the separated proteins (*see* **Note 17**) or electroblotted onto the appropriate type of membrane (*see* Chapter 4) for subsequent chemical characterisation. A typical separation of human myocardial proteins using this technique is shown in **Fig. 1**.

4. Notes

1. We routinely use IPG gels with an 18 cm pH gradient separation distance, but it is possible to use gels of other sizes (e.g., 7 cm, 11 cm, 13 cm, 24 cm pH gradient separation distance) *(18)*. Small format gels (e.g., 7 cm strips) are ideal for rapid screening purposes or where the amount of sample limited, whereas extended separation distances (e.g., 24 cm IPG strips) provide maximum resolution of complex protein patterns.

2. A wide-range, linear IPG 3.0–10.0 L pH gradient is often useful for the initial analysis of a new type of sample. However, for many samples this can result in loss of resolution in the region pH 4.0–7.0, in which the pI values of many proteins occur. This problem can be overcome to some extent with the use of a nonlinear IPG

pH

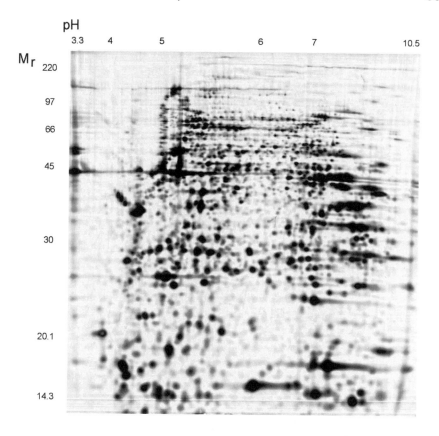

Fig. 1. Silver stained 2-D pattern of human myocardial proteins. A loading of 100 μg protein was used. The first dimension was pH 3.0–10.0 NL IPG IEF and the second dimension was 12% T SDS-PAGE.

3-10 NL pH gradient, in which the pH 4.0–7.0 region contains a much flatter gradient than in the more acidic and alkaline regions. This allows good separation in the pH 4.0–7.0 region while still resolving the majority of the more basic species. However, use of a pH 4.0–7.0 IPG IEF gel can result in even better protein separations within this range. With complex samples such as eukaryotic cell extracts, 2-DE on a single wide-range pH gradient reveals only a small percentage of the whole proteome because of insufficient spatial resolution and the difficulty of visualising low copy number proteins in the presence of the more abundant species. One approach to overcoming the problem is to use multiple, overlapping narrow range IPGs spanning 1–1.5 pH units; an approach that has become known as "zoom gels" *(23)*, "composite gels," or "subproteomics" *(24)*. Strongly alkaline proteins such as ribosomal and nuclear proteins with closely related pIs between 10.5 and 11.8 can be separated using narrow range pH 10.0–12.0 or pH 9.0–12.0 IPGs *(19)*.

3. The IPG Immobiline DryStrip reswelling tray (Amersham Pharmacia Biotech) is a grooved plastic tray with a lid designed for the rehydration of IPG DryStrip gels of any length from 7–24 cm in the presence of the solubilised protein sample.

4. We use the Multiphor II horizontal flatbed electrophoresis unit (Amersham Pharmacia Biotech). Any horizontal flat-bed IEF apparatus can be used for IPG IEF, but the Immobiline DryStrip kit (*see* **Note 5**) is designed to fit the Multiphor II. Another alternative is to use the IPGphor (Amersham Pharmacia Biotech), an integrated system dedicated to first-dimension IEF using IPG DryStrip with built-in temperature control unit and power supply.

5. The Immobiline DryStrip kit facilitates the sample application and running of IPG IEF gels in the first-dimension of 2-DE. The strip tray consists of a thin glass plate with a polyester frame. The frame acts as an electrode holder and the metal bars affixed to the frame conduct voltage to the electrodes. The electrodes, which are made of polysulphone, are moveable to accommodate gel strips of varying inter-electrode distance and have a platinum wire that rests against the electrode strip. It is also fitted with a bar, also made of polysulphone, which supports the sample cups (styrene-acrylnitrile). These cups can be used to apply sample volumes up to 100 µL as an alternative to the in-gel rehydration technique of sample application described here.

6. It is essential that the power supply can deliver less than 1 mA at 3,500 V, as these conditions are achieved during IEF of IPG gels. Powerpacks from some manufacturers are designed to cut out if a low current condition at high voltage is detected. The EPS 3501 XL power supply (Amersham Pharmacia Biotech) meets this requirement.

7. Urea should be stored dry at 4°C to reduce the rate of breakdown of urea with the formation of cyanate ions, which can react with protein amino groups to form stable carbamylated derivatives of altered charge.

8. Lysis buffer should be prepared freshly. Small portions of lysis buffer can be stored at –80°C, but once thawed they should not be frozen again.

9. We generally use the zwitterionic detergent CHAPS as this can give improved sample solubilisation compared with nonionic detergents such as Triton X-100 and Nonidet NP-40. However, the more hydrophobic membrane proteins are poorly solubilised under these conditions and it may be preferable to use a more powerful chaotropic agent such as thiourea and/or alternative linear sulphobetaine detergents such as SB 3-10 or 3-12 *(25)*.

10. For analytical purposes (e.g., silver staining) between 60 and 80 µg total protein from complex mixtures such as whole cell and tissue lysates should be applied. It is possible to obtain successful chemical characterisation on at least the more abundant protein spots using such a loading, but it is preferable for micro-preparative purposes for the sample to contain between 400 µg and 1 mg total protein.

11. The total volume for rehydration must be adjusted depending on the separation length of the IPG strip used; 175 µL for 7 cm, 275 µL for 11 cm, 325 µL for 13 cm, and 600 µL for 24 cm IPG strips.

12. The temperature at which IEF with IPG is performed has been shown to exert a marked effect on spot positions and pattern quality of 2-D separations *(26)*. Temperature control is, therefore, essential in order to allow meaningful comparison of 2-DE patterns. Focusing at 20°C was found to result in superior 2-D separations with respect to sample entry, resolution, and background staining compared with separations carried out at 10°C or 15°C *(26)*.

13. Exposure of the gel strips to the air should be as brief as possible to prevent the formation of a thin layer of urea crystals on the gel surface.

14. As a guide, for IPG IEF gel strips with an 18 cm pH gradient separation distance we use 60,000 Vh for micro-preparative purposes.

15. The second SDS-PAGE dimension can also be carried out using a horizontal flatbed electrophoresis apparatus. This method is described in **ref.** *(18)*.

16. For the isolation of proteins for chemical characterisation, it is essential to minimize the risk of chemical modification of the proteins during the various steps of 2-DE. The polymerisation efficiency of polyacrylamide is rarely greater than 90%, with the inevitable risk for modification of amino acid residues by free acrylamide. This area is reviewed in detail by Patterson *(27)*. The amino acid most at risk of acrylamide adduction has been found to be cysteine, resulting in the formation of cysteinyl-S-β-propionamide. In addition, the partial oxidation of methionine to methionine sulphoxide, presumably owing to the presence of residual persulphate in the gel, has also been demonstrated. Several approaches have been used to prevent gel electrophoresis-induced modification of proteins, including the use of scavengers such as glutathione or sodium thiogycolate for SDS-PAGE or free cysteine for IPG IEF gels *(27)*. However, we have not found it necessary to adopt these procedures for the successful chemical characterisation of proteins purified by 2-DE. This may be a consequence of the deionization step (*see* **Subheading 2.**, **step 13**) which we routinely employ and/or to the quality of the acrylamide and Bis obtained from the supplier (we use Electran grade reagents from Merck, Poole, Dorset, UK).

17. Organic dyes such as Coomassie blue R-250 and G-250 are compatible with most chemical characterisation methods, including mass spectrometry, but are limited by their relative insensitivity *(28)*. Silver staining allows the detection of low nanogram amounts of protein. However, standard silver-staining protocols almost invariably use glutaraldehyde and formaldehyde, which alkylate α- and ϵ-amino groups of proteins, thereby interfering with their subsequent chemical characterisation. To overcome this problem, silver-staining protocols compatible with mass spectrometry in which glutaraldehyde is omitted have been developed *(29,30)*, but these suffer from a decrease in sensitivity of staining and a tendency to a higher background. This problem can be overcome using postelectrophoretic fluorescent staining techniques *(28)*. The best of these at present appears to be SYPRO Ruby, which has a sensitivity approaching that of standard silver staining and is fully compatible with protein characterization by mass spectrometry *(31)*.

References

1. Fleischmann, R. D., et al. (1995) Whole-genome random sequencing and assembly of *Haemophilus influenzae* Rd. *Science* **269**, 496–512
2. Venter, J. C. et al. (2001) The sequence of the human genome. *Science* **291**, 1304–1351.
3. International Human Genome Sequencing Consortium (2001) Initial sequencing and analysis of the human genome. *Nature* **409**, 860–922.
4. Banks R., Dunn M. J., Hochstrasser D. F., et al. (2000) Proteomics: new perspectives, new biomedical opportunities. *Lancet* **356**, 1749–1756.
5. Merchant, M. and Weinberger, S. R. (2000) Recent advances in surface enhanced laser-desorption/ionization time-of-flight mass spectrometry. *Electrophoresis* **21**, 1165–1177.
6. Nelson, R. W., Nedelkov, D., and Tubbs, K. A. (2000) Biosensor chip mass spectrometry: a chip-based approach. *Electrophoresis* **21**, 1155–1163.
7. Link, A. J., Eng, J., Schieltz, D. M., et al. (1999) Direct analysis of protein complexes using mass spectrometry. *Nature Biotechnol.* **17**, 676–682.
8. Gygi, S. P., Rist, B., Gerber, S. A., et al. (1999) Quantitative analysis of complex protein mixtures using isotope-coded affinity tags. *Nature Biotechnol.* **17**, 994–999.
9. Rigaut, G., Shevchenko, A., Rutz, B., et al. (1999) A generic protein purification method for protein complex characterization and proteome exploration. *Nature Biotechnol.* **17**, 1030–1032.
10. Uetz, P., Giot, L., Cagney, G., et al. (2000) A comprehensive analysis of protein-protein interactions in Saccharomyces cerevisiae. *Nature* **403**, 623–627.
11. Dunn, M. J. and Görg, A. (2001) Two-dimensional polyacrylamide gel electrophoresis for proteome analysis, in *Proteomics, From Protein Sequence to Function* (Pennington, S. R. and Dunn, M. J., eds.), BIOS Scientific Publishers, Oxford, pp. 43–63.
12. Patton, W. F. (2001) Detecting proteins in polyacrylamide gels and on electroblot membranes, in *Proteomics, From Protein Sequence to Function* (Pennington, S. R. and Dunn, M. J., eds.), BIOS Scientific Publishers, Oxford, pp. 65–86.
13. Dunn, M. J. (1992) The analysis of two-dimensional polyacrylamide gels for the construction of protein databases, in *Microcomputers in Biochemistry* (Bryce, C. F. A., ed.), IRL Press, Oxford, pp. 215–242.
14. Wilkins, M. R. and Gooley, A. (1997) Protein identification in proteome analysis, in *Proteome Research: New Frontiers in Functional Genomics* (Wilkins, M. R., Williams, K. L., Appel, R. D. and Hochstrasser, D. F., eds.), Springer-Verlag, Berlin, pp. 35–64.
15. Patterson, S. D., Aebersold, R., and Goodlett, D. R. (2001) Mass spectrometry-based methods for protein identification and phosphorylation site analysis, in *Proteomics, From Protein Sequence to Function* (Pennington, S. R. and Dunn, M. J., eds.), BIOS Scientific Publishers, Oxford, pp. 87–130.
16. O'Farrell, P. H. (1975) High resolution two-dimensional electrophoresis of proteins. *J. Biol. Chem.* **250**, 4007–4021.

17. Klose, J. (1975) Protein mapping by combined isoelectric focusing and electrophoresis of mouse tissues. A novel approach to testing for induced point mutations in mammals. *Humangenetik* **26**, 231–243.

18. Görg, A., Obermaier, C., Boguth, G., et al. (2000) The current state of two-dimensional electrophoresis with immobilized pH gradients. *Electrophoresis* **21**, 1037–1053.

19. Görg, A., Obermaier, C., Boguth, G., et a;. (1997) Very alkaline immobilized pH gradients for two-dimensional electrophoresis of ribosomal and nuclear proteins. *Electrophoresis* **18**, 328–37.

20. Hanash, S. M., Strahler, J. R., Neel, J. V., et al. (1991) Highly resolving two-dimensional gels for protein sequencing. *Proc. Natl. Acad. Sci. USA* **88**, 5709–5713.

21. Bjellqvist, B., Sanchez, J.-C., Pasquali, C., et al. (1993) Micropreparative two-dimensional electrophoresis allowing the separation of samples containing milligram amounts of proteins. *Electrophoresis* **14**, 1375–1378.

22. Laemmli, U. K. (1970) Cleavage of structural proteins during the assembly of the head of bacteriophage T4. *Nature* **227**, 680–685.

23. Wildgruber, R., Harder, A., Obermaier, C., et al. (2000) Towards higher resolution: 2D-Electrophoresis of Saccharomyces cerevisiae proteins using overlapping narrow IPG's. *Electrophoresis* **21**, 2610–2616.

24. Cordwell, S. J., Nouwens, A. S., Verrils, N. M., et al. (2000) Sub-proteomics based upon protein cellular location and relative solubilities in conjunction with composite two-dimensional gels. *Electrophoresis* **21**, 1094–1103.

25. Santoni, V., Molloy, M., and Rabilloud, T. (2000) Membrane proteins and proteomics: un amour impossible? *Electrophoresis* **21**, 1054–1070.

26. Görg, A., Postel, W., Friedrich, C., et al. (1991) Temperature-dependent spot positional variability in two-dimensional polypeptide gel patterns. *Electrophoresis* **12**, 653–658.

27. Patterson, S. D. (1994) From electrophoretically separated protein to identification: Strategies for sequence and mass analysis. *Anal. Biochem.* **221**, 1–15.

28. Patton, W. F. (2000) A thousand points of light: The application of fluorescence detection technologies to two-dimensional gel electrophoresis and proteomics. *Electrophoresis* **21**, 1123–1144.

29. Shevchenko, A., Wilm, M., Vorm, O., and Mann, M. (1996) Mass spectrometric sequencing of proteins from silver-stained polyacrylamide gels. *Anal. Chem.* **68**, 85–858.

30. Yan, J. X., Wait, R., Berkelman, T., et al. (2000) A modified silver staining protocol for visualization of proteins compatible with matrix-assisted laser desorption/ionization and electrospray ionization-mass spectrometry. *Electrophoresis* **21**, 3666–3672

31. Yan, J. X., Harry, R. A., Spibey, C., and Dunn, M. J. (2000) Postelectrophoretic staining of proteins separated by two-dimensional gel electrophoresis using SYPRO dyes. *Electrophoresis* **21**, 3657–3665.

4

Electroblotting of Proteins from Polyacrylamide Gels for Chemical Characterization

Michael J. Dunn

1. Introduction

Since the first complete genome sequence, that of the bacterium *Hemophilus influenzae*, was published in 1995 *(1)*, a flurry of activity has seen the completion of the genomic sequences for more than 80 prokaryotic and eukaryotic organisms. Early in 2001 a major milestone was reached with the completion of the human genome sequence *(2,3)*. A major challenge in the post-genome era will be to elucidate the biological function of the large number of novel gene products that have been revealed by the genome-sequencing initiatives, to understand their role in health and disease, and to exploit this information to develop new therapeutic agents. The assignment of protein function will require detailed and direct analysis of the patterns of expression, interaction, localization, and structure of the proteins encoded by genomes; the area now known as "proteomics" *(4)*.

Techniques of polyacrylamide gel electrophoresis (PAGE) have an almost unrivaled capacity for the separation of complex protein mixtures. In particular, two-dimensional methods (2-DE) can routinely separate up to 2,000 proteins from whole cell and tissue homogenates, and using large format gels separations of up to 10,000 proteins have been described *(5,6)*. For this reason 2-DE remains the core technology of choice for protein separation in the majority of proteomics projects. Combined with the currently available panel of sensitive detection methods *(7)* and computer analysis tools *(8)*, this methodology provides a powerful approach to the investigation of differential protein expression. This has been complemented by the development over the last years of a battery of highly sensitive techniques of microchemical characterization, including N-terminal and internal protein microsequencing by automated

From: *Methods in Molecular Biology, vol. 211: Protein Sequencing Protocols, 2nd ed.*
Edited by: B. J. Smith © Humana Press Inc., Totowa, NJ

Edman sequencing, and amino acid compositional analysis *(9)*. More recently techniques based on the use of mass spectrometry for mass peptide profiling and partial amino acid sequencing have made this group of technologies the primary toolkit for protein identification and characterization in proteomics projects *(10)*.

A major obstacle to successful chemical characterization is efficient recovery of the separated proteins from the polyacrylamide gel as most procedures are not compatible with the presence of the gel matrix. The two major approaches used to overcome this problem for the recovery of intact proteins are electroelution and Western electroblotting. In the first method, protein zones are detected after PAGE by staining with Coomassie Brilliant Blue R-250. Protein-containing gel pieces are then excised and placed in an electroelution chamber where the proteins are transferred in an electric field from the gel into solution, and concentrated over a dialysis membrane with an appropriate molecular-weight cut-off. While this method can result in excellent protein recovery (>90%), it suffers from several disadvantages including:

1. The ability to handle only small numbers of samples at one time;
2. Contamination of the eluted protein with SDS, salts and other impurities which can interfere with subsequent chemical analysis;
3. Peptide chain cleavage during staining or elution; and
4. Chemical modification during staining or elution leading to N-terminal blockage *(11)*.

Although electroelution has largely been replaced by electroblotting, it is still occasionally successfully used, for example for protein mass analysis by matrix-assisted laser desorption mass spectrometry (MALDI-MS) *(12)*.

In Western blotting, proteins separated by 1-D or 2-D PAGE are blotted onto an appropriate membrane support, the total protein pattern visualized using a total protein stain, and the protein band or spot of interest excised. The protein, while still on the surface of the inert membrane support, can then be subjected to the appropriate microchemical characterization technique. The most popular method for the transfer of electrophoretically separated proteins to membranes is the application of an electric field perpendicular to the plane of the gel. This technique of electrophoretic transfer, first described by Towbin et al. *(13)*, is known as Western blotting. Two types of apparatus are in routine use for electroblotting. In the first approach (known as "tank" blotting), the sandwich assembly of gel and blotting membrane is placed vertically between two platinum-wire electrode arrays contained in a tank filled with blotting buffer. The disadvantages of this technique are that:

1. A large volume of blotting buffer must be used;
2. Efficient cooling must be provided if high current settings are employed to facilitate rapid transfer; and

3. The field strength applied (V/cm) is limited by the relatively large interelectrode distance.

In the second type of procedure (known as "semidry" blotting) the gel-blotting membrane assembly is sandwiched between two horizontal plate electrodes, typically made of graphite. The advantages of this method are that:

1. Relatively small volumes of transfer buffer are used;
2. Special cooling is not usually required although the apparatus can be run in a cold room if necessary; and
3. A relatively high field strength (V/cm) is applied due to the short interelectrode distance resulting in faster transfer times.

In the following sections both tank and semidry electroblotting methods for recovering proteins separated by 1-D or 2-D PAGE for subsequent chemical characterization will be described. In addition, a total protein staining procedure compatible with chemical characterization techniques is given. Electroblotting is ideal for the recovery of gel-separated proteins for automated Edman sequencing. It has also often been used (usually with trypsin) for subsequent peptide mass profiling by MALDI-MS. However, on-membrane digestion has now largely been superseded by methods of in-gel digestion as the latter process gives better overall sensitivity *(14)*.

2. Materials
2.1. Electroblotting

Prepare all buffers from analytical grade reagents and dissolve in deionized water. The solutions should be stored at 4°C and are stable for up to 3 mo.

1. Blotting buffers are selected empirically to give the best transfer of the protein(s) under investigation (*see* **Note 1**). The following compositions are commonly used:
   a. For characterization of proteins with pIs between pH 4.0 and 7.0: Dissolve 6.06 g Tris base and 3.09 g boric acid and make up to 1 L (*see* **Note 2**). Adjust the solution to pH 8.5 with 10 *M* sodium hydroxide *(15)*.
   b. For characterization of proteins with pIs between pH 6.0 and 10.0: Dissolve 2.21 g 3-(cyclohexyl-amino)-1-propanesulphonic acid (CAPS) and make up to 1 L (*see* **Note 2**). Adjust the solution to pH 11.0 with 10 *M* sodium hydroxide *(16)*.
2. Filter paper: Whatman 3MM filter paper cut to the size of the gel to be blotted.
3. Transfer membrane: FluoroTrans (Pall) cut to the size of the gel to be blotted (*see* **Note 3**).
4. Electroblotting equipment: A number of commercial companies produce electroblotting apparatus and associated power supplies. For tank electroblotting we use the Hoefer TE 42 Transphor II unit (Amersham Pharmacia Biotech), while

for semidry electroblotting we use the Multiphor II NovaBlot apparatus (Amersham Pharmacia Biotech).
5. Rocking platform.
6. Plastic boxes for gel incubations.

2.2. Protein Staining

1. Destain: 450 mL methanol, 100 mL acetic acid made up to 1 L in deionized water.
2. Stain: 0.2 g Coomassie Brilliant blue R-250 made up to 100 mL in destain.

3. Method

3.1. Electroblotting

3.1.1. Semidry Blotting

1. Following separation of the proteins by gel electrophoresis, place the gel in equilibration buffer, and gently agitate for 30 min at room temperature (*see* **Note 4**).
2. Wet the lower (anode) plate of the electroblotting apparatus with deionized water.
3. Stack 6 sheets of filter paper wetted with blotting buffer on the anode plate and roll with a glass tube to remove any air bubbles.
4. Place the prewetted transfer membrane (*see* **Note 5**) on top of the filter papers and remove any air bubbles with the glass tube.
5. Place the equilibrated gel on top of the blotting membrane and ensure that no air bubbles are trapped.
6. Apply a further six sheets of wetted filter paper on top of the gel and roll with the glass tube.
7. Wet the upper (cathode) plate with deionized water and place on top of the blotting sandwich.
8. Connect the blotter to power supply and transfer at 0.8 mA/cm^2 of gel area (*see* **Note 6**) for 1 h at room temperature (*see* **Note 7**).

3.1.2. Tank Blotting

1. Following separation of the proteins by gel electrophoresis, place the gel in equilibration buffer, and gently agitate for 30 min at room temperature (*see* **Note 4**).
2. Place the anode side of the blotting cassette in a dish of blotting buffer.
3. Submerge a sponge pad taking care to displace any trapped air and place on top of the anodic side of the blotting cassette.
4. Place two pieces of filter paper onto the sponge pad and roll with a glass tube to ensure air bubbles are removed.
5. Place the prewetted transfer membrane (*see* **Note 5**) on top of the filter papers and remove any air bubbles with the glass tube.
6. Place the equilibrated gel on top of the blotting membrane and ensure that no air bubbles are trapped.

7. Place a sponge pad into the blotting buffer taking care to remove any trapped air bubbles and then place on top of the gel.
8. Place the cathodic side of the blotting cassette on top of the sponge and clip to the anode side of the cassette.
9. Remove the assembled cassette from the dish and place into the blotting tank filled with transfer buffer.
10. Connect to the power supply and transfer for 6 h (1.5-mm thick gels) at 500 mA at 10°C (*see* **Note 7**).

3.2. Protein Staining

1. Remove the blotting membrane from the sandwich assembly.
2. Place the membrane into a dish containing the Coomassie blue staining solution for 2 min and agitate gently on the rocking platform.
3. Place the membrane into destaining solution and agitate for 10–15 min (or until the background is pale).
4. Wash the membrane with deionized water and place on filter paper and allow to air dry.
5. Place the membrane into a clean plastic bag and seal until required for further analysis. The membrane can be stored in this state at room temperature for extended periods without any apparent adverse effects on subsequent chemical characterization.
6. An example of a membrane stained by this method is shown in **Fig. 1**.

4. Notes

1. The use of transfer buffers containing glycine or other amino acids must be avoided for proteins to be subjected to microchemical characterization.
2. Methanol (10–20%, v/v) is often added to transfer buffers as it removes SDS from protein-SDS complexes and increases the affinity of binding of proteins to blotting membranes. However, methanol acts as a fixative and reduces the efficiency of protein elution, so that extended transfer times must be used. This effect is worse for high molecular-weight proteins, so that methanol is best avoided if proteins greater than 100 kDa are to be transferred.
3. Nitrocellulose membranes are not compatible with the reagents and organic solvents used in automated Edman protein sequencing. A variety of alternative (e.g., glass fiber-based and polypropylene-based) membranes have been used for chemical characterization *(17)*, but PVDF-based membranes (FluoroTrans, Pall; ProBlott, Applied Biosystems; Immobilon-P and Immobilon-CD, Millipore; Westran, Schleicher and Schuell; Trans-Blot, Bio-Rad) are generally considered to be the best choice for this application *(16)*. Nitrocellulose can be used as a support in applications such as internal amino acid sequence analysis and peptide mass profiling, where the protein band or spot is subjected to proteolytic digestion prior to characterization of the released peptides.
4. Gels are equilibrated in blotting buffer to remove excess SDS and other reagents that might interfere with subsequent chemical analysis (e.g., glycine). This step

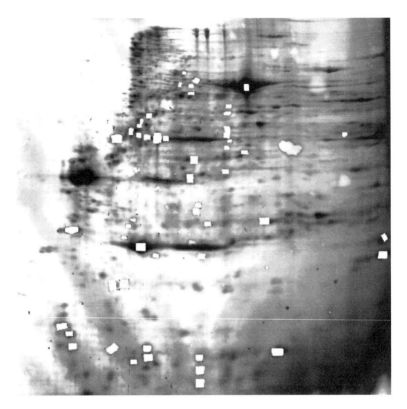

Fig. 1. PVDF (FluoroTrans) Western electroblot transfer of 300 µg human heart proteins separated by 2-DE and stained with Coomassie Brilliant blue R-250. The white areas indicate protein spots that have been excised for chemical characterization.

    also minimizes swelling effects during protein transfer. Equilibration may result in diffusion of zones and reduced transfer efficiencies of high molecular weight proteins. It is important to optimize the equilibration time for the protein(s) of interest.
5. Nitrocellulose membranes can be wetted with blotting buffer, but PVDF-based membranes must first be wetted with methanol prior to wetting with the buffer.
6. The maximum mA/cm² of gel area quoted applies to the apparatus we have used. This should be established from the manual for the particular equipment available.
7. Blotting times need to be optimized for the particular proteins of interest and according to gel thickness. Larger proteins usually need a longer transfer time, whereas smaller proteins require less time. Proteins will also take longer to be transferred efficiently from thicker gels. The transfer time cannot be extended indefinitely (> 3 h) using the semidry technique as the small amount of buffer used will evaporate. If tank blotting is used, the transfer time can be extended almost indefinitely (> 24 h) providing that the temperature is controlled.

References

1. Fleischmann, R. D., et al. (1995) Whole-genome random sequencing and assembly of *Haemophilus influenzae* Rd. *Science* **269**, 496–512

2. Venter, J. C., et al. (2001) The sequence of the human genome. *Science* **291**, 1304–1351.

3. International Human Genome Sequencing Consortium (2001) Initial sequencing and analysis of the human genome. *Nature* **409**, 860–922.

4. Banks R., Dunn M. J., Hochstrasser D. F., et al. (2000) Proteomics: new perspectives, new biomedical opportunities. *Lancet* **356**, 1749–1756.

5. Görg, A., Obermaier, C., Boguth, G., et al. (2000) The current state of two-dimensional electrophoresis with immobilized pH gradients. *Electrophoresis* **21**, 1037–1053.

6. Dunn, M. J. and Görg, A. (2001) Two-dimensional polyacrylamide gel electrophoresis for proteome analysis, in *Proteomics, From Protein Sequence to Function* (Pennington, S. R. and Dunn, M. J., eds.), BIOS Scientific Publishers, Oxford, pp. 43–63.

7. Patton, W. F. (2001) Detecting proteins in polyacrylamide gels and on electroblot membranes, in *Proteomics, From Protein Sequence to Function* (Pennington, S. R. and Dunn, M. J., eds.), BIOS Scientific Publishers, Oxford, pp. 65–86.

8. Dunn, M. J. (1992) The analysis of two-dimensional polyacrylamide gels for the construction of protein databases, in *Microcomputers in Biochemistry* (Bryce, C. F. A., ed.), IRL Press, Oxford, pp. 215–242.

9. Wilkins, M. R. and Gooley, A. (1997) Protein identification in proteome analysis, in *Proteome Research: New Frontiers in Functional Genomics* (Wilkins, M. R., Williams, K. L., Appel, R. D., and Hochstrasser, D. F., eds.), Springer-Verlag, Berlin, pp. 35–64.

10. Patterson, S. D., Aebersold, R., and Goodlett, D. R. (2001) Mass spectrometry-based methods for protein identification and phosphorylation site analysis, in *Proteomics, From Protein Sequence to Function* (Pennington, S. R. and Dunn, M. J., eds.), BIOS Scientific Publishers, Oxford, pp. 87–130.

11. Aebersold, R. (1991) High sensitivity sequence analysis of proteins separated by polyacrylamide gel electrophoresis, in *Advances in Electrophoresis, vol. 4* (Chrambach, A., Dunn, M. J., and Radola, B. J., eds.), VCH, Weinheim, pp. 81–168.

12. Patterson, S. D. (1994) From electrophoretically separated protein to identification: Strategies for sequence and mass analysis. *Anal. Biochem.* **221**, 1–15.

13. Towbin, H., Staehelin, T., and Gordon, G. (1979) Electrophoretic transfer of proteins from polyacrylamide gels to nitrocellulose sheets: procedure and some applications. *Proc. Natl. Acad. Sci. USA* **76**, 4350–4354.

14. Lahm, H. W. and Langen, H. (2000) Mass spectrometry: a tool for the identification of prtoeins separated by gels. *Electrophoresis* **21**, 2105–2114.

15. Baker, C. S, Dunn, M. J., and Yacoub, M. H. (1991) Evaluation of membranes used for electroblotting of proteins for direct automated microsequencing. *Electrophoresis* **12**, 342–348.

16. Matsudaira, P. (1987) Sequence from picomole quantities of proteins electroblotted onto polyvinylidene difluoride membranes. *J. Biol. Chem.* **262,** 10035–10038.

17. Eckerskorn, C. (1994) Blotting membranes as the interface between electrophoresis and protein chemistry, in *Microcharacterization of Proteins* (Kellner, R., Lottspeich, F., and Meyer, H. E., eds.), VCH Verlagsgesellchaft, Weinheim, pp. 75–89.

5

Enzymatic Cleavage of Proteins

Bryan John Smith

1. Introduction

Endoproteinases catalyse hydrolysis of polypeptide chains, most usefully at specific sites within the polypeptide, as described in **Table 1**. The number and nature of peptides generated by a proteinase of good specificity is characteristic of a protein, since it reflects the protein's sequence. The term "peptide map" is applied to the chromatogram or pattern of peptides resolved by a method such as high-performance liquid chromatography (HPLC) or capillary electrophoresis (*see* Chapter 8). Peptide mapping is widely used for quality control of recombinant proteins, where appearance of novel peptides indicates the presence of variant forms of protein (for example, *see* **refs. *1–4***). The mass spectroscopic equivalent of peptide mapping is called "mass mapping", whereby the masses of the products of proteolysis are characteristic of a given protein (*see* Chapters 17 and 18). Individual peptides may also be purified and subjected to various sequencing techniques as described elsewhere in this volume, the purpose being to identify a protein by its sequence, determine the partial sequence of a novel protein for cloning purposes, or identify sites of modification (for example, phosphorylation *[5]*).

Exopeptidases (carboxy- and aminopeptidases) also digest polypeptide substrates, but at their termini rather than at internal sites. Exopeptidases may be used to study C-terminal and blocked N-terminal sequences, for instance pyroglutamate aminopeptidase may remove an N-terminal pyroglutamate (*see* Chapter 29) Chemical methods of proteolysis have also been developed (*see* Chapter 6). They can usefully complement enzymatic methods because they have different specificity, but they may be unsuited to some purposes in that the harsh conditions employed may destroy biological activity.

From: *Methods in Molecular Biology, vol. 211: Protein Sequencing Protocols, 2nd ed.*
Edited by: B. J. Smith © Humana Press Inc., Totowa, NJ

It is usually preferable for the polypeptide cleavage event to be as specific as possible; cleavage at a multiplicity of different sites and in a variety of yields can lead to a complex mixture of peptides. The specificities of some proteinases are broad, but others are quite specific. Various of these are commercially available in especially pure from, usually obtained by extra rounds of chromatography. The purpose of the extra purification is to maximize specificity and minimize alternative cleavage owing to traces of contaminating enzymes. These especially pure enzyme preparations are given names such as "sequencer grade," and are recommended for more consistent peptide mapping and preparations. The specificities and other properties of various commercially available "sequencer grade" endoproteinases are summarized in **Table 1**. Trypsin (EC 3.4.21.4) is one of the most widely used proteinases, and the use of this enzyme is described in this chapter as an example of the approach to digestion of polypeptides by proteases.

Proteolysis is commonly carried out in solution but in recent years methods have been developed to allow proteolysis of samples that are bound to or trapped within a solid support, most notably nitrocellulose or polyvinylidene difluoride (PVDF) such as used in blotting of proteins from polyacrylamide gels (*see* Chapter 4), or the polyacrylamide gel itself. The aim of these developments has been to interface with the technique of polyacrylamide electrophoresis, a common separation method that has very high-resolving powers. In particular, two-dimensional electrophoresis is the method of choice for the resolution and analysis of complex protein mixtures and is a frequent starting point for the identification of proteins by chemical sequencing, amino acid analysis or mass-spectrometry. Proteolysis of samples resolved in gels can be achieved by digestion of the proteins in solution after they are eluted from the gel, but the recovery of proteins can be problematic in that it may be at low yield and with contamination by nonprotein components from the polyacrylamide gel. Protein digestions are therefore generally performed either within the gel itself (in-gel digestion) or after transfer of the sample to a membrane such as PVDF or nitrocellulose (*see* Chapter 4) and the digest performed *in situ*. These two approaches are generally equally successful. In contrast to digests in solution, however, digestion of proteins in gels and on blots may suffer from hinderance of access of enzymes to the protein substrate and possibly by poor retrieval of some peptides, which lead to incomplete peptide maps compared with digests in solution.

Optimization of the efficiency of proteolysis and the recovery of peptides from gels and blots is therefore of paramount importance.

Methods for digestion of polypeptides in these various states are presented separately.

Table 1
Characteristics of "Sequencer-Grade" Endoproteases

Enzyme	E C No.	Enzyme Class	Source	Approx Mol. Weight	Operating pH	Preferred Cleavage Site[a]	Example Digestion Buffer[b]	Inhibitor, Effective Concentration[b]	Notes
Chymotrypsin	3.4.21.1	Serine	Bovine pancreas	25 kDa	8–9	Y-X; F-X; W-X; (L-X; M-X; A-X; D-X; E-X)	Tris-HCl, 100 mM; CaCl$_2$, 10 mM, pH 7.8	AEBSF, 0.4–4 mM	Sites in brackets cleaved less rapidly. X may be amide or ester group
Endo Arg C	3.4.22.8	Cysteine	Clostridium histolyticum	50 kDa	8.0	R-X	Tris-HCl, 90 mM; CaCl$_2$, 8.5 mM	TLCK, 100–135 μM and DTT, 5 mM; pH 7.6	Reducing agent and Ca^{2+} for required activity (so oxidisin agents divalent metal ion chelators, e.g. EDTA, are alternative inhibitors) X may be amide or ester group. Alternative name: Clostripain.
Endo Asp N	–	Metallo	Pseudomonas fragi	27 kDa	6–8.5	X-D; X-C	Sodium phosphate, 50 mM; pH 8.0	EDTA, molar excess over divalent metal	
Endo Glu C	3.4.21.9	Serine	Staphylococcus aureus V8	27 kDa	7.8 4	E-X; D-X E-X	Ammonium carbonate; pH 7.8. Ammonium acetate; pH 4.0	3,4 dichloro-isocoumarin, 5–200 μM	Alternative name: Protease V8
Endo Lys C	3.4.99.30	Serine	Lysobacter enzymogenes	30 kDa	7–9	K-X	Tris-HCl, 25 mM; pH 8.5	TLCK, 100–135 μM	X may be amide or ester group. Apparent molecular weight increased to 33 KDa upon reducrion.
Trypsin	3.4.21.4	Serine	Bovine pancreas	23.5 kDa	8	K-X; R-X; Aminoethyl C-X	Tris. HCl, 100 mM; pH 8.5	AEBSF, 0.4–4 mM	X may be amide or ester group. Cleavage after Lys may be inhibited by succinylation or methylation of the Lys side chain.

[a] X = any amino acid. Susceptibility to proteolysis may be reduced or lost if the potentially cleavable bond is linked to P (e.g., K-P for chymotrypsin) or if between two like residues (e.g., E-E for Endo Glu C)

[b] Other conditions for reaction: enzyme:substrate (w/w):: 1:20 to 1:200, at 25–37°C for 2–18 h, optimized empirically for substrate in question. DTT = dithiothreitol. Abbreviations: AEBSF, 4-(2-aminoethyl)-benzenesulfonyl fluoride. EDTA, ethylenediaminetetra acetic acid. TLCK, L-1-chloro-3-[4-tosylamido]-7-amino-2-heptanone. *See* **Notes 7–9**, and **11**.

2. Materials
2.1. Proteolysis in Solution

1. Stock enzyme solution - Trypsin (EC.3.4.21.4) (*see* **Notes 1–3**): Available from various commercial suppliers. It is stable for periods of years as a dry solid kept at –20°C or –70°C. Make a stock solution of 1 mg/mL in 10 mM HCl (made in Milli-Q or HPLC-grade distilled water). Use fresh or divide into suitably sized aliquots and freeze to –70°C. A frozen stock solution may be thawed and refrozen several times without loss of the majority of the activity, but for consistent results thaw once only.
2. Reaction buffer: Ammonium bicarbonate (0.4 M) approximately pH 8.0 as prepared. For long-term storage, sterilize by filtration through a 0.2 mm filter and store refrigerated or frozen (*see* **Notes 4** and **5**)
3. Enzyme Inhibitor 4- (2-aminoethyl-benzenesulphonyl fluoride (AEBSF). 100 mM in water (Milli-Q or HPLC grade). Use fresh or store for 1 month at –20°C (*see* **Note 6**).

2.2. Proteolysis on Membranes

1. Destaining solution for stained proteins on membranes (see Note 10): dependent on the stain used, thus:
   a. For Coomassie brilliant blue R-stained PVDF: 70% (v/v) acetonitrile in water.
   b. For Amido Black-stained nitrocellulose or PVDF: 50% methanol, 40% water, 10% acetic acid (v/v/v).
   c. For Ponceau S-stained nitrocellulose or PVDF: 200 mM NaOH.
2. Digestion buffer-for trypsin: 10% methanol in ammonium carbonate (25 mM), pH 7.8 containing 1% (w/v) octyl ß glucoside. Make fresh or store for periods of days in the refrigerator to discourage microbial growth (*see* **Notes 11** and **12**).
3. Stock enzyme solution-for trypsin: dissolve the solid enzyme preparation in digestion buffer containing 10 % v/v methanol, to give a concentration of 1 mg/mL of enzyme. Stock solutions can be stored frozen in aliquots at –20°C, but should be thawed once only and then used immediately. Dilute to 0.1 mg/mL in digestion buffer prior to use (*see* **Note 13**).
4. Membrane extraction solution: 50% v/v formic acid in absolute ethanol. Prepare in advance and store at room temperature until used (*see* **Note 15**).

2.3. Proteolysis in Polyacrylamide Gel

1. Destain for Coomassie brilliant blue R-stained gel: 50% (v/v) acetonitrile in ammonium carbonate buffer (200 mM), pH 8.9.
2. Digestion buffer-for trypsin: 200 mM ammonium carbonate, pH 8.9, containing 0.02% (w/v) octyl β glucoside.
3. Stock enzyme solution - for trypsin: as in **Subheading 2.1.**, **step 1.**
4. Working enzyme solution: dilute the stock enzyme solution fourfold with digestion buffer to give an enzyme concentration of 250 µg/mL. Prepare immediately before use and discard excess solution after use (*see* **Note 13**).

5. Extraction solution: 60% (v/v) acetonitrile in 40% (v/v) trifluoroacetic acid (TFA) (0.1% v/v in water).

3. Methods

3.1. Proteolysis in Solution

1. Dissolve the sample protein in water (HPLC or Milli-Q grade) to a suitable concentration, such as 2 mg/mL. Add an equal volume of 0.4 M ammonium bicarbonate buffer i.e. final concentrations of 1mg/ml substrate, 0.2 M ammonium bicarbonate, pH 8.0. Add trypsin stock solution to a final enzyme : substrate ratio of 1:50 (w/w) i.e., to 1 mL of sample solution add 20 μL of 1 mg/mL trypsin solution. Mix thoroughly (*see* **Notes 1** to **5**).
2. Incubate the mixture at 37°C for 24 h.
3. Terminate proteolysis by addition of specific trypsin inhibitor, AEBSF, to a final concentration of 1 mM, i.e., add 1/100 volume of inhibitor solution. Store stopped reaction frozen (–20°C or –70°C) or immediately analyze or resolve by HPLC or other method (*see* **Note 6**).

3.2. Proteolysis on Membranes

1. Stained protein spots on PVDF membranes are excised and destained with 500 μL aliquots of the appropriate destaining solution (*see* **Subheading 2.2.**, **step 1**), refreshing the solution until all possible stain has been removed (*see* **Note 10**).
2. Dry the membrane pieces by laying them on filter paper, then cut each of them into 1–2 mm squares using a sharp scalpel. Transfer them to clean polypropylene microcentrifuge tubes.
3. Enzyme solutions are diluted to 0.1 μg/μL prior to use with the appropriate digestion buffer.
4. To each membrane piece add 1-4 μL of enzyme solution (diluted to 0.1 mg/mL in digestion buffer), the volume added depending on the size of the membrane piece. Note that PVDF is wetted by this solution without the need for prewetting in methanol. Allow the solution to absorb. Only add sufficient enzyme solution to wet the membrane. Centrifuge the tube briefly to deposit the membrane and digestion solution together in the bottom of the tube.
5. Cap tubes to restrict evaporation and incubate them in a water bath at 37°C for 3 h to overnight.
6. Extract peptides from each membrane piece by incubation with 50 μL of 50% formic acid/ethanol (v/v) for 1 h at 37°C. The extraction solution is then pipetted into a clean microcentrifuge tube and the membrane extracted with a further 50 μL of extraction solution for 1 h. Pool the two extraction aliquots and dry by centrifugal evaporation (*see* **Notes 14** and **15**).
7. Store the dried residue at 4°C until required for further analysis.

3.3. Proteolysis Within Polyacrylamide Gel

1. Excise protein spots of interest from the gel and then further destain each piece with two 500 μL aliquots of destain (50% acetonitrile in 200 mM ammonium

carbonate buffer, pH 8.9, *see* **Subheading 2.3.**, **step 1**) for 20–30 min at 30°C until colourless (*see* **Note 17**).

2. Place each gel slice on a clean sheet of sealing tissue (e.g., Parafilm) and air dry for 5–10 min until the gel has shrunk to less than half the original size (not to complete dryness), then cut into it small pieces (1–2 mm cubes) and place in a microcentrifuge tube (of small volume, to minimize the risk of gel drying by evaporation of liquid into the head space).

3. The slices are partially rehydrated with 5 μL of digestion buffer (ammonium carbonate, 200 m*M*, pH 8.9, containing 0.02% (w/v) octyl β glucoside, *see* **Subheading 2.3.**, **step 2**).

4. Rehydrate the gel further with 5 μL aliquots of digestion buffer containing trypsin (250 μg./mL, *see* **Subheading 2.3.**, **step 4**) until the gel has returned to its original size.

5. Seal and incubate the microcentrifuge tubes in a waterbath at 37°C overnight.

6. Extract peptides by addition of 250 μL extraction solution (*see* **Subheading 2.3.**, **step 5**). Incubate for 1 h at 37°C. Repeat the extraction and pool the extraction solutions. Dry by centrifugal evaporation. Store dried peptides at 4°C until required for analysis.

4. Notes

4.1. Digestion in Solution

1. Commercially available trypsin is prepared from bovine pancreas where it is synthesised as trypsinogen, the sequence of which is known (for example, **ref. [6]**). Active trypsin is generated in vivo by removal of the amino terminal hexapeptide. Its molecular weight is approximately 23,500 Daltons. Trypsin is optimally active at about pH 8.0. Stock solutions are made in 10 m*M* HCl at pH 2 (or 50 m*M* acetic acid is an acceptable alternative) and at –70°C, both of these conditions deterring autolysis. The trypsin regains activity when the pH is raised to above 4.0. Neutral or acidic buffers may be used (despite not providing the optimal pH for trypsin activity) where it is necessary to minimize the risk of disulphide bond interchange (as when isolating peptides to identify the positions of disulphide bridge in a sample protein).

   Trypsin has a serine at its active site and so belongs to the serine protease family. AEBSF, or phenylmethylsulphonyl fluoride (which is more unstable in water, and more toxic than is AEBSF) inhibit the enzyme by covalently modifying active site serine. Macromolecular trypsin inhibitors such as soybean trypsin inhibitor are not recommended because, being proteins, they may interfere with subsequent analyses.

   Trypsin displays good specificity, catalyzing the hydrolysis of the peptide bond to the COOH side of the lysyl and arginyl residues. Hydrolysis is slower if an acidic residue occurs to either side of the basic residue, and still slower if the residue to the COOH side is prolyl. Polylysine sequences may not be cleaved at every lysine. Bonds to the C-terminal side of methylated lysines may not be cleaved at all.

Trypsin available from some sources has been treated with L-1-chloro-3-tosylamido-4-phenylbutan-2-one in order to inhibit any chymotrypsin, which may be contaminating the preparation. Different trypsin preparations may vary in activity, so use only one batch of protease if reproducibility is important (e.g., for peptide mapping or for GMP work). As with other proteases, if trypsin is used at high concentration (say, trypsin:substrate::1:50 (w/w), or less) rare and unexpected cleavages may become apparent. This may be the case in digestion of samples on blots or in gels.

For some years it has been recognized that trypsin and other proteases may catalyze formation (rather than breakage) of peptide bonds, but for a significant level of this to occur addition of organic solvent to the buffer is generally required. To a small degree this transpeptidation reaction may occur in aqueous buffers neutral or acid pH conditions and this may produce small amounts of artificial polypeptides, detectable in peptide maps. As an example, Canova-Davis et al. *(7)* have reported that during digestion of relaxin by trypsin at pH 7.2 two (normally noncontiguous) peptides became linked by a peptide bond to a 10% level.

2. During digestion, autolysis of the trypsin occurs, to produce a background of trypsin peptides that is particularly noticeable when the protease is used at high concentration (e.g., for digestion of blots or in gels). These have been described in the literature *(8)* but it is always best to identify these in each experiment by inclusion of a control reaction of trypsin without substrate. To identify trace contaminants in the substrate (or buffers) include a control of substrate without trypsin. Reductive methylation of lysines (to ϵ-N,N-dimethyllysine) in the trypsin renders these sites insensitive to autolytic cleavage *(9)*. This modified trypsin is more stable and produces fewer interfering peptides. It is available commercially (for instance, from Promega).

3. Lysine is a common constituent of proteins and digestion with trypsin can generate a large number of peptides of small average size. This is a good point when peptide mapping, but some purposes, such as sequencing, may require longer peptides. The action of trypsin may be modified in an attempt to achieve this. It is done by modification of the side chains of lysyl or arginyl residues in the substrate, such that cleavage only occurs at unmodified residues. Perhaps the most common such method is succinylation of lysyl side chains, leading to tryptic cleavage at arginyl (and any remaining unmodified lysyl) residues (*see* Chapter 27, or **ref.** *[10]*). Introduction of additional sites of cleavage by trypsin may be achieved by conversion of cysteinyl residues, to aminoethyl cysteinyl residues by reaction with ethyleneimine as described in Chapter 27.

4. The method described is the basic procedure, and various of the conditions described for the method of digestion of substrates with trypsin in solution may be altered. Thus, the ammonium bicarbonate buffer (which is volatile and allows removal of salt by drying under vacuum) may be replaced by another buffer at pH 8.0. Again, 50 m*M* Tris-HCl will provide similar results but beware of the marked effect of temperature on the dissociation constant of Tris, the pH at 37°C being about one pH unit lower that at 4°C. Check the buffer pH at the temperature to be

used for the digestion. Addition of salt to high concentration (e.g., 0.5 M NaCl) favours compact folding of a structured polypeptide chain. Tightly folded domains are generally more resistant to proteolytic attack than are unstructured regions (though not indefinitely resistant), so high-salt conditions may be used to prepare structural domains which may retain biological function.

For complete digestion (as may be required for peptide mapping) prolonged incubation may be required, with further addition of trypsin (to replace trypsin which has been inactivated by digestion). Thus, conditions may be as follows: 37°C for 24 h at pH 8.0, then addition of further trypsin (similar to the first addition) followed by a further 24h incubation. On the other hand, preferential cleavage of particularly sensitive bonds or generation of partial cleavage products may be achieved by using less trypsin (say on enzyme:substrate ratio of 1:200 w/w, or more), shorter incubation times, lower incubation temperatures and/or a pH adjusted away from the optimum of pH8. Partial digestion products may be useful for determination of the order of neighbouring peptides in the parent sequence.

5. The condition of the substrate is important. First, the substrate should be soluble, or as finely divided as possible, in the digestion buffer. If a sample is not readily soluble in water or ammonium bicarbonate solutions, suitable solvents can be used initially and then adjusted by dilution or titration of pH to allow for trypsin action. If the polypeptide remains insoluble, the precipitate should be kept in suspension by stirring. Thus, 8 M urea may be used to solubilize a protein or disrupt a tightly folded structure and then diluted to 2 M urea for digestion by trypsin. Trypsin will also function in 2 M guanidinium chloride, in the presence of sodium dodecylsulfate (SDS, e.g., 0.1% w/v), or in the presence of acetonitrile (up to about 50%, v/v). Therefore fractions from reverse-phase HPLC (acetonitrile gradients in water/trifluoroacetic acid, 0.1%, v/v) may be readily digested after adjustment of pH by addition of ammonium bicarbonate or other buffer, and dilution (if necessary to lower the acetonitrile concentration).

Second, a native protein may be tightly folded, such as to markedly slow up or inhibit proteolytic attack. To remedy this, the substrate may be denatured and the structure opened out to allow for access of the proteinase. This may be done by boiling in neutral pH solution, or by use of such agents as urea, SDS, or organic solvent, as described above. Low concentrations of these agents (e.g., 5–10% acetonitrile, v/v) may give more rapid digestion than will a buffer without them, but high concentrations (e.g., 50% acetonitrile, v/v) will slow the digestion (also *see* **Note 1**, regarding the possibility of protease-catalyzed synthetic reactions in organic solvent).

An additional, and very common, technique is reduction and alkylation, i.e., permanent disruption of disulfide bonds. This treatment opens out the protein structure to allow for ready digestion and to minimize complications in peptide separation that are caused by pairs (or larger combinations) of peptides remaining connected by S-S bonds. This treatment is carried out as described in Chapter 27. Alternatively, if a sample has already been digested, S-S bonds may be reduced by simple addition of small amounts of dithiothreitol (as solid or as an aqueous solution) and incubation at room temperature (at pH 8.0) for 30 min or so. This treatment is followed immediately by HPLC to separate the various peptides. The amount of dithiothreitol required (i.e., slight molar excess over S-S bonds) may be calculated accurately if the cystine content of the protein concerned is known.

6. The digestion by trypsin may be stopped by addition of a serine protease inhibitor such as AEBSF as mentioned in **Note 1**. Alternatively, the solution's, pH may be adjusted far away from the optimal pH, e.g., lowered to pH 2.0 or so, by addition of acid (*see* **Note 1**). Keep the acidified solution cold, on ice, to minimize acid-catalyzed hydrolysis. Again, the reaction mixture may be immediately submitted to analysis (wherein trypsin is separated from substrate). Analysis and/or peptide preparation may be by reverse phase HPLC, capillary electrophoresis or polyacrylamide electrophoresis, or mass spectometry, as described in other chapters.

7. Other commercially available proteinases purified for sequencing purposes are summarized in **Table 1**. Use of these is essentially as described above for trypsin, except for use of buffers of appropriate pH and inclusion of divalent cations and reducing agents as required. Beware that complexing may occur between buffer salts and cations, thus affecting both cation concentration and buffering capacity for hydrogen ions. As an example calcium forms an insoluble phosphate in phosphate buffer.

    One of the most useful of these other proteinases is Endo proteinase Glu-C (Endo Glu C). It cuts to the COOH side of glutamyl residue. A lower frequency of cleavage to the COOH side of aspartyl residues may also occur at neutral pH, although at pH 4.0 this may not occur. Endoproteinase Glu-C functions in buffers containing 0.2% (w/v) SDS or 4 M urea. Its sequence is known *(11)*. It has been noted by a number of investigators that Endo Glu C can cause aspartamide formation by condensation of the side-chain of aspartate residues leading to a loss of 18 Da. Formation of aspartamide does not interfere with chemical sequencing, but may cause problems in peptide mass-fingerprinting

8. Other readily available proteinases are of broader specificity and may be affected by surrounding sequences. Their action is therefore difficult to predict. In particular instances, however, their observed action may prove beneficial by cleaving at one or a few particularly sensitive sites when incubated in suboptimal conditions (e.g., short duration digestion or nondenatured substrate).

    Good examples of this come from work on preparation of F(ab')$_2$, antigenbinding fragments of immunoglobulin IgG, that are bivalent and lack the constant Fc region of the molecule. Incubation of nondenatured IgG molecules with a proteinase of broad specificity can lead to proteolytic cleavage at a few sites or a single site in good yield. Pepsin has been used for this purpose (e.g., pH 4.2–4.5, enzyme:substrate::1:33, w/w, 37°C *[12]*). Different subclasses of mouse IgG were found to be digested at different rates, in the order IgG3 > IgG2a > IgG1. Different antibodies of the same subclass may also be degraded differently, some rapidly and without formation of F(ab')$_2$ *(12)*. Papain has been used to prepare F(ab')$_2$ fragments from the IgG1 subclass, which is the subclass that is most resistant to pepsin. The method described by Parham et al, *(13)* uses papain (which has been activated just before use by reaction with cysteine), at pH 5.5 (01.1 *M* acetate, 3 m*M* EDTA) 37°C, with an IgG concentration of about 10 mg/mL. The enzyme is added at time 0, and again later (e.g., at 9 h) to an enzyme:substate ratio of 1:20 (w/w). Digestion can be halted by addition of iodacetamide (30 m*M*) *(7)*. Rousseaux et al. *(14)* also described conditions for generating rat F(ab')$_2$, using papain (in the presence of 10 m*M* cysteine), pepsin or Endo Glu-C. Incubation of the IgG1 and 2a subclasses at pH 2.8 prior to digestion with pepsin improved the yields of F(ab')$_2$ fragments, presumably because the proteins thus denatured were effectively better substrates.

9. Two enzymes of broader specificities are worthy of further mention. The first is thermolysin, for its good thermostability that may prove useful when keeping awkward substrates in solution. Thermolysin remains active at 80°C or in 8 *M* urea. The second enzyme is pepsin, which acts at low pH. Disulfide bonds rearrange less frequently in acid than in alkaline conditions, so use of low pH buffers may not only help solubilize a substrate, but may also help preserve naturally disulfide bonded pairs of peptides. Endoproteinase Glu-C may also be used at low pH, having an optimum activity at pH 4.0.

4.2. Digestion on Blots

10. As a matter of course, destaining of blots removes what may be a source of interference in subsequent analyses. An alternative stain, sulforhodamine B, is compatible with mass spectrometry and need not be removed (*see* Chapter 18). If nitrocellulose membrane is used in place of PVDF, beware that it will dissolve in high concentrations of organic solvents and therefore care must be taken with staining and destaining steps (*see* **refs.** *15–17* for alternative formulations of stain and destain.

11. It is essential for the digestion of proteins on membranes to prevent adsorbtion of the enzyme to the membrane. This is achieved by using detergents, but the purity and stability of the detergents are important to prevent the formation or addition of reagent impurities, which will interfere with subsequent peptide analysis. For this reason detergents which are available in very high purity are necessary. Octyl β glucoside is one such detergent and it has little effect on subsequent analysis being both compatible with reversed-phase peptide fractionation and also with matrix-assisted laser desorption ionization mass-spectrometry (MALDI-MS). Octyl glucoside generates small peaks on reverse phase HPLC (monitored at 214 nm), but these are substantially less than those seen with Tween 20 or Triton X100, and are generally not a problem in analysis or peptide purification. Reduced Triton X-100 is also recommended owing to its low levels of UV absorbing contaminants compared with the unreduced detergent *(16)*.

Any enyzme-compatible buffers can be used for membrane digestions (*see* **Table 1**), but they should be supplemented with detergents and methanol to aid efficient digestion and peptide recovery. Volatile buffers such as ammonium carbonate and ammonium acetate are particularly useful as they do not interfere with subsequent biochemical procedures such as MALDI-MS. Buffers may also need to be supplemented with CaCl$_2$, dithiothreitol, or EDTA depending on the enzyme. The particular requirements of the enzyme in terms of pH and other co-factors is best determined by reference to the manufacturers technical bulletin which is normally supplied with the enzyme.

12. Access of the protease to substrate adsorbed to membrane is enhanced by the addition of detergents as mentioned in **Note 11**, which together with the addition of small quantities of organic solvents promotes wetting of the membrane. The compatibility of the membrane and enzyme with such solvents should however be ascertained before proceeding. The use of up to 50% (v/v) dimethyl sulfoxide (DMSO) in the digestion buffer has been reported to give much higher recoveries of hydrophobic peptides from digests on PVDF membranes. Generally, concentrations of up to 20% (v/v) methanol, n-propanol, or acetonitrile are tolerated by nitrocellulose and PVDF membranes, and are compatible with the proteases Lys-C, Glu C, Arg-C, and trypsin. However, the presence of solvents during proteolysis can cause transpeptidation (*see* **Note 1**), apparent in peptide maps and mass maps. A methanol concentration of 10% (v/v) can provide satisfactory results. If solvents are used they should be of HPLC-grade to prevent addition of contaminants that may interfere with subsequent analytical procedures. Likewise, water for all solutions should also be of HPLC grade or "polished" using a laboratory water-purification system (e.g., Milli-Q, Millipore).

13. A high concentration of protease is used for on-membrane digestions. This may encourage unusual cleavages and may also result in noticeable levels of autolysis products (especially if methylated trypsin is not used). These high concentrations of enzyme make high-purity enzymes (sequence-grade) essential for this type of analysis (*see* **Notes 1** and **4**).

14. Enzyme digestions of blotted samples are readily terminated by the addition of extraction solutions due to the high concentration of organic solvent and extreme pH. A double extraction is used to maximize the amount of peptides recovered and this is likely to be very important for hydrophobic peptides. Extraction of peptides from PVDF is an especial problem due to the high binding capacity and affinity of the membrane. Formic acid, which is a good solvent for peptides, can be used with ethanol and this gives good results with peptides digested on PVDF membranes and is also compatible with nitrocellulose.

15. After extraction, peptides can be dried down for subsequent analysis and centrifugal evaporation is the method of choice to reduce losses. However, with small quantities of peptides, drying completely can lead to irreversible binding of the peptides to the walls of tubes. In common with sample preparation for HPLC analysis, it may be advisable to only reduce the volume and remove organic solvents before further analysis. (*See also* Chapter 1.)

4.3. Digestion Within Gels

16. Successful digestion of small amounts of protein within polyacrylamide gels and their subsequent analysis is especially dependent on removal of residual SDS,

gel contaminants, and stain. This is most readily achieved by extracting the gel with 50% (v/v) acetonitrile *(18)*. Partial drying of the gel slices after destaining shrinks the gel and allows rapid entry of the enzyme solution into the gel as it rehydrates. Rosenfeld et al. have reported that complete drying of the slices leads to reduced recovery of peptides after digestion *(18)*. Hellman et al. *(19)*, on the other hand, have reported 50–85% yields of peptides after proteolysis in gels that have been completely dried. Jeno et al. *(20)* have described a further modification intended to suppress disulfide bond formation and so complexities in subsequent peptide maps. This is done by reduction (by DTT) and alkylation of Cys residues (by iodoacetamide) in the presence of 0.1% (w/v) SDS. The SDS is removed before HPLC analysis by addition of about 0.2 volume of 1 M guanidinium-HCl, followed by centrifugation.

17. Fixation or precipitation of protein in the gel during staining may reduce yields of extractable peptides. This can occur if the stain or destain is acidic or includes a fixation step such as with formaldehyde prior to some methods of silver staining. Sypro Ruby gel stain (Molecular Probes, Inc.) is a sensitive luminescent stain which has a pH of about 4.5 that is less likely to cause this problem. Equivalent Sypro Ruby blot stain is also available.

References

1. Jaquinod, M., Holtet, T. L., Etzerodt, M., Clemmensen, I., Thorgeson, H. C., and Roepstorff, P. (1999) Mass spectrometric characterisation of post-translational modificatin and genetic variation in human tetranectin. *Biol. Chem.* **380**, 1307–1314.

2. Gulati, D., Bongers, J., and Burman, S. (1999) RP-HPLC tryptic mapping of IgG1 proteins with post-column fluorescence derivatization. *J. Pharm. Biomed. Anal.* **21**, 887–893.

3. Samtora, L. C., Krull, I. S., and Grant, K. (1999) Characterization of recombinant human monoclonal tissue necrosis factor-alpha antibody using cation-exchange HPLC and capillary isoelectric focusing. *Anal. Biochem.* **275**, 98–108.

4. Bongers, J., Cummings, J. J., Ebert, M. B., Federici, M. M., Gledhill, L., Gulati, D., et al. (2000) Validation of a peptide mapping method for a therapeutic monoclonal antibody: what could we possibly learn about a method we have run 100 times? *J. Pharm. Biomed. Anal.* **21**, 1099–1128.

5. Xhou, W., Merrick, B. A., Khaledi, M. G., and Tomer, K. B. (2000) Detection and sequencing of phosphopeptides affinity bound to immobilized metal ion beads by matrix-assisted laser desorption/ionization mass spectrometry. *J. Am. Soc. Mass Spec.* **11**, 273–282.

6. Le Huerou, I., Wicker, C., Guilloteau, P., Toullec, R., and Puigserver, A (1990) Isolation and nucleotide sequence of cDNA clone for bovine pancreatic anionic trypsinogen. Structural identity within the trypsin family. *Eur. J. Biochem.* **193** 767–773.

7. Canova-Davis, E., Kessler, J., and Ling, V. T. (1991) Transpeptidation during the analytical proteolysis of proteins. *Anal. Biochem.* **196**, 39–45.

8. Vestling, M. M., Murphy, C. M., and Fenselau, C. (1990) Recognition of trypsin autolysis products by high-performance liquid chromatography and mass spectrmetry. *Anal. Chem.* **62**, 2391–2394.
9. Rice, R. H., Means, G. E., and Brown, W. D. (1977) Stabilization of bovine trypsin by reductive methylation. *Biochim. Biophys. Acta* **492**, 316–321.
10. Glazer, A. N., Delange, R. J., and Sigman, D. S. (1975) *Chemical Modification of Proteins.* Elsevier, North Holland, Amsterdam.
11. Drapeau, G. (1978) The primary structure of staphylococcal protease. *Can. J. Biochem.* **56**, 534–544.
12. Lamoyi, E. and Nisonoff, A. (1983) Preparation of F(ab')$_2$ fragments form mouse IgG of various subclasses. *J. Immunol. Meth.* **56**, 235–243.
13. Parham, P., Androlewicz, M. J. Brodsky, F. M., Holmes, N. J., and Ways, J. P. (1982) Monoclonal antibodies: Purification, fragmentation and application to structural and functional studies of class I MHC antigen. *J. Immunol. Meth.* **53**, 133–173.
14. Rousseaux, J., Rousseaux-Prevost, R., and Bazin, H. (1983) Optimal conditions for the preparation of Fab and F(ab')$_2$ fragments from monoclonal IgG of different rat IgG subclasses. *J. Immunol. Meth.* **64**. 141–146.
15. Sanchez, J.-C., Ravier, F., Pasquali, C., Frutiger, S., Paquet, N., Bjellqvist, B., et al. (1992) Improving the detection of proteins after transfer to polyvinylidene difluoride membranes. *Electrophoresis* **13**, 715–717.
16. Fernandez, J., DeMott, M., Atherton, D., and Mische, S. M. (1992) Internal protein sequence analysis: Enzymatic digestion of less than 10 mg of protein bound to polyvinylidene difluoride or nitrocellulose membranes. *Anal. Biochem.* **201**, 255–264.
17. Sutton, C. W., Pemberton, K. S., Cottrell, J. S., Corbett, J. M., Wheeler, C. H., Dunn, M. J., and Pappin, D. J. (1995) Identification of myocardial proteins from two-dimensional gels by peptide mass fingerprinting. *Electrophoresis* **16**, 308–316.
18. Rosenfeld., J., Capdevielle, J., Guillemot, J. C. and Ferrara, P. (1992) In-gel digestion of proteins for internal sequence analysis after one or two-dimensional gel electrophoresis. *Anal. Biochem.* **203**, 173–179.
19. Hellman, U, Wernstedt, C., Gonez, J and Heldin, C.-H. (1995) Improvement of an "in-gel" digestion procedure for the micropreparation of internal protein fragments for amino acid sequencing. *Anal. Biochem.* **224**, 451–455.
20. Jeno, P., Mini, T., Moes, S., Hintermann, E., and Horst, M. (1995) Internal sequences from proteins digested in polyacrylamide gels. *Anal. Biochem.* **224**, 75–82.

6

Chemical Cleavage of Polypeptides

Bryan John Smith

1. Introduction

Although proteins and peptides may be cleaved at various residues by use of endoproteolytic enzymes (*see* Chapter 5), they may be cleaved at still further sites by chemical methods. The most popular, best-yielding site for chemical cleavage is probably at the methionine residue. Cysteine is a significant residue, however, because it forms the disulfide bonds (by formation of cystine) which are so important in maintaining protein structure. Cleavage at cysteine may therefore be relevant to structural studies, and in any case can provide usefully large peptides because cysteine is a relatively uncommon amino acid. Likewise, cleavage of asparaginyl-glycyl bonds, and at tryptophanyl residues can provide large peptides. Further, since tryptophan is represented in the genetic code by a single codon, cleavage at that residue may be useful in cloning strategies in providing an unambiguous oligonucleotide sequence as part of a probe or primer. At the other extreme, aspartyl residues are relatively common, and cleavage of a protein there can generate a large number of small peptides (and partial cleavage products). Small proteins and peptides may be usefully cleaved at this point however, especially if other sites susceptible to other methods of cleavage are absent. Furthermore, partial hydrolysis can provide overlapping peptides, that can be analysed by mass spectrometric methods and used to order peptides in the sequence.

This chapter describes methods for chemical cleavage of Asn-Gly, Asp-X, Cys-X, Met-X, and Trp-X bonds, with the sample polypeptides being in solution. In recent years, however, methods have been adapted to suit samples on solid supports such as polyvinylidene difluoride (PVDF), as mentioned in the Notes section.

From: *Methods in Molecular Biology, vol. 211: Protein Sequencing Protocols, 2nd ed.*
Edited by: B. J. Smith © Humana Press Inc., Totowa, NJ

2. Materials

2.1. Cleavage of Asn-Gly Bonds

1. Cleavage buffer: 2 M hydroxylamine-HCl, 2 M guanidine-HCl, 0.2 M K_2CO_3, pH 9.0. Use Analar grade reagents and HPLC grade water. Beware the mutagenic, toxic, and irritant properties of hydroxylamine. Wear protective clothing. Clear wet spillages with absorbent material or clear dry spillages with a shovel, and store material in containers prior to disposal.
2. Stopping solution: trifluoroacetic acid (TFA), 2% (v/v) in water. Both of HPLC-grade.

2.2. Cleavage of Asp-X Bond

1. Dilute hydrochloric acid (approx 0.013 M) pH 2 +/-0.04: dilute 220 µL of constant boiling (6 M) HCl to 100 mL with distilled water.
2. Pyrex glass hydrolysis tubes.
3. Equipment includes a blowtorch suitable for sealing the hydrolysis tubes, a vacuum line, and an oven for incubation of samples at 108°C.

2.3. Cleavage of Cys-X Bond

1. Modification buffer: 0.2 M tris acetate, pH 8.0, 6 M guanidine-HCl, 10 mM dithiothreitol (DTT). Use Analar grade reagents and HPLC grade water.
2. NTCB: 2-nitro-5-thiocyanobenzoate. Commercially available (Sigma) as yellowish powder. Contact with skin, eyes, etc., may cause short-term irritation. Long term effects are unknown, so handle with care (protective clothing). Sweep up spillages. Store at 0–5°C.
3. NaOH: sodium hydroxide solution, sufficiently concentrated to allow convenient alteration of reaction pH. For example 2 M in HPLC grade water.
4. Deblocking buffer: 50 mM Tris-HCl, pH 7.0.
5. Raney nickel-activated catalyst:
   Commercially available (e.g., from Sigma as 50% slurry in water, pH >9.0). Wash in deblocking buffer prior to use. A supply of N_2 gas is also required for use with the Raney nickel.

2.4. Cleavage of Met-X Bond

1. 0.4 M Ammonium bicarbonate solution in distilled water. Stable for weeks in refrigerated stoppered bottle.
2. 2-Mercaptoethanol. Stable for months in dark, stoppered, refrigerated bottle.
3. TFA, HPLC- or sequencing grade.
4. Cyanogen bromide. Stable for months in dry, dark, refrigerated storage. Warm to room temperature before opening. Use only white crystals, not yellow ones. Beware of the toxic nature of this reagent.
5. Sodium hypchlorite solution (domestic bleach).
6. Equipment includes a nitrogen supply, fume hood and suitably sized and capped tubes (e.g., Eppendorf microcentrifuge tubes).

2.5. Cleavage of Trp-X Bond

1. Oxidizing solution: mix together 30 vol glacial acetic acid, 15 vol 9 *M* HCl, and 4 vol dimethylsulfoxide. Use best-grade reagents. Though each of the constituents is stable separately, mix and use the oxidizing solution when fresh.
2. 15 *M* Ammonium hydroxide.
3. Cyanogen bromide solution in formic acid (60% v/v): make 6 mL formic acid (minimum assay 98%, Aristar grade) to 10 mL with distilled water. Add white crystalline cyanogen bromide to a concentration of 0.3 g/mL. Use when fresh. Store cyanogen bromide refrigerated in the dry and dark, where it is stable for months. Use only white crystals. Beware of the toxic nature of this reagent.
4. Sodium hypochlorite solution (domestic bleach).
5. Equipment includes a fume hood and suitably sized capped tubes (e.g., Eppendorf microcentrifuge tubes).

3. Methods

3.1. Cleavage of Asn-Gly Bonds (see Notes 1–7)

1. Dissolve the protein sample directly in the cleavage buffer, to give a concentration in the range 0.1–5 mg/mL. Alternatively, if the protein is in aqueous solution already, add 10 volumes of the cleavage buffer (i.e., sufficient buffer to maintain pH 9.0 and high concentration of guanidine-HCl and hydroxylamine). Use a stoppered container (Eppendorf tube or similar) with small headspace, so that the sample does not dry out during the following incubation.
2. Incubate the sample (in stoppered vial) at 45°C for 4 h.
3. To stop reaction, cool and acidify by addition of 3 volumes of stopping solution. Store frozen (–20°C) or analyze immediately.

3.2. Cleavage of Asp-X Bond (see Notes 8–13)

1. Dissolve the protein or peptide in the dilute acid to a concentration of 1–2 mg/mL in a hydrolysis tube.
2. Seal the hydrolysis tube under vacuum, i.e., with the hydrolysis (sample) tube connected to a vacuum line, using a suitably hot flame, draw out and finally seal the neck of the tube.
3. Incubate at 108°C for 2 h.
4. To terminate the reaction, cool and open the hydrolysis tube, dilute the sample with water, and lyophilize.

3.3. Cleavage of Cys-X Bond (see Notes 14–20)

1. Dissolve the polypeptide to a suitable concentration (e.g., 2 mg/mL) in the modification buffer (pH 8.0). To reduce disulfides in the DTT, incubate at 37°C for 1–2 h.
2. Add NTCB to 10-fold excess over sulphydryl groups in polypeptide and buffer. Incubate at 37°C for 20 min.
3. To cleave the modified polypeptide, adjust to pH 9.0 by addition of NaOH solution. Incubate at 37°C for 16 h or longer.

4. Dialyse against water. Alternatively, submit to gel filtration or reverse phase HPLC to separate salts and peptides. Lyophilize peptides.

5. If it is necessary to convert the newly formed iminothiazolidinyl N-terminal residue to an alanyl group, dissolve the sample to, say, 0.5 mg/mL in de-blocking buffer (pH 7.0) and add to Raney nickel (10-fold excess, w/w, over polypeptide) and incubate at 50°C for 7 h under an atmosphere of nitrogen. Cool and centrifuge briefly to pellet the Raney nickel. Store supernatant at –20°C, or further analyze as required.

3.4. Cleavage of Met-X Bond (see Notes 21–30)

1. Reduction:
   a. Dissolve the polypeptide in water to between 1 and 5 mg/mL, in a suitable tube. Add 1 vol of ammonium bicarbonate solution, and add 2-mercaptoethanol to between 1 and 5% (v/v).
   b. Blow nitrogen over the solution to displace oxygen, seal the tube, and incubate at room temperature for approx 18 h.
2. Cleavage:
   a. Dry down the sample under vacuum, warming if necessary to help drive off all of the bicarbonate. Any remaining ammonium bicarbonate will form a salt on subsequent reaction with acid.
   b. Redissolve the dried sample in TFA to 1–5 mg/mL. Add water to make the acid 50% (v/v) finally.
   c. Add excess white crystalline cyanogen bromide to the sample solution, to between two- and 100-fold molar excess over methionyl residues. Practically, this amounts to approximately equal weights of protein and cyanogen bromide. To very small amounts of protein, add one small crystal of reagent. Carry out this stage in the fume hood.
   d. Seal the tube and incubate at room temperature for 24 h.
   e. Terminate the reaction by drying down under vacuum. Store samples at –10°C or use immediately.
   f. Immediately after use, decontaminate equipment (spatulas, tubes and so on) that has contacted cyanogen bromide, by immersion in hypochlorite solution (bleach) until effervescence stops (a few minutes).

3.5. Cleavage of Trp-X Bond (see Notes 31–42)

1. Oxidation: dissolve the sample to approx 0.5 nmol/μL in oxidizing solution (e.g., 2–3 nmol in 4.9 μL oxidizing solution). Incubate at 4°C for 2 h.
2. Partial neutralization: to the cold sample, add 0.9 vol of ice cold NH_4OH (e.g., 4.4 μL of NH_4OH to 4.9 μL oxidized sample solution). Make this addition carefully so as to maintain a low temperature.
3. Cleavage: add 8 vol of cyanogen bromide solution. Incubate at 4°C for 30 h in the dark. Carry out this step in a fume hood.
4. To terminate the reaction, lyophilize the sample (all reagents are volatile).
5. Decontaminate equipment, such as spatulas, that have contacted cyanogen bromide, by immersion in bleach until the effervescence stops (a few minutes).

4. Notes

4.1. Asn-Gly Cleavage

1. The reaction involved in the cleavage of the Asn-Gly bond is illustrated in **Fig. 1** (with more detail provided by **refs. *1*** and ***2***). The reaction of hydroxylamine actually is with the cyclic imide which derives from the Asn-Gly pair. Asp-Gly cannot form this succinimide, so that bond is resistant to cleavage by hydroxylamine. Kwong and Harris *(3)* have reported cleavage at an Asp-Gly bond, via a presumed succinimide at that site. Bornstein and Balian *(1)* have reported an Asn-Gly cleavage yield of about 80% but yields are somewhat dependent on the sequence of the protein. Other reactions may occur upon treatment of polypeptide with hydroxylamine. Cleavage is to the C-terminal side of the succinimide. The peptide to the C-terminal side is available for N-terminal sequencing. Because Asp-Gly is relatively rare (about 0.25% of dipeptide sequences), quite large peptides may result from cleavage by hydroxylamine.

   The succinimide residue is involved in spontaneous asparagine deamidation and aspartate racemisation and isomerization, for it can hydrolyze in neutral or alkaline conditions to aspartyl-glycyl and isoaspartyl-glycyl (or α aspartyl-glycyl and β aspartyl-glycyl). The isomerization of Asp to iso-Asp can affect immunogenicity and function (for instance, *see* **ref.** *[4]*). The succinimide is stable enough to be identified in proteins, the succinimidyl version being slightly more basic (by 1 net negative charge) than the aspartate version, which forms after incubation in neutral pH *(3)*. Assays are available for quantification of iso-Asp *(see* **ref.** *[5]* and refs. therein). Both iso-Asp and succinimide are detected as a termination of peptide sequencing, for both are refractory to Edman chemistry. Cleavage by hydroxylamine may be used to map the positions of succinimides and presumed iso-Asp that may arise from them (e.g., **ref.** *4*).

2. In addition to cleavage at Asn-Gly, there may be other, lower yielding cleavages. Bornstein and Balian *(1)* mention cleavage of Asn-Leu, Asn-Met, and Asn-Ala, while Hiller et al. *(6)* report cleavage of Asn-Gln, Asp-Lys, Gln-Pro, and Asn-Asp. Prolonged reaction times tend to generate more of such cleavages. Treatment with hydroxylamine may also generate hydroxamates of asparigine and glutamine, these modifications producing more acidic variants of the protein *(7)*.

3. Inclusion of guanidine-HCl as a denaturant seems to be a factor in improving yields. Kwong and Harris *(3)* reported that omission of guanidine-HCl eliminated Asn-Gly cleavage while still allowing cleavage at Asp-Gly. However, the literature does have examples of the use of buffers lacking guanidine-HCl. **References** *(6)* and *(7)* exemplify the use of a Tris-HCl buffer of approximate pH 9.0, with **ref.** *5* including 1 m*M* EDTA and ethanol (10% v/v). Other examples *(1,8,9)* describe the use of more concentrated (6 *M*) guanidine-HCl.

4. As when making peptides by other chemical or enzymatic cleavage methods, it may be advisable, prior to the cleavage steps, to reduce disulfide bonds and alkylate cysteinyl residues (*see* Chapter 27). This denatures the substrate and prevents formation of inter-peptide disulfide bonds. Alkylation and subsequent cleavage by hydroxylamine on a few-μL scale is described in **ref.** *(8)*.

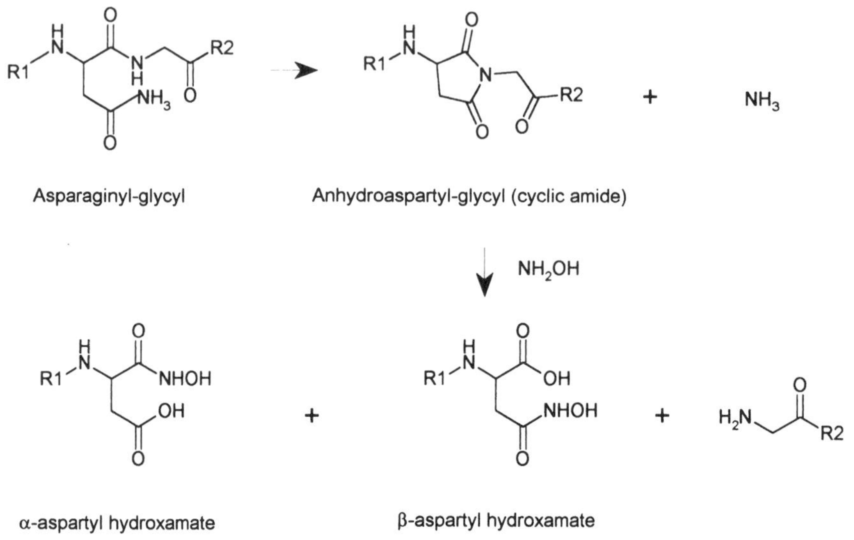

Fig. 1. Illustration of reactions leading to cleavage of Asn-Gly bonds by hydroxylamine.

5. After the cleavage reaction has been stopped by acidification, the sample may be loaded directly onto reverse-phase HPLC or gel filtration for analysis/peptide preparation. Alternatively, electrophoresis (*6*) will separate reactants and stop the reaction. Electrophoresis may be suitable for analysis of cleavage because large peptides are generally produced, but small peptides may be lost.

6. The hydroxylamine cleavage method has been adapted by Saris et al. (*10*) to cleave proteins in polyacrylamide gel pieces as follows:

   a. Wash gel piece(s) containing sample in 5% (v/v) methanol in order to remove SDS.
   b. Dry the gel pieces under vacuum.
   c. Submerge (and rehydrate) the gel pieces in cleavage solution, adding about 10–50 μL solution per 1 μL of gel piece. The cleavage solution is: 2 *M* hydroxylamine-HCl, 6 *M* guanidine-HCl. in 15 m*M* tris titrated to pH 9.3 by addition of 4.5 *M* lithium hydroxide solution. Preparation of the lithium hydroxide solution may generate insoluble carbonates, but these can be removed by filtration.
   d. Incubate at 45°C for 3 h.
   e. For analysis of cleavage, place the gel piece on the top of a second gel and undertake electrophoresis.

Saris et al. (*10*) reported that peptides of 10,000 Da or less could be lost during washes of the gel piece, while about 10% of the sample remained bound to the treated gel piece. Recoveries were about 60% in the second (analytical) gel, and cleavage yield was about 25%. Recovery may be adversely affected by fixing of protein in the staining procedure (by the use of acidic stain or destain solutions, for example).

7. In approximately neutral pH conditions, reaction of protein with hydroxylamine may cause esterolysis, and so may be a useful method in studying post-translational modification of proteins. Thus, incubation in 1 *M* hydroxylamine, pH 7.0, 37°C for up to 4 h cleaved carboxylate ester-type ADP-ribose-protein bonds (on histones H2A and H2B) and arginine-ADP-ribose bonds (in histones H3 and H4) *(11)*. Again, Weimbs and Stoffel *(12)* identified sites of fatty acid-acylated cysteine residues by reaction with 0.4 *M* hydroxylamine at pH 7.4, such that the fatty acids were released as hydroxamates. Omary and Trowbridge *(13)* adapted the method to release [³H] palmitate from transferrin receptor in polyacrylamide gel pieces, soaking these for 2 h in 1 *M* hydroxylamine-HCl titrated to pH 6.6 by addition of sodium hydroxide.

4.2. Asp-X Cleavage

8. The bond most readily cleaved in dilute acid is the Asp-X bond, by the mechanism outlined in **Fig. 2A**. The bond X-Asp may also be cleaved, in lesser yields (*see* **Fig. 2B**). Thus, either of the peptides resulting from any one cleavage may keep the aspartyl residue at the point of cleavage, or neither might, if free aspartic acid is generated by a double cleavage event. Any of these peptides is suitable for sequencing.

9. The method described is that of Inglis *(14)*. The amino acid sequence of the protein can affect the lability of the affected bond because the aspartic acid side chain can interact ionically with basic changes elsewhere in the molecule. Yields of cleavage are less than 100%, up to about 70% have been reported *(14)*.

   The aspartyl-prolyl bond is particularly labile in acid. Landon *(15)* has suggested that cleavage of Asp-Pro bonds may be maximised by minimizing the effect of intramolecular interactions, this being achieved by use of denaturing agent, as follows:

   a. Dissolve the sample in guanidine.HCl (7 *M*) in acetic acid (10% v/v, adjusted to pH 2.5 by addition of pyridine).
   b. Incubate at 37°C for a prolonged period (e.g., 24 h).
   c. Terminate by lyophilization.

   Because of the influence of protein sequence, the results of incubation of polypeptide in dilute acid are somewhat unpredictable and best investigated empirically.

10. The conditions of low pH can be expected to cause a number of side reactions: cleavage at glutamyl residues; deamidation of (and possibly some subsequent cleavage at) glutaminyl and asparaginyl residues; partial destruction of tryptophan; cyclization of N-terminal glutaminyl residues to residues of pyrolidone carboxylic acid; α-β shift at aspartyl residues. The last two changes create a blockage to Edman degradation. The short reaction time of 2 h is intended to minimize these side reactions. A small degree of loss of formyl or acetyl groups from N-termini *(14)* is another possible side reaction but is not recognized as a significant problem, generally.

11. The method described has the benefit of simplicity. It is carried out in a single reaction vessel, with reagents being removed by lyophilization at the end of reac-

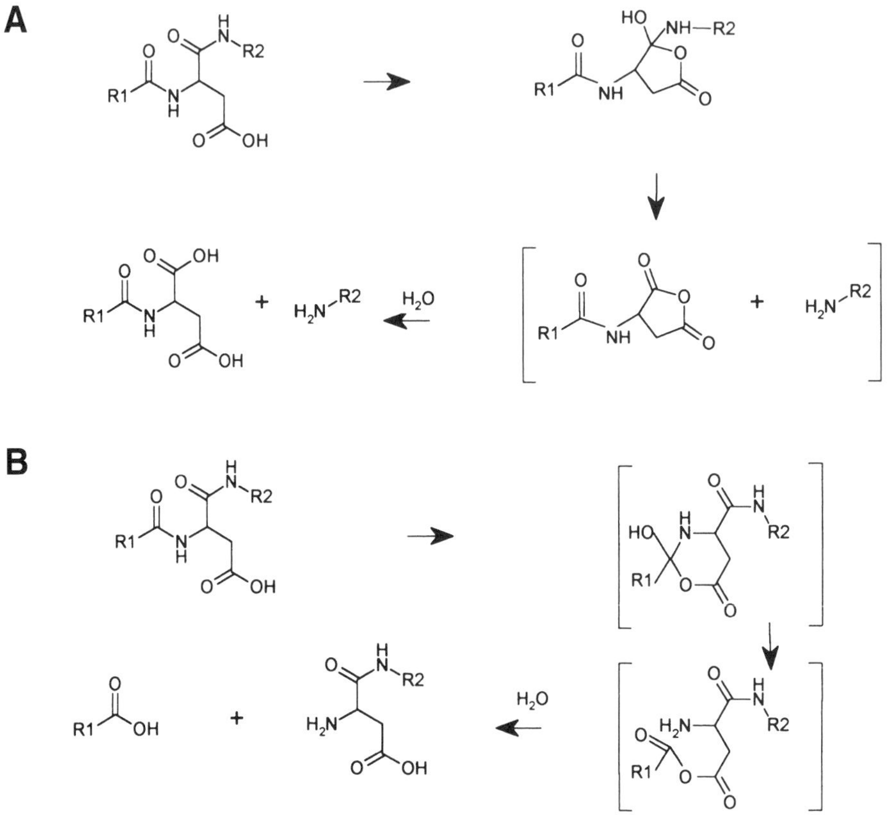

A

B

Fig. 2. Mechanisms of the cleavage of bonds (**A**) to the COOH side and (**B**) to the NH$_2$ side of aspartyl residues in dilute acid.

tion. Thus, sample handling and losses incurred during this are minimized. This makes it suitable for sub-nanomolar quantities of protein, though the method may be scaled up for larger amounts also.

12. A polypeptide substrate that is insoluble in cold dilute HCl may dissolve during the incubation at 108°C. Formic acid is a good protein denaturant and solvent and may be used instead of HCl as follows: Dissolve the sample in formic acid (minimum assay 98%, Aristar grade), then dilute 50-fold to pH 2.0; proceed as in method for HCl. Note, however, that incubation of protein in formic acid may result in formylation (detected as a 28 amu increase in mass [16]) and damage to tryptophan and tyrosine residues (altered spectral properties [17]).

13. Note that bonds involving aspartyl residues may also be cleaved by commercially available enzymes: endoproteinase Asp-N hydrolyses the bond to the N-terminal side of an aspartyl residue, but also of a cysteinyl residue; Glu-C cleaves the bond to the C-terminal side of glutamyl and aspartyl residues.

4.3. Cys-X Cleavage

14. The reactions involved in the method for Cys-X cleavage are illustrated in **Fig. 3**. The method described is basically that used by Swenson and Frederickson *(18)*, an adaptation of that of Jacobson et al. *(19*; also *see* **ref. 20**). The principle difference is that the earlier method *(19,20)* describes desalting (by gel filtration or dialysis) at the end of the modification step (**Subheading 3.3.**, **step 2**), followed by lyophilization and redissolution in a pH 9.0 buffer to achieve cleavage. Simple adjustment of pH as described in **Subheading 3.3.**, **step 3** has the advantages of speed and avoiding the danger of sample loss upon desalting.

    With conversion of the iminothiazolidinyl residue to an alanyl residue (in **Subheading 3.3.**, **step 5**), the peptide to the C-terminal side of the cleavage point is available for sequencing by Edman chemistry. If blockage of the N-terminal residue of the newly generated peptide to the C-terminal side of the cleavage point is not a problem (i.e. if sequencing is not required) **Subheading 3.3.**, **step 5** may be omitted.

15. Swenson and Frederickson *(18)* describe cleavage (**Subheading 3.3.**, **step 3**) at 37°C for 6 h, but report yields of 60–80%. Other references recommend longer incubations of 12 h or 16 h at 37°C to obtain better yields *(19–21)*.

16. Peyser et al. *(21)* have described a slightly modified procedure that may be more convenient for treating small samples. The procedure is as follows:

    a. Dissolve the sample to 1 mg/mL in a buffer of borate (20 mM) pH 8.0, urea (6 M).
    b. Add NTCB (0.1 M solution in 33% [v/v] dimethylformamide) at the rate of 40 μL of sample solution.
    c. Incubate at 25°C for 1 h.
    d. Adjust to pH 9.0 by addition of NaOH. Incubate at 55°C for 3 h. This brings about cleavage.
    e. Stop the reaction by addition of 2-mercaptoethanol to 80-fold excess over NCTB.

17. The conditions for reduction may be altered (**Subheading 3.3.**, **step 1**). Thus, if the sample contains no intramolecular or intermolecular disulfide bonds, the dithiothreitol (DTT) content of the modification buffer may be less, at 1 mM. Beware that nominally non-bonded cysteinyl residues may be involved in mixed disulfides with such molecules as glutathione or free cysteine. Reduction may be omitted altogether to allow reaction with native protein. Cys residues that remain protected in the native protein remain noncyanylated on reaction with 2-nitro-5-thiocyanobenzoate, and remain uncleaved upon alteration of pH to pH 9.0. Thus Cys residues that are buried within a native protein's structure or in a complex of proteins can be mapped within the proteins sequence, and regions involved in protein-protein interaction "footprinted" *(22)*.

18. Although Raney nickel is available commercially, Otiene *(23)* has reported that a more efficient catalyst may be obtained by the method he described, starting from Raney nickel-aluminium alloy. This is reacted with NaOH, washed, deionized and washed again (under H_2 gas).

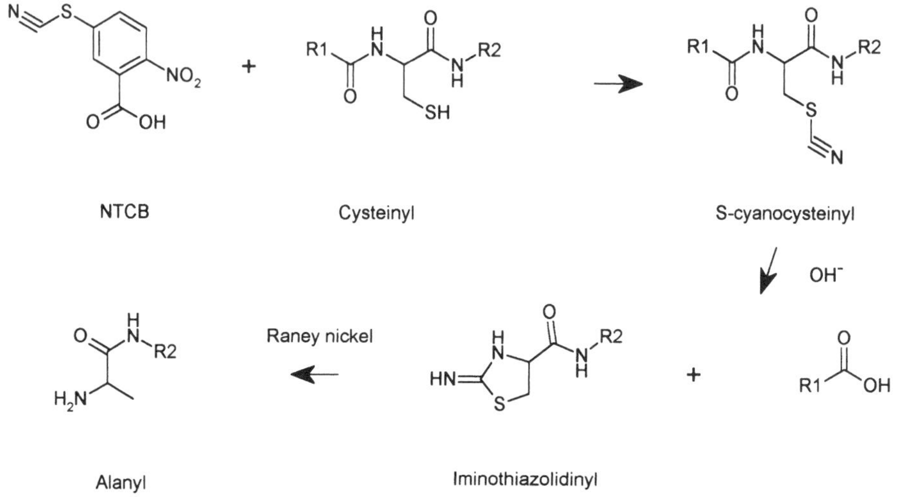

Fig. 3. Reactions in modification of, and cleavage at, cysteinyl residues by NTCB, and subsequent generation of alanyl N-terminal residue.

19. Treatment with Raney nickel (**Subheading 3.3., step 5**) causes desulfurization of residues. Thus, methioninyl residues are converted to 2-aminobutyryl residues. If the sample has not been reacted with NTCB, then remaining cysteinyl and cystinyl residues are converted to alanyl residues (without any cleavage). Otiene *(23)* has suggested that this modification might be used to study the dependence of protein function on Met and Cys content.

20. Endoproteinase Asp-N catalyses hydrolysis of bonds to the N-terminal side of either aspartyl or cysteinyl residues. Specificity for X-Cys can be generated by modification of aspartyl side chains *(24)*. Modification of cysteinyl residues to 2-aminoethylcysteinyl residues renders the Cys-X bond susceptible to cleavage by trypsin, and the X-Cys bond to cleavage by Lys-N, as discussed in **ref.** *(25)*.

4.4. Met-X Cleavage

21. The mechanism of the action of cyanogen bromide on methionine-containing peptides is shown in **Fig. 4**. For further details, see the review by Fontana and Gross *(26)*. The methioninyl residue is converted to homoseryl or homoseryl lactone. Peptides generated are suitable for peptide sequencing by Edman chemistry. Methionine sulfoxide does not take part in this reaction and the first step in the method is intended to convert any methionyl sulphoxide to methionyl residues, and so maximize cleavage efficiency. If the reduction is not carried out, the efficiency of cleavage may not be greatly diminished. If virtually complete cleavage is not necessary, partial cleavage products are desired (*see* **Note 26**), the sample is small and difficult to handle without loss, or speed is critical, the reduction step may be omitted.

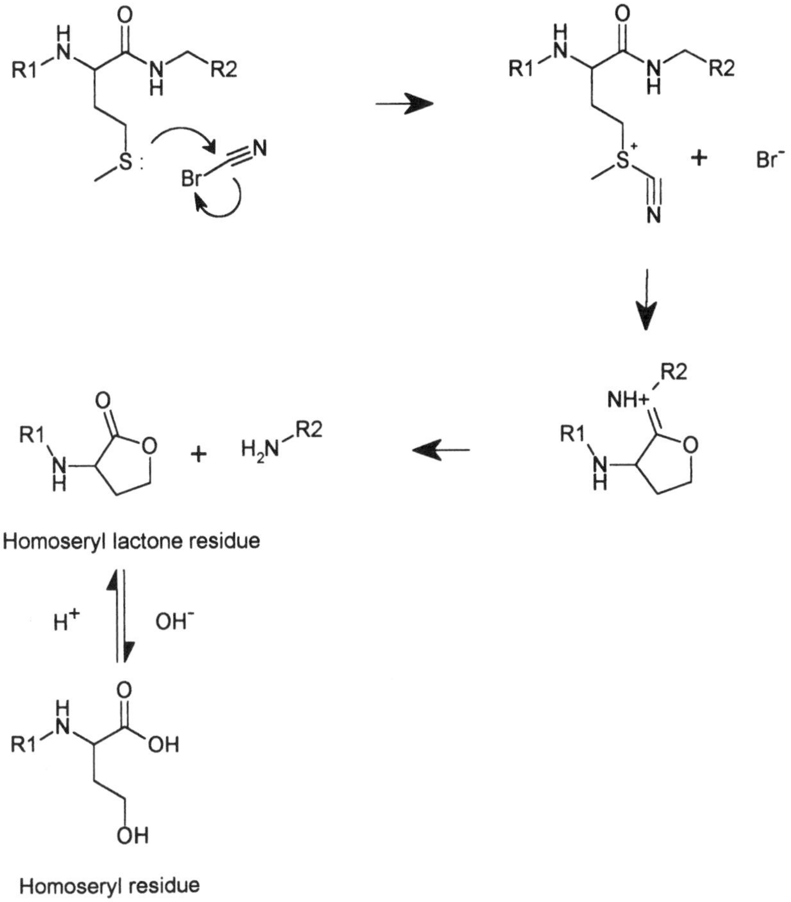

Homoseryl lactone residue

Homoseryl residue

Fig. 4. Mechanism of cleavage of Met-X bonds by cyanogen bromide.

An acid environment is required to protonate basic groups and so prevent reaction there and maintain a high degree of specificity. Met-Ser and Met-Thr bonds may give significantly less than 100% yields of cleavage and simultaneous conversion to methionyl to homoseryl residues within the uncleaved polypeptide. This is because of the involvement of the β-hydroxyl groups of seryl and threonyl residues in alternative reactions, which do not result in cleavage *(26)*. Morrison et al. *(27)* however, have found that use of 70% (v/v) TFA gives a better yield of cleavage of a Met-Ser bond in apolipoprotein A1 than does use of 70% formic acid *(see* **Note 2***)*. Using model peptides, Kaiser and Metzka *(28)* have analyzed the cleavage reaction at Met-Ser and Met-Thr and concluded that cleavage that efficiency is improved by increasing the amount of water present, and for practical purposes 0.1 M HCl is a good acid to use, giving about 50%

cleavage of these difficult bonds. Remaining uncleaved molecules contained either homoserine or methionyl sulfoxide instead of the original methionyl. Cleavage efficiency improved with increasing strength of acid, but there was an accompanying risk of degradation in the stronger acids.

22. Acid conditions are required for the reaction to occur. 70% (v/v) formic acid (pH 1.0) was formerly commonly used because it is a good protein solvent and denaturant, and also volatile. However, it may damage tryptophan and tyrosine residues (27) and also cause formylation of Seryl and Threonyl side chains (showing up during analysis by mass spectroscopy as an increase of 28 amu per modification [28,29]). Use of other acids avoids this problem. Trifluoroacetic acid (TFA) (also volatile) may be used in concentrations in the range 50% to 100% (v/v). The pH of such solutions is approx pH 0.5 or less. The rate of cleavage in 50% TFA may be somewhat slower than in 70% formic acid, but similar reaction times of hours, up to 24 h will provide satisfactory results. Caprioli et al. (30) and Andrews et al. (31) have illustrated the use of 60% and 70% TFA (respectively), for cyanogen bromide cleavage of proteins. Acetic acid (50%–100% v/v) may be used as an alternative but reaction is somewhat slower than in TFA. Alternatively, 0.1 M HCl has been used (28,29). To increase solubilization of proteins, urea, or guanidine-HCl may be added to the solution. Thus, in 0.1 M HCl, 7 M urea, for 12 h at ambient temperature, a Met-Ala bond was cleaved with 83% efficiency, and the more problematical Met-Ser and Met-Thr bonds with 56% and 38% efficiency (respectively) (28).

23. Although the specificity of this reaction is excellent, some side reactions may occur. This is particularly so if colored (yellow or orange) cyanogen bromide crystals are used, when destruction of Tyr and Trp residues may occur.

    The acid conditions employed for the reaction may lead to small degrees of deamidation of glutamine and asparagine side chains (which occurs below pH 3.0) and cleavage of acid-labile bonds, e.g., Asp-Pro. A small amount of oxidation of cysteine to cysteic acid may occur, if these residues have not previously been reduced and modified (e.g.. carboxymethylated). Occasional cleavage of Trp-X bonds may be seen, but this does not occur with good efficiency, as it does when the reduction step of this technique is replaced by an oxidation step (see **Subheading 4.5.** for cleavage of Trp-X bonds). Rosa et al. (32) cleaved both Met-X and Trp-X bonds simultaneously by treatment of protein with 12 mM cyanogen bromide in 70% TFA solution, plus 240 µM potassium bromide.

24. The protocol in **Subheading 4.5.** describes addition of solid cyanogen bromide to the acidic protein solution, to give a molar excess of cyanogen bromide over methionyl residues. This has the advantage that pure white crystals may be selected in favor of pale yellow ones showing signs of degradation (see **Note 23**). It does not allow accurate estimation of the quantity of reagent used, however. The work of Kaiser and Metzka (28) suggests that more than a 10-fold molar excess of cyanogen bromide over methionyl residues does not increase the extent of cleavage. If in doubt as to the concentration of methionyl residues, however, err on the side of higher cyanogen concentration.

If accurate quantification of cyanogen bromide is required, solid cyanogen bromide may be weighed out and dissolved to a given concentration by addition of the appropriate volume of 70% (v/v) TFA, and the appropriate volume of that solution added to the sample. The cyanogen bromide will start to degrade once in aqueous acid, so use when fresh. An alternative is to dissolve the cyanogen bromide in acetonitrile, in which it is more stable. Cyanogen bromide in acetonitrile solution is available commercially, for instance, at a concentration of 5 *M* (Aldrich). While such a solution may be seen to be degrading by its darkening color, this is not so obvious as it is with cyanogen bromide in solid form. For use, sufficient acetonitrile solution is added to the acidic protein solution to give the desired excess of cyanogen bromide over protein (e.g., 1/20 dilution of a 5 *M* cyanogen bromide solution to give a final 250 m*M* solution). The data of Kaiser and Metzka *(28)* indicate that high concentrations (70–100%) of acetonitrile can interfere with the cleavage reaction by decreasing the amount of water present, but below a concentration of 30% (in 0.1 *M* HCl) the effect is noticeable in causing a small decrease of Met-Ser and Met-Thr bond cleavage, but negligible for the Met-Ala bond.

25. The reagents used for Met-X cleavage are removed by lyophilization, unless salt has formed following failure to remove all of the ammonium bicarbonate. The products of cleavage may be fractionated by the various forms of electrophoresis and chromatography currently available. If analyzed by reverse-phase HPLC, the reaction mixture may be applied to the column directly without lyophilization. Since methionyl residues are among the less common residues, peptides resulting from cleavage at Met-X may be large and so in HPLC, use of wide-pore column materials may be advisable (e.g., 30-μ*M* pore size reverse-phase columns, using gradients of acetonitrile in 0.1% [v/v] TFA in water). Beware that some large peptides that are generated by this technique may prove to be insoluble (for instance if the solution is neutralized after the cleavage reaction) and so form aggregates and precipitates.

26. Incomplete cleavage that generates combinations of (otherwise) potentially cleaved peptides may be advantageous, for determination of the order of peptides within a protein sequence. Mass spectrometric methods are suitable for this type of analysis *(29)*. Such partial cleavage may be achieved by reducing the duration of reaction, even to less than 1 h *(29)*.

27. Methods have been described for cyanogen bromide treatment of low μg amounts of proteins in polyacrylamide gel *(33)*, on PVDF *(34)*, or on glass fiber, as used in automated protein sequencers *(31)*. The method described for treating a protein in polyacrylamide gel *(33)* is as follows:

    a. Lyophilize the piece of gel containing the protein of interest.
    b. Expose the gel piece to vapor from a solution of cyanogen bromide in TFA, for 24 h at room temperature, in the dark. The vapor is generated from a solution of 20 mg cyanogen bromide per mL of 50% (v/v) TFA, by causing it to boil under partial vacuum. A sealed container is used for this incubation.
    c. Lyophilize the treated piece of gel.

    d. Analyze cleavage by electrophoresis from the treated gel piece into a second gel.

28. For treatment of protein on PVDF *(34)* the method is as follows:

    a. Cut the protein band of interest from the PVDF, cutting closely around the band (since excess PVDF can reduce the final yield of peptide).

    b. Wet the dry PVDF piece with about 50 µL of cyanogen bromide solution in 50% (v/v) TFA (or 70% v/v TFA, or 70% v/v, formic acid - all of which can directly wet PVDF). Stone et al. *(34)* suggested application of cyanogen bromide at the rate of about 70 µg per 1 g of protein.

    c. Incubate in a sealed tube (to prevent drying out), at room temperature, 24 h, in the dark.

    d. Peptides generated may be extracted in the incubation solution itself, then successively in washes in 100 µL acetonitrile (40% v/v, 37°C, 3 h) and 100 µL TFA (0.05% [v/v] in 40% acetonitrile, 50°C). Pool extracts, dilute in water (to reduce acetonitrile concentration) and apply to reverse-phase HPLC, or dry down for analysis by PAGE.

    If the protein is run on PAGE prior to blotting onto PVDF, there is not a significant problem of methionine oxidation during electrophoresis; Stone et al. *(34)* reported approx 100% cleavage of myoglobin in these circumstances.

29. Protein may be treated with cyanogen bromide after having been subjected to Edman sequencing chemistry in an automated sequencer. This is useful for circumventing N-terminal blockage or for testing the alternatives of blockage or no sample in the event of failing to obtain any sequence. The method is similar to that described in **Note 28**, applicable either if the sample has been applied to a glass-fiber disk or to a piece of PVDF in the sequencer reaction cartridge. The method is as follows:

    a. Remove the glass fiber or PVDF from the sequencer, or leave in place in the reaction cartridge.

    b. Saturate the glass-fiber or PVDF piece with a fresh solution of cyanogen bromide in 50% (v/v) TFA (or 70% v/v formic acid). Make the cyanogen bromide solution to 100 mg/mL in the acid.

    c. Wrap the reaction cartridge, or loose glass fiber or PVDF in a small capped tube, in sealing film to prevent drying out. Incubate at room temperature in the dark, 24 h.

    d. Dry the sample under vacuum. Replace in the sequencer and start sequencing again. Yields tend to be poorer than the standard method described above for protein solutions; they may be down to 50% or less, and other, non-Met-X bonds may be cleaved at still lower yields. If the sample contains more than one methioninyl residue, more than one new N-terminus is generated. This may be simplified by subsequent reaction with orthophthalaldehyde, which blocks all N-termini except those bearing a prolyl residue *(35)*. For this approach to work, prior knowledge is required of the location of prolyl residues in the sequence, so that the orthophthalaldehyde reaction may be conducted at the correct cycle.

In order to test rapidly for the presence of any sample, a piece of CNBr-soaked PVDF may be treated at 65°C, for 1 h, followed by drying and sequencing. Yields, again, can be 30–50% with some (unpredictable) preference in Met-X bond cleavage, with somewhat greater levels of non-Met-X bond cleavage.

30. As described in **Note 21**, the peptide to the N-terminal side of the point of cleavage, has at its C-terminus a homoserine or homoserine lactone residue. The lactone derivative of methionine can be coupled selectively and in good yield *(36)* to solid supports of the amino type, e.g., 3-amino propyl glass. This is a useful technique for sequencing peptides on solid supports. The peptide from the C-terminus of the cleaved protein will, of course, not end in homoserine lactone (unless the C-terminal residue was methionine!) and so cannot be so readily coupled. Similarly, the C-terminal peptide carboxyl can react (if not amidated) with acidic methanol, to become a methyl ester (with a corresponding mass increment of 14 amu). Homoserine lactone, present as the C-terminal residue on other peptides in a cyanogen bromide digest, will react with acidic methanol and show a mass increase of 32 amu. With account made for side chain carboxyl residues, this is a means to identify C-terminal peptides by mass spectroscopy *(37)*.

4.5. Trp-X Cleavage

31. The method described for Trp-X bond cleavage is that of Huang et al. *(38)*. Although full details of the mechanism of this reaction are not clear, it is apparent that tryptophanyl residues are converted to oxindolylalanyl residues in the oxidation step, and the bond to the C-terminal side at each of these is readily cleaved in excellent yield (approaching 100% in **ref.** *[39]*) by the subsequent cyanogen bromide treatment. The result is seemingly unaffected by the nature of the residues surrounding the cleavage site.

   During the oxidation step, methionyl residues become protected by conversion to sulfoxides, bonds at these residues not being cleaved by subsequent cyanogen bromide treatment. Cysteinyl residues will also suffer oxidation if they have not been reduced and alkylated beforehand (*see* Chapter 27). Rosa et al. *(32)* cleaved both Trp-X and Met-X bonds simultaneously by omission of the oxidation step and inclusion of 240 µM potassium iodide in the reaction of protein with 12 mM cyanogen bromide.

   The peptide to the C-terminal side of the cleavage point has a free N-terminus and so is suitable for sequencing.

32. Methioninyl sulfoxide residues in the peptides produced may be converted back to the methioninyl residues by the action (in aqueous solution) of thiols (e.g., DTT, as described in **ref.** *[39]*, or as described in **Subheading 3.3., step 1**, or **Subheading 3.4., step 1**).

33. The acid conditions used for oxidation and cleavage reactions seem to cause little deamidation *(38)*, but one side reaction that can occur is hydrolysis of acid-labile bonds. The use of low temperature minimises this problem. If a greater degree of such acid hydrolysis is not unacceptable, speedier, and warmer alternatives to the reaction conditions described earlier can be used as follows:

a. Oxidation at room temperature for 30 min, but cool to 4°C before neutralization.
b. Cleavage at room temperature for 12–15 h.

34. As alternatives to the volatile base NH$_4$OH, other bases may be used (e.g., the nonvolatile potassium hydroxide or tris base).

35. As mentioned in **Note 22**, it has been found that use of 70% (v/v) formic acid can cause formylation of the polypeptide (seen as a 28 amu increase in molecular mass *[29]*) and damage to tryptophan and tyrosine (evidenced by spectral changes *[27]*). As an alternative to 70% formic acid, 5 *M* acetic acid may be used. Possibly, as in the use of cyanogen bromide in cleaving Met-X bonds 50% or 70% (v/v) TFA may prove an acceptable alternative *(32)*.

36. Samples of protein that have been eluted from sodium dodecylsulfate (SDS) gels may be treated as described, but for good yields of cleavage, Huang et al. *(32)* recommend that the sample solutions are acidified to pH 1.5 before lyophilization in preparation for dissolution in the oxidizing solution. Any SDS present may help to solubilize the substrate and, in small amounts at least, does not interfere with the reaction. However, nonionic detergents that are phenolic or contain unsaturated hydrocarbon chains (e.g., Triton, Nonidet P-40), and reducing agents are to be avoided.

37. The method is suitable for large-scale protein cleavage; this requires simple scaling up. Huang et al. *(38)* made two points, however:

a. The neutralization reaction generates heat. Since this might lead to protein or peptide aggregation, cooling is important at this stage. Ensure that the reagents are cold and are mixed together slowly and with cooling. A transient precipitate may be seen at this stage. If the precipitate is insoluble, addition of SDS may solubilize it (but will not interfere with the subsequent treatment).

b. The neutralization reaction generates gases. Allow for this by choosing a reaction vessel with reasonably large headspace.

38. At the end of the reaction, all reagents may be removed by lyophilization and the peptide mixture analyzed, for instance by polyacrylamide gel electrophoresis or by reverse-phase HPLC. Peptides generated may tend to be large, ranging up to a size in the order of 10,000 Da or more. Some of these large peptides may not be soluble, for instance if the solution is neutralized following the cleavage reaction, and consequently they aggregate and precipitate.

39. Note that all reactions are done in one reaction vial, eliminating transfer of sample between vessels, and so minimizing peptide losses that can occur in such exercises.

40. Various alternative methods for cleavage of the Trp-X bond have been described in the literature. The method that employs N-chlorosuccinimide is possibly the most specific, but shows only about 50% cleavage yield *(40)*. BNPS-skatole is a popular Trp-X-cleaving reagent whose reaction and products have been studied in some detail (for instance, *see* **refs.** *[41]* and *[42]*).

41. Methods have been described for Trp-X bond cleavage in small amounts (µg or less) of protein on solid supports or in polyacrylamide gel. These use N-chlorosuccinimide or BNPS-skatole (3-bromo-3-methyl-2-(2'-nitrophenyl-sulphenyl)-indolenine). For cleavage of protein in gel *(43)*:

a. Soak the gel piece for 30 min in a small volume of the solution: N-chlorosuccinimide, 0.015 M in urea (0.5 g/mL in 50% v/v acetic acid).

b. Wash the gel piece and electrophorese peptides from the treated gel into a second analytical gel.

42. Proteins bound to glass fiber (as used in protein sequencers) or to PVDF may be cleaved at Trp-X bond(s) by the method described in **ref.** *(44)*:

a. The glass-fiber disk, or PVDF, is wetted with a solution of BNPS-skatole (1 μg/mL in 70% v/v acetic acid).

b. Incubate in a sealed container to prevent drying out, at 47°C, 1 h in the dark.

c. Dry under vacuum. Replace in the sequencer and start sequencing.

References

1. Bornstein, P. and Balian, G. (1977) Cleavage at Asn-Gly bonds with hydroxylamine. *Methods Enzymol.* **47**, 132–145.

2. Blodgett, J. K., Londin, G. M., and Collins, K. D. (1985) Specific cleavage of peptides containing an aspartic acid (beta-hydroxamic) residue. *J. Am. Chem. Soc.* **107**, 4305–4313.

3. Kwong, M. Y., and Harris, R. J. (1994) Identification of succinimide sites in proteins by N-terminal sequence analysis after alkaline hydroxylamine cleavage. *Protein Sci.* **3**, 147–149.

4. Cacia, J., Keck, R., Presta, L. G., and Frenz, J. (1996) Isomerization of an aspartic acid residue in the complementarity-determining regions of a recombinant antibody to human IgE: identification and effect on binding affinity. *Biochemistry* **35**, 1897–1903.

5. Schurter, B. T. and Aswad, D. A. (2000) Analysis of isoaspartate in peptides and proteins without the use of radioisotopes. *Anal. Biochem.* **282**, 227–231.

6. Hiller, Y., Bayer, E. A., and Wilchek, M. (1991) Studies on the biotin-binding site of avidin. Minimised fragments that bind biotin. *Biochem. J.* **278**, 573–585.

7. Canova-Davis, E., Eng, M., Mukka, V., Reifsnyder, D. H., Olson, C. V., and Ling, V. T. (1992) Chemical heterogeneity as a result of hydroxylamine cleavage of a fusion protein of human insulin-like growth factor I. *Biochem. J.* **278**, 207–213.

8. Niles, E. G. and Christen, L. (1993) Identification of the vaccinia virus mRNA guanyltransferase active site lysine. *J. Biol. Chem.* **268**, 24986–24989.

9. Arselin, G., Gandar, J. G., Guérin, B., and Velours, J. (1991) Isolation and complete amino acid sequence of the mitochondrial ATP synthase ε-subunit of the yeast *Saccharomyces cerevisiae*. *J. Biol. Chem.* **266**, 723–727.

10. Saris, C. J. M., van Eenbergen, J., Jenks, B. G., and Bloemers, H. P. J. (1983) Hydroxylamine cleavage of proteins in polyacrylamide gels. *Anal. Biochem.* **132**, 54–67.

11. Golderer, G. and Gröbner, P. (1991) ADP-ribosylation of core histones and their acetylated subspecies. *Biochem. J.* **277**, 607–610.

12. Wiembs, T. and Stoffel, W. (1992) Proteolipid protein (PLP) of CNS myelin: Positions of free, disulfide-bonded and fatty acid thioester-linked cysteine residues and implications for the membrane topology of PLP. *Biochemistry* **31**, 12289–12296.

13. Omary, M. B. and Trowbridge I. S. (1981) Covalent binding of fatty acid to the transferrin receptor in cultured human cells. *J. Biol. Chem.* **256**, 4715–4718.

14. Inglis, A. S. (1983) Cleavage at aspartic acid. *Methods Enzymol.* **91**, 324–332.

15. Landon, M. (1977) Cleavage at aspartyl-prolyl bonds. *Methods Enzymol.* **47**, 132–145.

16. Beavis, R. C. and Chait, B. T. (1990) Rapid, sensitive analysis of protein mixtures by mass spectrometry. *Proc. Natl. Acad. Sci. USA* **87**, 6873–6877.

17. Morrison, J. R., Fiolge, N. H., and Grego, B. (1990) Studies on the formation, separation and characterisation of cyanogen bromide fragments of human A1 apolipoprotein. *Anal. Biochem.* **186**, 145–152.

18. Swenson, C. A. and Frederickson, R. S. (1992) Interaction of troponin C and troponin C fragments with troponin I and the troponin I inhibitory peptide. *Biochemistry* **31**, 3420–2427.

19. Jacobson, G. R., Schaffer, M. H., Stark, G., and Vanaman, T. C. (1973) Specific chemical cleavage in high yield at the amino peptide bonds of cysteine and cystine residues. *J. Biol. Chem.* **248**, 6583–6591.

20. Stark, G. R. (1977) Cleavage at cysteine after cyanylation. *Methods Enzymol.* **47**, 129–132.

21. Peyser, Y. M., Muhlrod, A., and Werber, M. M. (1990) Tryptophan-130 is the most reactive tryptophan residue in rabbit skeletal myosin subfragment-1. *FEBS Lett.* **259**, 346–348.

22. Tu, B. P. and Wang, J. C. (1999) Protein footprinting at cysteines: probing ATP-modulated contacts in cysteine-substitution mutants of yeast DNA topoisomerase II. *Proc. Natl. Acad. Sci. USA* **96**, 4862–4867.

23. Otiene, S. (1978) Generation of a free α-amino group by Raney nickel after 2-nitrothiocyanobenzoic acid cleavage at cysteine residues: applications to automated sequencing. *Biochemistry* **17**, 5468–5474.

24. Wilson, K. J., Fischer, S., and Yuau, P. M. (1989) Specific enzymatic cleavage at cystine/cysteine residues. The use of Asp-N endoproteinase, in *Methods in Protein Sequence Analysis* (Wittman-Liebold, B., ed.), Springer-Verlag, Berlin, 310–314.

25. Aitken, A. (1994) Analysis of cysteine residues and disulfide bonds, in *Methods in Molecular Biology, vol. 32: Basic Protein and Peptide Protocols* (Walker, J. M., ed.), Humana Press, Inc., Totowa, NJ, pp. 351–360.

26. Fontana, A. and Gross, E. (1986) Fragmentation of polypeptides by chemical methods, in *Practical Protein Chemistry: A Handbook* (Darbre, A., ed.), Wiley, Chichester, UK, pp. 67–120.

27. Morrison, J. R., Fidge, N. H., and Grego, B. (1990) Studies on the formation, separation, and characterisation of cyanogen bromide fragments of human A1 apolipoprotein. *Anal. Biochem.* **186**, 145–152.

28. Kaiser, R. and Metzka, L. (1999) Enhancement of cyanogen bromide cleavage yields for methionyl-serine and methionyl-threonine peptide bonds. *Anal. Biochem.* **266**, 1–8.

29. Beavis, R. C. and Chait, B. T. (1990) Rapid, sensitive analysis of protein mixtures by mass spectrometry. *Proc. Natl. Acad. Sci. USA* **87**, 6873–6877.

30. Caprioli, R. M., Whaley, B., Mock, K. K., and Cottrell, J. S. (1991) Sequence-ordered peptide mapping by time-course analysis of protease digests using laser description mass spectrometry in *Techniques in Protein Chemistry II* (Angeletti, R. M., ed.), Academic Press Inc., San Diego, CA, pp. 497–510.

31. Andrews, P. C., Allen, M. M., Vestal, M. L., and Nelson, R. W. (1992) Large scale protein mapping using infrequent cleavage reagents, LD TOF MS, and ES MS, in *Techniques in Protein Chemistry II* (Angeletti, R. M., ed.), Academic Press Inc., San Diego, CA, pp. 515–523.

32. Rosa, J. C., de Oliveira, P. S. L., Garrat, R., Beltramini, L., Roque-Barreira, M.-C., and Greene, L. J. (1999) KM+, a mannose-binding lectin from Artocarpus integrifolia: amino acid sequence, predicted tertiary structure, carbohydrate recognition, and analysis of the beta-prism fold. *Protein Sci.* **8**, 13–24.

33. Wang, M. B., Boulter, D., and Gatehouse, J. A. (1994) Characterisation and sequencing of cDNA clone encoding the phlorem protein pp2 of Cu curbita pepo. *Plant Mol. Biol.* **24**, 159–170.

34. Stone, K. L., McNulty, D. E., LoPresti, M. L., Crawford, J. M., DeAngelis, R., and Williams, K. R. (1992) Elution and internal amino acid sequencing of PVDF blotted proteins, in *Techniques in Protein Chemistry III* (Angeletti, R. M., ed.), Academic Press Inc., San Diego, CA, pp. 23–34.

35. Wadsworth, C. L., Knowth, M. W., Burrus, L. W., Olivi, B. B., and Niece R. L. (1992) Reusing PVDF electroblotted protein samples after N-terminal sequencing to obtain unique internal amino acid sequence, in *Techniques in Protein Chemistry III* (Angeletti, R. M., ed.), Academic Press Inc., San Diego, CA, pp. 61–68.

36. Horn, M. and Laursen, R. A. (1973) Solid-phase Edman degradation. Attachment of carboxyl-terminal homogenine peptides to an insoluble resin. *FEBS Lett.* **36**, 285–288.

37. Murphy, C. M. and Fenselau, C. (1995) Recognition of the carboxy-terminal peptide in cyanogen bromide digests of proteins. *Anal. Chem.* **67**, 1644–1645.

38. Huang, H. V., Bond, M. W., Hunkapillar, M. W., and Hood, L. E. (1983) Cleavage at tryptophanyl residues with dimethyl sulfoxide-hydrochloric acid and cyanogen bromide. *Methods Enzymol* **91**, 318–324.

39. Tseng, A., Buchta, R., Goodman, A. E., Loughman, M., Cairns, D., Seilhammer, J. et al. (1991) A strategy for obtaining active mammalian enzyme from a fusion protein expressed in bacteria using phospholipase A2 as a model. *Protein Expr. Purific.* **2**, 127–135.

40. Lischwe, M. A. and Sung, M. T. (1977) Use of N-chlorosuccinimide/urea for the selective cleavage of tryptophanyl peptide bonds in proteins. *J. Biol. Chem.* **252**, 4976–4980.

41. Vestling, M. M., Kelly, M. A., and Fenselau, C. (1994) Optimization by mass spectrometry of a tryptophan-specific protein cleavage reaction. *Rapid Commun. Mass Spectrom.* **8**, 786–790.

42. Rahali, V. and Gueguen, J. (1999) Chemical cleavage of bovine β-lactoglobulin by BNPS-skatole for preparative purposes: comparative study of hydrolytic procedures and peptide characterization. *J. Protein Chem.* **18**, 1–12.

43. Lischwe, M. A. and Ochs, D. (1982) A new method for partial peptide mapping using N-chlorosuccinimide/urea and peptide silver staining in sodium dodecyl sulphate-polyacrylamide gels. *Anal. Biochem.* **127**, 453–457.
44. Crimmins, D. L., McCourt, D. W., Thoma, R. S., Scott, M. G., Macke, K., and Schwartz, B. D. (1990) In situ cleavage of proteins immobilised to glass-fibre and polyvinylidene difluoride membranes, cleavage at tryptophan residues with 2-(2'-nitropheylsulfenyl)-3-methyl-3'-bromoindolenine to obtain internal amino acid sequence. *Anal. Biochem.* **187**, 27–38.

7

Enzymatic Preparation and Isolation of Glycopeptides

David J. Harvey

1. Introduction

The preparation and isolation of glycopeptides is the first step in the site-specific characterization of oligosaccharides attached to proteins. As discussed in Chapter 30, if the glycoprotein contains only one glycosylation site, the glycan profile can be obtained directly by mass spectrometry *(1)* providing that the mass of the glycoprotein is below about 30 kDa and the number of glycoforms is not large *(2)*. For larger glycoproteins and those containing multiple glycosylation sites, it is necessary to cleave the glycoprotein into smaller peptide fragments, either enzymatically or chemically (*see* Chapters 5 and 6, respectively), with the aim of isolating each glycosylation site into one glycopeptide. For some glycoproteins, it may be necessary to use a combination of cleavage methods in order to develop an approach for separating all of the glycosylation sites. The derived glycopeptides can then be isolated by high-performance liquid chromatography (HPLC) or lectin chromatography. The affinity of glycopeptides for different lectins provides preliminary information on the carbohydrates attached to the peptide. A number of strategies for serial lectin chromatography have been developed that result in isolation and characterization of individual glycopeptides *(3,4)*.

The proteases usually employed for digestion of glycoproteins, trypsin (which cleaves on the carboxy-terminal side of the basic amino acids lysine and arginine) and endoproteinase lys-C (which cleaves on the carboxy-terminal side of lysine), are available commercially to a very high standard and have been well-characterized. Other enzymes are discussed in Chapter 5. Both trypsin and Lys-C produce very low levels of autolytic products, which enables high ratios of sample protein to protease (1:10 to 1:40) to be used over long periods of time without significant loss of proteolytic activity. As both pro-

From: *Methods in Molecular Biology, vol. 211: Protein Sequencing Protocols, 2nd ed.*
Edited by: B. J. Smith © Humana Press Inc., Totowa, NJ

teases are active in 2 *M* urea, 8 *M* urea can be used to unfold target proteins and allow better access to the cleavage sites. Only a small dilution is then required for the proteases to be active while maintaining the protein at a high concentration (> 0.5 mg/mL) for digestion.

HPLC is the standard method for resolution of peptide and glycopeptide mixtures and is frequently used to provide specific profiles of digested proteins, for example in quality control of recombinant products. Other than a convenient method for separation and isolation of (glyco)peptides, HPLC does not provide any structural information on glycosylation without further investigation (*see* Chapter 30). Other methods for glycoprotein and glycopeptide fractionation are given in **ref.** *(5)*

As lectins are specific for particular sugars or discrete oligosaccharide structures, their ability to bind glycopeptides provides a means of characterizing the glycans present, in addition to providing a means of purifying them. Lectins immobilized on Agarose-based gels can be purchased from a number of commercial sources (*see* **Table 1**). However, the vast majority of lectins are available as highly purified preparations, which can readily be immobilised on CNBr-activated Sepharose or *N*-hydroxysuccinimide ester-activated Agarose *(6)*.

2. Materials
2.1. Trypsin and Endoproteinase Lys-C Digestion

1. Resuspension buffer: 8 *M* urea, 100 m*M* ammonium bicarbonate pH 7.8.
2. Reducing agent: 50 m*M* dithiothreitol (DTT) in distilled water (*see* **Note 1**).
3. Cysteine modifying reagent: 100 m*M* iodoacetamide in distilled water. Store in the dark or cover with aluminium foil.
4. Proteases (Boehringer Mannheim Ltd., Lewes, UK or Promega Ltd., Madison, WI):
   a. Sequence-grade, modified trypsin.
   b. Sequence-grade endoproteinase lys-C.
   Prepare each at 1 mg/mL in 2 m*M* hydrochloric acid. These solutions can be stored for up to 3 mo at –20°C.

2.2. Isolation of the Glycopeptide
2.2.1. Reversed-Phase HPLC

1. Suitable HPLC system capable of flow rates of 100 µL/min with a 200 µL sample loop, a C-18 or C-8 reversed-phase column (e.g., Beckman Ultrasphere 150 × 4.6 (id) mm C-18 from Alltech (Carnforth, Lancashire, UK), and with the detector set to a wavelength of 225 nm.
2. Eluant A: 0.1% trifluoroacetic acid (TFA) in water (500 mL) (*see* **Note 2**).
3. Eluant B: Acetonitrile containing sufficient TFA to give an absorbance at a wavelength of 225 nm equivalent to that of eluant A (approx 0.085% v/v) (500 mL) (*see* **Note 2**).
4. Fraction collector.

2.2.2. Lectin Chromatography

1. Sepharose- or Agarose-immobilized lectin (**Table 1**) (1–5 mL) packed into a PolyPrep (2 mL bed volume, 0.8 × 4.0 cm) or EconoPac (1.0–20 mL bed volume, 1 × 12 cm) disposable column (Bio-Rad Laboratories Ltd., Hemel Hempstead, UK).
2. Equilibration buffer: 20 mM phosphate buffer, pH 7.2, 150 mM sodium chloride (with 1 mM magnesium chloride and 1 mM calcium chloride for appropriate lectins; *see* **Table 1**) and 0.02% (w/v) sodium azide.
3. Elution buffer(s): equilibration buffer containing 100–500 mM of the appropriate sugar (*see* **Table 1**).

3. Methods

3.1. Trypsin and Endoproteinase Lys-C Digestion (see Note 3)

1. Solubilize the lyophilized protein (120 µg) in 50 µL of resuspension buffer to give a protein concentration of 2.4 mg/mL.
2. Add 5 µL of reducing agent (50 mM DTT).
3. Incubate at 50°C for 15 min.
4. Cool to room temperature and add 5 µL of 100 mM iodoacetamide. (All further steps should be performed in a foil-covered tube to exclude light.)
5. Incubate at room temperature for 15 min.
6. Add 140 µL of water.
7. Add 5 µL of trypsin or endoproteinase lys-C to give a protein:proteinase ratio of 24:1 (w/w).
8. Incubate at room temperature for 24 h.
9. Stop reaction by cooling to –20°C.

3.2. Isolation of the Glycopeptides

3.2.1. Reversed-Phase HPLC

1. Regenerate the C-8 column in 95% eluant B for 30 min at a flow rate of 100 µL/min.
2. Equilibrate the column in 5% eluant B for 30 min at a flow rate of 100 µL/min.
3. Diluted the digested protein mixture with eluant A to give a total volume of 200 µL.
4. Inject 200 µL of digested protein onto the column and elute using the following gradient:

Time (Min)	Eluant B (%)
0	5
25	5
95	50
97	95
102	95
105	5
120	5

Table 1
Lectins

Taxonomic name	Common name	Abbreviation	Specificity	Cation	Elution Sugar	Supplier[a]
Arachis hypogea	Peanut	PNA	Galβ1-3GalNAc	Ca^{2+}	Galactose	C, S
Artocarpus integrifolia	Jackfruit		Core Galβ1-3GalNAc Terminal αGalactose	Ca^{2+}	α-Methyl-D-galactoside	C
Canavalia ensiformis	Jack bean (Con-A)	Con A	α-Mannose, α-Glucose	Ca^{2+}	α-Methyl-D-mannoside	C, P, S
Datura stramonium		DSL	GlcNAcβ1-GlcNAc, LacNAc	Ca^{2+}	Chitobiose, Chitotriose	C
Dolichos biflorus	Horse gram	DBA	Terminal α-GalNAc	Ca^{2+}	N-Acetylgalactosamine	C, S
Glycine soja	Soybean	SBA	α- or β-GalNAc	–	N-Acetylgalactosamine	C, S
Griffonia simplicifolica			α-Galactose	Ca^{2+}	Raffinose, Galactose	C
Helix pomatia	Edible snail	HPA	O-Linked GalNAc		N-Acetylgalactosamine	S
Lens culinaris integrifolia	Lentil	LcH	Fucosylated antennary complex α-Mannose	Ca^{2+}	α-Methyl-D-mannoside plus α-Methyl-D-glucoside	C, P, S
Limax flavus	Slug	LFA	NeuAc, NeuGc	–	N-acetyl-neuraminic acid	C
Lotus tetragonolobus	Asapargus		Terminal α-fucose	Ca^{2+}	L-Fucose	C
Lysopersicon esculentum	Tomato		GlcNAcβ1-4GlcNAc oligomers	Ca^{2+}	Chitobiose and chitotriose	C
Maackia amurensis		MAL I	NeuNAcα2-3Galα1-4GlcNAc	Ca^{2+}	Lactose	C
Phaseolus vulgaris	Red kidney bean	PHA L4	Tri- and tetra-antennary complex with N-acetyl-lactosamine branches		N-Acetylgalactosamine	C, S

Table 1
Lectins (*continued*)

Taxonomic name	Common name	Abbreviation	Specificity	Cation	Elution Sugar	Supplier[a]
Phaseolus vulgaris	Red kidney bean	PHA E4	Bisected biantennary complex	Ca^{2+}	*N*-Acetylgalactosamine	C, S
Phytolacca americana	Pokeweed	PWM	GlcNAcβ oligomers complex		Chitobiose	S
Pisum Sativum	Garden pea	PSA	Bi- and tri-antennary with α1-6 fucose	Ca^{2+} Mg^{2+}	α-Methyl-D-mannoside α-Methyl-D-glucoside	C, S
Ricinus communis	Castor bean	RCA1, RCA60	Bi- and tri-antennary complex		Lactose	S
Sambucus nigra	Elderberry	SNA	NeuNAcα2-6Gal NewNAcα2-6GalNAc	–	N-Acetylgalactosamine, Lactose	C
Solanum tuberosum	Potato (STA)	STA	GlcNAcβ oligomers	Ca^{2+}	Chitobiose, Chitotetraose	C, S
Tetragonolobus purpureas	Winged pea	–	L-Fucose		L-Fucose	S
Triticum vulgaris	Wheatgerm	WGA	GlcNAcβ1⊗4GlcNAc (*N,N*-Diacetylchitobiose), NeuNAc	–	*N*-Acetylgalactosamine	C, P, S
Tritrichomonas mobilensis		TML	NeuNAc, NeuNGc	–	NeuNAc	C
Ulex europaeus	Gorse	UEA1 UEA2	Fucα1-2Galβ1-4GlcNAc	Ca^{2+}	L-Fucose	C, S
Wisteria floribunda	Wisteria		Terminal galNAcβ1-4-	Ca^{2+}	*N*-Acetylgalactosamine	C
Vicia villosa	Hairy vetch	VVA	GalNAcα1-O-serine	Ca^{2+}	*N*-Acetylgalactosamine	C

[a]C = Calbiochem, CN BioSciences, Nottingham, UK. S = Sigma-Aldrich, Poole, Dorset, UK. P = Pharmacia, Milton Keynes, UK.

5. Collect 2 min fractions (200 µL) fractions from time 0 to time = 100 min.

6. Lyophilize fractions and store at –20°C until required for analysis.

3.2.2. Glycopeptide Purification by Immobilized Lectin Chromatography

1. Equilibrate the column with five column volumes of equilibration buffer (including 1 mM divalent metal ions as appropriate; see **Table 1**).

2. Dilute the digested protein with equilibration buffer to a final volume of 1 mL and apply to the column.

3. Wash the column with five column volumes of equilibration buffer and collect the flow; this contains nonbound material.

4. Elute bound glycopeptides with five column volumes of elution buffer (see **Note 4**) and collect 1 mL fractions.

5. Store fractions at –20°C until required for analysis.

6. Measure the absorbance of the fractions spectrophotometrically at a wavelength of 225 nm or use methods outlined in Chapter 30.

7. Regenerate the column with 10 column volumes of equilibration buffer.

4. Notes

1. Distilled water is preferable to standard ion-free water, found in many laboratories, as the latter sources frequently contain residual organic material such as polyethylene glycol. This material can subsequently interfere with analyses performed by mass spectrometry.

2. Solutions should be sparged with helium for 10 min.

3. Protease digestion: Some flexibility is possible in the quantities and concentrations of the components used for digestion. The initial concentration on resuspension with 8 M urea, 100 mM ammonium bicarbonate, pH 7.8, in **step 1** should be between 2 and 10 mg/mL. Add a volume of 50 mM DTT that is between 1/10 and 1/25 the volume of suspended protein. DTT is present in at least 10-fold molar excess of the number of cysteine residues present in the protein to ensure complete reduction. Add an equal amount of iodoacetamide to ensure complete carboxamidomethylation of the reduced cysteine. The mixture is then diluted fourfold to reduce the urea concentration. Both trypsin and endoproteinase lys-C retain activity in 2 M urea. The final protein:proteinase ratio should optimally be between 20 to 40:1 (w/w).

4. Lectin chromatography: Fractionation of bound glycopeptides can be achieved by using a series of elution steps employing 0.1 M increments in specific sugar concentration up 0.5 M. Use five column volumes to elute bound material, collecting 1-mL fractions at each stage. Because of the selectivity of lectins for specific sugars, some glycopeptides will not bind and will appear in the void volume; others will elute in low sugar and others in high sugar concentrations.

References

1. Mock, K. K., Davy, M., and Cottrell, J. S. (1991) The analysis of underivatised oligosaccharides by matrix-assisted laser desorption mass spectrometry. Biochem. *Biophys. Res. Commun.* **177**, 644–651.
2. Tsarbopoulos, A., Bahr, U., Pramanik, B. N., and Karas, M. (1997) Glycoprotein analysis by delayed extraction and post-source decay MALDI-TOF-MS. *Int. J. Mass Spectrom. Ion Processes* **169/170**, 251–261.
3. Osawa, T. and Tsuji, T. (1987) Fractionation and structural assessment of oligosaccharides and glycopeptides by use of immobilized lectins. *Ann. Rev. Biochem.* **56**, 21–42.
4. Robertson, E. R. and Kennedy, J. F. (1996) Glycoproteins: a consideration of the potential problems and their solutions with respect to purification and characterisation. *Bioseparation* **6**, 1–15.
5. Montreuil, J., Bouquelet, S., Debray, H., Fournet, B., Spik, G., and Strecker, G. (1986) Glycoproteins, in *Carbohydrate Analysis: A Practical Approach* (Chaplin, M. F. and Kennedy, J. F., eds.), IRL Press, Oxford, pp. 143–204.
6. Sutton, C. W. (1989) Lectin affinity chromatography, in Protein Purification Protocols: A Practical Approach (Harris, E. L. V. and Angal, S., eds.), Oxford University Press, Oxford, pp. 268–282.

8

Analytical and Micropreparative Capillary Electrophoresis of Peptides

Alan J. Smith

1. Introduction

Capillary electrophoresis (CE) was developed as a high sensitivity, high resolution, quantitative, electrophoretic separation technique. Since its commercial introduction in 1987, it has proved to be a versatile analytical tool for the separation of both small molecules, e.g., inorganic cations and anions, and drugs; and large molecules, e.g., peptides, proteins, carbohydrates, and nucleic acids.

In the case of proteins and peptides CE has found utility both as an analytical technique and as a micropreparative separation technique. CE methods have been developed for monitoring the enzymatic digestion of proteins, for purity checks on both natural and synthetic peptides, for screening protein and peptide fractions from chromatographic separations, and for the micropreparative isolation of peptides from complex digestion mixtures.

The separation of peptides by CE is based on both size and charge considerations (1). It can be viewed as an orthogonal separation technique to reverse-phase high-performance liquid chromatography (RP-HPLC) which separates on the basis of hydrophobicity and to a lesser extent on size. A comparison between the two separation methods is shown in **Fig. 1**. The same tryptic digest of β lactoglobulin was separated by CE in **Fig. 1A** and RPHPLC in **Fig. 1B**. The high-sensitivity and high-resolution capabilities of CE are based on performing electrophoretic separations in long glass capillaries under conditions that keep diffusion to a minimum. Standard analytical capillaries have internal diameters of 50–75 micron and lengths of 20–50 cm. These glass capillaries are fragile and in order to improve their tensile strength of they are coated with a thin plastic (polyimide) film. The small internal volume of the capillaries (low

From: *Methods in Molecular Biology, vol. 211: Protein Sequencing Protocols, 2nd ed.*
Edited by: B. J. Smith © Humana Press Inc., Totowa, NJ

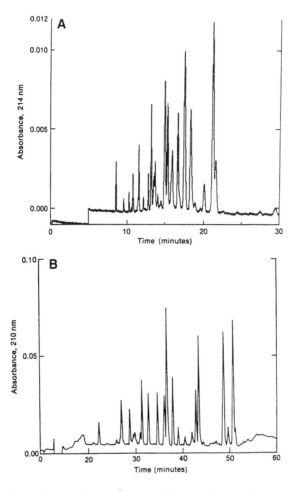

Fig. 1. (**A**) Analytical CE of β-lactoglobulin digest. (**B**) Narrowbore RP-HPLC anaytical separation of β-lactoglobulin digest.

microlitre) require very high voltages (10–25kV) to achieve optimal electro-phoretic separations, and for this reason they are incorporated into dedicated CE instruments. In its simplest form, (**Fig. 2**) a CE instrument consists of two electrode buffers joined by a glass capillary, and a UV detector. A third reser-voir containing the sample is also required. The sample is loaded on to the cap-illary from the sample reservoir either electrokinetically or mechanically (pressure or vacuum). The cathode end of the capillary is then removed from the sample reservoir, returned to the cathodic buffer reservoir, and electrophoresis is commenced.

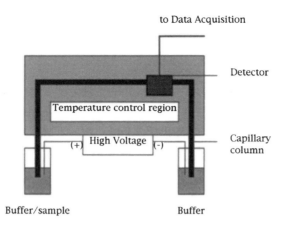

Fig. 2. Schematic of a CE instrument.

Once electrophoresis is started, an electroendosmotic flow (EOF) is set up within the capillary. The EOF is produced by the migration of cations and their associated water of hydration from the cathode towards the anode. Thus not only positively charge, but neutral, and negatively charged peptides are swept through the capillary to the anode. A small region of the polyimide coating is removed from the glass capillary and is inserted into the detector light path. Thus the capillary itself becomes the flow cell. The UV absorbing peptides are recorded and quantitated as they pass through the detector. The glass capillary can have a surface of native silica (open tube) or, it can be derivatized (coated) and can contain either electrolyte alone or, electrolyte and a suitable separation matrix.

One complication with the separation of peptides in open-tube capillaries is the potential for their irreversible adsorption to the silica wall. At low pH *(2)*, or high pH in the presence of a modifier *(3)*, this has not proved to be a significant problem. However, it is a serious problem with proteins. Recent work *(4)* from the Righetti group has shown that both peptides and proteins can be effectively separated by using uncoated capillaries and amphoteric, isoelectric buffers at low pH. The running buffer also contains hydroxyethylcellulose as a dynamic coating for the silica surface. This processes effectively eliminates the nonspecific adherence of peptides and proteins to the capillary walls. In the case of proteins, a denaturant such as 5 *M* urea or reduced Triton X-100 is added in order to keep them in their unfolded state. The use of amphoteric buffers such as aspartic acid at their isoelectric point means that higher electric fields and shorter capillary lengths can be used. In combination, these conditions can reduce the analytical separation times of protease digestion mixtures to less than 15 min. The definition as to what constitutes a protein and what constitutes a peptide becomes important when selecting an optimal separation protocol. In the context of CE, peptides should be viewed as containing less than 50 amino acids.

A second important aspect of CE is the definition of sensitivity. Because only nanoliter sample volumes are loaded on to the capillary for analytical separations, the absolute sensitivity is very high when compared to HPLC. However, in a practical sense, the sample must be at an approximate concentration of one microgram per microliter and a volume of at least 10 microliters even though only a very small percentage of the sample is used in the separation. For example, a 10-s injection would transfer approx 50 nl to the capillary, which would be equivalent to approx 50 ng of protein. At times these concentration ranges may not be readily available. It is possible to mitigate these difficulties to a certain extent by the preconcentration of dilute samples in the capillary by pressure loading a water slug prior to electrokinetic injection of the sample (*5*). More recently isotacophoretic injection methods have worked well for dilute samples (*6*).

Although CE was originally developed as an analytical technique, considerable interest has developed in utilizing the high-resolution capabilities as a preparative technique. The subsequent section will describe protocols that can be employed when micropreparative peptide separations are required.

Two major approaches have been developed for the successful micropreparative separation of peptides that have been generated from a protease digestion of a target protein. One method utilizes a single electrophoretic separation on a capillary of much larger diameter than those normally used for analytical separations (*7*). The other utilizes multiple separations on a single analytical capillary (*8*). Fractions with the same electrophoretic mobility are pooled in order to obtain sufficient material for further characterization.

In both cases, the amount of digest fractionated by micropreparative CE is in the 5–50 pmole range. In contrast, it is at these load levels that losses owing to adsorption become a significant problem for narrowbore (2.1 mm ID) RPHPLC separations. In this sense, the two techniques can be regarded as complimentary to each other.

2. Materials

2.1. Preconditioning Underivatized Capillaries (see Note 1)

1. Commercial CE instrument.
2. Sodium hydroxide: 100 m*M* in distilled water.
3. Sodium phosphate: 250 m*M* in distilled water pH 2.3.
4. Underivatized capillary: 75 micron × 50 cm.

2.2. Monitoring Protease Digestions

1. Commercial CE instrument equipped with a temperature controlled sample table.
2. Tris: 250 mM, pH 7.8, containing 8 M urea (*see* **Note 2**)
3. Protease: 1 μg/μL in distilled water (*see* **Note 3**).
4. Sodium phosphate: 50 mM in distilled water, pH 2.3.
5. Sodium phosphate: 250 mM in distilled water, pH 2.3.
6. Sodium hydroxide: 100 mM in distilled water (*see* **Note 4**).
7. Dithiothreitol (DTT): 50 mM in distilled water (*see* **Note 5**).
8. Preconditioned silica capillary.

2.3. Screening Fractions from Preparative Reverse Phase HPLC

2.3.1. Standard Protocol

1. HPLC fraction in actonitrile/TFA (*see* **Note 6**).
2. Vacuum centrifuge.
3. Ethylene glycol: 100% (*see* **Note 7**).
4. Sodium phosphate: 50 mM in distilled water, pH 2.3.
5. Preconditioned silica capillary.

2.3.2. Alternate Protocol (see **Note 5**)

1. HPLC fraction in actonitrile/TFA (*see* **Note 6**).
2. Vacuum centrifuge.
3. Ethylene glycol: 100% (*see* **Note 7**).
4. Sodium tetraborate: 100 mM with boric acid, pH 9.0.
5. Sodium dodecylsulphate (SDS): 0.2% in water.
6. Preconditioned silica capillary.

2.4. Micropreparative Separation

2.4.1. Single Separation Protocol

1. Commercial CE instrument equipped with active cooling (*see* **Note 8**).
2. Sodium phosphate: 50 mM in distilled water, pH 2.3.
3. Sodium phosphate: 250 mM in distilled water, pH 2.3.
4. Sodium hydroxide: 100 mM in distilled water.
5. Underivatized capillary: 50 micron × 57 cm. (Polymicro Inc.) (*see* **Note 9**).
6. Microvials for collection: 25 μL volume.
7. Ethylene glycol: 100% solution (*see* **Note 7**).
8. Protease digestion mixture: approx 1 μg/μL of substrate in Tris-urea buffer (*see* **Subheading 3.2.**).

2.4.2. Multiple Separations Protocol

1. Commercial CE instrument with forced air (passive) cooling.
2. Reagents as in **Subheading 2.4.1.**
3. Underivatized capillary: 75 micron × 57 cm (Polymicro Inc.).
4. Standard 100 μL vials.

3. Methods

3.1. Preconditioning Underivatized Capillaries

1. Place capillary in instrument.
2. Flush capillary with 10 column volumes of 100 mM sodium hydroxide at 0.5 psi.
3. Flush with 10 column volumes water.
4. Flush with 4 column volumes 250 mM sodium phosphate.
5. Store in same buffer until use.
6. When switching to a new buffer a 4 h equilibration is advised and **steps 2** and **3** eliminated.

3.2. Monitoring Protease Digestions

1. Dissolve 20 µg of the protein in 20 µL of Tris-urea buffer.
2. Add 5 µL of 50 mM DTT solution and incubate at 50°C for 15 min to reduce disulphide bonds in the protein.
3. Add 75 µL distilled water to dilute the tris buffer to 50 mM and the urea to 2 M.
4. Add protease 1/10 enzyme to protein ratio (w/w).
5. Load onto CE sample table that has been temperature equilibrated to 37°C.
6. Operate instrument in accordance with manufacturer's instructions.
7. Separate the peptide mixture using the following conditions:
    a. Electrolyte: 50 mM sodium phosphate pH 2.3.
    b. Sample table temperature: 37°C.
    c. Run Temp: 25°C.
    d. Voltage: 25kV.
    e. Sample injection: 10 s at 0.5 psi.
    f. Injection interval: 4 h.
    g. Detection: 200 nm.
    h. Run time: 37 min.
8. An example of a separation protocol for a Beckman P/ACE 5000 instrument with a 37-min run time from injection to injection is as follows:
    a. Screening of peptides (digests, HPLC fractions, and synthetic peptides).
    b. Sample vial = 11, Injection = 10 s, Voltage = 25kV, Separation = 30 min, 50 mM phosphate, pH 2.3.
    c. Vial contents:
        Position #7. Waste vial, water level just to contact capillary effluent.
        Position #9. Electrolyte: 50 mM phosphate, pH 2.3.
        Position #11. Sample.
        Position #29. Electrolyte 50 mM phosphate, pH 2.3.
        Position #32. 0.1 N NaOH for regeneration.
        Position #33. Water for rinse.
        Position #34. 0.25 M phosphate, pH 2.3.
9. The electropherograms are examined to determine the length of time required to produce a stable profile (*see* **Note 10**).

3.3. Screening HPLC Fractions for Peptide Purity

3.3.1. Standard Protocol

1. Concentrate the HPLC fractions in an evacuated centrifuge to approximately 20 μL in order to remove excess acetonitrile, but do not dry.
2. Add 5 μL of 100% ethylene glycol to reduce evaporation on the sample table and minimize adsorption to the walls of the tube.
3. Run same conditions as in **Subheading 3.2.** except that the sample table temperature is 25°C and only single injections are made.
4. Fractions that contain single or major components are suitable candidates for protein sequencing.

3.3.2. Alternate Protocol

1. Sample preparation is as in **Subheading 3.3.1.**, **steps 1** and **2**.
2. Prepare electrolyte by mixing equal volumes of sodium tetraborate and SDS solutions.
3. Separate the peptide mixture using the following conditions:
   a. Electrolyte: 50 m*M* sodium borate/0.1% SDS pH 9.0.
   b. Sample table temp: 25°C.
   c. Run temp: 25°C.
   d. Voltage: 30 kV.
   e. Sample injection: 15 s at 0.5 psi.
   f. Detection: 200 nm.
   g. Run time: 25 min.

3.4. Micropreparative Separation

3.4.1. Single Separation Protocol (see **Note 11**)

1. Place 10 μL of a 50 m*M* sodium phosphate/ethylene glycol (80:20 v/v) mixture in the microvials.
2. Program the Beckman Instruments P/ACE 5000 instrument to collect 3-min fractions.
3. All other operating conditions are as in **Subheading 3.2.** except that the operating voltage is 7.5 kV and the sample table temperature is 25°C.
4. Sample injection: 15 s at 0.5 psi.
5. Collect fractions over a 90-min separation period (*see* **Note 12**).

3.4.2. Multiple Separations Protocol (see **Note 13**)

1. The fraction collection tubes are filled with 10 μL of 50 m*M* sodium phosphate buffer to serve as the analyte electrodes.
2. Separation conditions and sample concentrations are as in **Subheading 3.2.** with the sample table temperature lowered to 25°C.
3. Fractions are collected every minute across the separation (*see* **Note 14**).

4. The CE instrument is programmed to perform 10 consecutive separations automatically.

5. The length of each separation is 25 min, which includes capillary regeneration.

4. Notes

1. Preconditioning of the capillary refers to the generation of ionized silanol (SiO⁻) groups on the inner wall of the capillary. This is essential for the generation of EOF.

   Preconditioning underivatized capillaries is essential when they are new or when it is necessary to change the separation buffer. Do NOT precondition coated capillaries as this will remove the coating and ruin the capillary. Buffers can be changed in these coated capillaries by pre-equilibration.

2. The proteins are dissolved in 8 M urea to facilitate unfolding and dissolution. This solution is diluted to 2 M urea with water prior to adding the protease. Most proteases are fully active in 2 M urea solutions.

3. Store the protease at −20°C.

4. The sodium hydroxide solution is used for regenerating the surface of the capillary before reequilibrating with the 250 mM and 50 mM phosphate buffers.

5. It is essential to reduce all the potential disulphide bonds in the protein prior to digestion and separation in order to obtain an accurate representation of the constituent peptides.

6. Fractions of approx 100 µL are normally collected from a narrowbore RP-HPLC separation of a protease digest that has been developed with an acetonitrile/water 0.5% (v/v) TFA gradient.

7. Obtained as a viscous solution from the supplier.

8. The large diameter capillary single-separation method requires that the CE instrument possesses refrigerated (active) cooling. The use of large (150–200 micron ID) capillaries produces significant quantities of Joule heating which can distort the electrophoretic separations unless they are adequately temperature controlled. Some CE instruments utilize forced ambient air (passive) cooling and this is inadequate to achieve effective temperature control when using these large diameter capillaries.

9. Much larger sample volumes can be loaded onto these capillaries (100–500 nl) and peptides can be recovered at the 5–50 pmole level from a single separation. These recoveries are adequate to allow direct protein sequencing of the fractions (7). The upper sample volume limit (500 nl) is dictated by the total volume of the capillary (approx 10 µL). A 15-s injection time would load approx 250 nl (0.5 µg protein digest) onto the capillary.

10. Overdigestion will result in the autodigestion of the protease, which will produce unwanted peptides.

11. The method suffers from the disadvantage that fractions have to be collected "blind" since it is not possible to obtain a continuous chromatographic profile during the run. This is due to the interruption in current that occurs when the end of the capillary is moved from one fraction collection tube to the next.

12. Fractions can be screened with either analytical CE or MALDI mass spectrometry prior to sequencing.
13. The multiple collection approach combines identical peaks from 10 consecutive analytical separations in order to obtain sufficient material for further characterization. Fractions are collected at 1-min time intervals across each separation. A single set of fraction collection tubes are used such that like-fractions are pooled in the same tube. The success of the method requires that the absolute electrophoretic migration times of the individual peaks must not vary by more than 0.1 min over the series of separations. This electrophoretic stability is not easily achieved with some commercial instruments.
14. As with the other micropreparative protocol, it is advisable to screen the fractions prior to selection for protein sequence analysis.

References

1. McCormick, R. M. (1994) Capillary zone electrophoresis of peptides, in *Handbook of Capillary Electrophoresis* (Landers, J. P., ed.), CRC, Boca Raton, FL, 287–324.
2. Strickland, M. and Strickland, N. (1990) Free solution capillary electrophoresis using phosphate buffer and acidic pH. *Am. Lab.* **22**, 60–65.
3. Chen, F.A., Kelly, L., Palmieri, R., Biehler, R., and Schwartz, H. (1992) Use of high ionic strength buffers for the separation of proteins and peptides with CE. *J. Liq. Chrom.* **15**, 1143–1150.
5. Burgi, D. S. and Chien, R. L. (1992) On-column sample concentration using field-amplification in CZE. *Anal. Chem.* **64**, 849–855A.
7. Kenny, J. W., Ohms, J. I., and Smith, A. J. (1993) Micropreparative capillary electrophoresis (MPCE) and micropreparative HPLC of protein digests, in *Techniques in Protein Chemistry IV* (Hogue Angeletti, R.A., ed.), Academic, San Diego, CA, pp. 363–370.
6. Bergman, T. and Jornvall, H. (1992) Capillary electrophoresis for preparation of peptides and direct determination of amino acids, in *Techniques in Protein Chemistry III* (Hogue Angeletti, R. A., ed.), Academic, San Deigo, CA, pp. 129–134.
7. Righetti, P. G., Gelfi, C., Perego, M., Stoyanov, A. V., and Bossi, A. (1997) Capillary zone electrophoresis of oligonucleotides and peptides in isoelectric buffers: theory and Methodology. *Electrophoresis* **18**, 2143–2153
8. Palmer, J., Burgi, D. S., Munro, N. J., and Landers, J. P. (2001) Electrokinetic injection for stacking neutral analytes in capillary and microchip electrophoresis. *Anal. Chem.* **73**, 725–731.

9

High-Performance Liquid Chromatography On-Line Derivative Spectroscopy for the Characterization of Peptides with Aromatic Amino Acid Residues*

Christoph W. Turck

1. Introduction

High-performance liquid chromatography (HPLC) is one of the most common separation techniques in today's protein chemistry. With the development of diode-array UV detectors for HPLC instruments on-line derivative spectroscopy has become possible and provides an extremely powerful tool for the analysis of peptides and proteins *(2)*. The method of on-line spectral analysis has been particularly useful in the analysis of peptides containing aromatic residues *(1,2)*. The formation of second derivatives of the absorption maxima leads to further increases in the resolution between spectral differences and allows one to distinguish between different aromatic residues in a peptide. Based on these findings we have utilized HPLC on-line derivative spectroscopy for the analysis of an important post-translational modification of peptides and proteins *(1)*.

Tyrosine phosphorylation has been shown to be a key step in the regulation of several cellular events *(3,4)* including signal-transduction mechanisms of stimulated growth factor receptors *(5,6)*. Traditional methods for direct mapping of phosphorylated tyrosines use biosynthetic radiolabeling procedures with [^{32}P] orthophosphate and subsequent isolation and cleavage of the protein of interest followed by peptide map analysis *(7)*. If the sequence of the studied protein is known, synthetic peptides can be prepared and compared to cleavage fragments carrying the phosphotyrosine residue. The mapping of protein phos-

*This article is a modified reprint of an article published in Peptide Research *(1)*.

From: *Methods in Molecular Biology, vol. 211: Protein Sequencing Protocols, 2nd ed.*
Edited by: B. J. Smith © Humana Press Inc., Totowa, NJ

phorylation sites is also possible with mass spectrometry *(8)*. However, owing to the low stoichiometries of protein phosphorylation and the unfavorable behavior of phosphopeptides during mass spectrometry analysis, this method is still not routine *(8)*. Also protein sequence analysis of phosphotyrosine containing peptides has been possible by several methods *(9)*.

We have developed a method for the detection of phosphotyrosine residues in peptides based on reversed-phase HPLC (RP-HPLC) on-line spectral analysis. It was found that tyrosine containing peptides show a hypsochromic shift of the aromatic absorbance maximum when the tyrosine is phosphorylated *(1)*. Subsequent second-order derivative spectra likewise reveal a hypsochromic shift of the corresponding minima of the phosphotyrosine residues compared to the unmodified tyrosine. This method allows mapping of tyrosine phosphorylation sites in proteins after cleavage into smaller peptides and separation and on-line spectral analysis of the latter by RP-HPLC. It furthermore provides a useful way for the characterization of synthetic phosphotyrosine containing peptides. The difference in absorption between phosphotyrosine and unmodified tyrosine can be exploited to determine phosphotyrosine residues in microgram amounts of polypeptides during their elution from reversed-phase columns with the help of an on-line scanning diode array detector and subsequent analysis of the second-order derivative spectra that exhibit characteristic minima at peaks and shoulders of the zero-order spectra *(2)*. This method can be used for the determination of tyrosine phosphorylation sites after isolation and cleavage of the protein of interest followed by on-line spectral analysis of the cleavage products during separation and subsequent Edman degradation or mass spectrometry analysis of phosphotyrosine containing peptides.

2. Materials

1. Analytical RP-HPLC is carried out with 1–5 µg of each amino acid or peptide in 25 µL 50% acetic acid using a Model 1090 Hewlett Packard instrument (Palo Alto, CA) equipped with a HP 1040 diode-array detector and HP 79996A data acquisition software (Hewlett Packard).
2. A narrow bore C_{18}-column from Vydac (2.1 mm × 25 cm) (Hesperia, CA) is used in 0.1% trifluoroacetic acid (TFA) with a gradient of acetonitrile from 0–90% in 60 min at a flow rate of 0.5 mL/min.
3. Effluents are monitored with a diode-array detector (flow cell: 6 mm pathlength, 8 µL volume). Peptide zero- and second-order derivative spectra are obtained between 230 and 300 nm and subsequently stored in a spectra library (HP 79996A).

3. Methods

1. Spectra of peptides collected in the above manner are analyzed for the presence of tryptophan, phenylalanine, tyrosine or phosphotyrosine by overlaying them with spectra of reference amino acids and peptides.

Fig. 1A and **1B** show the zero-and second-order derivative spectra of 5 μg of free tyrosine and phosphotyrosine, respectively. A hypsochromic shift of the aromatic ring absorption maximum of tyrosine can be observed when the phenolic hydroxyl group is modified with a phosphate moiety. Likewise in the second-order derivative spectra a hypsochromic shift of the corresponding two minima can be detected (281–272 nm and 273–264 nm, respectively).

2. In **Fig. 1C** and **1D** zero- and second-order derivative spectra for the other two natural aromatic amino acids, namely phenylalanine and tryptophan, are shown. The characteristic minima in the second-order derivative spectra of phenylalanine at 257 nm and 264 nm and for tryptophan at 268 nm and 278 nm are in accord with published data *(2)* and are used for subsequent comparison in the spectral analysis of the model peptides. A complete list of second-order derivative spectra minima of all four aromatic amino acids is shown in **Table 1**.

3. Results from on-line derivative spectroscopy of synthetic peptides (5 μg) with phosphorylated and unphosphorylated tyrosine residues are shown in **Table 2**. **Fig. 2A** and **2B** demonstrate that the hypsochromic shift of the absorption maximum between 250 and 300 nm is also observed for the tyrosine phosphorylated peptide. The method of real time spectral analysis hence can distinguish very clearly between phophotyrosine and unmodified tyrosine-containing peptides when no other aromatic residues are in the peptide sequence.

   Although additional minima are introduced into second-order derivative spectra when phenylalanine is present in the peptide sequence (**Fig. 2C** and **2D**) the hypsochromic shift of phosphotyrosine can still be detected (phosphotyrosine minima at 268 and 274 nm, phenylalanine minima at 261 and 268 nm).

   Less pronounced is the spectral difference between phosphotyrosine and tyrosine containing peptides when the sequence contains an additional tryptophan residue. **Fig. 3A** and **3B** show that the two peptide spectra are more similar than the ones without tryptophan but even in this case slight hypsochromic shifts of the second-order derivative minimum from 272 to 270 nm are observed. The tryptophan minimum at 290 nm is also present. A similar picture is obtained when all three amino acids, i.e., either phosphotyrosine or tyrosine and phenylalanine and tryptophan, are in the peptide sequence (minimum shift from 272–270 nm, **Fig. 3C** and **3D**). The results of a peptide containing a phosphotyrosine and an unmodified tyrosine residue (spectra not shown) are listed in the last row of **Table 2**. Its spectrum represents a combination of the spectra of the two peptides containing either only phosphotyrosine or only tyrosine (**Table 2**, rows 1 and 2). The second-order derivative minima at 261 nm and 282 nm are derived from the phosphotyrosine and tyrosine residues, respectively, whereas the third minimum at 273 nm represents two overlapping minima.

4. The spectral data of all the peptides listed in **Table 2** may be stored in a spectra library and can be used for the analysis of peptides derived from digests of proteins whose tyrosine phosphorylation sites are to be mapped.

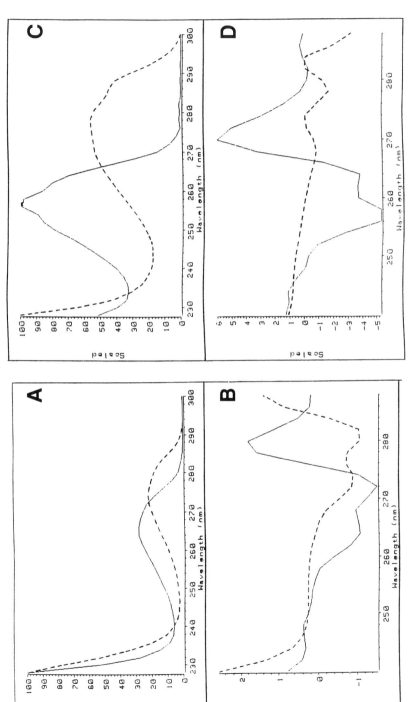

Fig. 1. Zero- (**A,C**) and second-order (**B,D**) derivative spectra of phosphotyrosine, tyrosine, phenylalanine and tryptophan. (**A,B**): phosphotyrosine (——) and tyrosine (- -) (**C,D**): phenylalanine (——) and tryptophan (- -).

Table 1
RP-HPLC On-Line Spectral Analysis of Free Aromatic Amino Acids

Amino acid		Second-order derivative spectra minima/nm
	*	
Phosphotyrosine	Y	264, 272
Tyrosine	Y	273, 281
Phenylalanine	F	257, 264
Tryptophan	W	268, 278, 290

Table 2
RP-HPLC On-Line Spectral Analysis of Synthetic Model Peptides

Peptide sequence	Aromatic amino acids	Second-order derivative spectra Minima/nm
*	*	
YVPML	Y	264,270
YVPML	Y	273,282
*	*	
YVPFL	Y,F	261,268,274
YVPFL	Y,F	258,264,274,282
*	*	
YVPWL	Y,W	262,270,280,290
YVPWL	Y,W	270,272,280,290
*	*	
YVFWL	Y,F,W	270,272,280,290
YVFWL	Y,F,W	272,280,290
*	*	
YVPYL	Y,Y	261,273,282

*: Phosphotyrosine.

4. Notes

1. A method has been developed that can detect the presence of phosphate groups on tyrosine residues with the help of on-line RP-HPLC spectral analysis, a method often used for purity check of peptides before sequence analysis (2). We have demonstrated that owing to the hypsochromic shift of the aromatic UV-absorption maximum between 250 and 300 nm, characteristic second-order derivative minima can be detected in tyrosine-phosphorylated vs -unphosphorylated peptides.

2. Although other factors, such as sequence context, pH, concentration of the organic solvent during elution (10), cause shifts in peptide absorption maxima, previous studies have shown that these differences are only in the range of 1–2 nm as compared to the corresponding free amino acids and thus are within the limit of resolution of the diode-array detector (11).

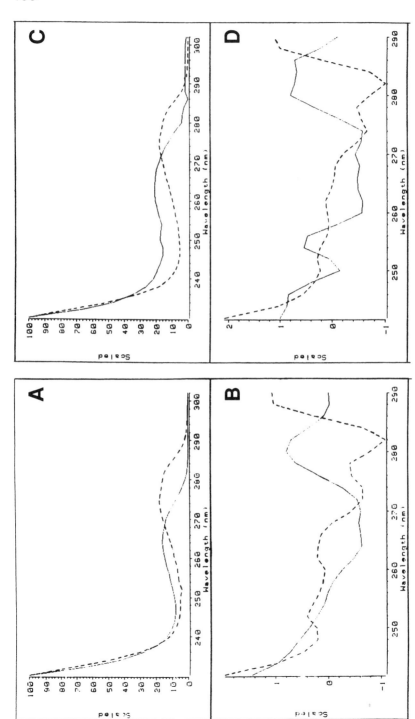

Fig. 2. Zero- (**A,C**) and second-order (**B,D**) derivative spectra of peptides *YVPML, YVPML, YVPFL and YVPFL. (**A,B**): *YVPML (—) and YVPML (- -) (**C,D**): YVPFL (—) and YVPFL (- -).

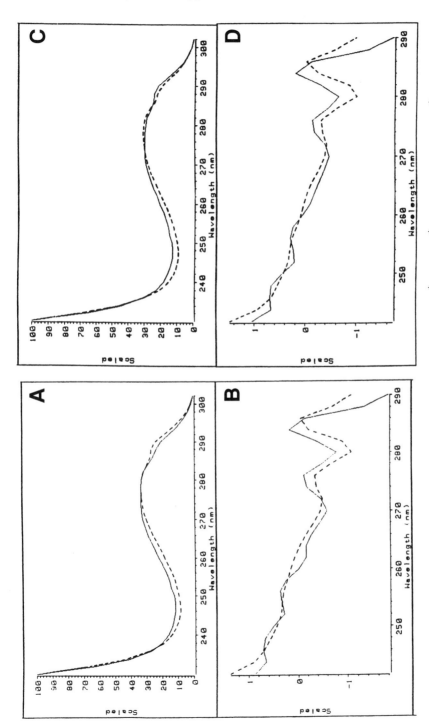

Fig. 3. Zero- (**A,C**) and second-order (**B,D**) derivative spectra of peptides YVPWL, YVPWL, YVFWL and YVFWL. (**A,B**): YVPWL (—) and YVPWL (- -) * (**C,D**): YVFWL (—) and YVFWL (- -).

3. With the exception of tryptophan-containing peptides where the hypsochromic shift of the absorbance maximum can be within the error of detection, tyrosine phosphorylation sites of microgram amounts of peptides can be detected with this method.

4. The established spectra library using the data from the peptides listed in **Table 2** can be expanded and should be useful in the analysis of peptide mixtures derived from tyrosine-phosphorylated natural as well as recombinant proteins during RP-HPLC separation.

5. The presented method for the identification of phosphotyrosine residues in peptides provides a simple and nondestructive way for the mapping of phosphorylation sites in proteins avoiding prior radiolabeling.

6. An additional application is the characterization of synthetic peptides containing phosphotyrosine. The synthesis of these compounds is associated with a high risk of phosphate ester hydrolysis in the assembled peptide chain during cleavage and deprotection steps leading to unphosphorylated peptide *(11)*. The on-line spectral analysis during RP-HPLC allows for a relatively simple way of checking for the presence of the phosphate group on tyrosine.

7. It is expected that other methods that use derivative spectroscopy to determine quantitatively the number of tyrosine and tryptophan residues in proteins *(12,13)* can be adapted for HPLC on-line analysis.

Acknowledgments

This work was supported by the Howard Hughes Medical Institute.

References

1. Turck, C. W. (1992) Identification of phosphotyrosine residues in peptides by high performance liquid chromatography on-line derivative spectroscopy. *Peptide Res.* **5**, 156–160.

2. Grego, B., Nice, E. C., and Simpson, R. J. (1986) Use of scanning diode array detector with reversed-phase microbore columns for the real-time spectral analysis of aromatic amino acids in peptides and proteins at the submicrogram level. *J. Chromatogr.* **352**, 359–368.

3. Cohen, P. (1982) The role of protein phosphorylation in neural and hormonal control of cellular activity. *Nature* **296**, 613–616.

4. Krebs, E. G. and Beavo, J. A. (1979) Phosphorylation-dephosphorylation of enzymes. *Ann. Rev. Biochem.* **48**, 923–939.

5. Ullrich, A. and Schlessinger, J. (1990) Signal transduction by receptors with tyrosine kinase activity. *Cell* **61**, 203–212.

6. Hunter, T. (2000) Signaling-2000 and beyond. *Cell* **100**, 113–127.

7. Martensen, T. M. (1984) Chemical properties, isolation, and analysis of O-phosphates in proteins. *Methods Enzymol.* **107**, 3–23.

8. Neubauer, G. and Mann, M. (1999) Mapping of phosphorylation sites of gel-isolated proteins by nanoelectrospray tandem mass spectrometry: potentials and limitations. *Anal. Biochem.* **71**, 235–242.

9. Meyer, H. E., Hoffmann-Posorske, E., Korte, H., Donella-Deana, A., Brunati, A.-M., Pinna, L. A., et al. (1990) Sequence analysis of phosphotyrosine-containing peptides. Determination of PTH-phosphotyrosine by capillary electrophoresis. *Chromatogr.* **30**, 691–695.
10. Donovan, J. W. (1969) Ultraviolet Absorption, in *Physical Principles and Techniques of Protein Chemistry, Part A* (Leach, S. J., ed.), Academic Press, New York, NY, pp. 101–105.
11. Valerio, R. M., Perich, J. W., Kitas, E. A., Alewood, P. F., and Johns, R. B. (1989) Synthesis of O-phosphotyrosine-containing peptides. II Solution-phase synthesis of Asn-Glu-PTyr-Thr-Ala through methyl phosphate protection. *Aust. J. Chem.* **42**, 1519–1525.
12. Bray, M. R., Carriere, A. D., and Clarke, A. J. (1994) Quantitation of tryptophan and tyrosine residues in proteins by fourth-derivative spectroscopy. *Anal. Chem.* **221**, 278–284.
13. Mach, H. and Middaugh, C. R. (1994) Simultaneous monitoring of the environment of tryptophan, tyrosine, and phenylalanine residues in proteins by near-ultraviolet second-derivative spectroscopy. *Anal. Biochem.* **222**, 323–331.

10

Hydrolysis of Samples for Amino Acid Analysis

Ian Davidson

1. Introduction

There is no single hydrolysis method that will effectively cleave all proteins to single amino acids completely and quantitatively. This is owing to the varying stability of the peptide bonds between the different amino acids and the amino acid side chains, which are themselves susceptible to the reagents and conditions used to cleave the peptide bonds (*see* **Table 1**). The classical hydrolysis conditions, to which all other methods are compared, is liquid-phase hydrolysis in which the protein or peptide sample is heated in 6 M hydrochloric acid under vacuum at 110°C for 18–24 h (*1*). The various methods of hydrolysis described here are summarized in **Table 2**.

2. Materials

1. Rotary evaporator, e.g., Savant "Speed Vac" equipped with a supply of high-purity nitrogen to the atmosphere inlet.
2. Rotary Vacuum pump, capable of evacuating the system to 50mTorr or better, equipped with a cold trap and high-purity nitrogen bleed valve. The cold trap should be kept at a temperature of –60°C or below. This can be achieved by cooling propan-2-ol in a Dewar Flask with dry ice or, e.g., Neslab (Waalresberg, Netherlands) Immersion Cryocool Type CC60 or by filling the Dewar flask with liquid nitrogen.
3. Argon gas supply, equipped with a needle valve to enable slow bubbling of gas.
4. Laboratory Oven with a variable temperature control up to 200°C
5. Vapor-phase hydrolysis tubes (Borosilicate glass 50 × 6 mm) (Sigma-Aldrich, Dorset, UK, Part no. Z144509). Before using the 50 × 6mm sample tubes they should first be cleaned thoroughly by inverting them in a borosilicate glass beaker, pyrolysing at 500°C for approx 4 h. Allow to cool, rinse with ultra high-quality water, dry, and store covered. Use these cleaned tubes only once.

From: *Methods in Molecular Biology, vol. 211: Protein Sequencing Protocols, 2nd ed.*
Edited by: B. J. Smith © Humana Press Inc., Totowa, NJ

Table 1
Stability of Amino Acid Residues and Peptide Bonds During Hydrolysis on 6 M Hydrochloric Acid at 110°C

Residue/bond	Stability/modification	Consequence	Remedy
Serine; Threonine	1. Side-chain hydroxyl group modified by dehydration, which is increased with increased hydrolysis time and temperature. 2. Ester formation with, e.g., glutamic acid can occur at the drying stage (3).	Serine and Threonine generated in low yield.	1. Hydrolyze protein samples for different times between 6–72 h. Calculate the yields, extrapolate results to time = 0 to compensate for losses (1,10). 2. Dry the hydrolysates rapidly in a rotary evaporator.
Tyrosine	The phenolic group (-C6H4OH) side chain is modified by traces of hypochlorite/chlorine radicals present in the acid (3).	Tyrosine is generated in low yield.	Incorporate phenol in acid to compete for hypochlorite/chlorine radicals (1,4,5).
Methionine	The thioether (-CH2-S-CH3) side chain is oxidized to the sulphoxide or sulphone (3).	Methionine, usually a less common residue anyway is converted to smaller peaks, more difficult to quantify on amino acid analyzers.	Add reducing agent (e.g., dodecanthiol or thioglycolic acid) to the acid/phenol mixture (5,6).
Cystine; Cysteine	The free sulfhydryl (-SH) and diusulphide (-S-S) side chain groups are oxidized (3).	Cystine and Cysteine recovered in low non-quantifiable yields.	Chemically modify prior to hydrolysis (7) (Chapter 27).

112

Table 1 (continued)
Stability of Amino Acid Residues and Peptide Bonds During Hydrolysis on 6 M Hydrochloric Acid at 110°C

Residue/bond	Stability/modification	Consequence	Remedy
Trytophan	The Indole group side chain is destroyed by oxidation under acid conditions (*3*).	Trytophan is not quantifiable under these conditions.	Add reducing agents (e.g., dodecanthiol or thioglycolic acid) to the acid mixture (*5,6,10*) or hydrolyze under alkaline conditions (*3,8,9*).
Aspargine; Glutamine	Asn and Gln are deaminated to form the respective acids (*3*).	Mixtures of Asp/Asn amd Glu/Gln are normally assigned as Asx and Glx respectively, in quantification data.	
Bonds between hydrophobic amino acids (e.g., Val-Val or any combination of Ala, Ile, Leu, Val).	Bonds are realtively stable (*3*).	Hydrolyze in poor yield. May be seen as dipeptides or similar, on amino acid analyzer or not at all.	Hydrolyze for longer time or elevated temperature, e.g., 165°C (*1,5*).
Phosphorylation	Phosphorylated amino acids are labile.	Destroyed under these extreme conditions.	Reduce time for hydrolysis to 1–4 h (*13*).
Glycosylation	Amino acid–sugar interactions produces complex secondary reaction products (*3*).	Complex reaction products are difficult to interpret even if they are seen on an analyzer.	Deglycosylate before hydrolysis as described in Chapter 30.

Table 2
Advantages and Disadvantges of Various Hydrolysis Techniques

Method of hydrolysis	Advantages	Disadvantages
Vapor phase under argon (165°C for 45 min) (*see* **Subheading 3.1.**).	High sensitivity. Relatively fast hydrolysis times. Samples can be processed in batches.	Owing to the high pressures the reaction vial and seals require regular inspection. Danger of exploding vials and eacaping hot acid can occur with defective vials and seals.
Vapor phase under vacuum (110°C for 18 h) (*see* **Subheading3.2.**).	Conditions not as extreme as in **Subheading 3.1.** Samples can be processed in batches.	Long duration time for hydrolysis. Most analyzers have relatively short derivatization and analysis times. The seals of the vial require regular inspection as above.
Liquid phase (110°C for 18 h) (*see* **Subheading 3.3.**)	Conditions not as extreme as in **Subheading 3.1.**	Long duration time for hydrolysis. Samples are processed individually over a number of manipulations, which is very time consuming.
Microwave irradiation (8 min) (*see* **Subheading 3.4.**).	Rapid hydrolysis times. Samples can be processed in batches.	More extreme conditions than 3.1.; thereefore more dangers from exploding vials. Potential contamination from the reusable, expensive tubes.
PVDF blots (*see* **Subheading 3.5.**).	Some useful composition data from samples that may not have been pure prior to SDS gel electrophoresis.	Low recoveries. Samples are difficult to remove successfully from the blot.
Liquid phase under alkaline conditions (*see* **Subheadings 3.6.** and **3.7.**).	Thrytophan is preserved throughout.	Nonvolatile reagents. Relatively high salt content. pH for derivatization difficult to control with such small volumes.

6. Liquid-phase hydrolysis tubes are made by cutting a piece of borosilicate glass tubing, 6 mm ID × 8 mm OD, 150–200 mm in length. The shortened tubing can be cleaned by immersing in 50% Nitric acid in water overnight, rinsing with distilled water, and finally with ultra high-quality water. Dry the tubes in an oven. Seal one end with a glass blowing torch, equipped with a gas and oxygen flame, to form a tube.

7. Preparation of acid hydrolysis mixture (add reducing agent as and when necessary).
    a. Hydrochloric acid (constant boiling) 6 *M* (Applied Biosystems, Foster City, CA, Part No. 400939 or Pierce, Rockland, IL, Part No.24309) (1 mL ampoules).
    b. Phenol (Sigma-Aldrich, Part No. 328111). (Store small amounts in a clean vial sealed under argon)
    c. Reducing agent, e.g., Dodecanethiol (Sigma-Aldrich, Part No. 62592) or Thioglycolic acid (Sigma-Aldrich, Part No. 88650).
    d. Gently heat the vial containing a little phenol, stored as above (**Note 7b**), on a hotplate set at 80°C in a fume cupboard to liquefy the phenol. Add 5 µL of the liquefied phenol before it cools, to 500 µL hydrochloric acid and mix thoroughly. Add 5 µL of the reducing agent and mix thoroughly. Allow the stock phenol to cool before purging with argon and storing.
    e. Alternative hydrolysis reagents for samples containing Tryptophan: 3 *M* mercaptoethanesulphonic acid (Pierce, Part no. 25555) or 4 *M* methanesulphonic acid (containing 0.2% 3-(2-Aminoethyl) indole) (Pierce, Part no. 25600).
    f. Alkaline hydrolysis reagent for samples containing tryptophan. 4.2 *M* sodium hydroxide (Sigma-Aldrich, Part no. 06213).
8. Reaction vial and valve (*see* **Fig. 1**) (e.g., Waters, Milford, MA, Part no. 07363). The Waters reaction vial is specified for vapour phase hydrolysis conditions at 110°C under vacuum. For hydrolysis conditions requiring elevated temperatures then a Reaction Vial (120 mm × 25 mm OD) consisting of a teflon lined Mininert valve (Pierce, Part no. 10150) fitted to a heavy walled, glass 25 mL universal container is required. A bulb is blown at 50 mm from the base of the universal container (a good glassblower will help with this) so that the tops of the sample tubes do not come into contact with the walls of the reaction vial. Any condensed hydrolysis mixture will therefore not run down the vial wall into the sample tube.
9. Microwave hydrolysis tubes, 100 ×10 mm borosilicate tubes with side arm and Teflon screw-on valve (Pierce, Part no. 29560). These tubes are reusable but should be cleaned thoroughly as described in **Subheading 2.6.**
10. Microwave Oven 650 Watt Full Power (Minimum requirement).
11. Norleucine as an internal standard is prepared by dissolving 32.75 mg of L-Norleucine (Pierce, Cat. no. 36323) in 100 mL of 0.1 *M* hydrochloric acid. Aliquots of approx 200 µL of this stock solution are frozen, almost indefinitely, at –20°C until required. A daily working solution is prepared by dissolving 20 µL of the stock solution up to 1 mL with water and vortexing thoroughly. A 5 µL aliquot of this solution contains 250 pmol Norleucine (*see* **Note 1**).
12. Digestion block (Pierce, Part no. 18970).
13. There are several different types of tips and packing material available and the choice is dependent on the molecular weight and type of peptide or protein of interest. (Supro tips, Nest Group, Southborough, MA, Part no. SSPUVC08F or Zip tips from Millipore, Bedford, MA, Part no. ZTC18S960. Alternatively, these tips can be prepared and adapted as required *(2)*. Constrict the end of a gel-loader tip (Eppendorf-Netherel-Hinz Gmbh, Hamburg, Germany, Part no. 0030

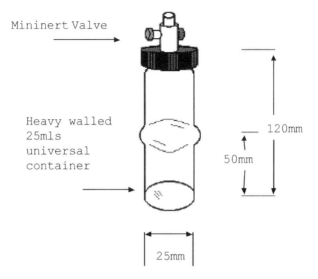

Fig. 1. Hydrolysis reaction vial.

001.222) by gently squeezing and pulling the end of the tip with a spatula so that liquid can still pass through the constricted tip but a few millimeters of a slurry of POROS R1 10 (Applied Biosystems, Framingham, MA, Part no. 1-1118-02) does not. The slurry is a mixture of the POROS material and 30% propan-2-ol in water. Use a 1-mL syringe to act as pump to push the POROS through but not out of the gel loader tip.

3. Methods

3.1. Manual Vapor-Phase Hydrolysis Under Argon

1. Pipet the sample aliquot, including any internal standard (*see* **Notes 1–5**) to a maximum volume of 200 μL, into the vapor-phase hydrolysis sample tube (50 × 6 mm borosilicate tube) as described in **Subheading 2.5.**, marked with a diamond-tipped pen as the ink from felt-tipped pens will run. Normally 50 ng to 1 μg is required depending on the derivatization method employed (*see* Chapters 11–15). Centrifuge so that the entire sample collects at the bottom of the tube and not on the sides.
2. Dry the sample. The samples should be dried in a rotary evaporator ensuring that there are no loose particles left in the bottom of the tube. Should loose particles be present after drying then add a little water and redry.
3. Add 500 μL of hydrolysis mixture to the reaction vessel (*see* **Fig. 1**).
4. Slowly purge the hydrolysis acid mixture in the reaction vessel and the mininert valve with argon for approx 2 min to displace atmospheric oxygen.

5. Add the sample tubes and purge with argon again. Purge each tube in turn taking care to avoid cross-contamination and displacing material from the tube. A gentle continuous flow from a 50 μL micropipet is a useful guide.

6. Screw the mininert valve securely onto the reaction vessel ensuring the valve is in the closed position.

7. Heat the reaction vessel to 165°C for 45 min (*see* **Note 6**).

8. **CAUTION!! The contents are under pressure.** Open the reaction vessel mininert valve very carefully in a fume cupboard. Use thermal gloves and face shield. Release the pressure very gradually to avoid the liquid boiling up into the sample tubes.

9. Remove the tubes from the reaction vessel while still hot. Wipe the outside of the tube with a clean tissue and dry off the sample tubes in a rotary evaporator without centrifugation to prevent any condensed acid vapor being drawn onto the sample at the bottom of the sample tube.

10. Store the samples dry at –20°C until required for derivatization.

3.2. Manual Vapor-Phase Hydrolysis Under Vacuum

1. The same as for **Subheading 3.1.**, **steps 1–3**.

2. Add the tubes to the reaction vial.

3. Screw the mininert valve securely onto the reaction vial ensuring the valve is in the closed position.

4. Attach the reaction vessel to the vacuum pump and open the valve slowly to avoid "bumping."

5. Evacuate for a few minutes with gentle agitation.

6. Close the valve.

7. Heat the reaction vessel to 110°C for 18 h (*see* **Note 6**).

8. Attach the reaction vessel to a nitrogen supply set at approx 1–5 psi and slowly open the mininert valve, taking care to avoid the samples being displaced from the tubes.

9. Remove the tubes from the reaction vial while still hot. Wipe the outside of the tube with a clean tissue and dry off the sample tubes in the rotary evaporator without centrifugation to prevent any condensed acid vapor being drawn onto the sample at the bottom of the sample tube.

10. Store the samples dry at –20°C until required for derivatization.

3.3. Manual Liquid-Phase Hydrolysis Under Vacuum

1. Pipet the sample aliquot including any internal standard, into a liquid phase (*see* **Notes 1–5**) hydrolysis borosilicate glass tube (200 mm × 6 mm ID) as described in **Subheading 2.6.** marked with a diamond tipped pen on the bottom one-third of the tube (the top two-thirds will be discarded later). Normally 50 ng to 1 μg is required depending on the derivatization method employed (*see* Chapters 11–15). Centrifuge so that the entire sample collects at the bottom of the tube and not on the sides.

2. Dry the samples in a desiccator under vacuum, slowly to avoid "bumping."

3. Add 200 µL of hydrolysis mixture to the hydrolysis tube.
4. Vortex thoroughly.
5. Centrifuge the hydrolysis tubes again so that all of the sample and hydrolysis mixture is on the bottom of the tube.
6. Cool the samples to –20°C or below to avoid the sample "bumping" when it is attached to the vacuum line. This can be achieved easily by dipping the tube into the –60°C cold trap of the vacuum pump for a few minutes.
7. Evacuate for a few minutes with gentle agitation.
8. Seal the tube while still under vacuum, approx 30–50 mm from the top, with a glass-blowing torch, equipped with a gas and oxygen flame.
9. Heat the reaction vessel to 110°C for 18 h (*see* **Note 6**).
10. Cool the tubes in ice. Centrifuge so that the samples are at the bottom of the tubes and not on the walls of the tubes.
11. Cut the tube open with a sharp glass knife approx 30–50 mm from the top by scoring around the tube, placing both thumbs either side of and close to the score and pulling both hands together against the thumbs.
12. Dry off the excess acid in a desiccator or rotary evaporator.
13. Store the samples dry at –20°C until required for derivatization.

3.4. Manual Liquid-Phase Hydrolysis by Microwave Irradiation

1. Pipet the sample aliquot (*see* **Notes 1–5**) as described in **Subheading 3.3.**, **step 1** into the specifically designed vacuum hydrolysis tube described in **Subheading 2.**, **item 9**.
2. Proceed as described **Subheading 3.3.**, **steps 2–7**.
3. Seal the tube by screwing down the Teflon plunger.
4. Place the tubes in a Microwave Oven on full power (650W) for 8 min.
   Extreme Caution!! High pressures of up to 140 psig. The tubes have been reported to explode at this point. (*11*)
5. Cool the hydrolysate tubes in ice, centrifuge to collect the entire sample on the bottom of the tube and not on the tube walls.
6. Pipet the hydrolyzed samples into a clean microcentrifuge tube.
7. Dry off the excess acid in a rotary evaporator.
8. Store the samples dry at –20°C until required for derivatization.

Note: A commercial Microwave Digestion System Type CEM-MDS-81D is available from CEM Corporation (Mathews, NC) (*10,11*).

3.5. Manual Hydrolysis of Samples Blotted onto Polyvinylidene Difluoride (PVDF) Membrane (see Note 6)

1. Using any of the described methods from **Subheadings 3.1.–3.3.**, cut the band from the membrane (with a scalpel) containing the sample and place the PVDF membrane into the bottom of a hydrolysis tube and proceed as normal taking care that the membrane is not dislodged during evacuation or when opening the tubes.
2. Proceed as described in **Subheadings 3.1.–3.3.**
3. Dry off the excess acid under vacuum in a rotary evaporator.

4. Extract the hydrolyzed amino acids in the sample from the hydrolyzed PVDF membrane by adding 100 μL of 70% 0.1 M hydrochloric acid in methanol (v/v).
5. Vortex thoroughly, by attaching the sample tubes to the mixer with a piece of parafilm, or similar, for 5 min.
6. Transfer the liquid to a clean microcentrifuge tube.
7. Repeat **steps 5** and **6**.
8. Dry the hydrolysed sample in a rotary evaporator.
9. Store the sample dry at –20°C until required for derivatization.

3.6. Alternative Reagents for Hydrolysis of Samples Where Tryptophan is to be Preserved (see Note 8)

1. Proceed as described in **Subheading 3.3.**, **steps 1–8**, substituting 30 μL 3 M mercaptoethanesulphonic acid (or 4 M methanesulphonic acid) for the acid hydrolysis mixture.
2. Heat the tube at 110°C for 22 h.
3. Cool the tube in ice and centrifuge so that the entire sample is collected at the bottom of the tube and not on the tube walls.
4. Neutralize with 50 μL of 1.38 M sodium hydroxide solution and mix thoroughly.
5. Store the samples at –20°C until required for derivatization.

3.7. Manual Alkaline Hydrolysis of Samples Where Tryptophan is to be Preserved (see Note 9)

1. Proceed as described in **Subheading 3.3.**, **steps 1–7** substituting the 200 × 60 mm Borosilicate glass tube with the specifically designed vacuum hydrolysis tube as described in **Subheading 2.9.** Note the tubes will not be cut at a later stage.
2. Add 200 mL of 4.2 M sodium hydroxide solution and vortex.
3. Seal the tube by screwing down the Teflon plunger.
4. Heat the hydrolysis tube to 110°C for 18 h (*see* **Note 5**) in the digestion block (*see* **Subheading 2.**, **item 12**).
5. Proceed as described in **Subheading 3.4.**, **steps 5–6**.
6. Neutralize to the correct pH for derivatization with approx 60 μL concentrated hydrochloric acid.
7. Alternatively store the hydrolyzed samples buffered at pH 4.25 at –20°C until required for derivatization.

4. Notes

1. The choice of internal standard depends on the following:
   a. Stability during hydrolysis.
   b. The derivatization procedure employed. The yield should be linear with concentration.
   c. The ability to be easily separated from other amino acids.
   d. Not occurring naturally.
   e. Commercially available and inexpensive.

2. Samples should ideally be dissolved in volatile solvent (i.e., no buffer salts present). Removal of buffer salts prior to hydrolysis may be required. This can be achieved by desalting either by HPLC as described in Chapter 1 or with the use of Supro tips or Zip Tips (see **Subheading 2.1.**, **step 3**). It should be noted that that the commercially available tips are for high sensitivity work and it may be necessary to apply several desalting cycles per sample to achieve sufficient material for amino acid analysis.

To equilibrate the tip:

a. Add 10 µL of a solution of Acetonitrile/0.1% Trifluoroacetic acid in water (9:1), Acetonitrile HPLC-grade, Trifluoroacetic acid (Sigma-Aldrich) to the zip tip. Attach the 1-mL disposable syringe to the above tip to act as a pump and gently expel the solution to waste. Remove the syringe. Repeat twice.

b. Add 10 µL of 0.1% TFA in water to the Zip Tip. Attach the 1-mL disposable syringe to the above tip and gently expel the solution to waste. Remove the syringe. Repeat twice.

c. Add the sample through the tip, the volume will depend on the concentration of the sample and the derivatization method for the analysis but a few picomoles should be sufficient. With some analyzers it may be necessary to repeat the procedure several times to obtain sufficiently pure material. Remove the syringe.

d. Remove the buffer salts from the protein or peptide sample by adding 10 µL of 0.1% TFA in water and expelling slowly to waste using the syringe. Repeat 2 or 3 times depending on the concentration of buffer salts expected.

e. Finally elute the purified peptide or protein from the column packing material with 10 µL of a solution of Acetonitrile/0.1% TFA in water (9:1 v/v) and collect the purified material.

f. The tips are cleaned and stored as in **step a** and may be used several times but this will depend on the level of impurities in the samples.

3. Powder-free gloves should always be worn when handling samples and associated glassware to avoid contamination. Powdered gloves can give rise to contamination from proteins, peptides, and amino acids adsorbing to the cornstarch in the powder. The particles of cornstarch can also accumulate in and eventually block narrow-bore tubing found in most modern analyzers.

4. The use of dedicated glassware is also recommended. The glassware should be cleaned, where possible, with 50% nitric acid, rinsed thoroughly with high-quality water and stored covered. Washing the glassware in a 3% solution of EDTA (Ethylenediaminetetraacetic acid, tetrasodium salt, Sigma-Aldrich, Part no. E26290) every couple of months will clean off and prevent metal ions, which can leach out from the glass, building up on the surfaces.

5. Reagents for all methods should be of the highest quality available. Water should be of ultra-high quality. (HPLC-grade or better).

6. A hydrolysis time-course experimental study is highly recommended to optimize laboratory conditions best suited to particular requirements. In a (1993) survey of

amino acid analysis test sites *(10)*, the times of hydrolysis averaged $111 \pm 2.7°C$ for 22 ± 2.4 h or $160 \pm 15°C$ for 1.4 ± 0.4 h.

7. PVDF blots are best analysed by manual liquid phase hydrolysis which requires excess acid but the recoveries are low, in the region of 25–30% with the total loss of methionine. Some useful composition information can be produced with care.

8. As with alkaline hydrolysis (*see* **Note 9**) the reagents are nonvolatile and have relatively high amounts of salts present. The hydrolyzed amino acids can be purified or determined by Ion Exchange Chromatography or derivatization followed by Ion exchange Chromatography for analysis.

9. Alkaline hydrolysis is dependent on the derivatization method to be used. The hydrolysis reagents are nonvolatile and have relatively large amounts of salts present, which is unsuitable for phenylisothiocyanate (PITC) derivatization for example.

Acknowledgment

My thanks to Lynne McKay for the illustration (**Fig. 1**).

References

1. Moore, S. and Stein, W. H. (1963) Chromatographic determination of amino acids by the use of automatic recording equipment. *Methods Enzymol.* **6**, 819–831.
2. Wilm, M., Shevchenko, A., Houthaeve, T., Brit, S., Schweigerer, L., Fotsis, T., and Mann, M. (1996) Femtomole sequencing of proteins from polyacrylamide gels by nano-electrospray mass spectrometer. *Nature* **379**, 466–469.
3. Hunt, S. (1985) Degradation of amino acids accompanying *in vitro* protein hydrolysis, in *Chemistry and Biochemistry of the Amino Acids* (Barrett, G. C., ed.) Chapman and Hall, London, pp. 376–398.
4. Bidlingmeyer, B. A., Tarvin, T. L., and Cohen, S. A. (1986) Amino acid analysis of submicrogram hydrolysate samples, in *Methods in Protein Sequence Analysis* (Walsh K., ed.) Humana Press, Totowa, NJ, pp. 229–244.
5. Dupont, D., Keim, P., Chui, A., Bozzini, M. L., and Wilson, K. J. (1988) Gas-phase hydrolysis for PTC-amino acids. *Appl. Biosys. User Bull.* **2**, 1–10.
6. Matsubara, H. and Sasaki, R. M. (1969) High recovery of tryptophan from acid hydrolysis of proteins. *Biochem. Biophys. Res. Commun.* **35**, 175–181.
7. Carne A.F. Chemical Modifications of Proteins, Methods in Molecular Biology, *Basic Protein and Peptide Protocols*, vol. 32 (Walker, J. M., ed.) Humana Press, Totowa, NJ, pp. 311–320.
8. Hugli, T. E. and Moore, S. (1972) Determination of the tryptophan content of proteins by ion exchange chromatography of alkaline hydrolysis. *J. Biol. Chem.* **247(9)**, 2828–2834.
9. Simpson, R. J., Neuberger, M. R., and Lui, T.-Y. (1976) Complete amino acid analysis of proteins from a single hydrolysate. *J. Biol. Chem.* **251**, 1936–1940.
10. Strydom, D. J., Anderson, T. T., Apostal, I., Fox, W. J., Paxton, R. J., and Crabb, J. W. (1993) Cysteine and tryptophan amino acid analysis of ABRF92-

AAA, in *Techniques in Protein Chemistry IV* (Angeletti, R.H., ed.) Academic Press, San Diego, CA, pp. 279–288.

11. Chiou, S. H. and Wang, K.-T. (1990) A rapid and novel means of protein hydrolysis by microwave irradiation using Teflon-Pyrex tubes, in *Current Research in Protein Chemistry, vol. 3* (Villafranca, J. J., ed.) Academic Press, San Diego, CA, pp 3–10.

12. Gilman, L. B. and Woodward, C. (1990) An evaluation of microwave heating for the hydrolysis of proteins, in *Current Research in Protein Chemistry, vol. 3* (Villafranca, J. J., ed.) Academic Press, San Diego, CA, pp. 23–26.

13. Capony, J.-P. and Demaille J. G. (1983) A rapid microdetermination of phosphoserine, phosphothreonine and phosphotyrosine in proteins by automatic cation exchange on a conventional amino acid analyser. *Anal. Biochem.* **128**, pp. 206–212.

11

Amino Acid Analysis

Precolumn Derivatization Methods

G. Brent Irvine and Ian Davidson

1. Introduction

The method described in this chapter is based on the derivatization of amino acids, produced by hydrolysis of peptides and proteins, with phenylisothiocyanate (PITC) under alkaline conditions. This forms the phenylthiocarbamyl (PTC-) amino acids, which are then separated by reverse-phase high-performance liquid chromatography (RP-HPLC) and quantified from their UV-absorbance at 254 nm.

Quantitative amino acid analysis, based on separation by ion-exchange chromatography followed by postcolumn derivatization using ninhydrin for detection, was developed during the 1950s (*1*), and remained the predominant method for 20 years. This method is described in Chapter 12. With the advent of RP-MPLC, however, rapid separation of amino acid derivatives became possible. Precolumn derivatization also avoids dilution of peaks and so increases sensitivity. Methods involving derivatization with fluorogenic reagents such as dansyl chloride (*2*), and o-phthaldialdehyde (*3*) were the first to be developed; these enable detection of less than 1 pmol of an amino acid. These methods have some disadvantages, however, including instability of derivatives, reagent interference, and lack of reaction with secondary amino acids. Derivatization using PITC, the reagent used in the first step of the Edman method for determining the amino acid sequence of proteins, avoids many of these problems. The reaction, shown in **Fig. 1**, is rapid and quantitative with both primary and secondary amino acids.

From: *Methods in Molecular Biology, vol. 211: Protein Sequencing Protocols, 2nd ed.*
Edited by: B. J. Smith © Humana Press Inc., Totowa, NJ

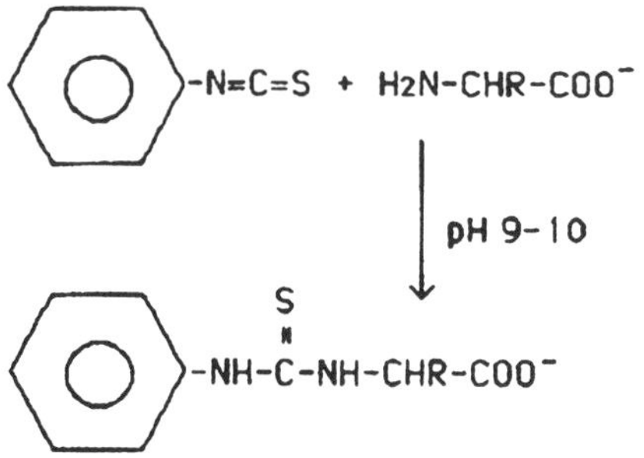

Fig. 1. Reaction of phenylisothiocyanate with an amino acid to form a phenylthiocarbamyl amino aicd derivative.

The products are relatively stable, and excess reagent, being volatile, is easily removed. Sensitivity, at about the level of 1 pmol, is more than adequate because it is difficult to reduce background contamination below this level.

Several commercial manufacturers have successfully automated the technique but although their analyzers are still in use in various types of laboratories they are no longer available as new, as other derivatization reagents have proved more popular and sensitive. In this chapter the manual PITC derivatization procedure will be described.

Quantitative analysis of phenylthiocarbamyl amino acids was first described by Tarr and coworkers *(4)*. Full details for the application of this method to the analysis of protein hydrolysates were published in 1984 *(5)*. Employees of the Waters Chromatography Division of Millipore Corporation *(6,7)* developed a system that was commercially available as the Waters Pico-Tag system (Millipore, Bedford, MA). Similarly, Applied Biosystems (Foster City, CA) employees developed the 420A and 421 Amino Acid Analyzers.

In addition to amino acids in protein hydrolysates, levels of free amino acids in foodstuffs *(8)*, unusual amino acids *(8)*, and amino acids in physiological fluids *(8–11)*, can also be determined.

Recent alternative systems have been introduced that utilize precolumn derivatization. Waters introduced a system with the reagent 6-aminoquinolyl-*N*-hydroxysuccinimidyl carbamate (AQC). The stability of the resulting derivatives is reported to give more reproducible analyses, a fluorescent detector is required for high-sensitivity analysis *(12)*.

Picometrics (Ramonville, France) have introduced a system with the reagent 3-(2-Furoyl)quinoline-2-carbaldehyde (FQCA). The technique is still to be fully developed but the sensitivity is claimed to be in the attomole level but a laser induced fluorescent detector (LIFD) is required *(13)*.

2. Materials

2.1. Apparatus

2.1.1. Equipment for Hydrolysis of Proteins and for Derivatization of the Resulting Amino Acids

1. Rotary evaporator (e.g., Savant "Speed Vac" equipped with a vacuum gauge, and a high-purity nitrogen bleed valve to the atmosphere inlet.
2. Rotary Vacuum Pump, capable of evacuating the system to 50mTorr or better, equipped with a cold trap. The cold trap should be kept at a temperature of –60°C or below. This can be achieved with liquid nitrogen contained in a Dewar flask or by cooling propan-2-ol in a Dewar flask with dry ice, or e.g., Neslab (Waalresberg, Netherlands) Immersion Cooler Type CC60.
3. Borosilicate glass tubes 50 × 6 mm (Sigma-Aldrich, Dorset, UK, Part no. Z144509).
4. Reaction vial and valve as described in Chapter 10.
5. Fan Oven adjustable to 200°C.

2.1.2. An RPLC System for Separation and Quantification of the Phenylthiocarbamyl Amino Acids

This requires two pumps, a gradient controller, an automated injector, a reverse-phase C18 column and column heater, a UV-detector set at 254 nm, and an integrator.

2.2. Chemicals (see Note 1)

2.2.1. Hydrolysis and Derivatization

1. HCl/phenol: Add melted crystalline phenol (10 µL) to hydrochloric acid (constant boiling at 760 mm) (1.0 mL) (Pierce, Rockford, IL).
2. Coupling Reagent: ethanol:water:triethylamine (2:2:1 by vol).
3. Derivatization Reagent: ethanol:water:triethylamine:phenylisothiocyanate (7:1:1:1 by vol). Vortex and allow to stand for 5 min before using. Use within 2 h. Phenylisothiocyanate (Pierce): Store at –20°C under nitrogen. After opening an ampoule, it can be divided into aliquots, which should be resealed under nitrogen. It is important to allow the container to come to room temperature before opening, since this reagent is sensitive to moisture. Each aliquot should be used within 3 wk of opening.
4. Amino Acid Standard H (Pierce): This contains a solution of 17 amino acids (2.5 mM each, except cystine, which is 1.25 mM) in 0.1M HCl. A fresh working standard solution is prepared daily by adding 10 µL of 2.5 mM stock solution up to 1 mL with water. A 10 µL-aliquot contains 250 pmol of each amino acid and 125 pmol cystine. Any internal standard should also be added.
5. Norleucine as an internal standard is prepared by dissolving 32.75 mg of L-Norleucine (e.g., Pierce, Part no. 36323) in 100 mL of 0.1 M Hydrochloric

acid. Aliquots of approx 200 μL of this stock solution are frozen at –20°C until required. Working standards are prepared fresh daily by dissolving 10 μL of the stock solution up to 1 mL with water. A 10 μL aliquot of this solution contains 250 pmol Norleucine.

6. Sodium Acetate Buffer 3 *M* solution pH 5.5, and pH 3.8 (Applied Biosystems, Foster City, CA, Part nos. 400471 and 400319).
7. Acetonitrile HPLC Grade S (Rathburn, Peebles, UK).
8. Tripotassium-EDTA solution 50 mg/mL (K₃EDTA)(Applied Biosystems, Part no. 400941).

2.2.2. Chromatography

1. Sample buffer: Dissolve 1.5 mL of Sodium Acetate 3 *M*, pH 5.5, in 150 mL of ultra high-quality water add 300 μL of Tripotassium-EDTA solution (50 mg/mL) and adjust the pH to 5.0 with 200 μL or so of sodium acetate 3*M* solution, pH 3.8. Filter through a 0.45-μm filter (Millipore type HA).
2. Eluent A: Dissolve 15.0-mL sodium acetate solution 3 *M* in water (900 mL). Adjust the pH to 5.38 with ammonium hydroxide (*see* **Note 2**) Filter through a 0.45-μm filter (Millipore type HA).
3. Eluent B: Dissolve 5.5 mL sodium acetate 3 *M* solution, pH 3.8, 5.0 mL sodium acetate 3 *M* solution, pH 5.5, in water and make up 300 mL with water. Mix this sodium acetate solution with acetonitrile (700 mL) Filter through a 0.45-μm filter (Millipore type HV).

3. Methods

3.1. Hydrolysis and Derivatization

1. To a Pyrex tube (50 × 6 mm), add an aliquot containing 200 ng–1 μg of a solution peptide or protein in a volatile solvent such as methanol or water (*see* **Notes 3** and **4**). Add 10 μL of a norleucine working solution as internal standard. Total volumes in excess of 200 μL should not be used. Place these tubes into the rotary evaporator.
2. Dry under vacuum until the pressure has fallen to 65 millitorr. Drying times using the rotary evaporator depend on the efficiency of the vacuum pump. With a good pump, this step should take less than 1 h.
3. To the bottom of the reaction vial (not into the tubes), add HCl/phenol (500 μL) and reducing agent.
4. Flush the vial with argon gas.
5. Place the vial in the oven at 110°C for 22 h (*see* **Note 5**).
6. Remove the vial and allow it to cool. Remove excess HCl from the outside of the tubes by wiping with a tissue.
7. Add two tubes, each containing an aliquot (10 μL) of Amino Acid Standard H and internal standard working solution. Dry under vacuum until the pressure has fallen to 65 millitorr (about 1 h).

Table 1
Gradient Table

Time, min	Flow μL/min	%A	%B
Initial	300	94	6
10	300	73	27
20	300	28	72
25	300	0	100
30	300	0	100
31	300	94	6
36	300	94	6

The steps from 20 min onwards are to clean and equilibrate the column ready for the next sample injection.

8. Clean out the cold trap to remove any traces of acid that might react with cyanate in the next stages.
9. To each tube, add coupling reagent (20 μL) and vortex.
10. Dry under vacuum in the rotary evaporator (about 30 min).
11. To each tube, add derivatization reagent (20 μL) and vortex. Leave at room temperature for 20 min.
12. Dry under vacuum in the rotary evaporator. After the pressure has dropped to 65 millitorr (about 2 h), leave for an additional 10 min to ensure complete removal of PITC.

3.2. Chromatography

1. To each tube, add sample buffer (60 μL) and vortex.
2. It may be necessary to filter some samples at this point if the samples are particularly dirty. This should be carried out as soon as possible, and certainly within a few hours if samples are left at room temperature (*see* **Note 6**).
3. Place the reverse phase C18 column 250 × 2.1 mm (e.g., Hichrom, Reading, UK, Part no. HIRPB-250AM) in the column heater at 38°C and commence flow of Eluent B (in which the column is stored) at a rate of 300 μL/min.
4. Run a linear gradient from 100% Eluent B to 100% Eluent A during 2 min. Allow the column to equilibrate for 30 min in the latter mobile phase.
5. Inject an aliquot (20 μL) of standard using the gradient shown in **Table 1**. Run time should be set at 20 min. The use of an automatic injector is highly recommended (*see* **Note 7**).
6. The second and third injections should be of the same solution (standard) as the first. Invariably, poor chromatography is obtained with the first injection but good separation should be achieved with second and third injections. These chromatograms should be checked to ensure that this is the case before proceeding with further samples.

7. Inject an aliquot (20 μL) of each unknown sample, recording the absorbance at 254 nm.
8. After the run, peaks are integrated. The amino acid composition of the unknown is determined by comparison with peak areas given by the Amino Acid Standard H. The peak areas of the internal standard from the standard chromatogram is compared with the peak areas of the internal standard from the sample chromatogram. The amounts of each amino acid are then adjusted accordingly.

Amount of unknown amino acid in the sample can be found from the calculation

$$A = \frac{\text{Peak Area of Internal standard from the standard} \times 250}{\text{Peak Area of Internal standard from the sample}}$$

$$B = \frac{\text{Peak Area of amino acid from the sample}}{\text{Peak area of amino acid from the standard}}$$

Amount of unknown amino acid = $A \times B$ pmol.

4. Notes

1. The highest quality of reagents should be used. Suitable water can be obtained using a Milli-Q purification system (Millipore, Bedford, MA) fed by a supply of tap water that has been distilled once. For other chemicals, suppliers of some suitable grades are suggested in **Subheading 2.** This is particularly important if high sensitivity is required.
2. The pH of Eluent A is increased to resolve other amino acids of interest such as hydroxyproline. At this lower pH the hydroxyproline is resolved between Asp and Glu without interfering with the resolution of other amino acids. At pH 5.7, the retention time of hydroxyproline is increased and appears between Glu and Ser in the chromatogram but the resolution between other amino acids are decreased (*see* **Fig. 2**).
3. Protein samples should be free of salts, amines, and detergents, although it has been reported that salts at concentrations up to 4 *M* do not affect derivatization or separation *(14)*. Although samples are normally introduced as protein solutions, it is also possible to carry out direct hydrolysis of protein bands that have been blotted from electrophoretic gels onto polyvinylidene difluoride (PVDF) membranes *(15,16)*. However, recovery of amino acids from PVDF membrane can be low.
4. The amount of each amino acid present in the chromatogram shown in **Fig. 2** is 250 pmol. This value is much higher than the sensitivity limit of the method. However, if adequate quantities of protein (200 ηg–1 μg) are available, working at this sensitivity will avoid problems owing to background contamination of samples. If limited quantities of protein are available, say less than 50 ηg, then all of the material (60 μL) can be injected but the injection and system peaks will also increase and may interfere with the resolution of some amino acids. A corresponding adjustment in the treatment of the standards will be essential particularly if internal standards have been used. If it is necessary to carry out analyses at low pmol range, special precautions must be taken, since background contamination (especially of serine and glycine, where it can reach several pmol per sample, 20–30 pmol is not uncommon) is a common problem. These precautions

UV Absorbance
at 254nm

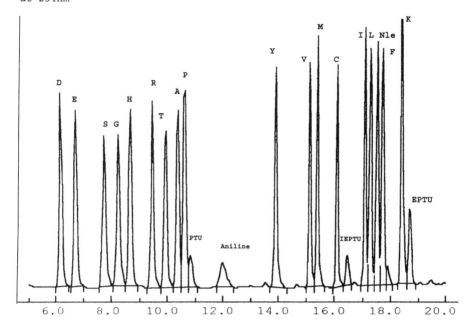

Fig. 2. Separation of a standard mixture of seventeen amino acids and norleucine (250 pmols) as internal standard after derivatization with Phenylisothiocyanate. The peak owing to each phenylthiocarbamyl amino acid is identified using the single-letter code for that amino acid. The peak marked C is for Cystine (125 pmols) rather than cysteine. The absorbance is at 254 nm with an attenuation of 32 mV.

include: pyrolysing glassware at 500°C overnight or steeping in sulfuric acid (250 mL) containing sodium nitrate (1 g); handling with clean gloves (nontalc) and forceps; and including a control blank that has been subjected to hydrolysis. Of course, it is good practice to use these procedures in any case, even for lower sensitivity analyses.

5. Rapid hydrolysis may be carried out at higher temperatures and the subject of peptide hydrolysis is covered in detail in Chapter 10. After hydrolysis, in order to prevent condensing HCl from running down the side of the vial into the tubes, keep the tubes upright. With fewer than 12 samples, add blank tubes for support.

6. Phenylthiocarbamyl amino acids in solution at neutral pH are relatively stable, with less than 10% loss of the least stable derivatives (Leu, Ile) during 10 h at room temperature. Losses are much reduced in the cold, with less than 5% loss during 48 h at 4°C.

7. The use of automatic injectors, apart from their laborsaving function, gives constant injection volumes and constant intervals between injections. An identical

interval between injections is an important criterion for obtaining reproducible retention times for each amino acid in different chromatograms, since the column is not given sufficient time to re-equilibrate in Eluent A.

References

1. Spackman, D. H., Stein, W. H., and Moore, S. (1958) Automatic recording apparatus for use in the chromatography of amino acids. *Anal. Chem.* **30,** 1190–1206.
2. Hsu, K. T. and Currie, B. L. (1978) High performance liquid chromatography of Dns-amino acids and application to peptide hydrolysates. *J. Chromatogr.* **166,** 555–561.
3. Lindroth, P. and Mopper, K. (1979) High performance liquid chromatographic determination of subpicomole amounts of amino acids by precolumn fluorescence derivatization with o-phthaldialdehyde. *Anal. Chem.* **51,** 1667–1674.
4. Knoop, D. R., Morgan, E. T., Tarr, G. E., and Coon, M. J. (1982) Purification and characterization of a unique isoenzyme of cytochrome P-450 from liver microsomes of ethanol-treated rabbits. *J. Biol. Chem.* **257,** 8472–8480.
5. Hendrickson, R. L., and Meredrith, S. C. (1984) Amino acid analysis by reverse phase high performance liquid chromatography; precolumn derivatization with phenylisothiocyanate. *Anal. Biochem.* **136,** 65–74.
6. Bidlingmeyer, B. A., Cohen, S. A., and Tarvin, T. L. (1984) Rapid analysis of amino acids using pre-column derivatization. *J. Chromatogr.* **336,** 93–104.
7. Bidlingmeyer, B. A., Cohen, S. A., and Tarvin, T. L. (1986) Amino acid analysis of submicrogram hydrolysate samples, *Methods in Protein Sequence Analysis* (Ken Walsh, ed.), Humana Press, Totowa, NJ, pp. 229–245.
8. Cohen, S. A. and Strydom, D. J. (1988) Amino acid analysis utilizing phenylisothiocyanate derivatives. *Anal. Biochem.* **174,** 1–16.
9. Lyndon, A. R., Davidson, I., and Houlihan, D. F., (1993) Changes in tissue and plasma free amino acid concentrations after feeding in atlantic cod. *Fish Physiol. Biochem.* **5,** 365–375.
10. Amezaga, M., Davidson, I., McLaggan, D., Verhaul, A., Abee, T., and Booth, I. (1995) The role of peptide metabolism in the growth of Listeria monocytogenes ATCC 23074 at high osmolarity. *Microbiology* **141,** 41–49.
11. Roe, A. J., McLaggan, D., Davidson, I., O'Byrne, C., and Booth, I., (1998) Pertubation of anion balance during inhibition of growth of *Escherichia coli* by weak acids. *J. Bacteriol.* **4,** 767–772.
12. Strydom, D. J. and Cohen, S. A. (1994) Comparison of amino acid analyses by phenylisothiocyanate and 6-aminoquinolyl-N-hydroxysuccinimidyl carbamate precolumn derivatization. *Anal. Biochem.* **222,** 19–28.
13. Beale, S. C., Hseih, Y., Wiesler, D., and Novotny, M., (1990) Application of 3-(2-Furoyl)quinoline-2-carbaldehyde as a Fluorogenic reagent for the analysis of primary amines by liquid chromatography with laser-induced fluorescence detection. *J. Chromatgr.* **499,** 579–587.

14. Hendrickson, R. L., Mora, R., and Maraganore, J. M. (1987) A practical guide to the general application of PTC-amino acid analysis, in *Proteins: Structure and Function* (L'Italein, J. J., ed.), Plenum, New York, pp. 187–195.

15. Tous, G. I., Fausnaugh, J. L., Akinyosoye, O., Lackland, H., Winter-Cash, P., Vitorica, F. J., and Stein, S.(1989) Amino acid analysis on polyvinylidene difluoride membranes. *Anal. Biochem.* **179**, 50–55.

16. Nakagawa, S. and Fukuda, T. (1989) Direct amino acid analysis of proteins electroblotted onto polyvinylidene difluoride membrane from sodium dodecyl sulfate-polyacrylamide gel. *Anal. Biochem.* **181**, 75–78.

12

Post Column Amino Acid Analysis

Alan J. Smith

1. Introduction

The quantitation of amino acids that are present in the acid hydrolysate of proteins and peptides was first developed into an automated procedure by Moore and Stein *(1)* in the mid-1950s. Quantitation was achieved by hydrindantin hydrate (ninhydrin) derivatization of the amino acids after ion-exchange chromatographic separation. Automated postcolumn amino acid analysis remains largely unchanged today in terms of the basic separation and derivatization chemistry. It is still the "gold standard" to which other methods are compared and is a testimonial to the robust nature of this chemistry.

Initially, amino acid analysis became the standard descriptor of the chemical content of a protein or peptide. More recently, the development of automated N-terminal Edman sequencing *(2)* and a variety of mass spectrometry techniques (*see* **Note 1**) have largely superceded amino acid analysis for this purpose. However, the introduction of recombinant proteins and synthetic peptides as therapeutics has rekindled interest in amino acid analysis as an accurate measure both of composition and absolute protein/peptide content. In addition, Hobohm *(3)* has demonstrated, with a high degree of success, the utility of amino acid analysis alone in identifying proteins. Adequate composition data can be obtained from a single hydrolysis timepoint and is relatively unaffected by the absence of the TRP and CYS values. The compositions are obtained from proteins that have been isolated as bands or spots on acrylamide gels and are compared to the theoretical compositions derived from proteins reported in the various databases. The molecular weight (from one dimensional [1D] gels) and/or isoelectric point (from two-dimensional [2D] gels) are required as data input. The searches can be performed via Internet at: http://www.embl-heidelberg.de/aaa.html. Since the compositions are obtained directly from stained proteins in

From: *Methods in Molecular Biology, vol. 211: Protein Sequencing Protocols, 2nd ed.*
Edited by: B. J. Smith © Humana Press Inc., Totowa, NJ

the gel, postcolumn, ninhydrin-based amino acid analysis is ideally suited to this purpose.

Obtaining the amino acid composition of a protein or peptide is not a single biochemical process, but rather the combination of several separate but interrelated processes. They are: sample preparation, sample hydrolysis, chromatography of the free or derivatized amino acids, detection (color development), quantitation, and data presentation in the finalized report. The suboptimal performance of any of these components can have a significant impact on the quality of the finished data. Amino acids can be quantitated using either precolumn or postcolumn derivatization chemistries. The difference refers to whether the amino acids are derivatized before or after the chromatographic separation. All of the other steps in the process, i.e., sample preparation, hydrolysis, quantitation (integration), and report preparation are common to both methods. This chapter will deal specifically with postcolumn derivatization with ninhydrin but will include comparisons with the precolumn method (*see* Chapter 11), where appropriate.

As stated earlier, postcolumn amino acid analysis using ninhydrin is very little changed from its inception. There are a number of very good reasons for this. The derivatization chemistry is very robust, reliable, and reproducible, and can readily detect a wide variety of amino acids and analogs. These same attributes also apply to the ion-exchange chromatography step, which carries the added advantage of separational versatility. The system can tolerate a wide variety of matrix components that might be present in the sample (e.g., buffers, salts, acrylamide gel, PVDF [polyvinylidine difluoride], etc.) since derivatization occurs after the amino acids have been chromatographically separated from these compounds. Proteins and peptides can be analyzed in most salt solutions or buffers at a concentration of 0.15 M or less, detergents (ionic and hydrophobic), or immobilized on resin, glass-fiber, PVDF (*see* **Note 2**), or acrylamide gels (*see* **Note 3**). Solutions containing ammonium ions or urea are not suitable for any method of amino acid analysis since they produce large quantities of ammonia upon acid hydrolysis. There are some disadvantages inherent in the ninhydrin system. It requires detection at two wavelengths, 440 nm and 590 nm, since amino acids (proline and hydroxyproline) form red-colored derivatives with ninhydrin rather than blue for the other amino acids. It is relatively insensitive having a lower detection range of 50–100 pmol of amino acid. It requires relatively long run times of approx 60 min.

The latest entrant into the postcolumn amino acid analysis arena requires no derivatization prior to the detection of the amino acids *(8)*. The acid hydrolysate is separated by ion-exchange chromatography and the free amino acids are quantitated by pulsed amperometric detection. The detector has been optimized for the detection of free amino acids and is minimally influenced by the presence of interfering substances in the sample matrix. Sensitivity appears to be at least an order of magnitude greater than ninhydrin detection.

Precolumn derivatization offers significant advantages over postcolumn derivatization in these areas, i.e., single wavelength detection, 10-fold greater sensitivity, and 45-min run times. However, the major disadvantage with precolumn derivatization is that there is much less flexibility with the composition of sample matrix. The actual derivatization occurs in the presence of the hydrolysis products of the matrix prior to chromatographic separation. This can lead to impairment of derivatization efficiency and can result in the variable recovery of some amino acids. Precolumn derivatization can be very effective when the proteins and peptides have been cleaned up from matrix components by high-performance liquid chromatography (HPLC) prior to hydrolysis and derivatization.

The most frequent hydrolysis conditions are 6 M HCl for 24 h at 110°C. However, there are a number of variations that have also been used. Some make use of the doubling rate for first-order reaction kinetics with every 10°C temperature increase, i.e., 150°C for 90 min vs 110°C for 24 h, while others use microwave heating *(4)* rather than conventional oven or heating block approaches. For a more detailed description of various hydrolysis protocols, *see* Chapter 10. The hydrolysis protocols described in this chapter have optimal application in postcolumn amino acid analysis. Using small amounts of oxygen scavengers such as phenol in addition to the physical exclusion of oxygen (i.e., vacuum) from the hydrolysis vessel, has greatly improved the recoveries of methionine. Although 6M HCl is the most common hydrolytic acid, it can be used in liquid or vapor phase *(5)* in order to achieve hydrolysis (*see* **Note 4**). Resin-bound peptides can be hydrolyzed with HCl/propionic acid (50:50 v/v) *(6)*, which can be used to monitor automated peptide syntheses or resin-loading levels (*see* **Note 5**).

Tryptophan is unstable in strong acidic conditions and 4 M methane sulphonic acid or 2 M sodium hydroxide are frequently used hydrolysis media (*see* Chapter 10, **Subheading 3.7.**). However, since these reagents are nonvolatile, they have to be neutralized prior to analysis (*see* **Note 6**). The quantities of protein or peptide that are required for successful amino acid analysis are dependent on molecular weight (*see* **Note 7**) but are generally in the low microgram to nanogram range.

Cysteine/cystine are also destroyed under normal hydrolytic conditions. However, performic acid oxidation or reduction and alkylation prior to hydrolysis *(7)* will give quantitative recoveries of these amino acids. Performic acid oxidation has the advantage of using volatile reagents and, after evaporation, the subsequent hydrolysis can be performed in the same tube.

After HCl hydrolysis, the samples are dried and the free amino acids are dissolved in an appropriate volume of a dilution buffer (*see* **Note 7**). If absolute sample amount is to be determined, then it is essential to include an internal standard of known quantity with the original sample, or more commonly, in the sample dilution buffer (*see* **Note 8**). The chromatographic separation of the free amino acids uses an ion-exchange, mixed-bed, sulphonated polystyrene resin. The hydrolysate is loaded on to the ion exchange column in a low pH, high ionic-strength buffer, and the amino acids are eluted with a combination of buffers using elevated pHs, elevated ionic strength, and elevated temperature.

After ion-exchange separation, the free amino acids are mixed with a detection reagent and quantitated by comparison to known quantities of amino acids in a standard calibration mixture (*see* **Note 9**). The most commonly used colorimetric reagent is ninhydrin, with detection at 440 nm and 570 nm. If higher sensitivity is required, orthophthalaldehyde (OPA) can be used as the detection reagent in place of ninhydrin. The OPA derivatives are detected by fluorescence, which results in an order of magnitude increase in sensitivity (i.e., 5–10 pmol of individual amino acids). OPA does not react with imines and consequently proline and hydroxyproline are not detected.

The final steps in the amino acid analysis process are quantitation and data presentation. The individual amino acids are identified by comparison to the retention times, and quantitated by area comparison, with a standard calibration mixture. The individual chromatograms must be inspected to confirm that the peaks have been correctly identified and quantitated (*see* **Note 10**).

As stated previously, when all the various components of the postcolumn amino acid analysis process are operating optimally, it is a highly accurate and reproducible method (*see* **Note 11**).

The final report format is dependent on the type of data required by the specific application. Proteins compositions can be presented as: mole percent of amino acids (*see* **Note 12**), weight percent of amino acids (*see* **Note 13**), molarity of amino acids (*see* **Note 14**), or absolute protein content (*see* **Note 15**).

Peptides can be presented as: absolute peptide content (*see* **Note 15**) or stoichiometry (*see* **Note 16**).

2. Materials

2.1. Hydrolysis

1. Hydrolysis tubes 10 × 75 mm Pyrex P/N 9820-10 (for cleaning, *see* **Subheading 3.1.1.**).
2. Muffle furnace capable of exceeding 500°C.
3. 6 *M* hydrochloric acid (Pierce Chemical Co., Rockford, IL).
4. Solid phenol (Baker AR, Phillipsburg, NJ).

5. PicoTag workstation (Waters Associates, Milford, MA).
6. Speedvac (Savant Instruments, Hicksville, NY).

2.2. Chromatography

1. Sample diluent NaS, 0.2 M sodium citrate, pH 2.2, (Beckman Instruments, Fullerton, CA).
2. Norleucine stock solution, 20 mM in NaS (store at –20°C).
3. Microfuge (Beckman Instruments, Fullerton, CA).
4. 7300 amino acid analyzer system (Beckman Instruments) or similar alternative (*see* **Note 17**), consisting of:
   a. Buffer NaE, 0.2 M sodium citrate, pH 3.28.
   b. Buffer NaF, 0.2 M sodium citrate, pH 4.25.
   c. Buffer NaD, 0.35 M sodium citrate, pH 6.40.
   d. Ninhydrin reagent NinRX.
   e. Column regenerant NaR, 2 M sodium hydroxide.
   f. System Gold data analysis software.

3. Methods

3.1. Hydrolysis

1. Pyrolyze the hydrolysis tubes by heating to 500°C for at least 2 h.
2. Store tubes inverted in a dust-free environment such as a parafilm covered beaker.
3. Place or weigh sample into a hydrolysis tube.
4. Add norvaline internal standard (optional, *see* **Note 7**).
5. Evaporate to dryness in the Speedvac. The presence of any liquid at this step will reduce the effective HCl concentration.
6. Add 0.5 mL of 6 M HCl.
7. Add 2 crystals solid phenol.
8. Place in PicoTag workstation hydrolysis vessel (*see* Chapter 10, Fig. 1) and evacuate and nitrogen flush. Repeat this step twice more, evacuate as the final step.
9. Place evacuated vessel in the workstation hydrolysis oven and heat for 24 h at 110°C.
10. Remove the hydrolysis vessel and air cool.
11. Break the vacuum seal carefully and remove the hydrolysis tubes. Some acid will have condensed inside the vessel and on the outside of the tubes. Wipe the outside of the tubes taking care not to spill the contents. The sample tubes will contain different levels of acid after hydrolysis even though they all initially contained 0.5 mL. Evaporate to dryness in an evacuated centrifuge.

3.2. Chromatography

1. Add 200 µL of norleucine solution to 400 mL of NaS to give a final concentration of 0.5 nmol/50 µL. This will be the dilution buffer that will be used for all sample manipulations. Store refrigerated and allow to equilibrate to room temperature before use.

2. Add a minimum of 120 µL of this diluent to each sample tube and vortex for 2 min to allow the free amino acids to dissolve (*see* **Note 18**). Insoluble material will frequently be present at this stage.
3. Remove contents with pasteur pipet and place in a plastic microcentrifuge tube.
4. Centrifuge at 7500*g* for 5 min in the Microfuge.
5. Remove supernatent with pasteur pipette and transfer to another plastic centrifuge tube (*see* **Note 19**).
6. Remove sufficient solution with the sample loading loop such that 50 µL of sample is loaded on to the analyzer column.
7. Operate the analyzer using the acid hydrolysis chromatography program according to the manufacturer's instructions.
8. Program the System Gold software such that the commencement and termination of peak integration are shown on all chromatograms and that baselines are drawn.
9. Inspect the chromatogram for correct integration of all peaks (*see* **Note 20**).
10. Reintegrate where necessary and correct data report file.
11. Use the System Gold format that reports the quantitation of amino acids in "nmol/50 µL."
12. Confirm that the norleucine value is 0.50 ± 0.01 nmole (*see* **Note 21**).
13. Normalize all quantities to 0.50 nmole norleucine for the measure-ment of absolute quantitation.
14. Run a standard calibration mixture every 15 samples or at least every other day to check for stability of the ninhydrin reagent. Each bottle of ninhydrin should be stable for up to 1 wk on the instrument before recalibration is necessary.

4. Notes

1. Fast atom bombardment mass spectrometry (FAB-MS) can be used for both mass analysis and sequence analysis of peptides, and electrospray (ES-MS) and matrix-assisted laser desorption (MALDI-MS) for the mass analysis of both peptides and proteins. These methods are discussed/described further in Chapter 17.
2. Successful postcolumn amino acid analysis of Coomassie Blue stained proteins on PVDF requires liquid-phase hydrolysis and the extraction of the amino acids with high-salt sample dilution buffer. The polar nature of the extraction buffer appears to facilitate extraction of the amino acids from the hydrophobic membrane. However, these buffer solutions are unsuitable for precolumn separations. The methanol/TFA solutions used in precolumn methods appear to give variable recoveries of some amino acids in the presence of PVDF.
3. Large amounts of ammonia are produced by the acid hydrolysis of polyacrylamide. It is essential that the analyzer is programmed to terminate the run immediately following the elution of lysine. High concentrations of ammonia can form an insoluble complex with ninhydrin, which can plug the instrument. Early termination allows the ammonia to flush from the column during the normal regeneration procedure without reacting with ninhydrin and risking precipitation.

4. Vapor-phase hydrolysis is preferred where high-sensitivity analysis is required since the acid does not come into contact with the sample and cannot therefore contribute any contaminants to the sample.

5. On-resin hydrolysis of synthetic peptides (150°C for 90 min), followed by amino acid analysis, can be a very effective and rapid means of troubleshooting an instrument problem or a difficult synthesis without waiting until the peptide has been cleaved and deprotected.

6. The high salt concentrations that result from the neutralization step have to be greatly diluted, i.e., to 0.2 M or less. This means that 10–20-fold more starting material is required for the analysis.

7. Protein quantities in the 1–2 µg range are normally adequate to give amino acid recoveries in the 1–2 nmole range for a 40kD protein. Smaller proteins will require proportionally more, and larger proteins proportionally less, protein. Peptides will require 10–100-fold less material, i.e., 10–100 ng. Alternatively, 10–100-fold greater dilutions can be used.

8. In general, there is little opportunity for sample loss during hydrolysis and the acid is subsequently removed by evaporative centrifugation. For this reason the internal standard is normally incorporated at the sample dilution step by inclusion in the sample dilution buffer. Two commonly used compounds are the non-naturally occurring amino acids, norleucine or norvaline.

9. The eluate-to-reagent mixing ratio is 2:1, which introduces a dilution factor of one-third and a proportional reduction in sensitivity.

10. Significant (>0.2 M) levels of salts or buffers in the sample will either increase or reduce the elution times of the early eluting amino acids. This will result in incorrect peak identification and subsequent quantitation. In the context of area integration, it is essential that the concentration of the standard that is used for setting the integration parameters is in the approximate concentration range of the sample. For example, integration parameters that are optimal for a separation of 5 nmol standard would be suboptimal for analyzing samples at the 100 pmol level, and vice versa. In some proteins (collagens for example), there is a huge (300-fold) range of amino acid content (e.g., glycine and tyrosine). In these instances, it is frequently necessary to perform analyses at two separate dilutions in order to obtain accurate amino acid quantitation.

11. Replicate analyses of the same sample will show a coefficient of variation (CV) of 1% at the 1 nmol detection level and 2% at the 100 pmol level. Replicate hydrolyses of the same sample will show a CV of 1–2% and 5% at these same levels, respectively. Absolute recoveries at the 1 nmol level, in the presence of an internal standard, should be accurate to 5% even without correction for the destruction of serine and threonine or the quantitation of tryptophan and cysteine/cystine.

12. This can be determined by summing the nanomole recoveries of the individual amino acids and calculating mole percentage (do not include ammonia or internal standards). *Remember*, Asp and Glu values represent a mixture of the free acids and their amides!

13. The individual amino acid recoveries are multiplied by their molecular weights minus water (18 Daltons) as listed in **Appendix 1**, and calculated as in **Note 12**.

14. This requires the molecular weight of the protein and subsequent normalization to the sum obtained in **Note 12**. For example, the values would be multiplied by 400 for a 40 kDa protein.

15. This requires the accurate initial weight of the sample (or OD_{280} if the extinction coefficient is required). The weights obtained from the **Note 13** calculation are corrected for both internal standard and sample dilution factors in order to obtain the absolute protein quantity.

16. Peptide stoichiometry can be obtained by dividing the amino acid recoveries by the value obtained for the least abundant amino acid. Care must be taken with this selection since it could be a contaminant or a poorly recovered amino acid (serine, threonine, methionine). In the final analysis, common sense and a knowledge of the approximate molecular weight of the peptide will dictate the selection of the appropriate common denominator.

17. The Beckman Instruments Model 7300 has been out of production since 1996. However, a significant number of these instruments are still in service. The Hitachi Model L-8800 (Tokyo, Japan) has proved to be a very capable alternative and is now the instrument of choice for postcolumn amino acid analysis.

18. Where large amounts of sample are used in the hydrolysis, it may be necessary to perform serial dilutions using the norleucine dilution buffer. It is essential to remain within the linear range of the ninhydrin reagent, i.e., less than 15 nmol of amino acid.

19. Care must be taken to remove only the clear supernatent. If particulates are loaded onto the separation column, they can plug the inlet frit.

20. On occasion any chromatography software can: delay integration, terminate integration early, or drop an inappropriate perpendicular to baseline, all of which can affect accurate quantitation.

21. If the norleucine value is low and the integration is correct, there are two possible explanations. There may have been less than 50 µL of sample transferred to the column, which can be corrected by normalization. Alternatively, the ninhydrin reagent may have deteriorated over time. In which case, a standard calibration mixture can be run to recalibrate the ninhydrin and recalculate the analysis.

References

1. Moore, S. and Stein, W. H. (1963) Chromatographic determination of amino acids by the use of automated recording equipment. *Methods in Enzymology*, vol. 6, (Colowick, S. P. and Kaplan, N., eds.), Academic, San Diego, CA, pp. 819–831.

2. Hewick, R. M., Hunkapiller, M. W., Hood, L. E., and Dryer, W. J. (1981) A gas-liquid solid phase peptide and protein sequenator. *J. Biol. Chem.* **256**, 7990–7997.

3. Hobohm, U., Houthaeve, T., and Sander, C. (1994) Amino acid analysis and protein database composition search as a fast and inexpensive method to identify proteins. *Anal. Biochem.* **222**, 202–209.

4. Gilman, L. B. and Woodward, C. (1990) An evaluation of microwave heating for the vapor phase hydrolysis of proteins, in *Current Research in Protein Chemistry* (Villafranca, J., ed.), Academic, San Diego, CA, pp. 23–36.
5. Knecht, R. and Chang, J. Y. (1986) Liquid chromatographic determination of amino acids after gas-phase hydrolysis and derivatization with (dimethyl-aminoazobenzenesulfonyl) chloride. *Anal. Chem.* **58**, 2375–2379.
6. Westall, F. and Hesser, H. (1974) Fifteen-minute acid hydrolysis of peptides. *Anal. Biochem.* **61**, 610–613.
7. Carne, A. F. (1994) Chemical modification of proteins, in *Methods in Molecular Biology, vol. 32: Basic Protein and Peptide Protocols* (Walker, J. M., ed.), Humana, Totowa, NJ, pp. 311–320.
8. Clarke, A. P., Jandik, P., Rocklin, R. D., Liu, L. , and Avdalovic, N. (1999) An integrated amperometry waveform for the direct, sensitive detection of amino acids and amino sugars following anion-exchange chromatography. *Anal. Chem.* **71**, 2774–2781.

13

Amino Acid Analysis Using Pre-Column Derivatization with 6-Aminoquinolyl-*N*-Hydroxysuccinimidyl Carbamate

Analysis of Hydrolyzed Proteins and Electroblotted Samples

Steven A. Cohen

1. Introduction

Over the past 20 years, amino acid analysis methods based on precolumn derivatization procedures have become a popular standard practice in many biochemistry laboratories as practical alternatives to the traditional older postcolumn derivatization techniques based on ninhydrin detection following the ion-exchange separation of the underivatized amino acids. The major reasons for this enthusiastic endorsement include greater sensitivity, faster analyses, and the ability to use less expensive, more flexible high-performance liquid chromatography (HPLC) instrumentation rather than dedicated amino acid analyzers.

Among the more widely employed derivatization reagents are ortho-phthalaldehyde (OPA) (*1*) and Edman's reagent, phenylisothiocyanate (PITC) (*2–5*), for which there are numerous publications illustrating their utility. OPA, however, has the significant drawback in that it does not react with secondary amino acids, and some of the derivatives, notably those with Gly and Lys, are unstable. PITC, while reactive with the secondary amino acids, also produces somewhat unstable derivatives, notably with Asp and Glu, which slowly cyclize from the desired phenylthiocarbamyl moieties to their respective phenylthiohydantion derivatives. In addition, excess PITC must be removed, usually by evaporation, prior to HPLC analysis to avoid column contamination and poor chromatographic separation.

From: *Methods in Molecular Biology, vol. 211: Protein Sequencing Protocols, 2nd ed.*
Edited by: B. J. Smith © Humana Press Inc., Totowa, NJ

More recently Cohen and Michaud introduced 6-aminoquinolyl-N-hydroxysuccinimidyl carbamate (AQC) for precolumn derivatization (6). This reagent reacts smoothly with primary and secondary amines to produce stable unsymmetrical urea derivatives that are highly fluorescent. Products are generated within seconds, and in a somewhat slower reaction, excess reagent is hydrolyzed to yield 6-aminoquinoline (AMQ), carbon dioxide, and N-hydroxysuccinimide (NHS). Key to the design and implementation of this method are the fluorescence properties of the derivatized amino acids and AMQ. While both have excitation maxima approx 248 nm, they have radically different emission maxima, with the AMQ near 520 nm, and the amino acid products at approx 395 nm. This shift in emission maximum allows the derivatization mixture to be injected onto the HPLC column without need for excess reagent removal. Since the initial publication, a number of papers have described various applications and extensions of the original method (7–11).

Fluorescence detection permits highly sensitive analyses with detection limits ranging from 50–300 fmol for the normal hydrolyzate amino acids. As little as 50 ng of protein hydrolyzate is sufficient for compositional analysis. Samples such as proteins recovered from two-dimensional (2D) gel electrophoresis, with typical spot amounts ranging from 100–1000 ng, can be readily analyzed (12).

The relevance of this ability to analyze single gel spots by amino acid analysis has been amplified with the increasing importance of the new proteomics studies (see ref. 13 for an excellent introduction). In these studies, protein mixtures from whole cell extracts or subcellular fractions are often resolved by 2D gel electrophoresis, typically an isoelectric focusing step followed by a sodium dodecylsulfate (SDS) gel-electrophoresis, size-separation step. Following the separation procedure, a major goal in proteomics is the identification and quantification of resolved species to establish trends in expression levels as a function of cellular condition, stress, or other perturbation of the cell or tissue. Identification of the gel spots is usually carried out on excised samples subjected to trypsin digestion and subsequent analysis by either matrix-assisted laser desorption ionization (MALDI-TOF) mass spectrometry (MS) (14–17) or liquid chromatography-tandem mass spectrometry (18). However, the ability to quantitate gel spots by MS is problematic and quantitative information is most commonly obtained on the gels themselves by densitometry following staining with Coomassie Blue, silver stain, or other dyes. Poor linearity and a lack of useful external standards limits the utility of quantitation on stained gels, however, and several laboratories have added quantitative amino acid analysis as a means of providing both quantitation and confirmation of identification via MS (12,19,20).

The techniques to derive useful amino acid analysis from stained gels are similar to those used for solubilized proteins, but must be modified to maximize recovery from either the gel spots themselves, or from a membrane surface following electrophoretic transfer of the proteins to a membrane. Most reported studies *(12,21–25)* have used polyvinylidene fluoride (PVDF), which provides high transfer efficiency *(26)*, stability to subsequent acid hydrolysis, good recovery of the amino acids following hydrolysis, and acceptably low background in blank samples. These samples do pose severe limitations on the method amino acid analysis employed. Sample amounts are at most one microgram, compositional accuracy sufficient to identify a protein from established protein databases is mandatory, and the samples often contain extraneous, nonamino acid components from the electrophoretic or transfer steps. The AQC chemistry is well-suited for analyzing gel spots given its high sensitivity, excellent compositional accuracy, and minimal effect of sample contaminants on derivative yield.

2. Materials

2.1. Membrane-Bound Samples

Gel-isolated proteins are more easily prepared for amino acid analysis after electrophoretic transfer to a PVDF membrane. Only a few studies have reported the hydrolysis of the gels themselves. The gel-to-membrane electrophoretic transfer method of Matsudaira *(26)* or modifications of this procedure are most popular and have demonstrated good recovery and robustness. Proteins absorbed onto PVDF can be visualized by common staining procedures, or can be observed under a strong light after wetting the membrane with methanol *(26a)*. Quantitative recovery of whole proteins absorbed to PVDF is difficult, but good recoveries of amino acids following on-membrane hydrolysis have been reported *(20,24)*. Minimum sample amounts are approx 50–100 ng, with higher amounts (e.g., 1 µg) giving higher reliability of identification and better reproducibility and accuracy *(12,24)*.

1. Coomassie Brilliant Blue Stain, R-250 (BioRad Inc.).
2. Acetic acid, analytical reagent grade.
3. HCl, constant boiling (Pierce Chemical Company or equivalent).
4. Phenol, crystal reagent grade (*see* **Note 1**).
5. Methanol, HPLC-grade.

2.2. Derivatization

1. AQC dry powder. Store in a dry place, up to 6 mo in a desiccator (Waters Corporation).
2. Borate derivatization buffer: 0.2 M sodium borate, pH 8.8, with 5 mM calcium disodium EDTA. Weigh 1.24 g of boric acid in a clean beaker. Add 100 mL of

water, stir to dissolve. Add 187 mg of calcium disodium EDTA. Titrate to pH 8.8 with sodium hydroxide (a solution made fresh from pellets). Use the best quality boric acid and EDTA available.

3. Acetonitrile for reagent dissolution. Only use the highest quality with low water and alcohol content. AQC dry powder, borate derivatization buffer, and acetonitrile are available as a kit from Waters Corporation.

4. Reagent Solution: Dissolve the AQC powder in pure, dry acetonitrile to provide a solution that is approx 3.0 mg/mL. Heat at 50°C for up to 10 min to complete solubilization. Store in a desiccator for up to 1 wk (*see* **Notes 2** and **3**).

5. Amino acid calibration stock solution (2.5 m*M* in 0.1*M* HCl) (Pierce Chemical Co. or Sigma Chemical Co.).

6. Working calibration mixture: Mix 40 µL of amino acid stock solution (2.5 m*M*) with 960 µL of water to give a solution with 100 pmol/µL of each AA.

7. 20 m*M* HCl: Mix 10 µL of constant boiling HCl with 3 mL of HPLC-grade water.

2.3. Chromatographic Analysis

1. Sodium acetate trihydrate, HPLC-grade.
2. Triethylamine, Pierce Sequenal-grade or equivalent.
3. Calcium disodium EDTA.
4. Acetonitrile (HPLC-grade).
5. Phosphoric Acid, 85%, reagent-grade.
6. Working phosphoric acid, 10% concentrated phosphoric acid in water.
7. Working EDTA solution. Weigh 100 mg calcium disodium EDTA and dissolve in 100 mL of water. Store at 4°C for up to 3 mo.
8. Eluent A: 140 m*M* sodium acetate, 17 m*M* triethylamine (TEA), pH 5.05, containing 1 m*M* calcium disodium EDTA. Weigh 19.04 g sodium acetate. Add 1 L water, stir. Bring pH to ca. 5.5 with working phosphoric acid (this reduces the odor from the TEA and improves its solubility in the aqueous buffer). Add 1 mL working EDTA. Add 2.37 mL of TEA (1.72 g) (*see* **Note 4**). Titrate to pH 5.05 with working phosphoric acid. Filter the buffer through 0.22- or 0.45-µm filters (*see* **Notes 5** and **6**). Use the HPLC-grade reagents and water. Store excess TEA under a blanket of inert gas.
9. Eluent B for ternary system, HPLC-grade acetonitrile.
10. Eluent C for ternary system, HPLC-grade water.
11. Alternate Eluent B for high-pressure mixing system, 60% acetonitrile. Mix 600 mL of acetonitrile with 400 mL of HPLC-grade water (note: measure these volumes separately).
12. AccQ-Tag reversed-phase column, 3.9 × 150 mm (*see* **Note 7**).
13. Guard column (3.9 × 20 mm) packed with reversed-phase packing, compatible with analytical column.
14. Ternary gradient HPLC system equipped with an in-line degasser or sparging system, including column heater, autosampler, and dual monochromator fluorescence detector (*see* **Notes 8–11**).

3. Methods

3.1. Hydrolysis and Extraction of Samples from PVDF Membranes

Protein samples are prepared by gel electrophoresis. One of the more common procedures using a Tris-glycine buffer has been described by Laemmli *(27)*. The Tricine buffer system of Schägger and von Jagow *(28)* is a common alternative to the Laemmli method. Typical 2D gel procedures are based on the work of O'Farrell *(29)* and Klose *(30)*. The protocol for extraction described here is a modified version of the one described Tous et al. *(21)*.

1. Stain the membrane using Coomassie brilliant blue R-250 blue *(27)*.
2. Destain the membrane with 5% acetic acid.
3. Rinse the membrane for 2–3 times with water. Alternatively membranes can be stained by Ponceau S as described by Gharahgaghi et al. *(23)*.
4. Excise visualized bands with a sharp, clean razor blade or Exacto knife blade.
5. Place the band in a clean 6 × 50 mm test tube (*see* **Note 12**) and add 200 µL of constant boiling (5.7 *M*) HCl. Place the tube into a sealable vacuum hydrolysis vial (e.g., a Pico-Tag® vacuum vial, Waters Corporation) and add a crystal of phenol into the external hydrolysis vial. Seal the vial after three alternate steps each of vacuum and nitrogen purging, finishing with a vacuum step.
6. Hydrolyze the sample for 22–24 h at 110–112° C or for 1 h at 150° C.
7. Cool the samples. Open the vial and remove the tubes with a forceps. Wipe the outside of the tubes with a Kimwipe. Remove excess HCl (present as a film on the inside of the tube) under vacuum (5 min is more than sufficient) or a stream of warm nitrogen (*see* **Note 13**). Caution: use nontalc gloves when handling strong acid!
8. To each tube, add 100 µL of 0.1 *M* HCl containing 30% MeOH and vortex vigorously.
9. Remove the extracted amino acids with a clean Pasteur pipet to a clean test tube and repeat the extraction with a second aliquot of 50 µL of the same extraction solvent.
10. Combine the extracts and dry the sample in a vacuum centrifuge (*see* **Note 14**).

3.2. Derivatization

This method of derivatization is suitable for either membrane extracted samples or solublilized proteins after hydrolysis.

3.2.1. Calibration Standard

1. Place 10 µL of dilute standard in a 6 × 50 mm tube using a syringe or micro disposable pipettor.
2. Add 70 µL of borate buffer; mix.
3. Add 20 µL of AQC, vortex immediately after addition (*see* **Notes 15** and **16**).
4. Transfer the contents of the tube to a conical-shaped HPLC sample vial and cap tightly with a silicone-lined septum.

5. Heat the vial in a reaction block or oven for 10 min at 50°C (*see* **Note 17**).
6. Inject 5 μL = 50pmol.

3.2.2. Hydrolyzed Samples

1. Add 20 μL of 20 m*M* HCl to the sample and vortex (*see* **Note 18**)
2. Add 60 μL of borate buffer and vortex.
3. Add 20 μL of AQC reagent solution; vortex again.
4. Transfer the contents of the tube to a conical-shaped HPLC sample vial and cap tightly with a silicone-lined septum.
5. Heat the vial in a reaction block or oven for 10 min at 50°C.
6. Inject up to 20 μL of sample (*see* **Note 19**).

3.3. Chromatographic Analysis

3.3.1. Instrument Settings

1. Set the column heater to 38°C.
2. Set the fluorescence detector excitation wavelength at 248 nm and the emission wavelength at 395 nm.

3.3.2. Chromatographic Operation

1. Equilibrate the system with 100% of eluent A for 10 min before beginning sample analysis.
2. Analyze samples using the following gradient shown in **Table 1**. The entire separation is completed in approx 35 min (*see* **Fig. 1**). Including column re-equilibration, the total run time for a typical HPLC system is 45 min (*see* **Notes 20–22**).

4. Notes

1. Crystalline phenol is preferred over liquefied phenol owing to lower contamination levels (unpublished data).
2. The AQC reagent can be slow to dissolve and may require brief sonication to solubilize. Make sure the vial is well-sealed with a cap and further sealed with Parafilm and limit the sonication to 60 s at a time. Be certain that no water gets in the reagent vial!
3. If the reconstituted reagent is kept dry it should last for a minimum of several weeks. Repeated use from the same vial will expose the reagent to atmospheric water and slowly degrade the reagent. Minor hydrolysis will form AMQ plus CO_2 and NHS, normal by-products of the derivatization procedure, and consequently will not contribute any significant interference. However, the AMQ produced will react slowly with another molecule of AQC, forming the symmetric bis-aminoquinoline urea. The urea has poor solubility and may form a precipitate in the vial. Any urea in solution will elute after any derivatized amino acid and will also not interfere. However, continued breakdown will deplete the reagent, and unidentified fluorescent components that do interfere with the analysis, do

Table 1
Gradient for Ternary Eluent System

Time	Flow rate	%A	%B	%C	Curve
Initial	1.0	100	0	0	*
0.50	1.0	99	1	0	6
18.0	1.0	95	5	0	6
19.0	1.0	91	9	0	6
29.5	1.0	83	17	0	6
33.0	1.0	0	60	40	11
36.0	1.0	100	0	0	11

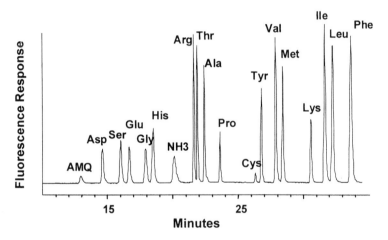

Fig. 1. Typical separation for a hydrolysate standard. The injected volume was 5 µL with a total of 50 pmol of each amino acid (25 pmol cystine) injected. Detection was accomplished by fluorescence with excitation at 248 nm and emission at 395 nm. The chromatographic conditions were those described in the methods section.

accumulate with continued use. For best results store unused reagent in the following manner.

a. Seal the vial as rapidly as possible following use.
b. Wrap Parafilm® tightly around the top of the vial.
c. Place the vial in a well-sealed desiccator.
d. Although it is not been proven necessary, it may be useful to protect the vial from light with aluminum foil.
e. If the vial is refrigerated or frozen, this may extend the life of the reagent. However it is imperative that the vial and the desiccator be warmed to room

temperature before opening to prevent condensation of moisture. In addition, storage at subambient temperature often results in significant precipitation that may require sonication to resolubilize.

4. We have found it to be much more reproducible to weigh the TEA. Tare a clean vial with a cap. Pipet the approximate volume of TEA into the vial in a hood. Weigh the vial and add or remove the proper amount of TEA to correct the weight (density = 0.7255). Final weight should be within 0.05 g the desired amount. Note: To increase eluent A shelf life, add 0.01% sodium azide. There will be no impact for fluorescence detection. UV-based systems will exhibit a gradually decreasing baseline. This can be offset by adding 100 µL of acetone to Eluent B (7).

5. Poor quality eluents are a common source of background gradient contamination. This is especially problematic for operation with UV detection.

6. This eluent may be made as an 11X concentrate. Weigh 1904 g of sodium acetate. Add 1 L of water, stir thoroughly, and adjust the pH to 5.5 with phosphoric acid. Add 17.2 g of TEA and stir thoroughly. Add 10 mL of working EDTA solution. Titrate to pH 5.05 with phosphoric acid. To make working eluent, mix 100 mL of concentrate with 1000 mL of HPLC grade water.

7. Other reversed-phase columns can be substituted, but optimal separation may require substantial changes to the Eluent A composition as well as the gradient profile.

8. It is advisable to use a small (e.g., 5 µL) flow cell in the detector. This will improve resolution compared to larger flow cells. Larger flow cells normally give a greater response.

9. Set the fluorescence emission bandwidth to the narrowest available setting. This will reduce the response from AMQ.

10. A UV detector set at 248 nm may be substituted for the fluorescence detector. Detection limits will be much higher, approx 1 pmol injected. There is no selectivity for detection. Thus, the AMQ peak is very large and can compromise the analysis of Asp. It may be necessary to adjust the gradient conditions and/or Eluent A pH to provide sufficient resolution (7). Lowering Eluent A pH to 4.95 will improve the AMQ/Asp resolution. Decreasing the gradient slope between 0.5 and 18 min will also improve the resolution. There may be some deterioration in the resolution of Gly and His, or between Arg and Thr.

11. Other HPLC systems can be used. To use a high pressure binary gradient, use the alternate Eluent B. A typical gradient is shown in **Table 2**. Use a total run time of 50 min to equilibrate the column.

12. For high-sensitivity analysis, it is necessary to have very clean tubes (e.g., fired at 500°C). Acid cleaning may be insufficient.

13. Excess residual acid may lower the derivatization pH below the optimal range (8.2–10.0). One potential source is nonvolatile acid formation during the hydrolysis. Phosphate salt or sodium dodecyl sulfate (SDS) will be converted to their respective acids, and will not be removed by vacuum drying. If excess acid is present, it may be neutralized with a TEA/ ethanol/water solution (2:2:1, v/v). The volume should be kept to a minimum (10–20 µL if possible) to avoid contamination problems. Vortex the dried sample with the TEA solution and vacuum

Table 2
Gradient for Binary Eluent System

Time	Flow rate	%A	%B	Curve
Initial	1.0	100	0	*
0.50	1.0	98	2	6
15.0	1.0	93	7	6
19.0	1.0	90	10	6
32.0	1.0	67	33	6
33.0	1.0	67	33	6
34.0	1.0	0	100	6
37.0	1.0	0	100	6
38.0	1.0	100	0	6

dry the sample again. Alternatively, a higher concentration borate derivatization buffer (up to 1 *M*) may be used instead of the 0.2 *M*. The background from the derivatization blank may increase.

14. AQC derivatization is compatible with other extraction protocols. Typical methods use an acidic solution of an aqueous-organic solvent mixture with the organic solvent ranging from 10–50% (*12,19,20,22–24*). In addition to HCl, acids such as TFA have also been used successfully. We have even used a solution containing up 1% SDS (unpublished results) as the derivatization can be carried out in the presence of detergents.

15. The amount of AQC added is sufficient to derivatize 5 µg of protein or peptide. Higher amounts may suffer lower yields for slow reacting amino acids, namely Asp, Glu, and Lys.

16. Less than quantitative derivatization of Lys results in production of either or both mono-derivatized species. Poor derivatization, often caused by an imbalance of sample and reagent or poor pH control of the derivatization mixture, can be monitored by observing the appearance of these peaks small "satellite" peaks on the front shoulder of Gly and the trailing shoulder of His. Poor mixing of the reagent with the buffered sample may also result in less than quantitative recovery of fully derivatized Lys.

17. The purpose of the 10 min heating step is to convert the phenolic side chain of Tyr to free phenol and yield a product labeled only at the on the amino group. During the derivatization at room temperature approx 30% of the phenol is labeled with AQC to form an unstable adduct. Heating the sample has no other significant effect.

18. The addition of HCl at this step helps improve the recovery of the basic amino acids Lys, His, and Arg.

19. Peak splitting for Asp and Ser and other early eluting components can be caused by excessive injection volume. To avoid this phenomenon, limit the injection volume to ≤ 20 µL.

20. Poor resolution between AMQ and Asp can be caused by too steep of an initial gradient slope or Eluent A pH being too high. Note that gradient delivery may vary with different HPLC systems.
21. Poor resolution of Asp and Ser or Glu and Gly can be caused by Eluent A pH being too low.
22. Poor resolution of Cys-Cys and Tyr can be caused by:
    a. column temperature too high;
    b. gradient slope from 19–28 min too shallow; or
    c. Eluent A pH too high.

References

1. Hill, D. W., Walters, F. H., Wilson, T. D., and Stuart, J. D. (1979) High performance liquid chromatographic determination of amino acids in the picomole range. *Anal. Chem.* **51**, 1338–1341.
2. Koop, D. R., Morgan, E. T., Tarr, G. E., and Coon, M. J. (1982) Purification and characterization of a unique isozyme of cytochrome P-450 from liver microsomes of ethanol-treated rabbits. *J. Biol. Chem.* **257**, 8472–8480.
3. Bidlingmeyer, B. A., Cohen, S. A., and Tarvin, T. L. (1984) Rapid analysis of amino acids using pre-column derivatization. *J. Chromatogr.* **336**, 93–104.
4. Heinrikson, R. L. and Meredith, S. C. (1984) Amino acid analysis by reversed-phase high-performance liquid chromatography: precolumn derivatization with phenylisothiocyanate. *Anal. Biochem.* **136**, 65–74.
5. Cohen, S. A. and Strydom, D. J. (1988) Amino acid analysis utilizing phenylisothiocyanate derivatives. *Anal. Biochem.* **174**, 1–16.
6. Cohen, S. A. and Michaud, D. P. (1993) Synthesis of a fluorescent derivatizing reagent, 6-aminoquinolyl-N-hydroxysuccinimidyl carbamate, and its application for the analysis of hydrolysate amino acids via high performance liquid chromatography. *Anal. Biochem.* **211**, 279–287.
7. Liu, H. J. (1994) Determination of amino acids by precolumn derivatization with 6-aminoquinolyl-N-hydroxysuccinimidyl carbamate and high performance liquid chromatography with ultraviolet detection. *J. Chromatogr.* **670**, 59–66.
8. Strydom, D. J. and Cohen, S. A., (1994) Comparison of amino acid analyses by phenylisothiocyanate and 6-aminoquinolyl-N-hydroxysuccinimidyl carbamate precolumn derivatization. *Anal. Biochem.* **222**, 19–28.
9. Strydom, D. J. and Cohen, S. A., (1993) Sensitive analysis of cystine/cysteine using 6-aminoquinolyl-N-hydroxysuccinimidyl carbamate derivatives, in *Techniques in Protein Chemistry IV* (Angeletti, R. H., ed.), Academic Press, San Diego, CA, pp. 299–306.
10. Cohen, S. A. and De Antonis, K. M. (1994) Applications of amino acid analysis derivatization with 6-aminoquinolyl-N-hydroxysuccinimidyl carbamate: Analysis of feed grains, intravenous solutions and glycoproteins. *J. Chromatogr.* **661**, 25–34.
11. Van Wandelen, C. and Cohen, S. A. (1997) Using quaternary high-performance liquid chromatography eluent systems for separating 6-aminoquinolyl-N-

hydroxysuccinimidyl carbamate-derivatized amino acid mixtures. *J. Chromatogr.* **763**, 11–22.

12. Grant, R. A. and Fieno, A. M. (1998) Use of the AQC precolumn method for determination of protein composition and identification from PVDF blots. Presented at ABRF'98: From Genomes to Function - Technical Challenges of the Post-Genome Era, http://www.abrf.org/ABRF/ABRFMeetings/abrf98/rgrant.pdf.

13. Wilkins, M. R., Williams, K. L., Appel, R. D., and Hochstrasser, D. F. (eds.), *Proteome Research: New Frontiers in Functional Genomics* (1997), Springer, Berlin.

14. Henzel, W. J., Billeci, T. M., Stults, J. T., Wong, S. C., Grimley, C., and Watanabe, C. (1993) Identifying proteins from two-dimensional gel slabs by molecular mass searching of peptide fragments in protein sequence databases. *Proc. Natl. Acad. Sci. USA* **90**, 5011–5015.

15. James, P., Quadroni, M., Carafoli, E., and Gonnet, G. (1993) Protein identification by mass profile fingerprinting. *Biochem. Biophys. Res. Commun.* **195**, 58–64.

16. Mann, M., Hojrup, P., and Roepstorff, P. (1993) Use of mass spectrometric molecular weight information to identify proteins in sequence databases. *Biol. Mass Spectrom.* **22**, 338–345.

17. Pappin, D. J. C., Hojrup, P., and Bleasby, A. J., (1993) Rapid identification of proteins by peptide-mass fingerprinting. *Curr. Biol.* **3**, 327–332.

18. Mann, M. and Wilm, M. (1994) Error tolerant identification of peptides in sequence databases by peptide sequence tags. *Anal. Chem.* **66**, 4390–4399.

19. Eckershorn, C., Jungblut, P., Mewes, W., Klose, J., and Lottspeich, F. (1988) Identification of mouse brain proteins after two-dimensional electrophoresis and electroblotting by microsequence analysis and amino acid composition. *Electrophoresis* **9**, 830–838.

20. Wilkins, M. R., Pasquali, C. Appel, R. D., Ou, K., Golaz, O., Sanchez, J.–C., et al. (1996) From proteins to proteomes: Large scale protein identification by two-dimensional electrophoresis and amino acid analysis. *Bio/Technology* **14**, 61–65.

21. Tous, G. I., Fausnaugh, J. L., Akinyosoye, O., Lackland, H., Winter-Cash, P., Vitorica, F. J., and Stein, S. (1989) Amino acid analysis on polyvinylidene difluoride membranes. *Anal. Biochem.* **179**, 50–55.

22. Ploug, M., Jensen, A. L., and Barkholt, V. (1989) Determination of amino acid compositions and NH2-terminal sequences of peptides electroblotted onto PVDF membranes from Tricine-sodium dodecyl sulfate-polyacrylamide gel electrophoresis: Application to peptide mapping of human complement component C3. *Anal.Biochem.* **181**, 33–39.

23. Gharahdaghi, F., Atherton, D., DeMott, M., and Mische, S. M. (1992) Amino acid analysis of PVDF bound proteins, in *Techniques in Protein Chemistry III* (Angeletti, R. H., ed.), Academic Press, San Diego, CA, pp. 249–260.

24. Hunziker, P., Andersen, T. T., Bao, Y., Cohen, S. A., Denslow, N. D., Hulmes, J. D., et al. (1999) Identification or proteins electroblotted to polyvinylidene

difluoride membrane by combined amino acid analysis and bioinformatics: an ABRF multicenter study. *J. Biomol. Tech.* **10**, 129–136.

25. Tyler, M. I. and Wilkins, M. R. (2000) Identification of proteins by amino acid composition after acid hydrolysis, in *Proteome Research: Two-Dimensional Gel Electrophoresis and Identification Methods* (Rabilloud, T., ed.), Springer, Berlin, pp. 143–161.

26. Matsudaira, P. (1987) Sequence from picomole quantities of proteins electroblotted onto polyvinylidene difluoride membranes. *J. Biol. Chem.* **262**, 10035–10038.

26a. Reig, J. A. and Klein, D. C. (1988) Submicrongram quantities of unstained proteins are visualized on polivinylidene difluoride membranes by transillumination. *Applied and Theoretical Electrophoresis* **1**, 59–60.

27. Laemmli, U. K. (1970) Cleavage of structural proteins during the assembly of the head of bacteriophage T4. *Nature* **227**, 680–685.

28. Schägger, H. and von Jagow, G. (1987) Tricine-sodium dodecyl sulfate-polyacrylamide gel electrophoresis for the separation of proteins in the range from 1–100 kDa. *Anal. Biochem.* **166**, 368–379.

29. O'Farrell, P. H. (1975) High resolution two-dimensional electrophoresis of proteins. *J. Biol. Chem.* **250**, 4007–4021.

30. Klose, J. (1975) Protein mapping by combined isoelectric focusing and electrophoresis of mouse tissues. A novel approach to testing for induced point mutations in mammals. *Humangenetik* **26**, 231–243.

14

Amino Acid Analysis in Protein Hydrolysates Using Anion Exchange Chromatography and IPAD Detection

Petr Jandik, Jun Cheng, and Nebojsa Avdalovic

1. Introduction

Following the commercial introduction as "AAA Direct," amino acid analysis by anion exchange chromatography and integrated pulsed amperometric detection (IPAD) *(1)* has been applied in many different types of amino acid assays. The new method does not require any derivatization. It is thus easier to use and less costly than derivatizaton-based methods. Additional savings are achieved by the relatively inexpensive single-component mobile phases and by the low flow rate (0.25 mL/min). The limits of detection are in the femtomol range and linear calibration plots extend over three orders of magnitude for the most amino acids *(1)*.

Various protocols for analyzing protein and peptide hydrolysates are outlined in a recent publication *(2)*. A Technical Note on the analysis of different types of peptide hydrolysates has become available from Dionex *(3)*. The results obtained for a collagen hydrolysate were compared with those generated by cation exchange/ninhydrin method *(1)*. Simultaneous analysis of carbohydrates and amino acids in cell cultures is discussed in a 1999 paper *(4)*. The use of the technique for on-line monitoring and control of metabolism in mammalian cell cultures has also been reported *(5)*.

In this chapter, we provide suitable protocols for the validation of reproducibility and accuracy. Such validation should be performed not only during a start up but also in regular intervals during the routine use. Evaluation of reproducibility is made easier by a new report format of Dionex software.

From: *Methods in Molecular Biology, vol. 211: Protein Sequencing Protocols, 2nd ed.*
Edited by: B. J. Smith © Humana Press Inc., Totowa, NJ

We expect many of the new users of AAA Direct to be experienced in one of the post- or precolumn derivatization techniques. In anticipation, we include guidelines for the selection of compatible buffers, surfactants, and reagents. Several of the compounds that are not compatible or difficult to use with the older methods do not cause any problems when utilized in conjunction with the AAA Direct (i.e., Triton X100, urea, ammonia from the hydrolysis of polyacrylamide, sodium hydroxide). On the other hand, some chemicals that are in widespread use in the pre- or postcolumn derivatization-based protocols do interfere, if utilized with our method. The most important examples of the latter category are tris(hydroxymethyl)aminomethane (TRIS) or carbohydrates when present in high concentrations. However, both TRIS and carbohydrates interfere only within the simplest format of AAA Direct. Recently, the authors have reported a modified version of AAA Direct overcoming TRIS and carbohydrate interference *(8,9)*.

2. Materials

2.1. Chromatography

1. Water 18 megohm (Millipore, Bedford, MA, Milli Q or equivalent, *see* **Note 1**).
2. Sodium hydroxide solution 50% (w/w) (Certified Grade, Fisher Scientific, Pittsburgh, PA).
3. Sodium acetate, anhydrous (Product No. 59326, Sunnyvale, CA).
4. Helium or argon gas cylinder, pressure regulation valve, Tygon tubing connecting cylinder to eluent containers.
5. Sterile filtration units from Nalge (0.2 μm Nylon filter in covered funnels, vol = 1 L, VWR, West Chester, PA).
6. Serological glass pipets, vol =10 mL (Fisher Scientific).
7. Eluent A: Water 18 megohm filtered through a 0.2-μm Nylon filter (*see* **Notes 2** and **3**).
8. Eluent B: Filter 1 L of water through a 0.2-μm Nylon filter; using the serological pipets (*see* **Note 4**), add 13.1 mL of 50% sodium hydroxide to the filtered 1 L-volume of water (*see* **Note 3**).
9. Eluent C: Dissolve 82 g of anhydrous sodium acetate in ca. 500 mL of water. Fill up to 1L and filter through the 0.2-μm Nylon filter.
10. AminoPac PA10 Guard Column (2 × 50 mm) and AminoPac PA10 analytical column (2 × 250 mm).
11. Ternary low-pressure gradient, microbore, high-performance liquid chromatography (HPLC) system, Dionex BioLC or comparable with an autosampler (e.g., Dionex AS50) and column thermostat. The system must include Dionex ED40 electrochemical detector.
12. AAA certified gold cell for use with the ED40 detector.
13. PC with installed copy of Dionex PeakNet 6.1 or equivalent (*see* **Note 5**).

2.2. Reproducibility Testing Using Injections of Standards

1. Amino Acids in 0.1 M HCl, Standard Reference Material 2389 (National Institute of Standards and Technology, Gaithersburg, MD).
2. Norleucine (Sigma, St. Louis, MO).
3. Sodium azide (Sigma).
4. HCl 0.1 M (reagent grade).
5. Diluent containing norleucine and sodium azide: Prepare 4 mM stock solution of norleucine (524.8 mg/L) in 0.1 M HCl. Dilute 500X with a solution of azide (20 mg/L).
6. Reproducibility Standard: pipet 320 µL of SRM 2389 into a 100-mL volumetric flask and fill up to volume with the norleucine diluent. If stored in a refrigerator, the Reproducibility Standard is stable for up to 30 d. If used at room temperature the Reproducibility Standard remains stable for up to 3 d (i.e., over a weekend).
7. The Reproducibility Standard from **step 6**, does not contain all of the standard compounds that are required for the analysis of certain types of hydrolysate samples. These standards can be purchased from Sigma and are listed below: δ-hydroxylysine, D(+)-galactosamine hydrochloride, D(+)-glucosamine hydrochloride, hydroxy-L-proline, phospho-L-arginine sodium salt, o-phospho-D,L-serine, o-phospho-D,L-threonine, o-phospho-D, L-tyrosine, L-cysteic acid hydrate, and L-tryptophan.
8. Prepare 1 mM, single component solutions of all standards from **step 7** in 0.1 M HCl and store them in a refrigerator. As required, add 100 µL of one or more of the above single-component stock solution to the Reproducibility Standard (**step 6**).
9. Disposable transfer pipets (glass or polyethylene).

2.3. Accuracy of Amino Acid Analysis in a BSA Hydrolysate

1. Bovine serum albumine (BSA) (7% solution) Standard Reference Material 927c (National Institute of Standards and Technology, Gaithersburg, MD).
2. Hydrolysate glass tubes with Teflon Plugs (Pierce, Rockford, IL).
3. Speedvac (Savant, Farmingdale, NY).
4. Heating Module with aluminum heating block (Pierce).
5. Constant Boiling HCl (Pierce).

2.4. Selecting Suitable Buffers, Surfactants and Reagents (see Note 6)

1. Buffers, surfactants, and reagents found to be compatible with AAA Direct: ACES (*see* **Note 7**); ADA; AMPSO; CAPS (*see* **Note 7**); CAPSO; CHES (*see* **Note 7**); citric acid (Fluka); EDTA (Aldrich); MES (Boehringer Manheim); MOBS; MOPS; MOPSO; n-octyl-β-glucoside (*see* **Note 8**); phenol; PIPES (*see* **Note 7**) sodium diphosphate; sodium monophosphate; sodium triphosphate; sodium azide; TES; TRITON X 100 (Aldrich).

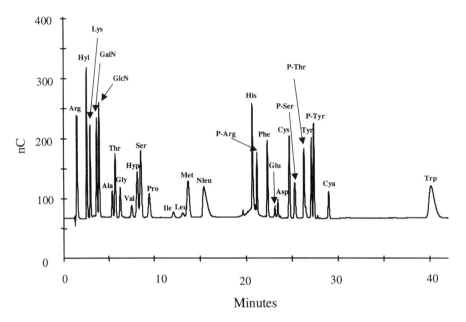

Fig. 1. Simultaneous separation of amino acids (including hydroxy-proline and hy-droxy-lysine, or Hyp and Hyl, respectively), amino sugars (GalN and GlcN), phospho amino acids (P-Arg, P-Ser and P-Thr) and cysteic acid (Cya) under the conditions of standard AAA Direct gradient.Fig. 1. Simultaneous separation of amino acids (including hydroxy-proline and hydroxy-lysine, or Hyp and Hyl, respectively), amino sugars (GalN and GlcN), phospho amino acids (P-Arg, P-Ser and P-Thr) and cysteic acid (Cya) under the conditions of standard AAA Direct gradient.

2. Buffers, surfactants, and reagents producing interfering peaks in AAA Direct: BES (BDH Chemicals, Poole UK); Bicine; BISTRIS (*see* **Note 9**); CHAPS; CHAPSO; DTE; EPPS; HEPES; imidazole; TEA (*see* **Note 8**); Tricine (*see* **Note 9**); TRIS (*see* **Note 9**).

3. Methods

3.1. Chromatography (see *Note 10*)

The chromatogram in **Fig. 1** was generated using the gradient conditions in **Table 1** and detection conditions in **Table 2**. We recommend using those conditions for all reproducibility and accuracy testing described below.

Figure 1 illustrates the capability of AAA Direct for protein and peptide hydrolysates. All common types of hydrolysates (HCl, HCl /propionic acid, MSA, NaOH,) can be analyzed under identical gradient and detection conditions. An important exception is the separation of methionine sulfone, that

Table 1
Gradient Conditions for Hydrolysates (Flow rate: 0.25 mL/min)

Time (min)	%A[a]	%B[a]	%C[a]	Curve
Init.	76	24	0	
0.00	76	24	0	
2	76	24	0	
8	64	36	0	8
11	64	36	0	
18	40	20	40	8
21	44	16	40	5
23	14	16	70	8
42	14	16	70	
42.1	20	80	0	5
44.1	20	80	0	
44.2[b]	76	24	0	5
75[c]	76	24	0	

[a] Eluents A,B,C: *See* **Subheading 2.1., steps 7–9**.
[b] Start of re-equilibration to initial conditions.
[c] Complete re-equilibration and column clean-up under normal conditions.

Table 2
Detection Conditions

Time (ms)	Potential (V)[a] vs pH	Integration
0	0.13	
40	0.13	
50	0.33	
110	0.33	
210	0.33	Begin
220	0.55	
460	0.55	
470	0.33	
560	0.33	End
570	−1.67	
580	−1.67	
590	0.93	
600	0.13	

[a] Sequence of potentials (or waveform) applied to the Au working electrode and referenced vs Glass/Ag/AgCl combination electrode.

can be achieved only at the column temperature of 35°C (*see* **ref. 2**). In contrast, the correct column temperature for all other separations (including that in **Fig. 1**) is 30°C.

1. Connect the fully assembled ED40 cell directly to the injector outlet. Running the initial gradient conditions (76%A, 24%B at 0.25 mL/min) switch the cell voltage on (Specify "Integrated Amperometry" and apply Waveform from **Table 2**). Verify that the signal is lower than 80 nC.
2. Install the AminoPac PA10 column set and check again that the signal background is at less than 80 nC.
3. Allow the column oven to reach 30°C and run a blank gradient (*see* **Table 1**) injecting 25 µL of water. Verify that gradient rise is not exceeding 30 nC.
4. Prepare an aliquot of Reproducibility Standard (**Subheading 2.1., step 6**) and place it into the autosampler. Follow proper procedures in executing the first standard injection (vol = 25 µL) by not allowing any delay between the end of the blank gradient run and first injection of standard.
5. Verify the absence of extraneous peaks in the blank gradient and a correct value of peak height for arginine (>100 nC). Verify the presence of peaks for all amino acids. P-amino acids, hydroxylysine, hydroxyproline, amino sugars, cysteic acid, and tryptophan are not present in the Reproducibility Standard. Verify that the separation between alanine and threonine in the Reproducibility Standard chromatogram is comparable to that in **Fig. 1**.

3.2. Reproducibility Testing Using Injections of Standards

1. Verify the system status by carrying out **steps 1–5, Subheading 3.1.**
2. Transfer 1.5 mL of Reproducibility Standard (**Subheading 2.2., step 6**) into a glass autosampler vial. Place the full vial into the autosampler. Make sure there is also a vial with water inside the autosampler.
3. Create a "PeakNet Sequence" for running a blank gradient (vol = 25 µL injection of water, **Table 1** gradient, **Table 2** waveform) followed by nine injections of Reproducibility Standard (vol = 25 µL "Full Loop," **Table 1**, **Table 2**). Do not perform more than nine vol = 25 µL Full-Loop-injections out of a single 1.5-mL vial.
4. Create or update a quantitation method (.qnt) and include it in the PeakNet sequence. A meaningful evaluation of reproducibility relies on correct identification and integration of all peaks of interest. The correct identification and integration is dependent on a correct user input of retention times and integration parameters (*see* **Note 11**). The retention times can be updated and integration parameters optimized during the first standard run. It is not necessary to input pmol amounts for amino acids and to carry out a calibration at this stage as reproducibility is evaluated from the relative standard deviations of peak areas.
5. Following the completion of all standard runs, verify correct identification and integration of all peaks by inspecting all nine standard chromatograms separately. Close the files you opened for identification and integration verification.

6. Select all nine standard chromatograms simultaneously (press Ctrl. and click on corresponding row numbers). In the menu that opens up by rightclicking on one of the selected rows, choose "compare>ECD1" and an overlay of all selected chromatograms appears on the screen.
7. Rightclick anywhere in the upper-right portion of the chromatographic overlay. In the menu that now appears choose "Load Report Format>%RSD Peak Areas" (*see* **Note 12**) to generate a similar report as the one shown in **Fig. 2**.
8. Rightclick in the upper-right portion of the screen, go to "Load Report Format" again and switch to %RSD Retention Times (actual report not shown).

3.3. Accuracy of Amino Acid Analysis in a BSA Hydrolysate

The following protocol utilizes a data reduction procedure producing three numerical QC values, "Average % Error," Recovery [g],"and "% Recovery," characterizing the accuracy of results for a protein of known amino acid composition.

Average % Error = (Σ % Error of 16 amino acids) / 16 *(1)*

Where:

% Error = 100* ABS (analytical result for *n* – true value for *n*)/true value for *n*
ABS (x): absolute value of x
n : number of each amino acid per molecule

Recovery [g] = {(*Corrected Average Ratio*) * molecular weight of protein} *(2)*

Where:

Corrected Average Ratio: mean value of those *First Ratios* values deviating by less than 20% from the Average of *First Ratios* (*Corrected Average Ratio* represents the analytical result for the mol-amount of protein injected)
First Ratio = analytical result in pmol of amino acid/ true value of *n*

% Recovery = {(**Recovery [g]** *8}/ g of Protein in 200 µL of hydrolysate) *(3)*

This data reduction approach has been used, for example, in the "multi-center" (round robin) studies organized by the Association of Biomolecular Resource Facilities (ABRF) *(6)*. Evaluation of accuracy is simplified in AAA Direct by an automatic calculation of QC parameters from pmol results after calibration. As in the reproducibility evaluation (*see* **Subheading 3.2.**), the parameters are calculated directly within the HPLC software. It is not necessary to export area or pmol results into another application such as Excel first (*see* **Note 14**).

As a protein of known amino acid composition, we have chosen the 7% BSA solution from the National Institute of Standard and Technology (**Subheading 2.3., step 1**). The amino acid composition of BSA can be obtained in the literature *(7)*. The NIST documentation certifies that the SRM 927c is from

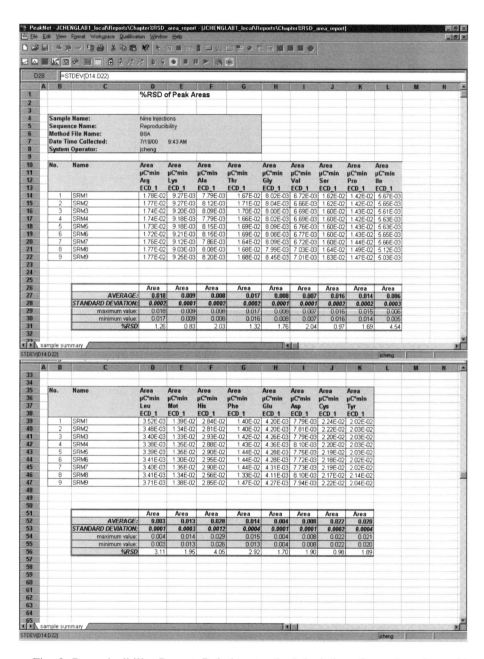

Fig. 2. Reproducibility Report. Relative standard deviations for each amino acid are listed in the bottom row. The report is linked to all data files in a given PeakNet Sequence. Any change of integration parameters produces instant updates of the corresponding area, average, and %RSD values.

bovine blood and has the correct molecular weight and spectral properties for BSA. It also states that the SRM 927c may be used for calibration in amino acid analysis. The SRM 927c documentation, however, does not certify the amino acid composition of the actual protein sample used to prepare the 7% solution.

1. Prepare a 500-fold dilution of SRM 927c in water containing 20 mg/L sodium azide.
2. Transfer 33.6 µL of 500X dilute SRM 927c into a clean microvial (vol = 1.5 mL) and carry out evaporative centrifugation (Speedvac, *see* **Subheading 2.3.**, **step 3**) to dryness.
3. Add to the dry residue 100 µL of constant boiling HCl (*see* **Subheading 2.3.**, **step 5**) and vortex thoroughly.
4. Using glass Pasteur pipets transfer the solution to a clean hydrolysis tube (*see* **Subheading 2.3., step 2**).
5. To the microvial emptied in **step 4**, add a new 50-µL aliquot of constant boiling HCl, vortex, and subsequently transfer as much as possible of the 50-µL aliquot to the sample in the hydrolysis tube from **step 4**. Use the same transfer pipet as in **step 4**.
6. Alternate vacuum and inert gas inside the hydrolysis tube three times. Place fully evacuated tube into a heating module (*see* **Subheading 2.3., step 4**). Carry out hydrolysis for 16–24 h at 110°C.
7. Remove the hydrolysis tube from the heating module and allow it to cool down to the ambient temperature.
8. Using a clean glass Pasteur pipet transfer the hydrolyzed sample from the hydrolysis tube into a clean microvial.
9. Pipet another 50-µL aliquot of constant boiling HCl into the hydrolysis tube from **step 8**. Rinse and subsequently transfer as much as possible of the 50-µL aliquot to the first aliquot of hydrolysate in the microvial. Use the same transfer pipet for **steps 8** and **9**.
10. Carry out evaporative centrifugation of the combined hydrolysate aliquots from **steps 8** and **9**.
11. To the completely dry residue from step 10, add 200 µL of norleucine/azide diluent (*see* **Subheading 2.2., step 5**) and vortex thoroughly.
12. Transfer an aliqot of the reconstituted hydrolysate into an autosampler vial (Polyethylene vial, (vol = 0.3 mL, with a split septum, Dionex). Make sure that vials with water and calibration standard have also been placed into the autosampler tray.
13. Create a "PeakNet Sequence" starting with an injection of water blank and followed by at least three injections of both, the calibration standard (*see* **Subheading 2.2., step 6**) and hydrolyzed sample. Use a program file (.pgm) that includes the gradient and waveform from **Tables 1** and **2**, respectively. Input correct pmol amounts (*see* **Note 15**) into the Amount Table of the quantitation method (.qnt, *see* **Subheading 3.2., step 4**). Use 25 µL Full-Loop injections

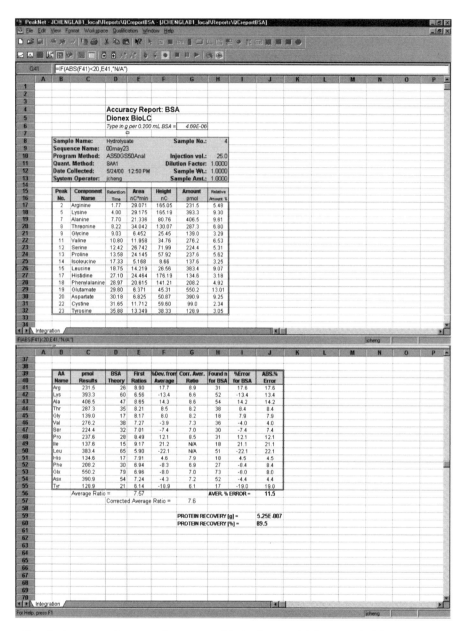

Fig. 3. Accuracy Report. The upper table is the usual "Amount Table" of PeakNet HPLC software. The pmol amounts in the sample are calculated from the pmol/area coefficients in the calibration file. The lower of the two tables evaluates equations *(1–3)* in the **Subheading 3.3.** This report is also linked to all data files of a given sequence. Any change of integration parameters produces an instant change of all calculated results.

for all samples and standards. In the quantitation method, specify the sequence line containing the calibration standard. Start the automatic execution of the entire sequence.

14. After executing all runs, inspect each standard and sample chromatogram. Verify correct identification and integration of all amino acid peaks. In a hydrolysate chromatogram, rightclick in the upper-right portion of the chromatogram. In the menu that now appears choose "Load Report Format>QCBSA (*see* **Note 14**)." A similar report as the one in **Fig. 3** appears on the screen.

15. Make sure to input the actual concentration of BSA in the hydrolysate as specified in the report subtitle (line 3). Switch to "Layout Mode" (Edit>Layout Mode) to enable input of BSA concentration. The calculated values of Average % Error, Protein Recovery [g], and % Protein Recovery appear instantaneously in the lower half of the report page. The BSA QC report is now ready to be saved or printed out.

16. Additional QC Reports for the other BSA hydrolysate injections can be generated by moving from one sample file to another using the corresponding two icons on the screen (left and right arrow, top bar, third row). To complete the documentation, it is also possible to generate the %RSD Area, %RSD Retention Times reports for the standard injections of the same sequence by carrying out the **steps 6–8** of the **Subheading 3.2.**

3.4. Selecting Suitable Buffers, Surfactants and Reagents

1. Generate chromatograms of 1 m*M* solutions of all buffers, surfactants and reagents present in your protein or peptide samples.

2. Hydrolyse (**Subheading 3.3.**, **steps 6–12**) 20 µL aliquots of 10 mM solutions of all buffers, surfactants, and reagents in 180 µL of constant boiling HCl.

3. Generate chromatograms of hydrolyzed and reconstituted (azide/norleucine diluent) samples from **steps 1** and **2** using the chromatographic conditions of **Subheadings 3.2.** and **3.3.**

4. Avoid using chemicals causing peaks interfering with amino acids or carry out desalting prior to hydrolysis of protein solutions containing such additives. *See* **Subheading 2.4.** for guidelines in the selection of new additives

5. Repeat **steps 1–4** for all new batches of additives.

4. Notes

1. Minimize extraneous plastic tubing connected to the laboratory water unit and use only filters supplied by the manufacturers. Polymeric surfaces support bacterial growth. Bacterial contamination of water used in preparation of eluents causes high backgrounds and spurious peaks in amino acid analysis.

2. The main purpose of eluent filtration is to eliminate bacterial contamination stemming from the water purification unit and from the atmosphere. Wear powder-free gloves whenever preparing or replenishing mobile phases to avoid contamination by skin contact.

3. Minimize carbonate contamination of the eluents. Follow the guidelines in the Installation Instructions and Troubleshooting Guide for AAA Direct Amino Acid Analysis System (Dionex Document No. 031481).

4. Most vol=10 mL pipets can be used for the prescribed volume of 13.1 mL. Usually this larger volume is at or near the markings for the maximum possible volume. Verify the correct height of 13.1 mL by pipeting and weighing out water. Do not re-use serological pipets.

5. Make sure that the "Report Publisher" option is part of the installed software and it is switched on. (In the Pull-down menu: Help/About PeakNet).

6. Except where indicated otherwise, all of the buffers, reagents, and surfactants were purchased from Sigma.

7. Four of the buffers that we evaluated (ACES, CAPS, CHES, PIPES) gave well-defined peaks, sufficiently separated from those of hydrolysate amino acids under the gradient conditions in **Table 1**. Users are advised to run control blanks when planning to utilize those buffers. Some of the peaks appear to be impurities unrelated to the actual buffer compound and may vary depending on the source and age of the chemical.

8. Glucoside-based surfactants produce a large interference peak when injected directly. However, they are completely hydrolyzed after 16–24 h and 110°C in constant boiling HCl and do not thus interfere in a completely hydrolysed sample.

9. Several of the nonanionic buffers (BISTRIS, TEA, TricineTRIS) produce a large peak interfering with arginine. This interference can be resolved using the modified AAA Direct technique described in **ref.** *(8)*.

10. A copy of the Installation Instructions and Troubleshooting Guide for the AAA Direct Amino Acid Analysis (Dionex Document No. 031481) is included with each AminoPac PA10 column shipped by Dionex. The text under Heading 3 describes only a subset of AAA Direct methodology used for reproducibility and accuracy testing.

11. Optimization of integration parameters and other general guidelines for reproducible quantitation are described in "PekNet 6 User's Guide." (Dionex Document No. 031630). Specify the version of PeakNet you are running when ordering the document from Dionex.

12. The "Excel-like" environment in the report part of Dionex HPLC software makes it possible for a user to create many different report formats executing simple or complex mathematical formulas. However, the user has to purchase the "Report Publisher" option to be able to generate a custom report format (*see* **Note 5** for verification of presence of Report Publisher in the installed software). The electronic files for reproducibility evaluation without installing the Report Publisher can be obtained from Dionex Customer Support. Specify "Reproducibility Evaluation of AAA Direct" when ordering the report format file.

13. A comprehensive evaluation of AAA Direct in conjunction with hydrolysis of peptides in HCl and MSA is presented in **ref.** *(3)*. **Reference** *(2)* contains detailed protocols for HCl, MSA (with and without performic acid oxidation), and NaOH hydrolysates. Also described in **ref.** *(2)* is a new method for using the

same set of chromatographic conditions for automatic analysis of carbohydrates, amino sugars, and amino acids in different types of TFA and HCl hydrolysates of glycoproteins.

14. As in the reproducibility evaluation (*see* **Subheading 3.2.** and **Note 12**), it is possible for a user to create highly customized report formats for the QC parameters of accuracy evaluation. The user has to have either the "Report Publisher" option installed on the PC (*see* **Note 5**) or obtain a suitable report format file from Dionex customer support. Specify "Accuracy of Amino Acid Analysis" when ordering the report format file from Dionex.

15. For a 312.5-fold dilution of SRM 2389 and 25 µL injection, divide all values in **Table 1** of SRM 2389 by 0.0125. For example: the original concentration of 2.50 mmol/L of aspartic acid yields 200 pmol amount to be entered into the Amount Table.

References

1. Clarke, A. P., Jandik, P., Rocklin, R. D., Liu, Y., and Avdalovic, N. (1999) An integrated amperometry waveform for the direct, sensitive detection of amino acids and amino sugars following anion-exchange chromatography. *Anal. Chem.* **71**, 2774–2781.

2. Jandik, P., Pohl, C., Barreto V., and Avdalovic, N. Anion exchange chromatography and integrated amperometric detection of amino acids, in *Methods in Molecular Biology, vol. 159: Amino Acid Analysis Protocols* (Cooper, C., Packer, N., and Williams, K., eds.), Humana Press, Inc., Totowa, NJ.

3. Dionex Corporation (2000) Determination of the amino acid content of peptides by AAA-Direct. Technical Note 50, pp.1–20.

4. Jandik, P., Clarke, A. P., Avdalovic, N., Andersen, D. C., and Cacia, J. (1999) Analyzing mixtures of amino acids and carbohydrates using bi-modal integrated amperometric detection. *J. Chromatogr. B.* **732**, 193–201.

5. Larson, T., Cacia, J., Wawlitzek, M., and Albers, U. (2000) On-Line HPLC for Monitoring and control of metabolism in mammalian cell cultures. P-1906, HPLC-2000, Seattle, WA.

6. Hunziker, P., Andersen, T. T., Bao, Y., Cohen, S. A., Denslow, N. D., Hulmes, J. D., et al. (1999) Identification of proteins electroblotted to polyvinylidene difluoride membrane by combined amino acid analysis and bioinformatics: An ABRF Multicenter Study. *J. Biomol. Tech.* **10**, 129–136.

7. Brown, J. R., (1975) Structure of Bovine Serum Albumine. *Fed. Proc.* **34**, 591.

8. Jandik, P., Cheng, J., Jensen, D., Manz, S., and Avdalovic, N. (2000) New technique for increasing retention of arginine on an anion-exchange column. *Anal. Biochem.* **287**, 38–44.

9. Jandik, P., Cheng, J., Jensen, D., Manz, S., and Avdalovic, N. (2000) Simplified in-line sample preparation for amino acid analysis in carbohydrate containing samples. *J. Chromatogr. B.* **758**, 189–196.

15

D-Amino Acid Analysis

Andrea Scaloni, Maurizio Simmaco, and Francesco Bossa

1. Introduction

In nature D-amino acids occur rarely as compared with their enantiomers. Their presence has been observed in several tissues and fluids of mammals *(1)*, molluscs *(2)*, invertebrates *(3)* and plants *(4)*, as well as in microbial fermented foods as result of bacterial metabolism *(5)*. Gram-positive and -negative bacteria contain D-amino acids as fundamental components of cell-wall peptidoglycans *(6)*. Similarly, they have been found inserted in polypeptide chains of several antibiotics synthesized by microbial multienzymatic complexes *(7)* or of bioactive peptides of ribosomal origin isolated from various vertebrates *(8)*. In the latter case, their presence, which is essential to exert the specific biological activity, is the result of a post-translational modification occurring on larger precursors encoded by mRNAs where the base triplet coding for the L-amino acid is present at the position occupied by the corresponding D-enantiomer in the mature product *(9)*. Furthermore, in proteins the presence of small quantities of D-amino acids is considered to be the result of a racemization process related to molecular aging *(10)*. Recently, the occurrence of D-serine in mammalian brain has been described and its role as neuromodulator has been proposed; subsequently, serine racemase, a glial pyridoxal phosphate-dependent enzyme synthesizing D-serine, has been cloned and expressed *(11)*. Thus, D-amino acids appear to be related to many important biological processes, but the poor diffusion in protein chemistry laboratories of methodologies and instruments dedicated to D-amino acid detection makes their occurrence in nature still underestimated.

Different liquid chromatogrphy (LC), gas chromatography (GC), and capillary electrophoresis (CE) techniques have been proposed for direct determination of D-amino acids based on the use of either chiral stationary phases or chiral mobile phases *(12–14)*. However, the recent introduction of specific

From: *Methods in Molecular Biology, vol. 211: Protein Sequencing Protocols, 2nd ed.*
Edited by: B. J. Smith © Humana Press Inc., Totowa, NJ

precolumn chiral derivatizing reagents made their diastereometric products more easily separable on conventional chromatographic supports *(15)*. Therefore, fully automated analyzers, not commercially available, have been developed based on the use of autosamplers and conventional high-performance liquid chromatography (HPLC) systems. Very recently a microfabricated capillary chip suitable for extraterrestrial environment analysis has been proposed for the analysis of few amino acid enantiomers *(16)*.

There are several precolumn chiral derivatizing reagents commercially available that allow the automated quantitative determination of D- and L-amino acids in protein hydrolysates through the simultaneous separation of their seventeen enantiomeric pairs. These are the *o*-phthaldialdehyde/N-isobutyryl-L-cysteine mixture *(17)*, (+)-1-(9-fluorenyl) ethyl-chloroformate *(18)* and 1-fluoro-2,4-dinitrophenyl-5-L-alanine amide, also known as Marfey's reagent *(19,20)*. The speed of analysis is similar for all three reagents but only the first two yield fluorescent derivatives that are easily detectable in the picomolar range. Even with the third reagent, however, routine analysis can be carried out with 300–500 pmol of sample. The reaction of 1-fluoro-2,4-dinitrophenyl-5-L-alanine amide, shown in **Fig. 1**, is rapid and quantitative with both primary and secondary amino groups. Moreover, this reagent has been used successfully for the determination of the chirality of the amino acid residues in the course of sequence analysis of peptides according to a variation of the subtractive-Edman degradation *(21)*. In this case, the Edman chemistry is used merely as a subtractive procedure. After each cycle of degradation, a suitable aliquot of the shortened peptide is subjected to end-group analysis by reaction with the chiral reagent, as reported in **Fig. 2**. It has to be mentioned that under the hydrolysis conditions used for the release of the resulting derivatized-amino acid, the 2,4-dinitrophenyl-5-L-alanine amide group is converted to 2,4-dinitrophenyl-5-L-alanine. High recoveries of each derivative are obtained after 5 h hydrolysis in vapor phase with negligible racemization. Therefore, the chirality of the N-terminal amino acid after each cycle can be determined through the reversed-phase HPLC (RP-HPLC) analysis of its 2,4-dinitrophenyl-5-L-alanine derivative. The separation of the corresponding derivatives has been optimized *(21)*. In fact, a reagent not containing the amide moiety (1-fluoro-2,4-dinitrophenyl-5-L-alanine) has been synthesized and successfully tested for the quantitative determination of D-amino acids in protein hydrolysate and for the determination of the chirality of the amino acid residues in the course of sequence analysis of peptides, but it is not commercially available. Early in 2000, a new chiral isothiocyanate was introduced for the simultaneous study of the sequence and the chirality of amino acids in peptides during Edman degradation; the promising properties of this method will be verified in future for routine analyses *(22)*.

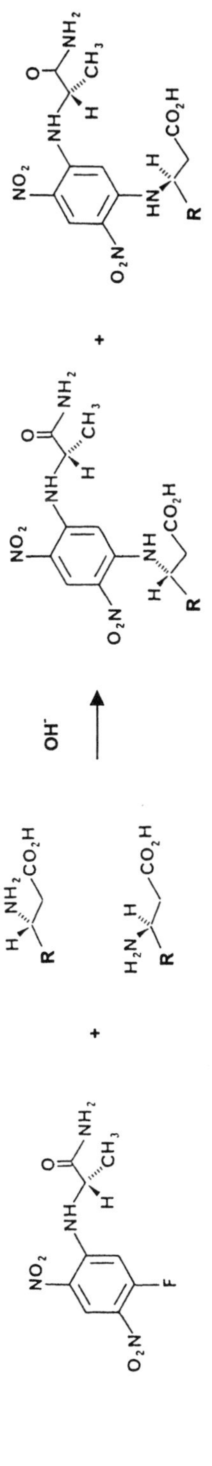

Fig. 1. Reaction of 1-fluoro-2,4-dinitrophenyl-5-L-alanine amide with D and L amino acids.

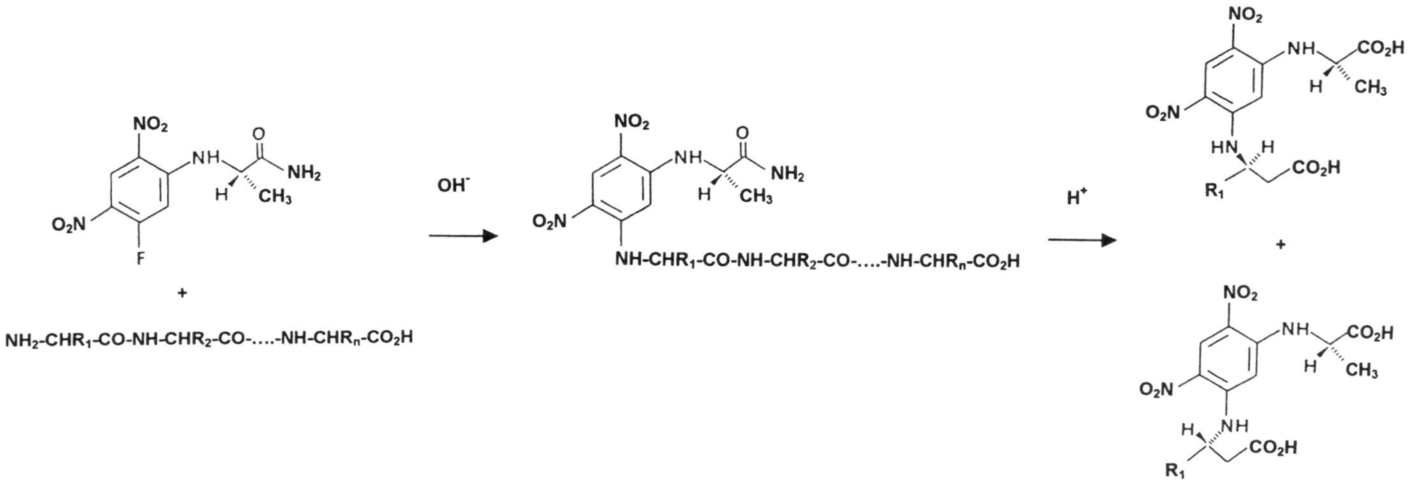

Fig. 2. Derivatization of the N-terminal residue of a peptide with 1-fluoro-2,4-dinitrophenyl-5-L-alanine amide and release of its amino acid derivative following acid hydrolysis.

2. Materials

2.1. Apparatus

1. Pyrolyzed Pyrex glass (40 × 6 mm ID) or polypropylene microtubes (e.g., Gilson P/N 450831).
2. Rotary evaporator, e.g., "Speed Vac" (Savant, Fullerton, CA), equipped with a rotary vacuum pump capable of evacuating the system to 50 mtorr or better and a cold trap kept at a temperature of –60°C.
3. Pyrex vessel for vapor-phase acid hydrolysis equipped with a Teflon screw cap and a vacuum/purge manifold, e.g., the Waters reaction vial and Mininert valve present in the Pico-Tag workstation (Waters P/N 07363).
4. Stoppered glass tubes for manual Edman degradation (3 mL).
5. Vacuum desiccator containing phosphorus pentoxide.
6. Automated RP-HPLC system for derivatization, separation, and quantification of the amino acid pairs. This requires two pumps; an automated sample manipulator and injector with a rack thermostated at 40°C (for the samples) and at 4°C (for the reagents), a column heater, a reversed-phase column, a UV-detector set at 340 nm and a database system for gradient control, quantitation of the chromatographic peaks, and data analysis (see **Note 1**).

2.2. Chemicals

1. Nitrogen gas supply.
2. 6 M hydrochloric acid (e.g., Pierce P/N 24309, Rockford, IL).
3. D- and L-amino acid standards (Sigma Chemical Co.).
4. 1-fluoro-2,4-dinitrophenyl-5-L-alanine amide (Novabiochem or Sigma Chemical Co.).
5. Sequencing-grade phenylisothiocyanate (PITC), pyridine, trifluoroacetic acid, triethylamine, 1,5-difluoro-2,4-dinitrobenzene (Fluka).
6. HPLC-grade solvents.

2.2.1. Chiral Derivatization and Subtractive Edman Degradation Solutions

1. Chiral derivatization buffer. Mix triethylamine: acetonitrile (2:3, by vol). Vortex and store at 4°C. Use within 4 wk.
2. Chiral derivatization reagent solution. Dissolve 1 mg of 1-fluoro-2,4-dinitrophenyl-5-L-alanine amide in 1 mL acetonitrile: acetone (1:2, by vol). Stored at –20°C, the reagent solution is stable for months.
3. Edman degradation buffer. Mix freshly distilled pyridine and water (1:1, by vol). Use within 24 h.
4. Edman degradation reagent solution. Mix PITC and freshly distilled pyridine (1:9, by vol). Use within 24 h.
5. Edman degradation extraction solution. Mix ethylacetate with water (1:1, by vol) under vigorous stirring. Let the two phases separate, store the bottle closed at room temperature and use within a week.

2.2.2. Chromatography

1. Loading solvent. Water: acetonitrile: ethylacetate (91:8:1, by vol).
2. Solvent A. 0.2% (by vol) trifluoroacetic acid (TFA) in MilliQ water. Filter through a 0.22-μm filter (Millipore) (*see* **Note 2**).
3. Solvent B. 0.1% (by vol) TFA in acetonitrile: 2-propanol (4:1, by vol).
4. Column. Waters Sperisorb S5 ODS2 (2.0 × 250 mm, 5 μm). An alternative is a Beckman Ultrasphere Octyl column (2.0 × 250 mm, 5 μm).

3. Methods

3.1. Quantitative Determination of the Chirality of Amino Acid Residues in Peptide/Protein Hydrolysates

1. Place the peptide/protein sample (0.5 nmol) dissolved in a volatile solvent (not more than 50 μL) into a Pyrex or polypropylene tube (*see* **Note 3**). Dry in the rotary evaporator as described for the acid hydrolysis procedures reported in Chapters 10–12.
2. Hydrolyze the dried samples by vapor-phase hydrolysis under vacuum as described in Chapters 10–12, placing the reaction vial at 110°C for 24 h (*see* **Note 4**).
3. Cool down the vial, open it carefully and clean the outside of the tubes from HCl excess.
4. Prepare additional tubes containing 1 nmol each of a mixture of D- or L-amino acids (calibration standard) or water (blank). Dry all samples in the rotary evaporator.
5. Add to each tube 10 μL of the chiral derivatization buffer and carefully vortex it. Add to each sample 10 μL of the chiral derivatization reagent solution and vortex again. Incubate at 40°C for 60 min without sealing the tubes. At the end of the incubation time, the sample is dry.
6. Dissolve each sample in 25 μL of the loading solvent and analyze it by using the gradient shown in **Table 1**. Calibration standards will give chromatographic profiles similar to that shown in **Fig. 3**.
7. After the run, qualitative and quantitative analyses of unknown samples is given by the integrator, on the basis of the comparison with amino acid standard peaks areas. Blank samples are used to identify reagent-related peaks (*see* **Note 5**).

3.2. Determination of the Chirality of Amino Acid Residues in Peptides in the Course of Subtractive Edman Degradation

1. Place each peptide sample to be sequenced (5–25 nmol) dissolved in a volatile solvent into a glass tube for manual Edman degradation and lyophilize (this procedure being time-consuming, the simultaneous processing of multiple samples is recommended) (*see* **Note 6**).
2. Dissolve the samples in 100 μL of the Edman degradation buffer.
3. After addition of 50 μL of Edman degradation reagent to each tube containing the solved sample, mix gently, flush with nitrogen, and incubate for 30 min at 50°C. Carefully seal the tubes.
4. Dry the samples in the vacuum desiccator.

Table 1
Gradient Parameters for the Separation of 2,4-
dinitrophenyl-5-L-alanine amide-1-amino Acids [a]

Time, min	Function	Value
0	Flow (mL/min)	0.2
0	%B	15
3	%B	15
110	%B	50
111	%B	85
117	%B	85
118	%B	15
125	End	

[a]Gradient curves are linear.

5. Add to each tube 100 µL of anhydrous TFA, flush with nitrogen, and incubate for 10 min at 50°C, and then dry again. Carefully seal the tubes.
6. Dissolve each sample in 250 µL of the aqueous phase from the Edman degradation extraction solution. Then, add 1 mL of the organic phase and vortex it. Let the two phases separate in each tube.
7. Carefully remove the organic phase from each sample and discard it; take a small sample from the aqueous phase for N-terminal analysis (*see* **Note 7**). Then dry the remaining aqueous phase and use it for the next cycle of Edman degradation starting from **step 2**. Repeat for the suitable number of cycles.
8. For N-terminal analysis of the aqueous phase sample, place each sample into a Pyrex or polypropylene tube, carefully label it and dry in the rotary evaporator as described in **Subheading 3.1.** Similarly, dry in a separate tube 0.5–2 nmoles of the intact peptide.
9. Prepare other tubes containing each 1 nmol of D- and L-amino acid (calibration standards) or water (blanks). Dry all samples in the rotary evaporator.
10. To each tube add 10 µL of the chiral derivatization buffer and carefully vortex. Add to each sample 10 µL of the chiral derivatization reagent solution and vortex again. Incubate at 40°C for 60 min without sealing the tubes. At the end of the incubation time, the sample is dry.
11. Place all samples in the reaction vial for vapor-phase hydrolysis under vacuum and hydrolyze the samples at 110°C for 5 h, as described at **step 2** of **Subheading 3.1.** (*see* **Note 8**).
12. Cool down the vials, open carefully, clean the outside of the tubes from HCl excess, and dry in the rotary evaporator.
13. Dissolve each sample in 25 µL of the loading solvent and analyze it as described in **Subheading 3.1., step 6** (*see* **Note 9**).
14. After the run, peaks are integrated and the identity of the N-terminal amino acid of the unknown is determined by comparison with peaks given by calibration standards. Blank samples are used to identify reagent-related peaks.

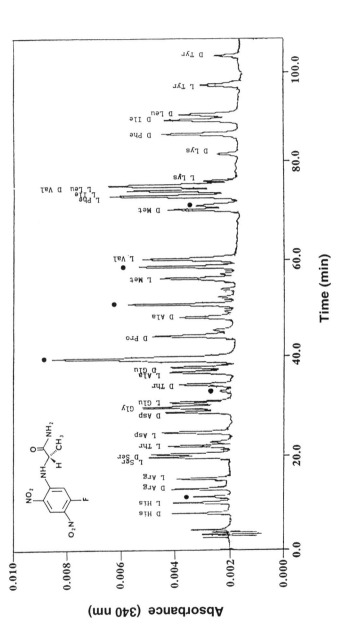

Fig. 3. Elution profile of a standard mixture of D- and L-amino acids plus glycine (150 pmol each) derivatized with 1-fluoro-2,4-dinitrophenyl-5-L-alanine amide. The peaks denoted with black circles are reagent related. Amino acids were derivatized and injected as described in **Subheading 3.1.** Injections were made using a Gilson autosampling injector consisting of a model 231 sample injector equipped with a code 9 modified rack, and a model 401 diluter, both controlled by a sample controller keypad. The HPLC system consisted of two LabFlow 4000 pumps (LabService Analytica), a model 430 UV-visible detector (Kontron AG, Zürich, Switzerland) equipped with a 3-µL flow cell, a Gilson model 811C dynamic mixer (65 µL mixing chamber), and a Rheodyne injection valve (model 7125) equipped with a 20 µL sample loop. Gradient formation is controlled by the pumps; quantitation of the peaks and data analysis is performed by a Kontron model 450 MT data system.

4. Notes

1. Traditionally, precolumn derivatizations are performed off-line from the liquid chromatograph. Therefore, they are time-consuming and errors inevitably occur when using manual procedures. Moreover, decomposition of sample derivatives may occur when analysis is delayed. We have set up a system where both derivatization and injection are completely automated *(20)*. During the analysis of a sample, the next one is derivatized. Different thermostating of the reagent and sample compartments allows storage of the reagents without degradation.

2. The use of volatile eluents instead of the previously reported 40 mM tri-ethylammine-phosphate buffer *(20,21)* will prevent damages to the HPLC system related to salt's corrosion and precipitation. Under these experimental conditions a column will allow at least a thousand analyses without significant loss of chromatographic resolution.

3. Samples should be ideally dissolved in a solvent not containing salts or detergents.

4. Do not perform acid hydrolysis of samples by using liquid-phase or microwave irradiation procedures. In the first case, background contamination is a frequent problem. In the second case, racemization of each amino acid during peptide/protein hydrolysis has been reported *(23)*. General considerations regarding the stability of amino acids during the acid hydrolysis are covered in details in Chapter 10.

5. The linearity of the response of all amino acids has been verified in the range 50–2000 pmol. The use of L-norvaline (eluting 0.8 min later than L-Val) as internal standard allows calculation of the recovery of each sample.

6. Reagents should be of the highest quality available. During Edman degradation, the use of freshly open sequencing grade reagents is strongly recommended.

7. The volume removed at each degradation cycle ranges from 10% to 100% of the total aqueous phase volume, depending on the number of cycles, the extent of degradation, and the sample size.

8. Highest recoveries of derivatized amino acids are obtained by adding to each acid hydrolysis vial a tube containing 0.5 mg of 1,5-difluoro-2,4-dinitrobenzene solved in 40 μL acetone. This compound has scavenger properties that can prevent diastereoisomer destruction *(20)*.

9. In this case, the elution profile of the calibration standards will be different from that reported in **Fig. 3**, since the amide moiety of each derivative has been converted to a carboxylic functionality *(see* **Fig. 2**). The chromatogram is similar to that reported in **Fig. 4**, where the nature of each peak was verified by using an *ad hoc* synthesized reagent (1-fluoro-2,4-dinitrophenyl-5-L-alanine) *(21)*.

10. As an alternative method for determination of the chirality of the N-terminal amino acid, the different peptides present in the aqueous phase at each step of Edman degradation can be analysed for their chiral amino acid composition as described in **Subheading 3.1.** The comparison of the amino acid content following each cycle will give indication on the occurrence of D-amino acids within the sequence *(24)*.

11. The procedure here reported for the identification of the amino acid chirality in the course of subtractive Edman degradation was successfully applied to the analysis of peptides from different sources. Its direct use during the analysis of

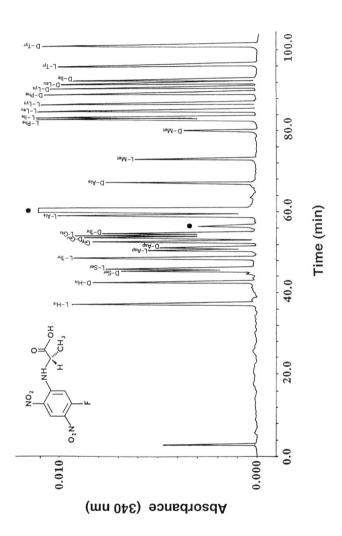

Fig. 4. Chromatographic profile of a standard mixture of D- and L-amino acids plus glycine (300 pmol each) derivatized with 1-fluoro-2,4-dinitrophenyl-5-L-alanine. Peaks denoted with black circles are reagent related. Amino acids were derivatized and injected as described in **Subheading 3.2.** Instrument details are reported in **Fig. 3**.

whole proteins could suffer of the use of organic solvent during the extraction step, therefore causing aggregation or precipitation of the sample and a resulting drop in the reaction yield.

References

1. Nagata, Y., Yamamoto, K., and Shimojo, T. (1992) Determination of D- and L-amino acids in mouse kidney by HPLC. *J. Chromatogr.* **575**, 147–152.
2. Felbeck, H. and Wiley, S. (1987) Free D-amino acids in marine molluscs. *Biol. Bull.* **173**, 252–259.
3. D'Aniello, A., Di Cosmo, A., Di Cristo, C., and Fisher, G. (1992) D-Aspartate in the male and female reproductive system of *Octopus vulgaris*. *Gen. Comp. Endocrinol.* **100**, 69–72.
4. Kullman, J. P., Chen, X., and Armstrong, D. W. (1999) Evaluation of the enantiomeric composition of amino acids in tobacco. *Chirality* **11**, 669–673.
5. Bruckner, H., Jaek, P., Langer, M., and Godel, H. (1994) Liquid chromatographic determination of D-amino acids in cheese and cow milk. Implication of starter cultures, amino acid racemases and rumen microorganisms on formation, and nutritional considerations. *Amino Acids* **2**, 271–284.
6. Quintela, J. C., Pittenauer, E., Allmaier, G., Arán, V., and de Pedro, M. A. (1995) Structure of peptidoglycan from *T. thermophilus* HB8. *J. Bacteriol.* **177**, 4947–4962.
7. Kleinkauf, H. and von Dohren, H. (1990) Nonribosomal biosynthesis of peptide antibiotics. *Eur. J. Biochem.* **192**, 1–15.
8. Kreil, G. (1997) D-amino acids in animal peptides. *Annu. Rev. Biochem.* **66**, 337–345.
9. Richter, K., Egger, R., and Kreil, G. (1987) D-Alanine in the frog skin peptide dermorphin is derived from L-alanine in the precursor. *Science* **238**, 200–202.
10. Bada, J. L. (1984) In vivo racemization in mammalian proteins. *Methods Enzymol.* **106**, 98–115.
11. Wolosker, H., Blackshaw, S., and Snyder S. H. Serine racemase: a glial enzyme synthesizing D-serine to regulate glutamate-n-methyl-D-aspartate neurotransmission (1999) *Proc. Natl. Acad. Sci. USA* **96**, 13409–13414
12. Hare, P. E. and Gil-Av, E. (1979) Separation of D and L amino acids by liquid chromatography: use of chiral eluants. *Science* **204**, 1226–1228.
13. Copper, C. L., Davis, J. B., Cole, R. O., and Sepaniak, M. J. (1994) Separation of derivatized amino acid enantiomers by cyclodextrin-modified capillary electrophoresis: mechanistic and molecular modelling studies. *Electrophoresis* **15**, 785–792.
14. Fukushima, T., Kato, M., Santa, T., and Imai, K. (1995) Enantiomeric separation and sensitive determination of D,L-amino acids derivatized with fluorogenic benzofuran reagents on Pirkle type stationary phases. *Biomed. Chrom.* **9**, 10–17.
15. Allenmark, S. G. (ed.) (1988) *Chromatographic Enantio-Separation: Methods and Applications.* Ellis Horwood, Chichester.

16. Hutt, L. D., Glavin, D. P., Bada, J. L., and Mathies, R. A. (1999) Microfabricated capillary electrophoresis amino acid chirality analyzer for extraterrestrial exploration. *Anal. Chem.* **71**, 4000–4006.
17. Bruckner, H., Haasmann, S., Langer, M., Westhauser, T., Wittner, R., and Godel, H. (1994) Liquid chromatographic determination of D and L-amino acids by derivatization with o-phtaldialdehyde and chiral thiols. *J. Chromatogr.* **666**, 259–273.
18. Einarrson, S., Josefsson, B., Moller, P., and Sanchez, D. (1983) Separation of amino acid enantiomers and chiral amines using precolumn derivatization with 1-(9-fluorenyl)ethyl-chloroformate and reversed phase liquid chromatography. *Anal. Chem.* **59**, 1191–1195.
19. Marfey, P. (1984) Determination of D-amino acids. II. Use of a bifunctional reagent, 1,5-difluoro-2,4-dinitrobenzene. *Carlsberg Res. Commun.* **49**, 591–596.
20. Scaloni, A., Simmaco, M., and Bossa, F. (1995) D-L amino acid analysis using automated precolumn derivatization with 1-fluoro-2,4-dinitrophenyl-5-L-alanine amide. *Amino Acids* **8**, 305–313.
21. Scaloni, A., Simmaco, M., and Bossa, F. (1991) Determination of the chirality of amino acid residues in the course of subtractive Edman degradation of peptides. *Anal. Biochem.* **197**, 305–310.
22. Toriba, A., Adzuma, K., Santa, T., and Imai, K. (2000) Development of an amino acid sequence and D/L-configuration determination method of peptide with new fluorescence Edman reagent, 7-methylthio-4-(2,1,3-benzoxadiazolyl) isothiocyanate. *Anal. Chem.* **72**, 732–739.
23. Weiss, M., Manneberg, M., Juranville, J. F., Lahm, H. W., and Fountoulakis, M. (1998) Effect of the hydrolysis method on the determination of the amino acid composition of proteins. *J. Chromatogr.* A **795**, 263–275.
24. Mignogna, G., Simmaco, M., Kreil, G., and Barra, D. (1993) Antibacterial and haemolytic peptides containing D-alloisoleucine from the skin of *Bombina variegata*. *EMBO J.* **12**, 4829–4832.

16

Validation of Amino Acid Analysis Methods

Andrew J. Reason

1. Introduction (*see* Notes 1–4)

This chapter presents a discussion of the points to consider during the validation of analytical methods and more specifically amino acid analysis.

Amino acid analyses are utilised in various areas of research for free amino acids (in, for example, foodstuffs) and for analysis of products and components of interest comprising peptides, polypeptides and peptide/protein conjugates. In this Chapter I have concentrated on the validation of amino acid analyses that are to be submitted in a package to obtain a regulatory authority licence for a peptide, protein, or conjugate product. In such studies, amino acid analyses can be used to provide various pieces of information generally required for biotechnological and/or biological products. For example quantitative amino acid analysis can be used to determine protein quality and quantity. Such data, combined with Optical Density (OD) measurements, can be used to determine the extinction coefficient for a protein.

Analyses that are to be submitted in a package to obtain a regulatory authority licence for a particular product should be validated in accordance with the International Conference on Harmonisation of Technical Requirements for Registration of Pharmaceuticals for Human Use (ICH) Harmonised Tripartite Guidelines "Validation of Analytical Procedures: Definitions and Terminology" *(1)* and "Validation of Analytical Procedures: Methodology" *(2)* except where there are specific issues for unique tests used for analyzing biotechnological and biological products *(3)*. The terms and definitions within these documents are meant to bridge the differences that often exist between the regulators in Europe, Japan and USA.

The validation criteria discussed below can be assessed using both precolumn and postcolumn amino acid analysis derivatization methods. The available techniques are described in detail in other chapters of this book (10–15).

From: *Methods in Molecular Biology, vol. 211: Protein Sequencing Protocols, 2nd ed.*
Edited by: B. J. Smith © Humana Press Inc., Totowa, NJ

Requirements of speed and sensitivity have led M-Scan to utilize a method involving acid hydrolysis followed by precolumn derivatization using phenylisothiocyanate and separation by reverse-phase high-performance liquid chromatography (RP-HPLC) for amino acid analysis. This method is used as a background to the discussion of validation of amino acid analysis methods.

2. Materials

2.1. Apparatus (see Note 5)

2.1.1. Equipment for Hydrolysis of Proteins and Precolumn Derivatization of the Released Amino Acids

1. An RP-HPLC system equipped with a Supelcosil LC-18-DB (250 × 4.6 mm, 5 µm; Supelco, Poole, UK) HPLC column, and capable of tertiary solvent/buffer supply, gradient control, autoinjection, UV detection at 254 nm and data acquisition (*see* **Note 6**).
2. Vacuum hydrolysis tubes (Pierce, Rockford, IL) 5 mL internal volume.
3. Vacuum pump system capable of evacuating hydrolysis tubes to 1–2 Torr.
4. Oven or dry heating block capable of heating vacuum hydrolysis tubes to hydrolysis temperature (usually 110°C).

2.2. Chemicals (see Note 5)

2.2.1. Chemicals Required for Hydrolysis of Proteins and for Precolumn Derivatization of the Released Amino Acids

1. Constant boiling 6 *M* HCl (Pierce).
2. Coupling buffer, acetonitrile:pyridine:water:triethylamine (10:5:3:2 v/v).
3. Phenylisothiocyanate (Pierce).
4. Amino Acid Standard H (Pierce), containing a mixture of 17 amino acids at 2.5 m*M* (cystine at 1.25 *M*) in 0.1 *M* HCl (*see* **Note 7**).
5. Internal standard, L-Norleucine (Pierce).

2.2.2. Chemicals Required for RP-HPLC Separation of Derivatized Amino Acids

1. Solvent A, Ultra High Quality (UHQ) water (1 L).
2. Solvent B, Acetonitrile (HPLC grade):UHQ water (80:20 v/v), 1 L.
3. Buffer C, Weigh out 28.7g of anhydrous sodium acetate and dissolve in 500 mL of UHQ water ensuring that all the solid dissolves. Add 1.25 mL of triethylamine and adjust the pH of the buffer to 6.4 using glacial acetic acid.
4. Sample loading solvent, Solvent A:Solvent B:Buffer C (75:5:20 v/v).

3. Methods

3.1. Preparation of Amino Acid Standard Mixture

1. Weigh out a known amount of norleucine internal standard (~1 mg).

2. Dissolve the weighed internal standard (Norleucine) in 5% (v/v) acetic acid in water to generate a 1 nanomole/µL solution. This solution is stable for up to 12 mo at –20°C ± 5°C and can be used as the internal standard bulk solution.
3. An aliquot (50 µL) of the resulting solution, i.e., 50 nmoles is added to each sample being prepared for amino acid analysis and to the tube to be used for the amino acid standard mixture.
4. Add 20 µL of Amino Acid Standard H to 50nmoles of the norleucine (internal standard) in the tube set aside for the standard mixture.
5. The standard mixture of amino acids including the internal standard (Norleucine) is derivatized in parallel with the samples to be studied (*see* **Subheading 3.3.**).

3.2. Hydrolysis (see Notes 8 and 9)

1. Aliquot the desired amount of peptide, protein or conjugate (ideally 50–100 µg) to be hydrolyzed into a clean vacuum hydrolysis tube and label the tube(s) to identify the sample.
2. Add a known amount (ideally 50 nmoles) of norleucine (internal standard) to the sample tube.
3. Lyophilize resulting sample under vacuum using a Savant or similar drying device.
4. Add 200 µL of 6 *M* (constant boiling) hydrochloric acid.
5. Evacuate the vacuum hydolysis tube(s) for 2 min with agitation.
6. Incubate samples in a heating block or oven at 110°C for 24 h.
7. Allow samples to cool and lyophilize the products.
8. Derivatize released amino acids with phenylisothiocyanate.

3.3. Derivatization of Released Amino Acids and Amino Acid Standard Mixture

1. Add 100 µL of coupling reagent to the dried hydrolyzed or amino acid standard sample (*see* **Note 10**).
2. Lyophilize (*see* **Note 10**).
3. Add a further 100 µL of coupling reagent and 5 µL of phenylisothiocyanate (PITC) to the dried sample.
4. Incubate the mixture at room temperature for 5 min.
5. Lyophilize the products.
7. Add 100 µL UHQ water to the dried products (*see* **Note 11**).
8. Lyophilize (*see* **Note 11**).
9. Resuspend products in 200 µL of sample loading buffer and analyze an aliquot by RP-HPLC as described in **Subheading 3.4.**

3.4. RP-HPLC Separation of Derivatized Amino Acids

1. Connect the flow to the Supelcosil LC-18-DB HPLC column within the column heater compartment equilibrated at 45°C.
2. Wash the column with solvent B until a stable baseline is observed (*see* **Note 12**).
3. Load the amino acid analysis gradient parameters and allow the system to equilibrate at the initial conditions (*see* **Table 1** and **Note 13**).

Table 1
Reversed Phase-High Performance Liquid Chromatography
Conditions for Separation of PTC-Derivatized Amino Acids

Time (min)	% Solvent A	% Solvent B	% Solvent C	Flow rate (mL/min)
0	75	5	20	1.0
25	30	50	20	1.0
28	20	80	0	1.0
33	20	80	0	1.0
34	75	5	20	1.0

4. Once the baseline has stabilized inject 10 μL of the sample loading buffer and run the gradient shown in **Table 1**.
5. Visually examine the chromatogram obtained and ensure that there is no significant UV absorption in the region of interest (between 5 and 25 min following injection; *see* **Fig. 1**). If the chromatogram contains no significant absorbance in the region of interest proceed to **step 6**. If UV peaks are observed the column should be cleaned or a new column used and a second aliquot of sample loading buffer should be analyzed.
6. Inject sequentially six separate 10 μL aliquots of the derivatized amino acid standard mixture.
7. Examine the resulting chromatograms for peak area variation, elution time variation, peak resolution, and tailing to determine that the system is suitable for use. Peak resolution should be calculated using the equation:

$$R = \frac{2(t_2 - t_1)}{W_2 + W_1}$$

Where, t_2 and t_1 are the retention times of the two closest eluting components and, W_2 and W_1 are the corresponding widths at the bases of the peaks obtained by extrapolating the relatively straight sides of the peaks to the baseline (*see* **Fig. 2**). Peak tailing should be calculated using the equation:

$$T = \frac{W_{0.05}}{2f}$$

Where $W_{0.05}$ is the width of the peak at 5% height and, f is the distance from the peak front to the peak center at 5% peak height (*see* **Fig. 3**). If the system is suitable for use (*see* **Note 14**) analysis of samples may continue. If the data do not pass the set criteria, a test regime must be established to determine the reason for the failure (*see* **Note 15**).
8. Inject sequentially duplicate aliquots (10 μL) of the samples of interest.
9. Following each run sample peaks should be integrated taking into account the internal standard. The response ratio for each amino acid in each sample run is

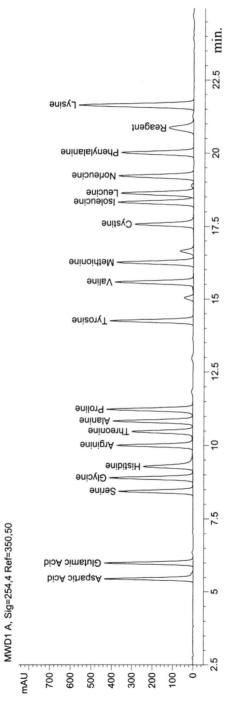

Fig. 1. UV chromatogram (254 nm) obtained from analysis of a standard mixture of derivatized amino acids.

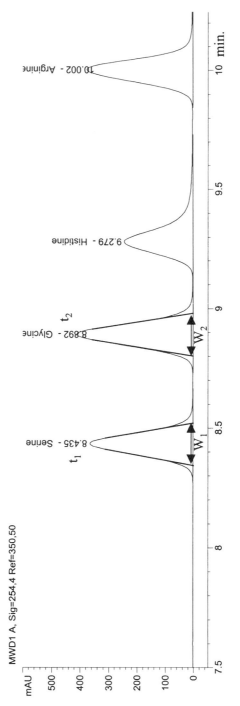

Fig. 2. Peak resolution measurement for derivatized serine.

calculated by dividing the peak area for each amino acid by the peak area obtained for the internal standard (norleucine) in the same run.

The response ratio is in turn divided by the response ratio for the same amino acid in the amino acid standard mixture. The figure generated is multiplied by the known amount of the relative amino acid present in the standard mixture (i.e., 50 nmoles) to produce the molar amount of that amino acid in the sample. This may be summarized in the equations,

$$R_r = R_s/R_{is} \text{ (in the sample data)}$$
$$R_m = R_s/R_{is} \text{ (in the standard mixture data)}$$
$$R_c = R_r/R_m$$
$$\text{nmoles of amino acid} = R_c \times 50$$

Where R_r is the response ratio for a particular amino acid in the sample or standard run of interest having a peak area R_s and an internal standard peak area (in the same run) R_{is} and, R_m is the response ratio for the same amino acid in the relevant standard mixture having a peak area R_s and an internal standard peak area (in the same run) R_{is}. and, R_c is the overall response ratio.

3.5. Validation Parameters (see Note 16)

The validation parameters that should be assessed during a full validation to ICH guidelines are outlined in **Subheadings 3.5.1.–3.5.9.** Samples should be prepared as outlined in **Subheadings 3.1.–3.4.** or using another suitable amino acid analysis procedure.

3.5.1. Specificity

Specificity is defined as the ability to assess unequivocally the analyte(s) in the presence of other components.

A suitable study should be designed to establish retention times and responses of known amino acids released from the study sample and the data obtained should be compared to that obtained from a standard mixture containing derivatized amino acids.

Generally, a standard aliquot of the peptide, protein, or conjugate for which the validation is being carried out is hydrolyzed and analyzed in duplicate and the data generated is compared with the data obtained from derivatized phenylthiocarbamyl (PTC)-amino acid standards.

The retention time of each of the PTC-amino acids released from the peptide, protein, or conjugate sample should normally be within ±3% of the same PTC-amino acid present in the standard mixture analyzed in the same session.

3.5.2. System Suitability Test (see Note 17)

System suitability is a test designed to ascertain the effectiveness of the operating system.

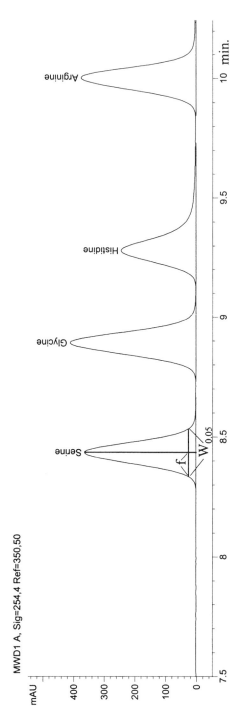

Fig. 3. Peak tailing measurement for derivatized serine.

System suitability can be determined in a number of ways. Generally a series of replicate injections (n = 6) of the derivatized amino acid standards is performed to assess the relative standard deviations for peak area measurements and retention times. In addition, peak tailing (T) and peak resolution (R) can be determined (*see* **Subheading 3.3.**).

The retention time variance should normally be within 3% and the relative standard deviation of peak areas taken from the six sets of data should normally be less than 5%. The peak tailing factor (T) should be less than or equal to two and the resolution factor (R) should be greater than one for the amino acids of interest.

3.5.3. Linearity

Linearity is defined as the ability (within a given range) of the analytical method to obtain results that are directly proportional to the concentration of analyte(s) in the sample.

Linearity may be demonstrated on the peptide, protein, or conjugate and standard mixture of amino acids via dilution of a standard stock solution or on separate weighings out of the peptide, protein, or conjugate.

Linearity of response of the amino acid standard mixture can be demonstrated by choosing at least 5 points in duplicate in a range to cover that anticipated for the peptide, protein, or conjugate.

This study should be repeated with aliquots of the peptide, protein, or conjugate (5 points in duplicate; 0, 50, 75, 100, and 150% of the usual aliquot analyzed).

Linearity should initially be determined using a visual inspection of a plot of peak area ratio (area of amino acid signal over the area of internal standard signal) against relative amount of the peptide, protein, or conjugate taken. If a linear relationship exists the plot obtained should be statistically evaluated. For example the sum of least squares around the regression line (r^2) should be calculated.

From the final plot, the correlation coefficient, y-intercept, slope of the regression line, and residual sum of squares should be determined and entered in the validation document.

The linearity data obtained should obey the equation $y = mx + C$, where C is zero within 95% confidence limits, and the sum of least squares around the regression line (r^2) is greater than 0.98.

3.5.4. Range

The range is derived from the linearity study and is established by confirming that the analytical procedure provides an acceptable degree of linearity, accuracy, and precision when applied to samples containing amounts of peptide, protein, or conjugate within or at the extremes of the specified range of the analytical procedure.

Following the linearity procedure described above, the working range for the analysis would be 50–150% of the concentration of the batch used for the analysis.

If the working range required is likely to be wider, additional concentrations may be added to the linearity study to take into account the future expected concentration range.

3.5.5. Accuracy/Recovery

An assay method should provide a true (accurate) result for a test peptide, protein, or conjugate sample.

Accuracy should be established across the specified range of the analytical procedure and can often be inferred once precision, linearity, and specificity have been established.

Generally percentage recoveries for the amino acids of interest should fall within the range of 90–110% and the percentage correlation of variance (%CV; standard deviation $\times$ 100 / mean) for multiple measurements should normally be less than 10%.

3.5.6. Repeatability

Repeatability is measured by carrying out analysis (in six replicates) in one laboratory by one operator, using one instrument over a relatively short time span.

A standard aliquot of the peptide, protein or conjugate is hydrolyzed and the derivatized products are injected six times onto the RP-HPLC system.

In general, the percentage correlation of variance (%CV) for the released amino acids should be less than or equal to 5%.

3.5.7. Intermediate Precision

Intermediate Precision is defined as long term variability of the measurement process, which may be determined for a method run on different days and or by different operators.

Six separate aliquots of the peptide, protein, or conjugate should be hydrolyzed, derivatized and analyzed by RP-HPLC on different days. The analyses should be carried out using different operators and different instruments in the same facility (*see* **Note 18**).

In general, the percentage correlation of variance (%CV) for the released amino acids should be less than or equal to 10%.

3.5.8. Reproducibility

Reproducibility expresses the precision between laboratories and can be assessed using an extension of the intermediate precision study described in **Subheading 3.5.7.** to include analyses at different laboratories.

Ideally, the percentage correlation of variance (%CV) for the released amino acids should be less than or equal to 10%.

3.5.9. Robustness/Ruggedness

Robustness/ruggedness is defined as a review of the critical parameters in the method and the steps taken to minimise method variability.

Generally experiments should be designed to assess the following criteria:

1. Stability of analytical solutions and derivatized products.
2. Influence of variations of pH in a mobile phase.
3. Influence of variations in mobile-phase composition.
4. Different columns (different lots and/or suppliers).
5. Temperature.
6. Flow rate.

For most validation studies it is impractical to carry out analyses to study all the factors previously outlined. The use of experimental matrix software can often be used to design a set of experiments to test the robustness/ruggeddness of the method being validated. Acceptance criteria are set prior to the start of the analyses and should adhere to the criteria set for linearity and percentage correlation of variance (%CV) described earlier. Unfortunately there are no actual in-depth guidelines to aid us in designing robustness/ruggeddness analyses.

3.5.10. Detection and Quantification Limit (i.e., Minimum Amount of Amino Acid Needed for Reliable Detection and Quantification)

Several approaches for determining the detection limit are possible *(2)*. However, during a validation of amino acid analysis detection limits for each amino acid can be most easily determined based on the standard deviation of the response and the slope of the linearity plot.

The detection limit (DL) may be expressed as:

$$DL = \frac{3.3\alpha}{S}$$

where α = the standard deviation of the response about the line of best fit

S = the slope of the calibration curve

The quantification limit (QL) may be determined in the same way and is expressed as:

$$QL = \frac{10\alpha}{S}$$

where α =   the standard deviation of the response, and
        S =   the slope of the calibration curve.

4. Notes

For other methods of amino acid analysis some details may differ, according to equilibration times, column lifetime, sensitivity (e.g., the amount of standard required), and so forth, but the general principles of validation outlined
above apply. Ideally, the HPLC system used should meet the basic criteria
specified in the liquid chromatography Pharmacopoeias *(4–6)*.

1. It is important to remember that the main objective of validation of an analytical
   procedure is to demonstrate that the procedure is suitable for its intended purpose.
2. All of the analytical data collected during a validation and any calculations made
   should be discussed in the resulting validation document as appropriate.
3. Well-characterized reference materials, with documented purity, should be used
   throughout a validation study.
4. The experimental work should be designed so that the appropriate validation characteristics can be considered simultaneously to provide a sound, overall knowledge of the capabilities of the analytical procedure.
5. Analyst information, performance validation, and/or calibration certificates for
   the equipment used should also be recorded and stored along with the raw data.
   All reagents should be tested for content and suitability for use prior to any validation study that is to be included in a regulatory authority application. For regulatory submissions, this includes records of supplier, catalog number, and lot
   number of reagents used.
6. A Hewlett-Packard HP 1050, HP 1090, or HP 1100 with quaternary switching pump
   is ideal. Reagents of the highest quality should be used during these analyses.
7. If validation is being carried out for a protein or amino acid mixture that contains unusual amino acids or amino acids that are not present in the standard
   mixture, the unusual amino acid should be purchased, accurately weighed, and
   added to the standard mixture. If the amino acid is not commercially available,
   the amino acid can either be synthesized, rigorously characterized, accurately
   weighed, and added to the standard mixture or quantitated relative to a similar
   amino acid.
8. Ideally prior to hydrolysis peptide, protein and conjugate samples should be salt-,
   amine-, and detergent-free, because these may interfere with hydrolysis or subsequent chromatography.
9. Inter-amino acid bonds differ in their susceptibility to acid hydrolysis. For
   example, longer acid hydrolyses are more likely to yield accurate results for
   hydrophobic amino acid residues, such as Valine, Isoleucine, and Phenylalanine. Other factors also have to be taken into account when utilizing amino acid
   analysis data. Asparagine and Glutamine are converted to their corresponding
   acids, Aspartic acid and Glutamic acid, and therefore cannot be directly quantified. Losses of Serine, Threonine, and to an extent Tyrosine, can be expected

during acid hydrolysis through hydrolytic breakdown. More accurate estimates of Serine and Threonine can often be determined either by assuming 10% and 5% losses, respectively, over 24 h at 110°C, or by performing a series of hydrolyses for various times and extrapolating the amounts of these amino acids detected back to zero time. Determination of Cysteine is also problematic when using standard hydrolysis conditions, due to oxidation. Tryptophan is completely destroyed during normal hydrolyses and is not detected. Also, the quantitation of Aspartic and Glutamic acid, in precolumn derivatization methods, can also be compromised by the presence of buffer salts. Proteins are routinely hydrolyzed at 110°C for 24 h. Higher recoveries of more hydrophobic amino acids from membrane proteins can be achieved using higher-temperature hydrolyses for example 160°C for 6 h. This matter is discussed further in Chapter 10.

10. Addition of coupling buffer and subsequent lyophilization ensures neutralization of the hydrolysed sample and the amino acid standard mixture.

11. The addition of UHQ water to the derivatized amino acids and subsequent lyophilization ensures optimal removal of derivatizing reagents and prevents variable chromatography.

12. The column routinely needs to be flushed for 30 min with solvent B to ensure that the column and UV lamp have equilibrated.

13. Equilibration at the initial conditions for amino acid analysis normally requires 5–10 min.

14. The system can be considered suitable if the retention time variance is ±3.0% and the relative standard deviation in peak areas for the six runs is < 5%. Peak tailing (T) should be less than or equal to two and Resolution (R) should be greater than one.

15. If the system fails a suitability test, the liquid chromatography system being used including hardware (pump, UV detector, switching valves, column, etc.) and solvents should be fully examined, for instance for leaks or partial blockages. Columns have a finite lifetime. For instance the type of RP-HPLC column used in the system described in **Subheading 2.1.1.** would normally be expected to allow 150–300 sample injections before peak broadening and loss of resolution result in excessively poor performance.

16. Validation of amino acid analysis for a particular sample should only be contemplated following optimization of the method or methods to be employed. In general the optimized method, the sample formulation and, of course, amino acid sequence for each peptide, protein, and conjugate will differ. Therefore, amino acid analysis should be validated for each product rather than using a generic validated method for analysis of various products.

17. Following validation the system suitability test should ideally be performed prior to each analysis of peptide, protein, or conjugate, providing that the data obtained passes the criteria set the system is deemed to be suitable for analysis.

18. Precision across different instruments need only be assessed if more than one instrument is to be used for analysis of the sample of interest.

References

1. The European Agency for the Evaluation of Medicinal Products Human Medicines Evaluation Unit ICH Topic Q2A Validation of Analytical Methods: Definitions and Terminology (CPMP/ICH/381/95) (1995).
2. The European Agency for the Evaluation of Medicinal Products Human Medicines Evaluation Unit ICH Topic Q2B Validation of Analytical Procedures: Methodology (CPMP/ICH/281/95) (1997).
3. The European Agency for the Evaluation of Medicinal Products Human Medicines Evaluation Unit ICH Topic Q6B Specifications: Test Procedures and Acceptance Criteria for Biotechnological/Biological Products (CPMP/ICH/365/96) (1999).
4. High Pressure Liquid Chromatography USP (1995) Ninth Supplement, USP-NF Section 621 pp. 4648–1805.
5. Liquid Chromatography BP (1993) A100 Appendix IIIc – A102 Appendix IIIe.
6. Liquid Chromatography Ph. Eur (1997) 2.2.29, pp. 32–34.

17

Electrospray Mass Spectrometry of Peptides, Proteins, and Glycoproteins

Fiona M. Greer and Howard R. Morris

1. Introduction

Mass spectrometry (MS) has now become the method of choice for identifying post-translational modifications in proteins, for protein mapping and identification studies, and for peptide and protein sequence analysis. In the early 1980s, two major advances in mass spectrometry combined to allow its more widespread application to protein analysis. These were, first, the development of high field magnet instrumentation *(1)* in the mid-1970s, allowing routine high-mass applications for the first time and, subsequently, the introduction of the fast atom bombardment (FAB) technique in 1981 *(2)*. FAB-MS enabled previously intractable, thermally labile, and highly polar compounds such as peptides, to be analyzed with minimal sample preparation and no chemical derivatisation. Since then, both methods and instrumentation for characterizing protein and carbohydrate structure by MS have advanced rapidly. In particular, these developments were complemented in the late 1980s with the introduction of another "soft ionization" method, Electrospray ionization (ES-MS) *(3,4)*, allowing both the analysis of very large molecules on instruments of conventional mass range (due to multiple charge states), and importantly, on-line liquid chromatography mass spectrometry (LL/ES-MS) analysis of purified proteins or peptide digest mixtures.

Although FAB-MS was used extensively on high field magnet instruments in the 1980s and early 1990s to solve real sequencing problems and to "map" and identify proteins and their post-translational modifications *(5)*, in recent years the majority of applications have centered on ES (sometimes called ionspray) ionization studies on quadrupole or tandem mass spectrometers. Whatever the ionization technique chosen, the strength of MS techniques in

From: *Methods in Molecular Biology, vol. 211: Protein Sequencing Protocols, 2nd ed.*
Edited by: B. J. Smith © Humana Press Inc., Totowa, NJ

general is the ability to detect any changes in a peptide or protein structure caused, for example, by insertion, deletion or substitution of single amino acids. Indeed, any structural alteration that changes the mass of the molecule can not only be detected, but also identified. These changes can include the addition of blocking groups at the amino (N-) or carboxy (C-) terminus, oxidation, deamidation, carboxylation (such as Gla), disulphide bridge formation; or nonprotein modifications including phosphorylation, glycosylation, sulphation, and so on.

1.1. Principle of ES-MS

ES-MS is a method by which a stream of liquid containing the sample is injected via a capillary into the atmospheric pressure ion source of a mass spectrometer. A spray of microdroplets is produced using a potential difference between the capillary and a counter-electrode; this flows through a series of "skimmer" electrodes, encountering a drying gas in some designs. The net effect is the creation of multiply charged molecular species, devoid of solvent and ready for analysis either by a magnetic sector, or more commonly, a less expensive quadrupole analyzer. Large molecules such as proteins and glycoproteins tend to form a distribution of multiple charge states and, because the mass spectrometer measures mass to charge ratios (m/z), rather than mass, even molecules exceeding 100kDa are amenable to analysis on a 3000 mass range quadrupole. ES-MS is particularly valuable as a method of observing whether a protein is post-translationally modified by observation of any mass difference between the observed signal and that calculated from the sum of the amino acids present in the sequence.

Electrospray provides a powerful method for examining solutions of peptides or glycopeptides, which should nevertheless be as pure as possible when using direct injection analysis (as opposed to LC-MS), as deconvolution and assignment of charge states is complex and sometimes impossible in mixture analysis (*see* **Note 1**). Mixtures are best analyzed by on-line LC/ES-MS, which provides a method for concomitant purification and mass analysis of protein/ glycoprotein digest components. ES-MS gives reported sensitivities in the low fmol-1 nmol range.

2. Materials

2.1. Equipment

2.1.1. Mass Spectrometers

The mass spectrometer may be viewed as a three-component system: an ion source, an analyzer, and a detector. The detector technology, except for special high-sensitivity systems such as focal plane array *(6)*, is not usually the limiting or defining feature of the instrument. In contrast, the analyzer defines the quality of the mass spectrometric data based on mass accuracy, resolution, mass range, and discrimination effects. As already mentioned, the majority of protein/glycoprotein MS applications have utilized either magnetic sector or quadrupole analyzer mass spectrometers. Double-focusing magnetic-sector instruments of the high field magnet type normally have an upper mass range of 3000–15,000 Dalton and possess the capability to produce high-quality mass data with routine mass accuracy of better then 0.3 Dalton and isotopically resolved signals, if required. These features are important, particularly in circumstances in which 1 Dalton mass differences must be assigned, e.g., differentiation between acids and amides or sugar composition masses, e.g., NeuAc and 2xFuc. The mass range is entirely appropriate and adequate for FAB-MS analysis of permethylated carbohydrates released from glycoproteins either enzymatically (by PNGase F for example) or by reductive elimination, and this method still has some advantages over matrix-assisted laser desorption ionization (MALDI) analysis due to the production of highly informative nonreducing end "A" type ions in FAB ionization *(7,8)*. Quadrupole analyzers give poorer resolution, which can affect mass accuracies. Mass discrimination effects normally restrict their use to m/z 4000 or less. These instruments are more commonly utilized for the study of multiply charged ions in electrospray ionization, which falls into the quadrupoles' effective mass range (e.g., a 20kDa protein carrying 15 charges would be observed at m/z 1333.3).

Time-of-flight (TOF) analyzers, while having excellent transmission sensitivity, have relatively poor resolution unless they incorporate a reflectron focusing device. Ionization energy-spread problems, together with metastable decomposition, can also lead to poor mass accuracy in some simpler designs. However, a proper combination of a quadrupole analyzer followed by a collision cell followed by orthogonal injection of ions into a reflectron time-of-flight analyzer, a Q-TOF mass spectrometer *(9)*, gives an instrument with excellent resolution, mass accuracy, and sensitivity, and this device is currently one of the most powerful tandem MS/MS instruments available for proteomics and glycobiology research *(8)*.

2.1.2. On-Line LC/ES-MS

An HPLC system is chosen (analytical, microbore or nanoflow, depending on the problem under study) to allow gradient elution of biopolymer digests with flow rates of nanoliters-1000 µL/min. To allow further study and manipulation of biologically important samples, a stream-splitting device can be introduced between the LC UV detector and the Electrospray introduction capillary, allowing fraction collection where appropriate.

2.2. Reagents

Reagents used should be the highest quality available and should be free from contaminants such as salts. Solvents should be high-performance chromatography (HPLC) grade.

2.2.1. For ES-MS and LC/ES-MS

1. Acetic acid.
2. Trifluoroacetic acid (TFA) or formic acid.
3. Triethylamine.
4. Methanol.
5. Acetonitrile.
6. Isopropanol.
7. Methoxyethanol.
8. Calibration compounds: e.g., horse heart myoglobin (Cat. no. M-1882, Sigma, St. Louis, MO), Glu Fibrinopeptide (Sigma F-3261), Caesium Iodide (Aldrich 20,213-4). or polypropylene glycol (Aldrich various) as appropriate.

3. Methods

3.1. Sample Application in Electrospray

The following schemes are used routinely with triple-quadrupole instruments and hybrid tandem instruments of the Q-TOF type.

3.1.1. Direct Injection

1. Dissolve the sample in a minimum volume of an appropriate solvent, preferably 5% (v/v) aqueous acetic acid or 0.1–0.2% formic acid for work in the positive ion mode, or a mild base such as ammonium acetate or triethylamine for negative ionization studies.
2. Dilute the sample with organic solvent, usually methanol or acetonitrile, and with the carrier solution (*see* **step 3**) where necessary to provide an approx 50% (v/v) organic solution and a final sample concentration of 1–100 pmol/µL.
3. Inject 10 µL of sample solution from **step 2** into a flow of approx 5 µL/min of carrier solution made up from 0.1% (v/v) aqueous formic acid: acetonitrile (1:1, v/v) for positive ion work, or 0.1% aqueous triethylamine (or 10 mM aqueous ammonium acetate):acetonitrile (1:1, v/v) for negative ion work (*see* **Note 2**).
4. The mass spectrometer should be tuned, calibrated, and ready to acquire data through its data system as specified in the manufacturer's manual.
5. Collect data over the lifetime of the sample; approx 2 min, under the aforementioned conditions measured from the first appearance of the spectrum.

3.1.2. Nanospray

For proteomics studies including in-gel tryptic digest samples from one-dimensional (1D) or two-dimensional (2D) gels, the gel extract containing the peptide/glycopeptide mixture should be pre-cleaned by elution from a C-18 microcartridge in 10–20 microliters of 30% or 60% acetonitrile in 0.1% TFA or 0.1% formic acid. Some ES sources are particularly susceptible to suppression by TFA, and hence on those, formic acid will be preferable. Beware, however, of the possibility of peptide formylation over prolonged contact periods. One to two microliters of solution is then loaded into a nanospray capillary, and the end "primed" by bending or stubbing allowing the spray to be produced. Typically this volume of sample will last for 15–30 min, allowing good MS acquisition and several MS/MS analyses at high sensitivity, particularly on Q-TOF type instruments.

3.1.3. On-Line LC/ES-MS

1. Dissolve the sample or digest in its own best solvent, 5% (v/v) aqueous acetic acid, 0.1% (v/v) TFA, or 0.1% formic acid for analysis in the positive ion mode; centrifuge and inject onto a reverse-phase HPLC column of appropriate dimensions (nanoflow, microbore or analytical). For digest mapping and/or removal of detergent, a gradient elution from the LC should be chosen (*see* **Note 3**).
2. If a TFA containing HPLC solvent system is used, then the suppressive effect of TFA on Electrospray ionization may be mitigated by the introduction of a 1:1 isopropanol:2-methoxyethanol solution between the UV detector and the stream splitter if present. In a typical microbore (2-mm column) system, an LC flow rate of 50-µL/min should be used, with an addition of 50-µL/min anti-TFA suppression solvent, with a 1:10 (MS:collection) stream splitter for ES sources limited to introduction rates of 10-µL/min.
3. The mass spectrometer should be tuned, calibrated, and ready to acquire data through its data system as specified in the manufacturers manual. Data are collected utilizing 5 or 10 s scans or acquisitions over the period of the HPLC elution profile, normally 60–90 min.
4. Sophisticated MS/MS type instruments will allow on-line MS/MS data to be collected, including the possibility of data-dependent switching from MS to MS/MS mode when peaks above a certain threshhold are encountered in the LC-MS trace. Of course, the acquisition time for data collection is limited in on-line experiments, even where stop-flow techniques are available, and data may be compromised compared to nanospray experiments.

3.2. Interpretation

3.2.1. Calculation of Molecular Weight

The spectrum obtained from a sample studied by one of the soft ionization methods will contain minor background signals at each mass owing to chemical noise and fragment ions, together with a major signal "cluster" for the singly charged quasimolecular ion, $[M+H]^+$ and/or multiply charged ions.

The cluster is caused by the existence of isotopes. For instance, ^{13}C exists at a natural abundance of 1.1% per carbon atom, therefore for every **n** Carbon atoms in a molecule there will be an **n** $\times$ 1.1% relative intensity for the ^{13}C isotope signal at $(M+1+H)^+$. The relative heights of the different peaks in the cluster will reflect this principle for the naturally occurring isotopes in any sample. The instrument resolution parameters may be set to resolve these individual isotope signals in a cluster, but at higher mass, the desire for maximum sensitivity may lead to a reduction in resolving power, causing unresolved signals which are seen as broad gaussian-like peaks.

It is important then to know whether we are dealing with accurate or average masses.

The molecular weight of a peptide can be calculated by adding together the masses of the appropriate amino acid residues (*see* **Appendix 1**) plus the masses of the terminating ("capping") groups. Normally, this would involve the addition of H at the N-terminus and OH at the C-terminus. The same principle applies to the addition of sugar residues or other post-translational modifications. The three methods of calculating mass are:

1. Nominal mass: A molecular weight calculated from nominal, integer masses i.e. C = 12, O = 16,and so on. For example, ARG = 156, LEU = 113 as amino acid residue masses. This is only useful as a rough guide to the expected peptide masses observed in an MS experiment.

2. Accurate mass: The exact monoisotopic mass values of the elements, i.e., O=15.9949146, yield a mol wt of so-called "accurate" mass for use when the individual isotopic peaks in a cluster can be resolved and identified allowing assignment of the ^{12}C isotopic mass. This may prove difficult for large peptides or glycopeptides (>3000Dalton) where high resolutions may be needed and the ^{12}C isotope represents only a small proportion of the overall distribution.

3. Average mass: For large peptides or glycopeptides, the average mass value, incorporating a weighting for the intensities of all the elemental isotopes added together, gives an average chemical molecular weight corresponding to the centroid of the total unresolved isotopic distribution.

3.2.2. Interpretation of ES-MS Spectra

A typical ES-MS spectrum will consist of a distribution of multiply-charged ions over the scanning mass range of approx 500–2,500 Dalton. The differing number of charges corresponds to the distribution of a differing number of protons added to basic sites in the molecule, such as the amino groups. (Note: Carboxylate anions will be protonated in the positive ion mode, rendering them neutral.) The maximum point in the distribution normally corresponds to something less than the total predicted number of basis sites, summing the amino-terminus plus LYS, ARG, and HIS on a protein molecule. With glycopeptides and glycoproteins the carbohydrate portion will also be protonated, increasing the overall charge state distribution of the molecule. The appearance of the raw data for Myoglobin, a commonly used calibration compound is illustrated in **Fig. 1A**. The data show a distribution of signals carrying varying numbers of net positive charges (A10-21) on protonable basic sites in the molecule (multiply charged ions). Because the series of signals are all separated by just one proton, each signal is a data point for the protein molecular weight and the presence of a number of data points as shown in **Fig. 1A** will normally yield a measured mass with a low standard deviation (SD) (accuracy of +/- 0.01%). The peak top data shown represent the average chemical mass (there is insufficient resolution in the quadrupole to see isotopic contributions). These charge distribution data can be "transformed" using a simple computer algorithm provided with most instruments, to produce a more normal MS profile of intensity against mass as shown in **Fig 1B**. Such a transformed ES-MS spectrum can be used to "diagnose" glycosylation or other post-translational modifications and to give some indication of the glycoforms present, since the increased mass difference compared to the protein mass can be correlated with sugar composition *(8)*.

3.3. Mass Mapping

A number of years ago *(10)*, it was realized that the determination of the molecular weights of peptides generated from proteolytic or chemical digestion of a protein (*see* Chapters 5 and 6) would constitute an excellent proof of structure or identity replacing the necessity for sequence analysis in many applications. The concept of "mixture analysis" of peptides by mass spectrometry had been introduced and proven some years earlier *(11)*, and the "MAPPING" of protein and glycoprotein structures by soft ionization methods such as FAB, MALDI and Electrospray has now become an important strategy for protein identification and post-translational modification analysis. Not every component of a digest will be necessarily observed, depending on the complexity and suppression owing to matrix effects and depending on the hydrophobicity of component peptides (*see* **Note 4**), or the mass range scanned.

1. The protein/glycoprotein is first normally reduced and alkylated (*see* Chapter 27) to protect thiols (unless disulfide bridge analysis is required).
2. A suitable digestion strategy is chosen, e.g., trypsin or cyanogen bromide, for good specificity.
3. The sample digest is then analyzed either by direct ES or on-line LC/ES-MS in mapping experiments. The mass data acquired are then screened using simple computer programs to allow assignment of peptides/glycopeptides to regions of the protein structure utilizing not only mass but the specificity of the digest used. For example, one mass may map onto two regions of the protein structure, but only one of these may contain a C-terminal LYS or ARG corresponding to a tryptic cleavage site. Mapping of multiple peptide masses is such a powerful method regarding the definitive (unambiguous) nature of the many data points produced that it can be used not only to define and confirm many sequences but

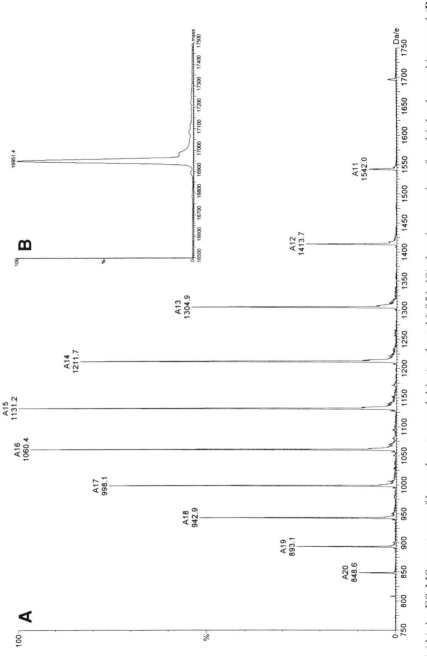

Fig. 1 (**A**) An ES-MS spectrum of horse heart myoglobin (mol-wt 16,951.48) showing a series of multiply charged ions and (**B**) the corresponding transformed spectrum with peak top mass 16,951.4.

also to detect protein impurities and assign errors, variant substitutions, or post-translational processing of proteins, including defining glycosylation, S-S bridges, phosphorylation, and so on *(5)*.

4. Glycopeptides are detected in such data either by looking for low molecular weight fragment ions formed by "cone voltage" effects, for example m/z 204 (HexNAc) and 366 (HexNAc.Hex), or by examining the higher-mass, higher charge state regions of the spectra for sugar mass differences. The carbohydrate structure of a glycoprotein is often heterogeneous and the mass difference correlation can be easily made if the charge state of principal ions is known. For example, if we are examining a fourthly charged ion in a region of the spectrum, then a Hexose mass difference would show up as a satellite ion separated by 40.5 Daltons (162/4) (*see* **Note 5**).

3.4. MS/MS

Internal energy is imparted to an ion by any ionization process. In some cases, it is insufficient to cause much fragmentation, but in others (including FAB and ES-MS), some natural inherent fragmentation is observed as a minor total of the ions created. This fragmentation, originally defined via protonation of the amide bond and concomitant fragmentation by one of two principle routes *(12)* can give rise to both N- and C-terminal structure information via the N-terminal (b) and C-terminal (y) sequence ions (*see* **Note 6**).

In situations where insufficient internal energy is imparted to molecular ions for effective fragmentation, this may be enhanced by collisional activation (collisionally activated decomposition [CAD], sometimes known as collisionally induced decomposition [CID] with an appropriate gas such as helium or argon, using a collision cell *(13)*. Briefly, in the so-called MS/MS experiment, a primary ion beam selected using the first mass analyzer (usually the quasimolecular ion $(M + H)^+$ in FAB or MALDI, or the $(M + 2H)^{2+}$ in ES-MS) is passed through a collision cell where fragment ions are produced. These fragments are then separated and analyzed by a second mass analyzer prior to detection, to produce a fragment ion mass spectrum from which the structure can be determined. Many types of MS/MS experiment exist depending on instrument type, e.g., 4 sector high-energy MS/MS on a tandem magnetic sector instrument or low-energy triple quadrupole tandem ES-MS/MS analysis. High quality MS/MS data for the most demanding high sensitivity studies, such as proteomics, is probably best achieved using the Quadrupole Orthogonal Acceleration Time of Flight or Q-TOF type instrument *(8,9,13–15)*. High sensitivity MS/MS data (low femtomole and attomole range) show isotopically resolved daughter ions (allowing easy charge-state recognition), mass accuracies of better than 0.1 Dalton (5ppm on the new Q Star and Q-TOF Ultima instruments), and excellent signal/noise characteristics as shown in **Fig. 2** for a seminal plasma Glycodelin S derived peptide sequenced as VL/IVEDDEL/IMQGFL/IR.

De novo sequencing, where deduced sequences are not corroborated in data base searches, can be achieved with confidence on Q-TOF type instruments, and the resulting data is used to generate primers for a gene-sequence approach to determining the full protein sequence *(14,15)*. In glycopeptide MS/MS, the glycosidic bonds are more readily labile at lower collision energies. A combination of low- and high-energy data will often be needed to allow not only peptide and carbohydrate sequence analysis but also identification of the site(s) of carbohydrate substitution *(8)*. This is illustrated in **Fig. 3** showing the discovery and structure determination of a novel hydroxy proline linear pentasaccharide (Hex-Hex-Fuc-Hex-HexNAc)-linked glycopeptide NDFTP(OH)EEEEQIRK determined by MS/MS analysis of the triply charged *m/z* 829.4 ion from a tryptic digest of *Dictyostelium* SKP1 protein *(13)*.

4. Notes

1. Sample Purification and Preparation. For successful analysis, samples should be free from contaminants such as detergents, salts, or substances which ionize more readily than the sample itself as these will suppress sample ionization and in many cases completely obliterate spectra. Small quantities of impurities having similar chemical properties to the sample may not pose problems. Although it is important to remove foreign impurities, it is not necessarily required to have one single biopolymer species. Mixtures of related substances are amenable to analysis; however, it should be recognized that selective ionization may occur and not all components of a mixture may be seen.

   Most of the purification techniques used by chemists or biochemists are compatible with ES-MS. Ideally the final purification step should be HPLC (using volatile elution solvents and, in particular, avoiding salt buffers). A convenient alternative to HPLC is a Sep-Pak cartridge (available from Waters) or a reverse phase microcartridge (Jones Chromotography), which are disposable cartridges containing HPLC packings. When handling small quantities of biological substances, careful and clean experimental techniques are required. Biological activity may be lost (and the sample altered chemically) on exposure to air or high temperatures and the sensitivity of the sample to routine handling procedures should always be borne in mind.

   In general, for high sensitivity work, sample dry-down should be avoided. Otherwise irreversible adsorption effects may occur. Samples should be analysed directly by LC-MS or after micro clean-up by direct injection or nanospray keeping volumes to a minimum, and thus relative concentration to a maximum.

2. Solvent systems other than the ones suggested for ES here are possible, provided they assist protonation (or deprotonation) and allow microdroplet formation at the end of the capillary injector (a function of the surface tension of the liquid as well as the voltage gradient applied). A major consideration is the solubility of the sample in the aqueous/organic mixture. Samples with poor solubility may block the capillary inlet or prevent adequate spray formation. If poor quality data

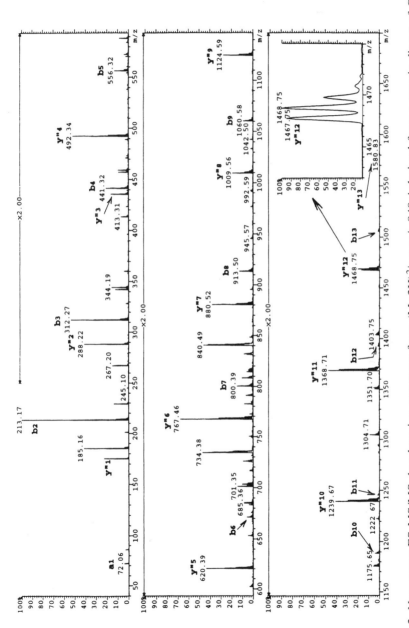

Fig. 2. Nanospray ES-MS/MS daughter ion spectrum from $(M+2H)^{2+}$ at $m/_2$ 840.4 derived from a tryptic digest of Gds (**9**). Insert shows the resolution of the fragment ion at $m/_2$ 1467.75.

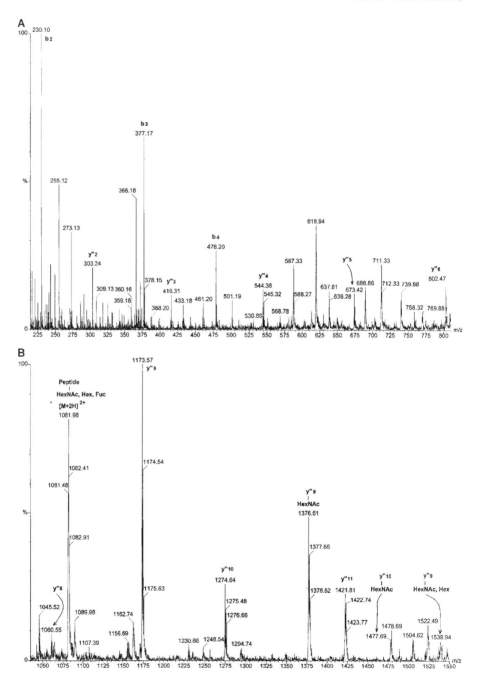

Fig. 3. Nanospray ES-MS/MS spectrum of a triply-charged (M+3H)³⁺ ion at 829.4 in a tryptic digest of SKP1 protein. Interpretation allowed the discovery of a novel cytosolic protein glycosylation *(13)*.

are obtained, check the pump pressure readout (for direct injection) or the nanospray tip to identify blockage. If this is not the case, redissolve the sample in stronger solvents, e.g., formic acid, prior to repeating the previous steps. If the sample fails to produce a good spectrum, it may require further purification, particularly to remove detergents and so on.

3. It is often useful to incorporate an on-line valve (e.g., rheodyne) taking the void "salt" peak to waste prior to on-line operation of the LC/ES-MS system.

4. The suppression of ionization or preferential ionization of some components of complex mixtures, first described in FAB, is in fact observed in all ionization methods, including ES and MALDI, and is best minimized by LC or other purification where possible.

5. The carbohydrate portions of glycoproteins can be mapped very effectively as the released permethylated products either by FAB on a magnetic sector instrument or by MALDI-TOF *(8)*.

6. Fragmentation. A schematic presentation of this for a singly charged ion is shown in **Fig. 4** for both positive and negative ion FAB-MS/ES-MS. (For doubly charged ion MS/MS the "neutral" lost may also be charged and therefore observed.) This shows typical sequence ions resulting from cleavage of the peptide backbone. These so-called low energy fragmentation processes are particularly useful since interpretation is easy (no side-chain losses or more complex high-energy mechanisms).

A general fragmentation nomenclature including these amide bond cleavages together with other commonly observed subfragments is in common use *(16)*.

Factors determining which sequence ions will predominate include the nature of the residues in the chain (particularly at the termini), the ionization technique and the method of collisional activiation (if any). The three principle cleavage points are labeled a, b, and c when the charge is retained on the N-terminal fragment of the peptide, and x, y, and z when the charge is retained by the C-terminal fragment. A subscript indicates which peptide bond is cleaved counting from the N or C terminus respectively. The number of hydrogens gained by the fragment is indicated by apostrophes to the right of the letter, e.g., y.

N-terminal sequence ions: cleavage of CO-NH with charge retention on the N-terminal fragment i.e., acylium ion can be accompanied by another 28 Dalton lower due to loss of CO. These correspond to types (b) + (a).

C-terminal sequence ions: cleavage at CO-NH with charge retention on the C-terminal fragment (y"). The equivalent alkyl fragment is 15 Dalton lower in mass (z').

The sequence of a peptide is simply built up by examining the mass differences between the sequence ion fragments in the spectrum. With a few exceptions (e.g., LEU/ILE or LYS/GLN), the mass differences will correspond to a unique amino acid residue mass. An example of this interpretation is shown in **Fig. 5**.

The same principles apply to carbohydrate sequencing, where glycosidic bond cleavage and beta elimination are the main mechanisms of fragmentation (*see* **ref. 8** and references therein).

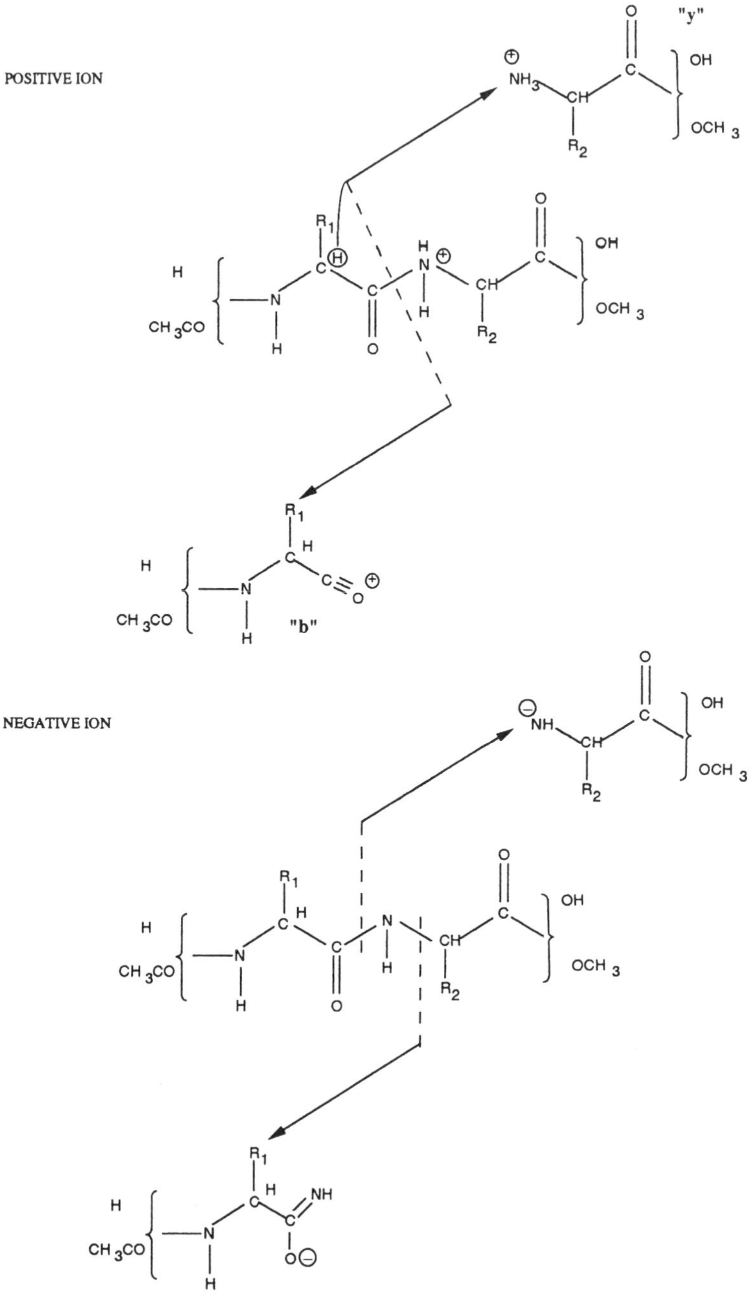

Fig. 4. Fragmentation patterns observed in positive and negative ion modes of FAB-MS and ES-MS. Derivatives can be made as shown (esters or acyls) to allow assignment and differentiation of N- and C-terminal sequence ions by mass shift.

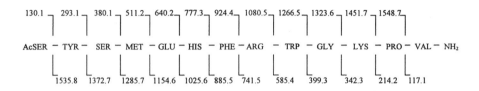

N-terminal

b ions

$[M+H]^+ = 1663.8$ "accurate"

1664.9 "average"

130.1	293.1	380.1	511.2	640.2	777.3	924.4	1080.5	1266.5	1323.6	1451.7	1548.7

AcSER — TYR — SER — MET — GLU — HIS — PHE — ARG — TRP — GLY — LYS — PRO — VAL — NH_2

	1535.8	1372.7	1285.7	1154.6	1025.6	885.5	741.5	585.4	399.3	342.3	214.2	117.1

y″ ions

C-terminal

Fig. 5. Example of potential sequence ions; both N- and C- terminal for the peptide α MSH.

References

1. Morris, H. R., Dell, A., and McDowell, R. A. (1981) Extended performance using a High Field Magnet mass spectrometer. *Biomed. Mass Spectrom.* **8**, 463–473.
2. Barber, M., Bordoli, R. S., Sedgwick, R. D., and Tyler, A. N. (1981) Fast Atom Bombardment of solids (FAB): a new ion source for mass spectrometry. *J. Chem. Soc. Chem. Commun.* **7**, 325–327.
3. Fenn, J. B., Mann, M., Meng, C. K., Wong, S. F., and Whitehouse, C. M. (1989) Electrospray ionization for mass spectrometry of large molecules. *Science* **246**, 64–71.
4. Alexandrov, M. L., Gall, L. N., Krasnov, N. V., Nikolaev, V. I., Pavlenko, V. A., and Shkurov, V. A. (1984) *Dokl. Akad. Nauk S. S. S. R.* **227**, 379.
5. Morris, H. R. and Greer, F. M. (1988) Mass spectrometry of natural and recombinant proteins and glycoproteins. *Trends Biotechnol.* **6**, 140–147.
6. Khoo, K.-H., Dell, A., Morris, H. R., Brennan, P. J., and Chatterjee, D. (1995) Inositol phosphate capping of the non-reducing termini of lipoarabinomannan from rapidly growing strains of Mycobacterium: mapping the non-reducing terminal motifs of LAMS. *J. Biol. Chem.* **270**, 12380–12389.
7. Sutton-Smith, M., Morris, H. R., and Dell, A. (2000) A rapid mass spectrometric strategy suitable for the investigation of glycan alterations in knockout mice. *Tetrahedron Assymetry* **11**, 363–369.
8. Dell, A. and Morris, H. R. (2001) Glycoprotein structure determination by mass spectrometry. *Science* **291**, 2351–2356.
9. Morris, H. R., Paxton, T., Dell, A., Langhorne, J., Berg, M., Bordoli, R. S., et al. (1996) High sensitivity collisionally-activated decomposition tandem mass spectrometry on a novel quadrupole/orthoganol acceleration time of flight mass spectrometer. *Rapid Comm. Mass Spectrom.* **10**, 889–896.

10. Morris, H. R., Panico, M., and Taylor, G. W. (1983) FAB-MAPING of recombinant DNA derived protein products. *Biochem Biophys. Res. Comm.* **117**, 299–305.
11. Morris, H. R., Williams, D. H., and Ambler, R. P. (1971) Determination of the sequences of protein-derived peptides and peptide mixtures by mass spectrometry. *Biochem. J.* **125**, 189–201.
12. Morris, H. R., Panico, M., Barber, M., Bordoli, R. S., Sedgwick, R. D., and Tyler, A.N. (1981) Fast atom bombardment: a new mass spectrometric method for peptide sequence analysis. *Biochem Biophys. Res. Comm.* **101**, 623–631.
13. Teng-umnuay, P., Morris, H. R., Dell, A., Panico, M., Paxton, T., and West, C. M. (1998) The cytoplasmic F-box binding protein SKP1 contains a novel pentasaccharide linked to hydroxyproline in dictyostelium. *J. Biol. Chem.* **273**, 18242–18249.
14. Calabi, E., Ward, S., Wren, B., Paxton, T., Panico, M., Morris, H., et al. (2001) Molecular characterization of the surface layer proteins from Clostridium difficile. *Molecular Microbiology* **40**, 1187–1199.
15. van der Wel, H., Morris, H.R., Panico, M., Paxton T., North, S. J., Dell, A., et al. (2001) A non-golgi alpha1,2-fucosyltransferase that modifies Skp1 in the cytoplasm of dictyostelium. *J. Biol. Chem.* **276**, 33952–33963.
16. Roepstroff, P. and Fohlman, J. (1984) proposal for a common numenclature for sequence ions in mass spectra of peptides. *Biomed. Mass Spectrom.* **11**, 601.

18

Peptide Mass Fingerprinting
Using MALDI-TOF Mass Spectrometry

Darryl J. C. Pappin

1. Introduction

Large-format 2D gel electrophoresis systems in routine operation are capable of resolving several thousand cellular proteins in 1 or 2 d *(1,2)*. For the last decade, a combination of Edman microsequence analysis and identification of proteins by staining with specific antibodies has been used to systematically identify proteins and establish cellular databases *(3–5)*. There are, however, significant problems associated with these approaches. Most proteins are only present in the low- to upper-femtomole range, which is significantly below the level at which automated sequencers can reliably operate *(6,7)*. The relatively slow speed of the Edman process also means that the number of proteins is too great to permit large-scale characterization within any useful period of time. The use of monoclonal antibodies, while both rapid and sensitive, requires the ready availability of a large pool of specific antibody probes.

New methods have been recently developed using a combination of protease digestion, matrix-assisted laser-desorption ionization (MALDI) mass spectrometry (MS), and screening of peptide-mass databases that offer significant increases in the speed at which proteins can be identified **(8–12)**. Practical sensitivity extends into the low-femtomole range, and experience has shown that proteins can be reliably identified using as few as five to six determined peptide masses when screened against databases derived from more than 100,000 proteins. Mass errors of 1–2 Daltons are well tolerated by the scoring algorithms, permitting the use of inexpensive time-of-flight (TOF) mass spectrometers. Search discrimination can be improved by combining results obtained using more than one proteolytic enzyme. Simple chemical modifications (such as esterification) can also be performed on

From: *Methods in Molecular Biology, vol. 211: Protein Seqeuencing Protocols, 2nd ed.*
Edited by: B. J. Smith © Humana Press Inc., Totowa, NJ

whole-digest mixtures to yield important compositional information that increases search specificity.

The major practical problems are the development of protocols that allow multiple samples to be processed in parallel with the minimum of sample handling. Several procedures have been described for enzymatic cleavage of proteins bound to nitrocellulose or PVDF transfer membranes (13–15). In general, all these procedures require pretreatment of the membranes with polymers, such as PVP-40, to prevent nonspecific adsorption and denaturation of the proteolytic enzyme. Postdigest extraction of peptides also requires pooling of several separate washes to ensure efficient recovery of peptides. These methods considerably increase both the level of background contamination and the time and manipulation required per sample. A significant improvement was made in the later work of Fernandez et al. (16), where the use of hydrogenated Triton X-100 made blocking with PVP-40 redundant. One major drawback to the use of heteropolymeric detergents (such as Triton, Tween, or Emulphogene) for peptide-mass fingerprinting is that residual detergent interferes significantly when analyzing peptides by MALDI MS.

In the case of in-gel digestion protocols, there are valid arguments that digestion and extraction of peptides may be more efficient. An excellent review by Patterson and Aebersold (17) discusses the relative merits of in-gel vs on-blot digestions. Additional effort is nearly always required, however, to remove excess detergent, buffer, and low-mol-wt acrylamide polymers that cause significant problems for direct MALDI MS analysis of the digest mixture. This usually involves some form of reverse-phase chromatography, which is difficult to perform at high sensitivity and significantly increases the amount of work involved in preparing samples for analysis. We have found that in spite of the problems associated with electrotransfer and enzymatic digestion from the membrane surface (including incomplete transfer, digestion, and lower recovery of digest fragments) the overall simplicity and cleanliness of the protocols described herein permit reliable peptide maps to be obtained from subpicomole quantities of material.

2. Materials (*see* Note 1)

2.1. PVDF Membranes and Staining

1. Sulforhodamine B stain (Kodak or Aldrich). Prepare solutions of 0.005% w/v in 30% v/v aqueous MeOH (50 mg/L) containing 0.1% v/v acetic acid. The solution may be stored at room temperature for several months and used repeatedly.
2. Immobilon P PVDF membranes (Millipore; *see* **Note 2**).
3. Whatman 3MM filter paper (sheets).
4. Laboratory bench-top orbital shaker.
5. Laboratory UV light-box.

2.2. Enzymatic Digestion and Peptide Recovery

1. Trypsin (Promega, modified) or S. aureus V8 protease (Boehringer, Sequencing grade). Stock (200 ng/μL) solutions of both enzymes are prepared in 2 mM HCl, and 2–5 μL aliquots are immediately frozen on dry ice and stored at –20°C (stable for months).
2. 50 mM ammonium bicarbonate (pH 7.8) containing 1% w/v octyl-β-D-gluco-pyranoside (OBG or octyl glucoside; Millipore).
3. 0.5 mL polypropylene Eppendorf tubes (Treff AG).
4. 98% v/v formic acid and absolute ethanol (1:1 v/v), prepared immediately before use (*see* **Note 3**).
5. Fine-gauge needle.
6. Laboratory vacuum source capable of operation at 10–50 mTorr.
7. Laboratory incubator or convection oven capable of stable temperature operation at 30°C.

2.3. Esterification of Peptide Mixtures (see Note 4)

1. Thionyl chloride (Aldrich, 99%+). Care: This reagent reacts violently with water and decomposes in air to yield HCl vapor. Restrict use to fume-hood.
2. Methanol, anhydrous (Aldrich 99%+, Sure-seal bottle).
3. 100 μL tapered glass tubes and PTFE-lined crimp caps (Chromacol Ltd). Tubes are rinsed briefly in 6M HCl, rinsed thoroughly with distilled/deionized water, and dried at 110°C before storing in presence of desiccant.
4. 10-mL glass syringe. Stored dry in presence of desiccant.
5. Laboratory heat-block capable of temperatures of at least 50°C.
6. Laboratory vacuum source.

2.4. MS Materials

1. α-Cyano-4-hydroxycinnamic acid (Aldrich, 97%). Recrystallize from boiling 50% v/v aqueous acetic acid to remove Na and K salts. Store dry at room temperature.
2. 50% v/v aqueous acetonitrile containing 0.1% v/v TFA.
3. Oxidized insulin B-chain (Sigma; mol wt 3496.9 Dalton) and [Glu1] fibrinopeptide B (Sigma; mol wt 1570.6 Dalton) for use as internal mol-wt standards. Stock solutions (2 pmol/μL) of these peptides are prepared in 50% aqueous acetonitrile/0.1% TFA and stored frozen at –20°C.

3. Methods

Certain simple precautions should be observed for all the following procedures. Contamination of samples is a very real problem when working in the subpicomole range. Whenever possible, clean plastic surgical gloves should be worn. Blotting membranes should only be handled using clean forceps and all surfaces (e.g., polypropylene cutting boards) should be rinsed with aqueous ethanol before use.

3.1. Staining of Electroblotted Proteins

In general terms the precise method of electrotransfer is unimportant and is of little overall consequence. Users are encouraged to employ whatever method they are comfortable with, whether using wet or semi-dry electroblotting procedures or any electroelution buffers from conventional Tris-glycine to the higher pH CAPS-type buffers (*see* **Note 5**).

1. Following electrotransfer (by whatever method) the PVDF membrane should be removed from the apparatus and washed thoroughly with distilled or deionized water to reduce residual transfer buffer and detergent. This is best achieved by immersing the membrane sheet in several hundred milliliters of distilled or deionized water in a shallow tray with gentle agitation (orbital shaker). Change the water at least three times at 30-min intervals.
2. Dry the membrane sheet thoroughly, first between sheets of filter paper, then in vacuo for 10–15 min. This drying step is critical for the success of the staining method (*see* **Note 6**).
3. Immerse the dry membrane in the staining solution. Gently agitate for 30–60 s to ensure that the staining solution covers all parts of the membrane.
4. Remove the membrane and rinse the surface with distilled/deionized water for 30 s.
5. Dry the membrane between paper filters.
6. Stained protein bands or spots are pink in visible light. For higher sensitivity, the membrane can be illuminated with short-wave UV light to exploit the fluorescent properties of the stain. Care: Please ensure that the UV source is enclosed (laboratory light-box) or that eyes are adequately shielded. Faint bands or spots may be marked using a fine needle to define the margin.
7. The dry, stained membranes can be stored in sealed plastic bags at –20°C almost indefinitely.

3.2. Enzymatic Digestion and Elution of Peptides

The main features of this procedure are very small digestion volumes (to maximize the digestion kinetics for a given amount of enzyme) and no requirement to destain or block the PVDF surface with polymers, such as PVP 40.

1. Excise the stained protein spots or bands with a clean scalpel or razor blade. Cut the membrane further into small pieces (<1 mm) and transfer to the bottom of a 0.5-mL Eppendorf tube. This is most easily accomplished using a fine needle, although some care is needed as the small membrane pieces become highly charged with static.
2. As a vital control, excise an equivalent area of blank membrane and treat with enzyme in exactly the same fashion (*see* **Note 7**).
3. Prepare the enzyme solution as follows: Thaw one aliquot of the frozen enzyme stock solution (200 ng/µL in 2 mM HCl) and dilute fivefold with 50 mM ammonium bicarbonate/0.1% w/v octyl glucoside buffer to give a final enzyme concentration of 40 ng/µL.

4. Immediately add a sufficient amount of this buffered enzyme solution to just wet the PVDF membrane pieces, with minimal residual volume. For a protein spot covering a few square millimeters of PVDF membrane surface this may be as little as 3–5 μL of solution. It is not necessary to prewet the membrane pieces or block with polymers, such as PVP40 (*see* **Note 8**).

5. Seal the Eppendorf and incubate overnight at 30°C in an incubator oven. Even after overnight incubation, the membrane pieces should remain wet (*see* **Note 9**).

6. Add 5–20 μL of freshly prepared formic acid:ethanol (1:1 v/v) and allow to stand for 30 min to allow cleaved peptides to diffuse from the PVDF surface.

7. Sample 0.1- and 0.5-μL aliquots directly from the supernatant and apply to the MS target slide or probe (*see* **Note 10**). Allow to air dry for a few seconds, then dry in vacuo for 15–20 min to remove residual ammonium buffer salts. For a typical digest, these 0.1–0.5 μL aliquots represent some 2–15% of the total digest material.

8. The residual supernatant may be collected, evaporated in vacuo, and stored at –20°C. We have found that the tapered, acid-washed glass tubes are preferable to polypropylene (Eppendorf) tubes for storage purposes. The presence of residual OBG detergent seems to protect against significant loss of peptide through non-specific adsorption to the tube walls. The stored peptides can be redissolved in formic acid:ethanol:water (1:1:1 v/v) for reanalysis (*see* **Note 11**).

3.3. Esterification of Peptide Mixtures

This procedure is used to quantitatively esterify all peptide carboxylic acids to the corresponding methyl esters. Each peptide carboxyl group increases in mass by +14 Dalton. Cross correlation of peptide mass-spectra before and after esterification thus allows for the assignment of acidic residue composition (number of Asp or Glu residues) in any given peptide.

1. Sample additional 0.5–1.0 μL aliquots from the digest supernatant and dry in vacuo in the 100 μL tapered glass vials.

2. Prepare a solution of dry 1% v/v thionyl chloride in anhydrous methanol (using the dry glass syringe).

3. Immediately add 10–15 μL of this solution to the dry peptide, seal the vial with a PTFE crimp-cap, and heat to 50°C for 30 min in a laboratory heat-block.

4. Unseal the vial and pipet 0.1–0.3 μL of the solution onto an MS target slide or probe. Air-dry for a few minutes.

3.4. MS Analysis Using MALDI

1. Matrix solution is prepared by dissolving 100 mg recrystallized α-cyano-4-hydroxycinnamic acid in 10 mL 50% v/v aqueous acetonitrile containing 0.1% v/v TFA (final concentration 1% w/v). Fresh solution is prepared daily, then discarded.

2. Aliquots of the peptide standard solutions (Insulin or Glu-fibrinopeptide) are diluted 20-fold with matrix solution to give final concentrations of 100 fmol/μL.

3. 0.5-μL aliquots of this "spiked" matrix solution are now applied to the peptide samples already dried onto MS targets or probes, and allowed to air-dry for 30–60 s.

4. Peptide spectra are collected at just above the laser threshold intensity and calibrated using the appropriate internal standard.

Important summary points are as follows: Using the above digest and analysis procedures, peptide spectra are obtained directly from the whole digest mixture with no chromatographic separation or "clean-up" of samples required.
There is no requirement to wet or block the membranes with polymers, such as
PVP, and the use of Sulforhodamine or Ponceau S removes the requirement to
destain. The enzyme solution is added directly to the dry membrane pieces
and, following digestion, peptides are efficiently eluted form the surface with
formic acid:ethanol. Sample manipulation is thus kept to a bare minimum. The
choice of OBG is critical in that it does not interfere with the digestion or
subsequent analysis by MALDI.

3.5. Database Searches

There are now several freely accessible World Wide Web (WWW) sites that
offer database search facilities for screening of MS fingerprint data. In addition
to the ability to search using peptide masses alone, most sites permit the use of
peptide mass information in conjunction with sequence or compositional data.
Full instructions for use are available on-line from the respective sites.

1. MOWSE search program at SEQNET facility, Daresbury, UK: http://www.dl.ac.uk/
   SEQNET/mowse.html
2. MS-Fit and MS-Tag programs from the UCSF Mass Spectrometry Facility: http://
   rafael.ucsf.edu/
3. Peptide Search program from the EMBL Protein and Peptide Group: http://
   www.mann.embl-heidelberg.de/Services/PeptideSearch/PeptideSearchIntro.html
4. MassSearch program from ETH Switzerland: http://cbrg.inf.ethz.ch/subsection
   3_1_3.html

4. Notes

1. All reagents used should be of the highest quality available. All water should be
   deionized to >15 MΩ resistivity and filtered through 0.45-μm filters.
2. Although PVDF blotting membranes are available from a variety of vendors, the
   very fine surface finish of the Immobilon P membrane allows retention of fine
   detail in stained bands or spots. This becomes important for adequate resolution
   of close-running bands.
3. After standing at room temperature for several hours, significant amounts of ethyl
   formate can be formed by reaction of the acid and alcohol. This is a powerful
   formylating agent and may, under certain conditions, react with peptide primary amines.
4. It is very important that all reagents, tubes, and syringes are anhydrous to prevent
   formation of aqueous HCl. Glass tubes and syringes are preferable to polypropy-

lene. The described method has significant advantages over the more conventional methanolic-HCl procedure in that esterification occurs via an active thioacyl chloride intermediate. With this route, preparation of esters from longer-chain alcohols (e.g., ethyl- or propyl alcohol) is also facile.

5. It is a simple truth that any one individual protein has an optimum set of transfer characteristics (time, applied voltage, and buffer pH). As a general rule, experimenters are urged to use small quantities of sample to optimize transfer conditions (visualized by silver staining or autoradiography for intrinsically labeled preparations). In all cases the rule is simple: Perform the transfer experiment using "best guess" initial conditions and stain both the transfer membrane and gel to assess transfer efficacy.

6. This staining procedure relies entirely on the hydrophobic nature of the PVDF surface *(18)*. As a general rule, Immobilon P membranes require at least 50% aqueous methanol or acetonitrile solutions to "wet." Where proteins are bound to the surface, however, the local properties of the surface are altered such that the surface will wet much more readily. At concentrations of 25–30% v/v aqueous methanol, therefore, the surface wets only where protein is adsorbed, allowing any water-soluble stain to complex with the protein. The rest of the membrane surface does not wet at these low methanol concentrations. As a consequence, the general membrane surface does not take up the dye and requires no destaining. The whole process can thus be completed in 2–3 min, but it is essential that the membrane surface be thoroughly dry and free of residual transfer buffer and detergent before immersion in the stain. One inherent property of this procedure is that any water-soluble stain may be used. For MALDI MS purposes. Sulforhodamine B or Ponceau S are the stains of choice as they show little or no suppressive effect. More traditional stains such as Coomassie blue should be avoided at all costs as they cause massive suppression of signal under conditions of laser desorption.

7. This control piece of membrane is necessary to allow the experimenter to identify and eliminate those peptide fragments resulting from autolytic cleavage of the enzyme used. We prefer the use of the Promega modified trypsin, where autolysis is significantly reduced by partial methylation of exposed lysine residues. In our experience, inclusion of 2–5 mM $CaCl_2$ has little effect on the rate of autolysis of native trypsin.

8. The 1% w/v octyl glucoside detergent included in the digest buffer is sufficient to wet the dry pieces of PVDF membrane in aqueous solution. Once the surface is wetted, there is sufficient detergent to block any remaining hydrophobic character of the surface, obviating the requirement to treat with polymers, such as PVP 40. Tests have shown that 1% w/v OBG has no observable effect on the properties of trypsin or V8 protease. Unlike any of the heteropolymeric nonionic detergents (Triton, Tween, Nonidet, Emulphogene), octyl glucoside also has little suppressive effect on laser desorption.

9. It is important that the entire tube be heated to equilibrium temperature. If the tubes are placed in a laboratory heat block at 30°C, the exposed lids act as the

cooling plate for what is essentially a miniature reflux tower. In this situation, the very small volumes of liquid (usually only a few microliters) will evaporate completely and condense on the tube lid within 2–3 h (end of digest). If the entire tube is heated (as is the case in a convection oven or incubator) sufficient vapor pressure is maintained within the 0.5-mL tube to prevent total evaporation of even 2–3 μL of the digest buffer.

10. One of the most difficult practical problems associated with MALDI analysis of peptide digests of unknown amount is in preparing solutions of appropriate concentration for adequate spectra. In our experience, this is generally where the peptides are in the range of 50–500 fmol, although good spectra can sometimes be obtained using lower amounts. In practice, the easiest way to achieve this is to prepare different loadings from one sample (e.g., sample 0.1-, 0.3-, and 0.5-μL aliquots) and examine all to obtain the best spectrum. In some cases (i.e., for more densely stained bands or spots) it may be necessary to further dilute the sample by the addition of more formic acid:ethanol.

11. We have found that samples may be redissolved in this manner for repeated analysis several times without effect, and after storage for more than 24 mo. We have seen no evidence of N-terminal formylation of peptides after at least four cycles of storage and reconstitution using the formic acid:ethanol solvent.

References

1. O'Farrell, P. (1975) High resolution two-dimensional electrophoresis of proteins. *J. Biol. Chem.* **250**, 4007–4021.

2. Patton, W. F., Pluskal, M. G., Skea, W. M., et al. (1990) Development of a dedicated two-dimensional gel electrophoresis system that provides optimal pattern reproducibility and polypeptide resolution. *Biotechniques* **8**, 518–527.

3. Celis, J. E., Gesser, B., Rasmussen, H. H., et al. (1990) Comprehensive two-dimensional gel protein databases offer a global approach to the analysis of human cells: the transformed amnion cells (AMA) master database and its link to genome DNA sequence analysis. *Electrophoresis* **11**, 989–1071.

4. Garrels, J. and Franza, B. (1989) The REF52 protein database: methods of database construction and analysis using the QUEST system and characterization of protein patterns from proliferating and quiescent REF52 cells. *J. Biol. Chem.* **264**, 5283–5298.

5. Rasmussen, H. H., Van Damme, J., Bauw, G., et al. (1991) Protein electroblotting and microsequencing in establishing integrated human protein databases, in Methods in Protein Sequence Analysis (Jornvall, H., Hoog, J. O., and Gustavsson, A. M., eds.), Birkhauser Verlag, Basel, pp. 103–114.

6. Hewick, R. M., Hunkapiller, M. W., Hood, L. E., and Dreyer, W. J. (1981) A gas-liquid solid phase peptide and protein sequenator. *J. Biol. Chem.* **256**, 7990–7997.

7. Totty, N. F., Waterfield, M. D., and Hsuan, J. J. (1992) Accelerated high-sensitivity microsequencing of proteins and peptides using a miniature reaction cartridge. *Protein Sci.* **1**, 1215–1224.

8. Henzel, W. J., Billeci, T. M., Stults, J. T., et al. (1993) Identifying proteins from 2-dimensional gels by molecular mass searching of peptide-fragments in protein-sequence databases. *Proc. Natl. Acad. Sci. USA* **90**, 5011–5015.
9. Pappin, D. J. C., Hojrup, P., and Bleasby, A. J. (1993) Rapid identification of proteins by peptide-mass fingerprinting. *Curr. Biol.* **3**, 327–332.
10. Mann, M., Hojrup, P., and Roepstorff, P. (1993) Use of mass-spectrometric molecular-weight information to identify proteins in sequence databases. *Biol. Mass Spectrom.* **22**, 338–345.
11. Yates, J. R., Speicher, S., Griffin, P. R., and Hunkapiller, T. (1993) Peptide mass maps—a highly informative approach to protein identification. *Anal. Biochem.* **214**, 397–408.
12. James, P., Quadroni, M., Carafoli, E., and Gonnet, G. (1993) Protein identification by mass profile fingerprinting. *Biochem. Biophys. Res. Commun.* **195**, 58–64.
13. Aebersold, R. H., Leavitt, J., Saavedra, R. A., et al. (1987) Internal amino acid sequence analysis of proteins separated by one- or two-dimensional gel electrophoresis after in situ protease digestion on nitrocellulose. *Proc. Natl. Acad. Sci. USA* **84**, 6970–6974.
14. Bauw, G., Van Damme, J., Puype, M., et al. (1989) Protein electroblotting and microsequencing strategies in generating protein databases from two-dimensional gels. *Proc. Natl. Acad. Sci. USA* **86**, 7701–7705.
15. Fernandez, J., DeMott, M., Atherton, D., and Mische, S. M. (1992) Internal protein sequence analysis: enzymatic digestion for less than 10 micrograms of protein bound to polyvinylidene difluoride or nitrocellulose membranes. *Anal. Biochem.* **201**, 255–264.
16. Fernandez, J., Andrews, L., and Mische, S. M. (1994) An improved procedure for enzymatic digestion of polyvinylidene difluoride-bound proteins for internal sequence-analysis. *Anal. Biochem.* **218**, 112–117.
17. Patterson, S. D. and Aebersold, R. (1995) Mass-spectrometric approaches for the identification of gel-separated proteins. *Electrophoresis* **16**, 1791–1814.
18. Coull, J. M. and Pappin, D. J. C. (1990) A rapid fluorescent staining procedure for proteins electroblotted onto PVDF membranes. *J. Protein Chem.* **9**, 259,260.

19

Nanoelectrospray Tandem Mass Spectrometry and Sequence Similarity Searching for Identification of Proteins from Organisms with Unknown Genomes

Anna Shevchenko, Shamil Sunyaev, Adam Liska, Peer Bork, and Andrej Shevchenko

1. Introduction

Recent developments in technology and instrumentation have made mass spectrometry the method of choice for identification of proteins in a sequence database [reviewed in **refs.** *(1–4)*]. Regardless of what mass spectrometric technology is applied, it is ultimately required that acquired mass spectra be accurately matched to a protein sequence from the corresponding database entry.

If the sequence of the analyzed protein is not present in a database, the further possible strategy depends on the sequence similarity between the protein of interest and homologous proteins from other species. If the sequence similarity is high and proteins share 5–10 identical tryptic peptides, it would be possible to identify the "unknown" protein of interest by cross-species matching of its peptide mass map to a known sequence of its homolog *(5)*. If the similarity is low, usually the peptides have to be sequenced *de novo* and the gene of the unknown protein has to be cloned by polymerase chain reaction (PCR) *(6,7)*.

How is it possible to determine sequences of tryptic peptides? Continuous series of fragment ions containing the C-terminus (Y" ions) *(8)*, which are usually observed in tandem mass spectra of tryptic peptides, have been successfully recognized for *de novo* sequencing *(7)*. Partial peptide sequence can be readily obtained by considering precise mass differences between the adjacent Y"- ions in the m/z region above the multiply charged precursor *(9)*. However, in order to determine a complete peptide sequence it is necessary to identify Y"-ions in the low m/z region of the spectrum, where ions of other series and ions originating from chemical noise are abundant. Therefore it is necessary to obtain additional

From: *Methods in Molecular Biology, vol. 211: Protein Sequencing Protocols, 2nd ed.*
Edited by: B. J. Smith © Humana Press Inc., Totowa, NJ

evidence that the particular fragment ion indeed belongs to the Y"-series. Masses of the ions of Y" series are shifted upon peptide esterification with methanol *(7,10)* and could be identified by comparing tandem mass spectra acquired from esterified and native peptides. Alternatively, C-terminal carboxyl group of the peptide could be selectively labeled with ^{18}O isotope by digesting the protein in a buffer containing 50% of $H_2^{16}O$ and 50% $H_2^{18}O$ (v/v). Y" ions could be distinguished by a characteristic isotopic pattern: a doublet of peaks, split by 2 mass units *(11)*. Using any of the approaches, complete and accurate sequences of tryptic peptides can be determined *(12–16)*. However, *de novo* sequencing remains laborious and time consuming, and requires significantly higher amount of protein compared to conventional protein identification.

Because of the rapid growth of sequence databases, it is becoming increasingly possible to identify unknown proteins via sequence similarity searching. Importantly, it is possible to utilize peptide sequences of much lower quality, compared to the ones required for designing degenerate oligonucleotide primers for PCR *(17–20)*. Partially redundant peptide sequence candidates determined by automated or rapid manual interpretation of tandem mass spectra are combined (regardless of their completeness, confidence and length) and used in a single search by Mass Spectrometry driven BLAST (MS BLAST) *(19)*, which is performed over the web on a high computational capacity server. A few thousands candidate sequences can be submitted in a single search, which takes a few minutes to complete. Therefore MS BLAST searching could be coupled with high throughput protein identification methods as MALDI quadrupole TOF *(21)*, MALDI TOF/TOF *(22)* and LC MS/MS *(23)* by straightforward script-based automation.

2. Materials

2.1. Chemicals

All chemicals should be of analytical grade or better.

1. Water, acetonitrile methanol, and formic acid high-performance liquid chromatography (HPLC) grade from Merck (Darmstadt, Germany) (*see* **Note 1**).
2. Perfusion sorbent POROS 50 R2 (PerSeptive Biosystems, Framingham, MA) (*see* **Note 2**)
3. Borosilicate glass capillaries GC120F-10 1.2 mm OD × 0.69 mm ID (Harvard Apparatus Ltd, Edenbridge, UK). Needles for nanoelectrospray and columns for micropurification could be purchased from Protana (Odense, Denmark) or manufactured as described in *(24,25)*.

2.2. Equipment and Software

1. Mass Spectrometers. A QSTAR Pulsar *i* quadrupole time-of-flight mass spectrometer and an API III triple quadrupole mass spectrometer (both from MDS Sciex, Concord, ON, Canada) each equipped with a nanoelectrospray ion source

(24) (Protana, Odense, Denmark and EMBL, Heidelberg, respectively), tuned for peptide sequencing as described in **Note 3**.

2. Benchtop mini-centrifuge (as "PicoFuge," Stratagene, Palo Alto, CA).

3. Micropurification holder purchased from Protana (Odense, Denmark) or manufactured as described in *(25)*.

4. Software for automated interpretation of tandem mass spectra (*see* **Note 4**). Tandem mass spectra acquired on an API III mass spectrometer were interpreted by PredictSequence routine (a part of the BioMultiview software package from MDS Sciex, Concord, Canada). Spectra acquired on a QSTAR Pulsar *i* were interpreted using BioAnalyst v.1.0 supplied by the same company.

   MS BLAST web interface is located at: http://dove.embl-heidelberg.de/Blast2/msblast.html

3. Methods

3.1. Desalting and Concentration of Peptides for NanoES MS/MS Sequencing (see Note 5)

1. Pipet ca. 5 μL of POROS R2 slurry (*see* **Note 2**) into the pulled glass capillary (*see* **Subheading 2.1.3.**), here and further down referred as a "column."

2. Spin the beads down and then open the pulled end of the column by gentle touching against a bench top. The amount of POROS R2 resin in the column should be within the tapered region only.

3. Wash the beads with 5 μL of 5% formic acid and make sure the liquid can flow out of the column if gentle centrifuging is applied. Open the column end wider if necessary. Mount the column into the micropurification holder.

4. Dissolve the dried protein digest in 10 μL of 5% formic acid and load onto the column. Pass the sample through the beads layer by centrifuging.

5. Wash adsorbed peptides with another 5 μL of 5% formic acid.

6. Align the column and the nanoelectrospray needle in the micropurification holder and elute peptides directly into the needle with 1 μL of 60% of methanol in 5% formic acid by gentle centrifuging. Add the eluent in three applications of 0.3 μL or two applications of 0.5 μL for best results.

7. Mount the spraying needle with the sample into the nanoelectrospray ion source and acquire mass spectra.

3.2. Sequencing on a Triple Quadrupole Mass Spectrometer (see Note 6)

1. After desalting and concentration of peptides (*see* **Subheading 3.2.**), initiate spraying and acquire Q1 spectrum of the peptide mixture. Turn on collision gas and acquire the spectrum in the precursor scan mode with the fragment mass set at *m/z* 86 (*see* **Note 7**).

2. Stop spraying by dropping spraying voltage to zero. Drop the air pressure applied to the spraying capillary to zero. Move the spraying capillary away from the inlet of the mass spectrometer.

3. Examine the acquired spectra and compare them with the spectra acquired from

the control sample (*see* **Note 6**). Select precursor ions for subsequent tandem mass spectrometric sequencing.

4. Add 0.3–0.5 µL of 60% of methanol in 5% formic acid directly to the spraying capillary if the remaining sample volume is less than 0.5 µL. Re-establish spraying and acquire tandem mass spectra from selected precursor ions.

3.3. Sequencing on a Quadrupole Time-of-Flight Mass Spectrometer

Sequencing is performed as described in **Subheading 3.2.**, but no precursor ion scanning is applied and the collision cell is always filled with gas. High resolution of a quadrupole time-of-flight mass spectrometer allows to distinguish genuine multiply charged peptides ion as characteristic sharp isotopic features superimposed on chemical background, which is mostly observed as singly charged broad irregular peaks. Collision energy could be adjusted by monitoring the intensity of the residual precursor ion during acquisition of the tandem mass spectrum *(26)*.

3.4. Identification of Proteins by MS BLAST Searching (see **Note 8**)

MS BLAST is a specialized BLAST-based *(27)* tool for identification of proteins by sequence similarity searching that utilizes peptide sequences produced by the interpretation of tandem mass spectra *(19)* (*see* **Note 9**).

1. Obtain, edit, and assemble peptide sequences.
   a. Automated interpretation of MS/MS spectra: If tandem mass spectra were interpreted by *de novo* sequencing software, disregard relative scores and use the entire list of candidate sequences (or some 50–100 top scoring sequence proposals per fragmented peptide precursor).
   b. Manual interpretation of MS/MS spectra: Try making the longest possible sequence stretches, although their accuracy may be compromised. For example, it is usually difficult to interpret unambiguously fragment ion series at the low *m/z* range because of abundant peaks of chemical noise and numerous fragment ions from other series. In this case, it is better to include many complete (albeit low confidence) sequence proposals into the query rather than using a single (although accurate) three–four amino acid sequence stretch deduced from a noise-free high *m/z* segment of the spectrum.
2. Gaps and ambiguities in peptide sequences. Some *de novo* sequencing programs may suggest a gap in the peptide sequence that can be filled with various isobaric combinations of amino acid residues. For example:

ASDF[...]FGTR, [...] = [L,T] or [D,V]

If one or two combinations were suggested, include all variants into a searching string:

-ASDFLTFGTR-ASDFTLFGTR-ASDFDVFGTR-ASDFVDFGTR-

If more combinations were possible, the symbol X can be used instead to fill the gap. Zero score is assigned to X symbol in PAM30MS scoring matrix (*see* **Subheading 3.4.5.**) and therefore it matches weakly any amino acid residue:

-ASDFXXFGTR-

Note that MS BLAST is sensitive to the number of amino acid residues that are filling the gap. If the gap could be filled by a combination of two and three amino acid residues, consider both options in the query:

-ASDFXXFGTR-ASDFXXXFGTR-

3. Isobaric amino acids. L stands for Leu and Ile; Z stands for Gln and Lys, if undistinguishable in the spectrum. Use Q or K if the amino acid residue could be determined.

4. Generic trypsin cleavage site. If the proposed sequence is complete, a putative trypsin cleavage site symbol B is pasted prior to the peptide sequence:

...-BASDFLTFGTR-

It is often difficult to determine two amino acid residues located at the N-terminus of the peptide. In this case present them as:

...-BXXDFLTFGTR-...

MS BLAST will then consider BXX residues in possible sequence alignments.

5. MS BLAST options and settings:

   a. –NOGAP: absolutely essential, it turns off gapped alignment method so that only high scoring pairs (HSPs) with no internal gaps are reported

   b. –SPAN1: absolutely essential, it identifies and fetches the best matching peptide sequence among similar peptide sequences in the query. Therefore the query may contain multiple partially redundant variants of the same peptide sequence without affecting the total score of the protein hit.

   c. –HSPMAX 100 limits the total number of reported HSPs to 100. Set it to higher number (for example, 200) if a large query is submitted and complete list of protein hits including low confidence hits) is required in the output.

   d. SORT_BY_TOTALSCORE places the hits with multiple high scoring pairs to the top of the list. Note that the total score is not displayed, but can be calculated, if necessary, by adding up scores of individual HSPs.

   e. EXPECT: It is usually sufficient to set EXPECT at 100. Searching with higher EXPECT (as, 1000) will report many short low-scoring HSPs thus increasing the sequence coverage by matching more fragmented peptides to the protein sequence. Note that low scoring HSPs do not increase statistical confidence of protein identification. EXPECT setting also does not affect the scores of retrieved HSPs.

   f. MATRIX: PAM30MS is a specifically modified scoring matrix. Do not use it for conventional BLAST searching!

   g. PROGRAM: blast2p; DATABASE: nrdb95 are default settings of MS BLAST interface. In principle, MS BLAST could be also used as a ***tblastn*** program for searching EST or genomic databases.

   h. FILTER: By default, filtering is set to "none." However, if a query contains many low-complexity sequence stretches, (as, ...EQEQEQ...), filtering should be set to "default."

6. MS BLAST searching. Space all candidate sequence proposals obtained from all fragmented precursors with "–" (minus) symbol and merge them into a single text string that can be pasted directly into the query window at the MS BLAST

web interface. The query may contain space symbols, hard returns, numbers, and so on since they are ignored by the server. For example, it is convenient to keep masses of precursor ions in the query since it makes retrospective analysis of data much easier.

7. Statistical evaluation of MS BLAST hits. Statistical evaluation is a very important element of MS BLAST protocol since the query typically comprises many incorrect and/or ambiguous peptide sequences. Note, that statistics of conventional BLAST searching is not applicable and therefore ignore reported E-values and P-values.

Thresholds of statistical significance of MS BLAST hits are set conditionally on a number of reported HSPs and were estimated in a computational experiment (*see* **Table 1**). To evaluate the significance of MS BLAST hits:

a. Check the number of fragmented precursor ions from which sequences for MS BLAST searching were obtained. Accordingly, pick up appropriate number of expected unique peptides from **Table 1**.

b. Consider the top hit protein in MS BLAST output and check the list of matching HSPs. Pick up the HSP with the highest score and compare this score with the threshold value reported in the table for a single reported HSP.

c. If the score of the HSP is higher than the threshold score in **Table 1**, the match is statistically significant. If not, pick up the score of the second ranked HSP, add it to the score of the first ranked HSP and compare the sum with the threshold score reported in Table 1 for two reported HSPs. Again, if the combined score exceeds the threshold, the identification is positive. Otherwise, add the score of the third ranked HSP, compare the sum with the threshold expected for three reported HSPs and so forth.

Always start the evaluation from the highest ranked HSP reported for the given protein hit!

If necessary, repeat the procedure for other protein hits in the MS BLAST output.

4. Example Analysis: Identification of a Protein from African Cloned Frog

Currently sequences of less than 3,500 *Xenopus* proteins are available in a public database. Therefore the characterization of proteins isolated from *Xenopus* cells usually requires *de novo* sequencing and cloning of the genes (*14*). Direct identification of proteins via sequence similarity searching provides a shortcut to their functional characterization by skipping time-consuming and laborious cloning experiments (*28*).

Identification of 120 kDa protein isolated from *Xenopus* oocytes by affinity chromatography and one-dimensional electrophoresis is presented as an example. A Coomassie stained band was excised from the gel, the protein was in-gel digested with trypsin and recovered tryptic peptides were sequenced on a QSTAR Pulsar *i* quadrupole time-of-flight mass spectrometer. Tandem spectra were

Table 1
Threshold Scores for Statistical Evaluation of MS BLAST Hits[a]

Number of reported HSPs	Number of unique peptides in the query		
	10	20	50
1	68	72	75
2	102	106	111
3	143	146	153
4	177	< 208[b]	< 180[b]
5	< 238[b]	n.o.[c]	< 212[b]
6	n.o.	n.o.	< 275[b]
7	n.o.	n.o.	< 285[b]

[a]Threshold scores were estimated in computational experiment (*19*) and are provided for guidance rather than as a stringent identification criterion.

[b]The calculated value is statistically unreliable because just a few hits matching with the specified number of HSPs were observed. In those cases the maximal score from the ones observed is presented.

[c]n.o. No random hits with specified number of HSPs were observed.

acquired from 21 multiply charged peptide precursor ions (*see* **Fig. 1**) and were subjected to manual and/or automated interpretation. For example, in the MS/MS spectrum acquired from a doubly charged ion with m/z 883.44 three amino acid residues (…DYT…) were confidently called in the m/z region above the precursor, which covers the N-terminal part of the peptide sequence (*see* **Fig. 2**). The sequence stretch could be extended a few residues further down to the lower m/z region (towards the C-terminus of the peptide). However, the sequence became ambiguous and could not be anchored at C-terminal arginine or lysine residues, expected from the cleavage specificity of trypsin. Automated interpretation produced a few ambiguous candidate sequences that covered C-terminal part of the peptide, but they did not extend far enough towards the N-terminus and did not overlap with the sequence obtained by manual interpretation. All sequences were then edited and merged according to MS BLAST requirements (*see* **Subheading 3.4.**) and combined with sequences obtained by the interpretation of other tandem mass spectra. Altogether, 793 peptide sequences of varying quality, completeness, and length were merged and the text string was submitted to MS BLAST searching at the EMBL server. A few homologous catalytic subunits of DNA polymerase delta from various species were hit with the statistical significance exceeding the threshold scores (*see* **Table 2**) (*see* **Note 10**). Altogether, 10 peptide sequences matched to bovine DNA polymerase delta (although only two tryptic peptides matched its sequence exactly) and additionally two peptides matched to homologous polymerases from other species.

Thus a combination NanoES tandem mass spectrometry and MS BLAST sequence similarity searching allowed rapid and confident identification of the

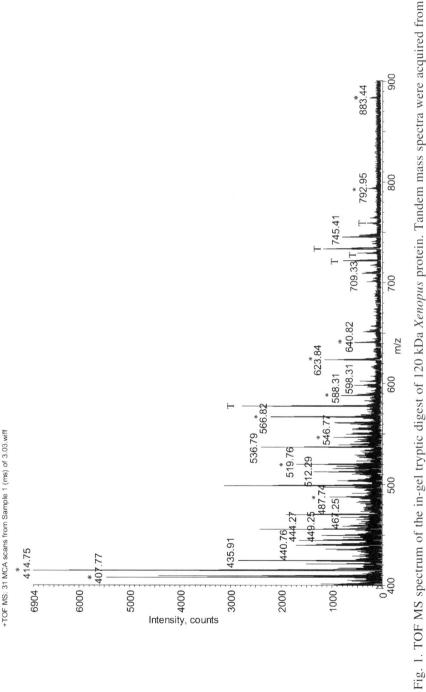

Fig. 1. TOF MS spectrum of the in-gel tryptic digest of 120 kDa *Xenopus* protein. Tandem mass spectra were acquired from peaks designated with *m/z*. Peaks originating from trypsin autolysis products are designated with T. Peaks of the peptides that matched the sequence of bovine DNA polymerase delta are labeled with asterisks.

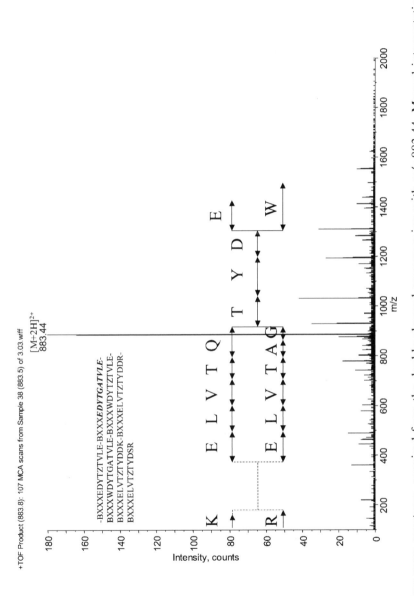

Fig. 2. Tandem mass spectrum, acquired from the doubly charged precursor ion with m/z 883.44. Manual interpretation of spectra considered precise mass difference between adjacent Y" ions starting from the m/z segment above the precursor ion (correspondent peaks and amino acid residues are designated by arrows). Automated interpretation resulted in a few partially redundant ambiguous sequences, covering the C-terminus of the peptide (inset).

Table 2
Identification of a Xenopus protein by MSBLAST Sequence Similarity Searching

m/z	z	Precursor mass	Automated interpretation	Manual interpretation	Bovine	Mouse	Rat	Human	Yeast	Arabidopsis	Score
407.77	2	813.52	/	**BVVSZLLR**	+	+	+	+			47
414.75	2	827.49	/	**BLLEZGLR**	+	+	+	+			46
435.91	3	1304.70	YES	BXXXTAVLZD			+				-
440.76	2	879.50	YES	BLAVYD							-
444.27	3	1329.79	YES	BXXAHFNTAVLK, BXXAHFNTAVLK							-
449.25	3	1344.73	NONE	BXXXXADLL, BXXXXXADLL							-
467.25	2	932.49	NONE	BTPTPT							-
487.74	2	973.46	YES	**BYTLDDGYK**	+	+	+	+			42
512.29	2	1022.57	YES	BVSTFPG							-
519.76	2	1037.52	YES	**BTPTGDZV**	+	+	+	+			41
536.79	2	1071.56	YES	BLALZDPFLR							-
546.77	2	1091.52	YES	BLZDLSDFZK							-
566.82	2	1131.62	YES	**BLFEPLL**	+	+	+	+			51
588.31	2	1174.60	NONE	**BVLSFDLE**	+	+	+	+		+	53
598.31	2	1194.62	YES	BYGLNPEDFLK		+	+				-
623.84	2	1245.66	YES	**BXXSZLSALEEK**	+	+	+	+			54
640.82	2	1279.65	/	**BVLSFDLEE**	+	+		+			54
709.33	2	1416.65	NONE	BXXEVPDZ							-
745.41	2	1488.80	YES	BXXXTVAEA, BXXXXTVAEA							-
792.95	2	1583.89	YES	BXADSVYGFT, **BXXADSVYGFT**,	+	+	+	+	+	+	58
883.44	2	1764.86	YES	**BXXXEDYTGATVLE** BXXXEDYTZTVLE	+	+	+	+	+		72

aBovine DNA Polymerase delta was the top hit; scores of matched HSPs are presented. Sequence stretches in bold matched the bovine sequences exactly. Peptides for which the software predicted no sequences are labeled "NONE." Peptides for which the automated sequence predictions were made are labeled "YES." For peptides labeled "/," no automatically predicted sequences were included because high-quality sequences were retrieved directly from Y-ion series in the spectrum.

protein using only rapid and "rough" interpretation of tandem mass spectra (*see* **Note 11**).

5. Notes

1. All chemicals should be of the highest degree of purity available. Plastic ware (pipet tips, gloves, dishes, etc.) may acquire a static charge and attract dust thus leading to massive contamination of samples with human and sheep keratins during in-gel digestion. Any polymeric detergents (Tween, Triton) should not be used for cleaning the laboratory dishes and tools.

2. Methanol (1 mL) is added to *ca.* 30 µL of POROS R2 resin to prepare a slurry. A submicrometer-sized fraction of the resin beads, whose presence increases the resistance to liquid flow, is removed by repetitive sedimentation. Vortex the test tube containing the slurry and then let it stay in a rack until the major part of the resin reaches the bottom of the tube. Aspirate the supernatant with pipet and discard it. Repeat the procedure 3–5 times if necessary.

3. Make sure that the settings controlling resolution of the first quadrupole (Q1) allow good transmission of precursor ions. On the other hand, unnecessary low resolution of Q1 results in the transmission of too many background ions, which may densely populate the low *m/z* region of the MS/MS spectra. The third quadrupole (Q3) should likewise be operated at a low resolution settings in order to improve its transmission and to achieve acceptable ion statistics in the MS/MS spectra. In our experience, resolution of Q3 as low as 250 (FWHM) still allows accurate readout of peptide sequences. The Q1 and Q3 resolution settings can be tuned in tandem mass spectrometric experiment using synthetic peptides.

   Resolution of the first quadrupole (Q1) should be tuned in a similar way as described for a triple quadrupole mass spectrometer. External calibration using masses of [Glu]-fibrinogen peptide fragments allows better than 20 ppm mass accuracy for both TOF MS and product ion operation modes, if calibration and sequencing experiments are performed within *ca.* 2 h.

4. *De novo* sequencing software is often included in the software packages shipped together with mass spectrometers: BioMultiview, BioAnalyst (both are from MDS Sciex, Canada); BioMassLynx (Micromass Ltd, UK), BioTools (Bruker Daltonics, Germany). Lutefisk program *(20,29)* can be downloaded from: http://www.immunex.com/researcher/lutefisk/. Automated interpretation of tandem mass spectra often requires adjusting of settings that affect scoring of candidate sequences. It is therefore advisable to test the settings in advance using digests of standard proteins and to adjust them if necessary. Note that the settings may depend on a charge state of the fragmented precursor ion. Use only standard one-letter code for amino acid residues (*see* **Appendix 1**). If the software introduces special symbols for modified amino acid residues, replace them with standard codes.

5. In-gel digestion of proteins with trypsin was described in detail in *(30–32)*.

6. In-gel digestion of proteins with unmodified trypsin is accompanied by trypsin autolysis. Therefore it is necessary to acquire the spectrum of a control sample

(blank gel pieces processed using the same in-gel digestion method) in advance. Spectra should be acquired both in conventional single MS mode (Q1 scanning on a triple quadrupole or TOF MS mode on a quadrupole TOF machines) and in precursor ion scanning mode (triple quadrupole instruments only).

7. Scanning for precursor ions that produce the characteristic fragment ions with m/z 86 (immonium ion of leucine or isoleucine) helps to distinguish genuine peptide ions from chemical noise *(33)*.

8. It is not known in advance if the sequence of the analyzed protein is already present in a database. Therefore conventional database searching routines based on stringent matching of peptide sequences should be applied first *(9,34)*. Only if the protein is unknown and no convincing cross-species matches were obtained, it is recommended to proceed with *de novo* interpretation of tandem mass spectra and sequence similarity searching.

9. The algorithm and principles of BLAST homology searching are discussed in detail in *(27,35)*. A useful list of BLAST servers accessible on the web is provided in *(36)*.

10. The top peptide matched with the score 72, which as such is not sufficient for positive identification (the threshold score for a single matched peptide is 72, *see* **Table 1**). Taken together with the score of the second ranked peptide matched (58), the total score of two matched peptides is 72 + 58 = 130, which is higher than the threshold score for two matched peptides that is 106 (*see* **Table 1**) and the protein was positively identified. More matching peptide (*see* **Table 2**) additionally validated the hit.

11. The success of MS BLAST identification depends on the number of sequenced peptides, on the quality of peptide sequences and on the sequence similarity between the protein of interest and its homologues available in a database. On average, candidate sequences determined for five tryptic peptides should be submitted to MS BLAST searching to identify the protein by matching to a homolog from the organism within the same kingdom.

References

1. Chalmers, M. J. and Gaskell, S. J. (2000) Advances in mass spectrometry for proteome analysis. *Curr. Opin. Biotechnol.* **11**, 384–390.
2. Lahm, H. W. and Langen, H. (2000) Mass spectrometry: a tool for the identification of proteins separated by gels. *Electrophoresis* **21**, 2105–2114.
3. Pandey, A. and Mann, M. (2000) Proteomics to study genes and genomes. *Nature* **405**, 837–846.
4. Mann, M., Hendrickson, R. C., and Pandey, A. (2001) Analysis of proteins and proteomes by mass spectrometry. *Annu. Rev. Biochem.* **70**, 437–473.
5. Podtelejnikov, A. V., Bachi, A., and Mann, M. (1998) Cross-species identification of proteins by high mass accuracy MALDI peptide mapping, in *Proceedings 46th ASMS Conference on Mass Spectrometry and Allied Topics*, Orlando, FL, Abstract no. 212.

6. Wilm, M., Shevchenko, A., Houthaeve, T., Breit, S., Schweigerer, L., Fotsis, T., et al. (1996) Femtomole sequencing of proteins from polyacrylamide gels by nanoelectrospray mass spectrometry. *Nature* **379**, 466–469.
7. Shevchenko, A., Wilm, M., and Mann, M. (1997) Peptide sequencing by mass spectrometry for homology searches and cloning of genes. *J. Protein Chem.* **16**, 481–490.
8. Roepstorff, P. and Fohlman, J. (1984) Proposed nomenclature for sequence ions. *Biomed. Mass Spectrom.* **11**, 601.
9. Mann, M. and Wilm, M. (1994) Error tolerant identification of peptides in sequence databases by peptide sequence tags. *Anal. Chem.* **66**, 4390–4399.
10. Hunt, D. F., Yates, J. R., Shabanowitz, J., Winston, S., and Hauer, C. R. (1986) Peptide sequencing by tandem mass spectrometry. *Proc. Natl. Acad. Sci. USA* **83**, 6233–6237.
11. Shevchenko, A., Chernushevich, I., Ens, W., Standing, K. G., Thomson, B., Wilm, M., et al. (1997) Rapid 'De Novo' peptide sequencing by a combination of nanoelectrospray, isotopic labeling and a quadrupole/time-of-flight mass spectrometer. *Rapid Commun. Mass Spectrom.* **11**, 1015–1024.
12. McNagny, K. M., Petterson, I., Rossi, F., Flamme, I., Shevchenko, A., Mann, M., et al. (1997) Thrombomucin, a novel cell surface protein that defines thrombocytes and multipotent hematopoetic progenitors. *J. Cell Biol.* **138**, 1395–1407.
13. Lingner, J., Hughes, T. R., Shevchenko, A., Mann, M., Lundblad, V., and Cech, T. R. (1997) Reverse transcriptase motifs in the catalytic subunits of telomerase. *Science* **276**, 561–567.
14. Chen, R. H., Shevchenko, A., Mann, M., and Murray, A. W. (1998) Spindle checkpoint protein Xmad1 recruits Xmad2 to unattached kinetochores. *J. Cell Biol.* **143**, 283–295.
15. Muzio, M., Chinnaiyan, A. M., Kischkel, F. C., Rourke, K. O., Shevchenko, A., Jian N., et al. (1996) FLICE, a novel FADD-homologous ICE/CED-3-like protease, is recruited to the CD95 (Fas/APO-1) death-inducing signaling complex. *Cell* **85**, 817–827.
16. Aigner, S., Lingner, J., Goodrich, K. J., Grosshans, C. A., Shevchenko, A., Mann, M., et al. (2000) Euplotes telomerase contains an La motif protein produced by apparent translational frameshifting. *EMBO J.* **19**, 6230–6239.
17. Taylor, J. A . and Johnson, R. S. (1997) Sequence database searches via de novo peptide sequencing by tandem mass spectrometry. *Rapid Commun. Mass Spectrom.* **11**, 1067–1075.
18. Huang, L., Jacob, R. J., Pegg, S. C., Baldwin, M. A., Wang, C. C., Burlingame, A. A., et al. (2001) Functional assignment of the 20S proteasome from *Trypanosoma brucei* using mass spectrometry and new bioinformatics approaches. *J. Biol. Chem.* **17**, 28327–28339.
19. Shevchenko, A., Sunyaev, S., Loboda, A., Shevchenko, A., Bork, P., Ens, W., et al. (2001) Charting the proteomes of organisms with unsequenced genomes by MALDI- Quadrupole Time-of-Flight mass spectrometry and BLAST homology searching. *Anal. Chem.* **73**, 1917–1926.
20. Johnson, R. S. and Taylor, J. A. (2000) Searching sequence databases via de novo peptide sequencing by tandem mass spectrometry. *Methods Mol. Biol.* **146**, 41–61.

21. Shevchenko, A., Loboda, A., Shevchenko, A., Ens, W., and Standing, K. G. (2000) MALDI Quadrupole Time-of-Flight mass spectrometry: a powerful tool for proteomic research. *Anal. Chem.* **72**, 2132–2141.

22. Medzihradszky, K. F., Campbell, J. M., Baldwin, M. A., Falick, A. M., Juhasz, P., Vestal, M. L., et al. (2000) The characteristics of peptide collision-induced dissociation using a high-performance MALDI-TOF/TOF tandem mass spectrometer. *Anal. Chem.* **72**, 552–558.

23. Link, A. J., Eng, J., Schieltz, D. M., Carmack, E., Mize, G. J., Morris, D. R., et al. (1999) Direct analysis of protein complexes using mass spectrometry. *Nat. Biotechnol.* **17**, 676–682.

24. Wilm, M. and Mann, M. (1996) Analytical properties of the nano electrospray ion source. *Anal. Chem.* **66**, 1–8.

25. Jensen, O. N., Wilm, M., Shevchenko, A., and Mann, M. (1999) Peptide sequencing of 2-DE gel-isolated proteins by nanoelectrospray tandem mass spectrometry. *Methods Mol. Biol.* **112**, 571–588.

26. Shevchenko, A., Chernushevich, I., and Mann, M. (1998) High sensitivity analysis of gel separated proteins by a quadrupole – TOF tandem mass spectrometer, in *Proceedings 46th ASMS Conference on Mass Spectrometry and Allied Topics*, Orlando, FL, Abstract no. 237.

27. Altschul, S., Madden, T., Schaffer, A., Zhang, J., Zhang, Z., Miller, W., et al. (1997) Gapped BLAST and PSI-BLAST: a new generation of protein database search programs. *Nucleic Acids Res.* **25**, 3389–3402.

28. Tournebize, R., Popov, A., Kinoshita, K., Ashford, A. J., Rybina, S., Pozniakovsky, A., et al. (2000) Control of microtubule dynamics by the antagonistic activities of XMAP215 and XKCM1 in *Xenopus* egg extracts. *Nat. Cell Biol.* **2**, 13–19.

29. Taylor, J. A. and Johnson, R. S. (2001) Implementation and uses of automated de novo peptide sequencing by tandem mass spectrometry. *Anal. Chem.* **73**, 2594–2604.

30. Shevchenko, A., Wilm, M., Vorm, O., and Mann, M. (1996) Mass spectrometric sequencing of proteins from silver-stained polyacrylamide gels. *Anal. Chem.* **68**, 850–858.

31. Shevchenko, A., Chernushevich, I., Wilm, M., and Mann, M. (2000) *Methods Mol. Biol.* **146**, 1–16.

32. Shevchenko, A. (2001) Evaluation of the efficiency of in-gel digestion of proteins by peptide isotopic labeling and MALDI mass spectrometry. *Anal. Biochem.* **296**, 279–283.

33. Wilm, M., Neubauer, G., and Mann, M. (1996) Parent ion scans of unseparated peptide mixtures. *Anal. Chem.* **68**, 527–533.

34. Perkins, D. N., Pappin, D. J., Creasy, D. M., and Cottrell, J. S. (1999) Probability-based protein identification by searching sequence databases using mass spectrometry data. *Electrophoresis* **20**, 3551–3567.

35. Altschul, S. F., Boguski, M. S., Gish, W., and Wootton, J. C. (1994) Issues in searching molecular sequence databases. *Nat. Genet.* **6**, 119–129.

36. Gaeta, B. A. (2000) BLAST on the Web. *Biotechniques* **28**, 436–440.

20

Direct Identification of Proteins in Ultracomplex Mixtures

Applications to Proteome Analysis

David M. Schieltz and John R. Yates, III

1. Introduction

Tandem mass spectrometry has been used for many years as a technique for sequencing peptides *(1)*. In conjunction with liquid chromatography and database searching, this technique has developed into a very powerful means of elucidating protein identities in complex mixtures *(2–5)*. These mixtures may result from an affinity enrichment or immuno-precipitation of an extract with a "bait" protein. Whole protein complexes may be enriched, such as the ribosome, which contains over 70 proteins *(3)*. With recent improvements, greater mixtures of proteins have been identified. It is now possible to directly identify proteins from organelles and whole-cell extracts containing several thousand proteins using chromatography, tandem mass spectrometry, and database searching *(6)*.

Generally, the approach for identification of proteins, consists of digesting the mixture with a site-specific enzyme after the complex is isolated *(1)*. The resulting peptide mixture is loaded onto a C_{18} reverse-phase high-performance liquid chromatography (RP-HPLC) column. Depending on the complexity of the sample, the duration of the reverse-phase gradient can be adjusted. As the gradient increases, peptides will be released from the solid phase based on hydrophobicity, where hydrophilic peptide release under low concentrations of organic phase and hydrophobic peptides elute later. This provides a method of concentration and separation so that the entire mixture is not introduced into the mass spectrometer all at once, as in the case with nano-spray *(7)*. The column is coupled to the tandem mass spectrometer through an electrospray ion source *(8–10)*. This serves to ionize the peptides by placing a high voltage on

From: *Methods in Molecular Biology, vol. 211: Protein Sequencing Protocols, 2nd ed.*
Edited by: B. J. Smith © Humana Press Inc., Totowa, NJ

the eluting solution, providing a voltage drop between exit of the electrospray ion source and the entrance to the mass spectrometer. The solvent exiting the source is nebulized into fine droplets, which retain the positive charges. The charged droplets are drawn towards the entrance of the mass spectrometer owing to the lower potential. As the droplets evaporate, positive charges do not leave with the evaporating solution; rather, they condense onto the peptide, thereby providing a method of "soft" ionization (9).

To acquire sequence information from a peptide, it undergoes two stages of mass measurement. This is done by quadrupole mass filters or ion traps (11–13). First, as peptides enter the mass spectrometer, the mass to charge (m/z) values of all peptides at that moment are recorded (MS scan). Then a single m/z value is selected and all other ions are removed by mass filtering. This is done in triple quadrupole instruments by setting the first quadrupole to pass only a window of ± 1.5 Da around the specified m/z. In ion-trap instruments, a notch waveform is generated that cause all other ions to be expelled from the trap except for the selected m/z value. This peptide ion is then vibrationally activated by collisions with an inert gas that causes the peptide to fragment at points along the peptide bond (14). Two series of ions are generated, fragments from the carboxy terminus (y-type ions) and fragments from the amino terminus (b-type ions) (1,15). Subsequently, the m/z values of the fragments are recorded. This represents the second step of mass spectrometry or "tandem" mass spectrometry (MS/MS scan). By subtracting neighboring y-ion peaks or b-ion peaks from each other, the mass of the amino acid at that position in the sequence can be determined. Therefore, a sequence can be obtained and protein can be identified. Current technology allows mass spectrometers to be programmed to set a threshold, acquire the top three to five most abundant ions per mass spectrum, and populate an exclusion list so that once an ion has been selected for MS/MS it will not be selected again. Mass spectrometers can be set to cycle from MS to MS/MS to acquire as many unique tandem mass spectra as possible. In a single experiment it is possible to record over a one thousand spectra in a 60-min period. With available database search programs such as SEQUEST (16) and MASCOT (17) that are able to match un-interpreted tandem mass spectra to sequences in protein, nucleotide, or expressed sequence tag (EST) databases, protein identification in complex mixtures can be very rapid. Using nano-LC integrated ion sources, protein identification is possible with high sensitivity (18). The power of this technique comes from the ability to separate peptides through chromatography and use the mass spectrometer to select single peptide ions for fragmentation.

Recently, this method was improved to provide a greater capacity for identification of proteins. A common problem with analysis of very complex mixtures on a single dimension of reverse phase material is co-elution of peptides.

When this occurs, the mass spectrometer cannot acquire tandem mass spectra quickly enough to sample all of the peptides within a peak before they have completely eluted. This inefficiency leads to incomplete identification of all the proteins in the mixture. To combat this, a second dimension of chromatography is added to the existing material to form a two-dimensional chromatography column (2D-LC). Capillary pulled needles are filled with reverse-phase material then filled with strong cation-exchange (SCX) material *(3)*. This system separates peptides first by charge then hydrophobicity. Increasing concentrations of salt are used to displace a certain population of peptides from the SCX, which then bind to the reverse-phase bed. After the salt is flushed from the column, a reverse-phase gradient is used to elute the peptides into the mass spectrometer. This cycle is repeated, each time with a higher concentration of salt until all the peptides have been removed. The SCX acts as a holding area for the peptides. Portions of the mixture can be sequentially parsed out to the reverse phase column. In this manner, the reverse phase is not overwhelmed by the entire mixture and can efficiently separate peptides.

Using this improved technique, known as Multi-Dimensional Protein Identification Technology (MuDPIT), has allowed for the high-throughput and high-sensitivity identification of proteins in the yeast ribosome *(3)* and the yeast whole-cell proteome *(6)*. Another advantage of the technique is the discovery of a large number of protein-protein interactions in complexes. This was demonstrated with the discovery of new members of the yeast proteasome *(5)*. The following outline describes the method for the MuDPIT analysis of the proteomes of organelles or whole cells.

2. Materials
2.1. Instruments/HPLC Hardware

1. Tandem mass spectrometer: ThermoFinnigan LCQ Classic™, Deca™ or EXP™ tandem mass spectrometer (Finnigan MAT, San Jose, CA); Micromass QTOF1™, QTOF2™ tandem mass spectrometer (Micromass, Inc., Beverly, MA).
2. HPLC system: HP1100™ quaternary pump (Agilent Technologies, Palo Alto CA).
3. Laser Puller: A Sutter Instruments P2000 laser puller (Sutter Instruments, Novato, CA).
4. Nano-LC ion sources: (ThermoFinnigan, San Jose, CA), (The Scripps Research Institute, LaJolla, CA).
5. Stainless steel pressurization device (pneumatic "bomb") (The Scripps Research Institute).
6. Helium Tank 99.99% purity and regulator to deliver 1000 psi (Airgas Inc., Radnor, PA)
7. LabquakeTM shaker (Barnstead/Thermolyne, Dubuque, IA).

2.2. HPLC buffers/Chemicals/Enzymes/Materials

1. Buffers: Deionized water, HPLC grade Acetonitrile (ACN), HPLC-grade Methanol (MeOH), (Fisher Scientific, Fair Lawn, NJ); Heptafluorobutyric acid (HFBA), (Pierce Chemical Co., Rockford, IL); Glacial acetic acid (Mallinckrodt Baker, Inc., Phillipsburg, NJ); 98% formic acid (Fluka Chemicals, Buchs, Switzerland).
2. Chemicals: Sodium vanadate ($NaVO_3$), sodium fluoride (NaF), sodium pyrophosphate ($Na_4P_2O_7$), cyanogen bromide (CNBr), (Aldrich Chemicals, Milwaukee, WI); Ammonium acetate (AmOAC), calcium chloride, Urea, ammonium bicarbonate (AmBic), ethylenediamine tetraacetate (EDTA), Tris-HCl, dibasic sodium phosphate (Na_2HPO_4), dibasic potassium phosphate (KH_2PO_4), (Sigma Chemicals, St. Louis, MO).
3. Proteolytic Enzymes: Trypsin (sequencing grade) working concentration of 0.5 µg/µL, Endoproteinase Lys-C (sequencing grade) working concentration of 0.1 µg/µL, (Roche Diagnostics, Indianapolis, IN); Trypsin, Poroszyme™ bulk immobilized (Applied Biosystems, Framingham, MA).
4. 100 micron inner diameter (ID) by 360 micron outer diameter (OD) fused silica capillary, 50 micron ID by 360 micron OD fused silica capillary (Agilent Technologies), (Polymicro Technologies, Phoenix, AZ).
5. Partisphere™ 5 µm strong cation exchange (SCX), (Whatman, Clifton, NJ); Zorbax™ XDB 5 µm C_{18} reverse-phase packing material (RP) (Agilent Technologies).
6. SPEC Plus PT C18 solid-phase extraction pipet tips (Ansys Diagnostics, Lake Forest, CA); 1-mL plastic syringe (Becton-Dickinson, Franklin Lakes, NJ); Finger-tight PEEK cross, finger-tight sleeves (UpChurch Scientific, Oak Harbor, WA); Gold wire (Scientific Instrument Services, Inc., Ringoes, NJ).

2.3. Solutions for HPLC: Strong Cation-Exchange/Reversed-Phase Two-Dimensional HPLC

1. Buffer A: 5% acetonitrile/0.02% HFBA.
2. Buffer B: 80% acetonitrile/0.02% HFBA.
3. Buffer C: 5% acetonitrile/0.02% HFBA/containing 250 mM ammonium acetate.
4. Buffer D: 5% acetonitrile/0.02% HFBA/containing 1 M ammonium acetate.

2.4. Digestion Buffers

1. Endoproteinase Lys-C Digestion: 8 M urea, 100 mM ammonium bicarbonate, pH 8.5.
2. Trypsin Digestion: 2 M urea, 100 mM ammonium bicarbonate, pH 8.5, 1 mM $CaCl_2$.
3. Cyanogen Bromide digestion: 90% formic acid.

3. Methods

Once the tissue, cell, or organelle has been isolated, the proteins are extracted in the presence of protease inhibitors and split into two fractions, soluble and insoluble. Generally the soluble extract is generated in the presence of 8 M

urea/100 m*M* ammonium bicarbonate, pH 8.5, or the extract can be lyophilized and buffer exchanged into 8 *M* urea/100 m*M* ammonium bicarbonate, pH 8.5. This solution is centrifuged to pellet cellular debris and insoluble material. The resulting membrane or insoluble fraction is brought up in 90% formic acid for identification of insoluble proteins *(6)* (*see* **Note 1**).

The digestion of the soluble fraction consists of adding between 0.5 μg and 1.5 μg of endoproteinase Lys-C depending on the concentration of the solution. The enzyme is allowed to incubate overnight at 37°C in the dark. The following day the solution is diluted to 2 *M* urea with 100 m*M* ammonium bicarbonate, pH 8.5. CaCl$_2$ is added to give a final concentration of 1 m*M*. Depending on the volume of the solution, liquid trypsin or the Poroszyme™ immobilized trypsin can be used. Poroszyme trypsin is more effective for digestion of large volume (500–750 μL) solutions with high protein concentrations (several milligrams/milliliter). When using the immobilized trypsin, the digestion is carried out at 37°C in a rotating shaker to spread the beads throughout the solution. When digestion is complete, the sample is buffer exchanged and concentrated using an Ansys Diagnostics SPEC Plus PT C18 solid-phase extraction pipet tip (*see* **Note 2**).

The digestion of the insoluble fraction consists of resuspending the insoluble pellet in 90% formic acid. Cyanogen bromide crystals are added to digest the proteins, where cleavage of the proteins occurs at methionines. This step generates long peptides that should be soluble in 8 *M* urea, where Lys-C and trypsin can be used to further complete the digestion prior to mass spectrometry. The sample is incubated overnight in the dark at room temperature. After digestion with CNBr the pH is adjusted using ammonium bicarbonate and then vacuum concentrated. The sample is brought up in 8 *M* urea and digested with Lys-C followed by immobilized trypsin (*see* **Note 3**).

3.1. Digestion of Soluble Fraction

1. Solubilize sample in 8 M urea/ 100 mM ammonium bicarbonate pH 8.5.
2. Add 1–5 μL of a 0.1 μg/μL endoproteinase Lys-C solution.
3. Incubate overnight at 37°C in the dark.
4. Dilute the sample to 2 *M* urea with 100 m*M* ammonium bicarbonate, pH 8.5.
5. Add 100 m*M* CaCl$_2$ to give a final volume of 1 m*M* CaCl$_2$.
6. Add 5–15 μL of Poroszyme immobilized trypsin slurry to sample.
7. Place microfuge tube, containing the sample in a Labquake™ rotating shaker and incubate the sample overnight at 37°C in the dark.

3.2. Digestion of Insoluble Fraction

1. Re-suspend insoluble pellet in 100–300 μL of 90% formic acid.
2. Add 100 mg of cyanogen bromide and incubate the sample in the dark at room temperature, overnight.

3. Adjust the pH of the sample to 8.5 using saturated ammonium bicarbonate (volume will increase by 3× or 4×, make sure the solution is transferred to a larger tube prior to starting this step).
4. Lyophilize the solution to near dryness using a vacuum concentrator.
5. Re-suspend the sample in 8 *M* urea/100 m*M* ammonium bicarbonate, pH 8.5.
6. Follow **steps 2–7** in **Subheading 3.1.**

3.3. Buffer Exchange

Before the sample is loaded onto the 2D nano-LC column, it must be desalted. The salts present in the sample will prevent peptides from binding to the strong cation exchange material, effectively reducing the 2D system to one dimension. To remove the urea and salts that are present, an Ansys Diagnostics SPEC Plus PT C18 solid-phase extraction pipet tip is used. A 1-mL plastic syringe is needed to move the solutions through the solid phase disk (*see* **Note 4**).

1. Activate the SPEC Plus PT C18 tip by flowing 200 µL of buffer B through slowly. Do not let the disk dry out.
2. Equilibrate the disk with 400 µL of buffer A.
3. Load the sample by either introducing it from the top or pulling it through the bottom.
4. Pass the sample through the disk three times to ensure the disk binds as much material as possible.
5. Wash the disk with 200 µL of buffer A to remove salts.
6. Elute the sample with 50–100 µL of buffer B.
7. Vacuum concentrate the sample to near dryness in a lyophilizer.
8. Resuspend the sample in 5–15 µL of 5% formic acid/5% acetonitrile.

3.4. Sample Loading

The peptide mixture is separated online using high-sensitivity 2D HPLC. Typical conditions for high-sensitivity analysis require the use of a microcolumn capillary needles with 100 micron ID or less and flow rates between 100 and 300 nL/min. Loading of the sample is performed by removing the 2D-nano-LC column from the source, and placing it in a pneumatic bomb *(19)*. The microfuge tube containing the sample is seated into the bomb and the lid is fastened down. The entrance of the column is threaded through a Teflon ferrule in the lid of the bomb and is tightened down once the column is placed into the correct position. The bomb is pressurized to ~ 800 psi, which forces liquid up through the column. The amount loaded onto the column is measured by displacement using a graduated glass capillary. The column is then placed back into the microelectrospray source and equilibrated with 0% buffer B (*see* **Note 5**).

3.5. Two-Dimensional HPLC Separation
of Ultra-Complex Peptide Mixtures

By combining strong cation exchange with reversed-phase chromatography, peptides are separated orthogonally, first by charge, then by hydrophobicity. This greatly improves the resolution of the chromatographic separation. The sample is loaded onto the column and the peptide mixture binds to the SCX. Any peptides that do not bind will pass through the SCX bed and bind to the reverse-phase bed. After the column is washed with 0% B for several minutes a 90-min reverse-phase gradient is initiated, which runs from 0–60% B. Peptides are eluted, ionized and analyzed in the tandem mass spectrometer. Once the gradient is finished the column is re-equilibrated with 0% B for several minutes. Then a 2-min plug of X concentration of 250 mM ammonium acetate is allowed to flow through the column to displace a portion of the peptides from the SCX bed to the RP bed. The column is equilibrated again followed by a 90-min 0–60% B reverse phase gradient. X is the percent concentration of 250 mM ammonium acetate, which is used in each of the salt bumps (X = 10, 20, 30, 40, 50, 60, 70, 80, 90, 100). When all of the 250 mM ammonium acetate salt steps have been performed, a final salt step of 1 M ammonium acetate is used to strip all peptides from the SCX bed.

3.6. Tandem Mass Spectrometry

The mass spectrometers are programmed to perform data-dependant tandem mass spectrometry as peptides elute from the column. In the case of the ThermoFinnigan LCQ DECA the scan range was set from m/z 400 to 1400. The DECA was programmed to acquire one scan of MS followed by three rounds of MS/MS in which the first, second, and third most intense ions from the MS scan are selected individually. This cycle was repeated throughout the duration of the gradient and Dynamic Exclusion was activated to prevent ions from being re-selected while they eluted.

3.7. Parameters for Data Dependant MS/MS Scanning
for ThermoFinnigan LCQ Deca™

1. Scan range for MS, m/z 400–1400, 7 microscans, maximum injection time 450 ms.
2. Minimum MS signal for triggering to MS/MS 1E+05.
3. Default charge state of peptide MS/MS is 2.
4. Three most intense ions are selected for MS/MS analysis, 7 microscans for each tandem mass spectrum.
5. Isolation width for selection of precursor for fragmentation 2.5 Da.
6. Electrospray needle voltage 1.5–1.8 kV.
7. Dynamic Exclusion duration, 10 min for a 90-min gradient.

3.8. Protein Identification Using Database Searching

To identify proteins in mixtures, database searching is used, provided that the organism's genome has been sequenced. Several software programs are available that match un-interpreted tandem mass spectra to peptide sequences in databases (*16,17,20*). This allows for the rapid identification of proteins in mixtures. In a typical MuDPIT analysis, anywhere from 70,000–90,000 tandem mass spectra are acquired. By using a database search algorithm such as SEQUEST (*16*), proteins in whole cell lysates or organelles can be identified quickly in an automated fashion.

To ensure that as many proteins in a mixture can be identified using the SEQUEST algorithm, indexed databases are not used and the database search is not constrained to look only for tryptic peptides. In very complex mixtures such as whole-cell lysates it is possible for nonspecific cleavages to occur and by constraining the search only to look for tryptic peptides these nontryptic peptides would be missed. When tryptic peptides are matched in the search, this gives added assurance that the peptide is present, because trypsin was used. Thus, it is important to use protease inhibitors during the cell lysis, as this helps reduce the number of nontryptic cleavages. In general high-confidence protein identifications usually follow these criteria, a high SEQUEST correlation score (X_{corr}), a high SEQUEST δC_n score, tryptic peptides, and several unique peptides belonging to the same protein. Several unique peptides generally assure a given protein is present. Usually one peptide sequence from the group is manually confirmed against the tandem mass spectrum to determine, if the fragment ion signals are above background and that there is continuity with the *y*-ion and *b*-ion series (*3*). If a single peptide identifies a protein, then manual confirmation is required (*see* **Note 6**).

3.9. Filtering SEQUEST Output for Positive Identifications

1. For +1 peptides look for an X_{corr} above 1.5.
2. For +2 peptides look for an X_{corr} above 2.4.
3. For +3 peptides look for an X_{corr} above 3.7 and full or partially tryptic.
4. For all peptides look for δC_n values above 0.08.
5. Look for several unique peptides that match the same protein.
6. Look for continuous or near continuous matched series of *y*- or *b*-ions.

4. Notes

1. It is important to use protease inhibitors when whole cell or organelle extracts are being generated to suppress proteolytic enzymes from within the cell. This should help reduce, to some extent, the number of non-tryptic peptides present. This will improve identification of proteins, which is discussed in **Subheading 3.8.**

2. To ensure that a high degree of specific tryptic cleavage of proteins occurs. The 8 *M* urea is used to solubilize the proteins in the extract. This should denature tightly folded protein and proteins that are associated in complexes. Endoproteinase Lys-C cleaves at lysines and remains active in 8 *M* urea. By using this enzyme most of the tertiary structure of the proteins is eliminated which prevent proteins precipitating when the solution is diluted to 2 *M* urea prior to trypsin digestion. Using Lys-C in 8 *M* urea followed by trypsin in 2 *M* should effectively solubilize, denature, and digest the protein into a large percentage of tryptic peptides *(3)*.

3. To gain access to peripheral and integral membrane proteins, 90% formic acid is used to solubilize the remaining pelleted material and solid CNBr is added to digest the protein into peptides. These peptides are then buffer exchanged into a more convenient buffer (8 *M* urea) where additional digestion can generate tryptic peptides *(6)*.

4. The 5% formic acid/5% acetonitrile solution is added to the lyophilized sample so that during loading on the 2D LC system, the peptides will effectively bind to the SCX bed.

5. The ion-source consists, of an integrated nano-flow capillary needle column and micro-electrospray interface according to the design of Gatlin et al. *(18)*. The column is constructed by pulling a 3–5 µm diameter tip, from 365 µm OD by 100 µm ID fused silica capillary. First 12 cm of 5 µm Zorbax XDB C18 reverse-phase material is loaded into the capillary followed by 10 cm of 5 µm Partisphere strong cation-exchange material. An UpChurch fingertight PEEK cross is used to connect the transfer line from the HPLC pump, the flow splitting line, the gold wire, and the column. The transfer line uses 365 µm × 50 µm capillary to quickly move the gradient from the HPLC pump to the cross. The flow-splitter line consists of 365 µm × 50 µm capillary. The length of this capillary depends on the length of the column. Generally, the flow splitter line is cut to a length that will give a flow-rate from the column of 100 nL/min to 250 nL/min. Most of the flow from the pump exits the splitter and a small amount of solvent flows through the column. The gold wire is used to place a high voltage on the solution in the column and produces electrospray ionization.

6. To generate a meaningful list of the protein identification data, *SEQUEST-summary* is sorted according to locus or gene name, this allows viewing of the number of peptides (unique and nonunique) per protein. This is exported to Microsoft ExcelTM where the ion exchange fraction in which the peptide was found, is listed along with an abbreviation of the cross-correlation score. The SEQUEST software is commercially available from ThermoFinnigan.

References

1. Hunt, D. F., Yates, J. R., 3rd, Shabanowitz, J., Winston, S., and Hauer. C. R. (1986) Protein sequencing by tandem mass spectrometry. *Proc. Natl. Acad. Sci. USA* **83**, 6233–6237.
2. McCormack, A. L., Schieltz, D. M., Goode, B., Yang, S., Barnes, G., Drubin, D., and Yates, J. R., 3rd (1997) Direct analysis and identification of proteins in mix-

tures by LC/MS/MS and database searching at the low-femtomole level. *Anal. Chem.* **69**, 767–776.

3. Link, A. J., Eng, J., Schieltz, D. M., Carmack, E., Mize, G. J., Morris, D. R., et al. (1999) Direct analysis of protein complexes using mass spectrometry. *Nat. Biotechnol.* **17**, 676–682.

4. Meeusen, S., Tieu, Q., Wong, E., Weiss, E., Schieltz, D., Yates, J. R., and Nunnari, J. (1999) Mgm101p is a novel component of the mitochondrial nucleoid that binds DNA and is required for the repair of oxidatively damaged mitochondrial DNA. *J. Cell Biol.* **145**, 291–304.

5. Verma, R., Chen, S., Feldman, R., Schieltz, D., Yates, J., Dohmen, J., and Deshaies, R. J. (2000) Proteasomal proteomics: identification of nucleotide-sensitive proteasome-interacting proteins by mass spectrometric analysis of affinity-purified proteasomes. *Mol. Biol. Cell* **11**, 3425–3439.

6. Washburn, M. P., Wolters, D., and Yates, J. R., 3rd (2001) Large-scale analysis of the yeast proteome by multidimensional protein identification technology. *Nat. Biotechnol.* **19**, 242–247.

7. Wilm, M. and Mann, M. (1996) Analytical properties of the nanoelectrospray ion source. *Anal. Chem.* **68**, 1–8.

8. Whitehouse, C. M., Dreyer, R. N., Yamashita, M., and Fenn, J. B. (1985) Electrospray interface for liquid chromatographs and mass spectrometers. *Anal. Chem.* **57**, 675–679.

9. Fenn, J. B., Mann, M., Meng, C. K., Wong, S. F., and Whitehouse, C. M. (1989) Electrospray ionization for mass spectrometry of large biomolecules. *Science* **246**, 64–71.

10. Covey, T. R., Huang, E. C., and Henion, J. D. (1991) Structural characterization of protein tryptic peptides via liquid chromatography/mass spectrometry and collision-induced dissociation of their doubly charged molecular ions. *Anal. Chem.* **63**, 1193–1200.

11. Yost, R. A. and Boyd R. K. (1990) Tandem mass spectrometry: quadrupole and hybrid instruments. *Methods Enzymol.* **193**, 154–200.

12. Louris, J. N., Brodbelt Lustig, J. S., Cooks, R. G., Glish, G. L., van Berkel, G. J., and McLuckey, S. A. (1990) Ion isolation and sequential stages of mass spectrometry in a quadrupole ion trap mass spectrometer. *Int. J. Mass Spectrom. Ion Proc.* **96**, 117–137.

13. Jonscher, K. R. and Yates J. R., 3rd (1997) The quadrupole ion trap mass spectrometer—a small solution to a big challenge. *Anal. Biochem.* **244**, 1–15.

14. McLafferty, F. (1983) *Tandem Mass Spectrometry.* Wiley-Intersciences, New York, NY.

15. Roepstorff, P. and Fohlman J. (1984) Proposal for a common nomenclature for sequence ions in mass spectra of peptides. *Biomed. Mass Spectrom* **11**, 601.

16. Eng, J., McCormack, A., and Yates, J. R., 3rd (1994) An approach to correlate tandem mass spectral data of peptides with amino acid sequences in a protein database. *J. Am. Soc. Mass Spectrom.* **5**, 976–989.

17. Perkins, D. N., Pappin, D. J., Creasy, D. M., and Cottrell J. S. (1999) Probability-based protein identification by searching sequence databases using mass spectrometry data. *Electrophoresis* **20**, 3551–3567.

18. Gatlin, C. L., Kleemann, G. R., Hays, L. G., Link, A. J., and Yates J. R., 3rd (1998) Protein identification at the low femtomole level from silver-stained gels using a new fritless electrospray interface for liquid chromatography-microspray and nanospray mass spectrometry. *Anal. Biochem.* **263**, 93–101.

19. McCormack, A., Eng, J., and Yates, J. R., 3rd (1994) Peptide sequence analysis on quadrupole mass Spectrometers, in *Methods: A Companion to Methods in Enzymology*, vol. 6, (Shively, J., ed.), Adademic Press, San Diego, CA, pp. 274–283.

20. Clauser, K. R., Baker, P., and Burlingame, A. L. (1999) Role of accurate mass measurement (+/- 10 ppm) in protein identification strategies employing MS or MS/MS and database searching. *Anal. Chem.* **71**, 2871–2882.

21

Identification of PTH-Amino Acids by HPLC

Gregory A. Grant and Mark W. Crankshaw

1. Introduction

Protein sequence analysis employing Edman degradation chemistry commonly uses high-performance liquid chromatography (HPLC) separation as the means for identification of the PTH-amino acid (phenylthiohydantoin amino acid) produced at each cycle. In fact, all modern automated protein sequencing instruments come with "on-line" HPLC detection as standard equipment. This method of identification of PTH-amino acids is dependent on essentially two factors: resolution of one PTH-amino acid from another and coincidence of elution position of the unknown PTH-amino acid with that of a known standard. Because the identification of the PTH-amino acid is based almost solely on its elution position, the ideal situation is that each PTH-amino acid elute at a unique position in the chromatogram. This is accomplished by the choice of column, mobile phase, and elution program. However, even if conditions are found to separate every PTH-amino acid from one another, identification still depends on matching the elution position with a standard of known structure. The method is straightforward for the genetically coded amino acids but becomes more problematic for modified amino acids that are not routinely encountered. Because the method does not provide a direct identification of the PTH-amino acid, additional analysis by chemical or physical means may often be necessary to provide unequivocal identification of nonstandard PTH-amino acids. However, it is often helpful to have some knowledge of where known modified PTH-amino acids elute in these systems. This provides a starting point for the investigator and provides an additional level of knowledge upon which to proceed.

Applied Biosystems is presently the only active manufacturer of automated protein sequencers. Therefore, this article will focus mainly on these instru-

From: *Methods in Molecular Biology, vol. 211: Protein Sequencing Protocols, 2nd ed.*
Edited by: B. J. Smith © Humana Press Inc., Totowa, NJ

ments. The Hewlett-Packard Model G1005A protein sequencer is still in use and still being supported by the manufacturer, so sections dealing with this instrument have been included. However, since the Beckman Model 2000 and 3000 series (previously Porton) have not been manufactured for many years, they are not included here. Pertinent information about these instruments can be found by consulting the original edition of this article *(1)*.

The reverse phase HPLC separation of PTH-amino acids provided as standard equipment on the different automated protein sequencers are similar but not identical (*see* **Note 1**). Thus a discussion of each is presented with information on optimizing the separation of the PTH-amino acids as well as the elution positions of some modified PTH-amino acids on each system. The data on the elution positions of modified PTH-amino acids have mostly been taken from a survey conducted by the Association of Biomolecular Resource Facilities (*see* **Note 2**).

2. Materials

2.1. Applied Biosystems Sequencers

Applied Biosystems Inc. was the first company to manufacture modern protein sequencers and has produced several different models (i.e., 470, 473, 475, 476, 477, and the Procise (49x) series, where x designates the number of reaction cartridges). The company has undergone several name changes since its inception, including Perkin Elmer, Perkin Elmer/Applied Biosystems, and PE Biosystems. At the present time they have come back to the original name of Applied Biosystems. Presently, only the 477 and Procise 49x series are being supported by the company and the 49x series is the only one being manufactured. It can be purchased with a single or multiple (1, 2, or 4) reaction cartridges and with a capillary HPLC system, referred to as the Procise cLC. The chromatography development protocols used on these instruments do not differ significantly, but several different solvent compositions and gradient programs have been used. With the exception of the elution position of PTH-histidine, PTH-arginine, and PTH-pyridylethyl cysteine, the standard chromatograms using either system are similar. Only the 477 and 49x series will be discussed here. Readers interested in information on earlier models are referred to the first edition *(1)*.

2.1.1. Premix Solvent System

The premix buffer system is the most recent one to be utilized on the Applied Biosystems sequencers and virtually all of the 49x series instruments utilize this system. Solution A is prepared by adding a buffer solution containing an ion pairing reagent to a solution of 3.5 % (v/v) tetrahydrofuran in water so that

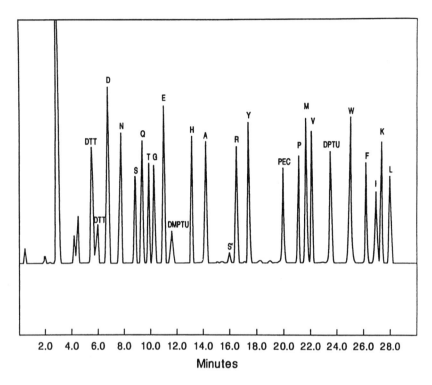

Fig. 1. Elution profile of standard PTH-amino acids on an Applied Biosytems Model 477 Sequencer. 10 pmol of each PTH-amino acid is used. The one-letter code designates standard PTH-amino acids. Gradient conditions are described in **Table 3**.

protonatable PTH-amino acids such as PTH-arginine, PTH-histidine, and PTH-pyridylethyl cysteine are well-separated from the other PTH amino acids (compositions are given in **Subheading 3.**). Solution B is acetonitrile or acetonitrile/isopropanol, often with dimethylphenylthiourea (DMPTU) added as a scavenger in all systems prior to the Procise series. Premix buffer is supplied by Applied Biosystems as are the other solutions and HPLC columns. Standard chromatograms developed with the premix system for the Model 477, Model 49x, and Model 49x cLC are shown in **Figs. 1–3**. As you can see, they are very similar, differing mainly in elution time and sensitivity. The elution positions of modified amino acids relative to the common genetically coded amino acids as they would run on a 477 sequencer are given in **Table 1**. This table also lists the relative elution positions of the commonly encountered phenylthiocarbamyl (PTC)-amino acids, which are partial conversion products, as well as common sequencing artifact peaks. **Table 2** lists relative positions of modified amino acids often encountered in synthetic peptides when they are analyzed by

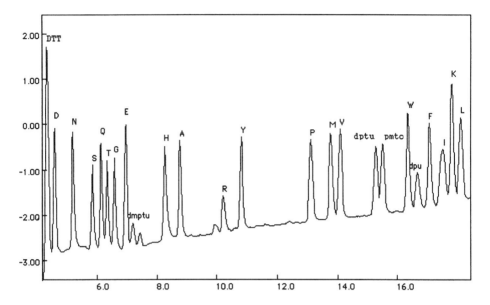

Fig. 2. Elution profile of standard PTH-amino acids on an Applied Biosytems Model 494 Sequencer. 10 pmol of each PTH-amino acid is used. The one-letter code designates standard PTH-amino acids. Gradient conditions are described in **Table 4**.

on-resin sequencing (*see* **Note 4**) with a Model 477 sequencer. Note that these tables were generated with a Model 477 sequencer and that the elution times are different for the 49x Models. Relative elution positions on the 49x sequencers can be estimated by comparing the positions of the common genetically coded amino acids.

2.1.2. Sodium Acetate Solvent System

The original sodium acetate solvent system for PTH-amino acid separation is still being used by a number of investigators on the Model 477 instruments. In this system, the pH and ionic strength of solution A are adjusted by adding various amounts of pH 3.8 and pH 4.6 3 *M* sodium acetate solutions to 3.5 % (v/v) tetrahydrofuran in water. The ratio of the two sodium acetate solutions determines the final pH and the total volume of sodium acetate determines the ionic strength. As with the premix system, solution B is acetonitrile or acetonitrile/isopropanol, often with DMPTU added. Buffers are supplied by Applied Biosystems as are the other solutions and HPLC columns. This system gives essentially the same separation profile as the premix system with the exception of the elution positions of the protonatable PTH-amino acids. In this case, PTH-histidine and PTH-arginine generally elute after PTH-alanine and PTH-

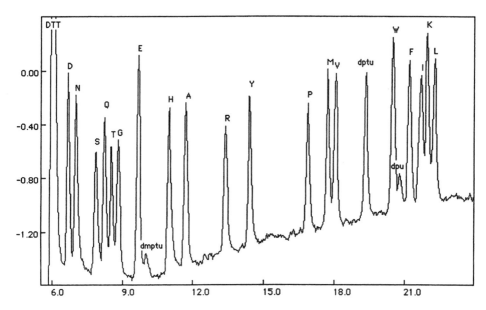

Fig. 3. Elution profile of standard PTH-amino acids on an Applied Biosytems Model 494 cLC Sequencer. 1 pmol of each PTH-amino acid is used. The one-letter code designates standard PTH-amino acids. Gradient conditions are described in **Table 4**.

tyrosine, respectively, and PTH-pyridylethyl cysteine elutes with or just before DPTU.

2.2. Hewlett Packard Sequencer

The Hewlett Packard Model G1005A Protein Sequencing System is the latest instrument to be developed but it is also no longer being manufactured. It utilizes a three solvent system for the PTH-amino acid separation shown in **Fig. 4** *(2)*. All solutions and the HPLC column are supplied by Hewlett-Packard. The composition of the solvents is proprietary. Although the relative elution positions of most of the PTH-amino acids are the same as in other systems, there are some differences, particularly early in the chromatogram. One characteristic of the separation is that the two allomers of PTH-isoleucine are completely resolved. The standard chromatogram shown in **Fig. 4** shows only one of these. The other, which usually appears in a ratio of approx 40 % to the one shown, co-elutes with PTH-phenylalanine. This does not usually present a problem unless a mixture of peptides is being sequenced. In some cases it may be difficult to tell if both PTH-isoleucine and PTH-phenylalanine appear in the same cycle. The elution positions of several cysteine derivatives are described in the legend to **Fig. 4**.

Table 1
Relative Elution Positions of PTH-Amino Acids on HPLC[a]

Retention time (min)	PTH-amino acid	Notes
5.8	Tyr(P)	Stable
5.9	PTC-Gly	Stable, partial conversion product of ATZ-Gly
6.0	γ-carboxyglutamic acid (Gla)	Broad peak. Some decarboxylation to Glu
6.2	Asp	Stable, slight conversion to Asp derivative
6.3	phenylthiourea (PTU)	Reaction product of PITC and ammonia
7.0	PTC-Ala	Stable, partial conversion product of ATZ-Ala
7.1	Asn	~ 5–10% recovered as PTH-Asp
7.6	aniline	
8.0	O-fucosyl threonine	Deglycosylates with successive cycles
8.2	Ser	Typically ~ 0–20% recovered, see Ser derivative
8.5	S-carboxymethyl cysteine	Partially unstable
8.8	Gln	~ 5–20% recovered as PTH-Glu
9.3	Homoserine	
9.3	Thr	Typically ~ 5–25 % recovered, see Thr derivative
9.3	Tris buffer	
9.7	Gly	Stable
10.5	Glu	Stable, slight conversion to Glu derivative
11.0	Dimethylphenylthiourea (DMPTU)	Edman chemistry by-product
11.0	Cysteine S-β-propionamide	
11.2	Tris buffer	
11.4	S-carboxamidomethyl cysteine	
11.7	Carboxamidomethyl methionine	
12.0	Methionine sulfone	
12.5	N-ε-succinyl lysine	Partially unstable

Table 1 (*continued*)
Relative Elution Positions of PTH-Amino Acids on HPLC[a]

Retention time (min)	PTH-amino acid	Notes
12.8	His[b]	Stable but may not be completely extracted from sequencer
12.9	5-hydroxy lysine derivative	Not always seen
13.0	Hypro (cis)	Hydroxyl can be cis or trans, Edman chemistry produces botl
13.6	N-ε-acetyl lysine	
13.8	Ala	Stable
14.0	Hypro (trans)	Hydroxyl can be cis or trans, Edman chemistry produces botl
14.2	Methyl Histidine	
15.1	PTC-Met	Stable, partial conversion product of ATZ-Met
15.2	Ser derivative	Breakdown product of serine due to Edman Chemistry
16.5	PTC-Val	Stable, partial conversion product of ATZ-Val
16.7	Arg[b]	Stable but may not be completely extracted from sequencer
17.0	O-methyl threonine	
17.7	Cystine	
17.7	Tyr	Stable
17.9	O-methyl glutamic acid	
18.0	Threonine-related peak	
18.2	N-ε-methyl lysine	*See* text **Note 5**
18.4	N-ε-dimethyl lysine	*See* text **Note 5**
18.6	N-ε-trimethyl lysine	*See* text **Note 5**
18.7	Cystine related peak	
18.8	Ser derivative	Breakdown product of serine due to Edman Chemistry
18.9	Canavanine	
19.1	α-aminobutyric acid	
19.2	Methyl arginine	

Table 1 (continued)
Relative Elution Positions of PTH-Amino Acids on HPLC[a]

Retention time (min)	PTH-amino acid	Notes
19.3	S-methyl cysteine	
19.6	DL-homocysteine	
19.9	PTC-Lys	Stable, partial conversion product of ATZ-Lys
19.9	Threonine-related peak	
20.0	S-pyridylethyl cysteine	
20.4	S-carboxamidopropyl cysteine	
20.5	Methanol PITC	
20.6	R4 contamination	
20.7	PTC-Leu	Stable, partial conversion product of ATZ-Leu
20.7	PTC-Ile	Stable, partial conversion product of ATZ-Ile
20.9	DTT/PTC adduct	
21.1	Pro	Stable
21.9	Met	Usually stable but susceptible to oxidation
22.3	Val	Stable
22.8	5-hydroxylysine	
23.7	Asp derivative, aniline amide of β carboxyl	Associated with conversion to PTH-Asp
23.9	Diphenylthiourea (DPTU)	Edman chemistry by-product
24.0	Glu derivative, aniline amide of γ carboxyl	Associated with conversion to PTH-Glu
24.0	Iodotyrosine	
24.1	Trp derivative, kynurenine	
24.1	Methanol PITC	Elutes with DPTU on systems using neat acetonitrile as solvent B
25.0	α,γ-diaminobutyric acid	
25.3	Trp	Typically only partially recovered at inconsistent levels
25.5	Diphenylurea	

Table 1 (continued)
Relative Elution Positions of PTH-Amino Acids on HPLC[a]

Retention time (min)	PTH-amino acid	Notes
25.6	Ornithine	
25.9	O-methyl tyrosine	
26.1	Phe	Stable
26.3	Coomassie blue	
26.5	Oxidized DTT/PITC adduct	
26.7	Ile	Stable
27.1	Lys (ptc)	Stable, both amino groups are derivatized with PITC
27.1	Lanthionine	Reports of minor peaks between ser and gln and at dehydroalanine
27.4	Leu	Stable
27.6	S-aminopropyl cysteine	
28.2	Norleucine	Stable
28.4	p-chlorophenylalanine	
29.0	Diiodotyrosine	
29.0	N-ε-methyl lysine	*See* text **Note 5**
29.0	N-ε-dimethyl lysine	*See* text **Note 5**
29.4	Canavanine-related peak	

[a]Elution positions are relative and may vary somewhat depending on system and column. Some of these derivatives were originally done with different conditions and are now represented on a typical sequencing cycle. Table composed from material taken from references

[b]Position is variable depending on ionic strength and elution buffer composition.

Table 2
Relative Elution Positions of PTH-Amino Acids on HPLC using REZ Cycle[a]

Retention time (min)	PTH-amino acid	Notes
5.8	Tyr(P)	Stable
5.9	PTC-Gly	Stable, partial conversion product of ATZ-Gly
6.2	Asp	Stable, slight conversion to Asp derivative
7.1	Asn	~ 5–10% recovered as PTH-Asp
7.0	PTC-Ala	Stable, partial conversion product of ATZ-Ala
8.2	Ser	Typically ~ 0–20% recovered, see Ser derivative
8.8	Gln	~ 5-20 % recovered as PTH-Glu
9.3	Thr	Typically ~ 5–25% recovered, see Thr derivative
9.7	Gly	Stable
10.5	Glu	Stable, slight conversion to Glu derivative
11.0	Dimethylphenylthiourea (DMPTU)	Edman chemistry by-product
12.5	Cys-acetamidomethyl (Acm)	Stable
12.8	His[b]	Stable but may not be completely extracted from sequencer
13.0	Hypro (cis)	Hydroxyl can be cis or trans, Edman chemistry produces both
13.8	Ala	Stable
14.0	Hypro (trans)	Hydroxyl can be cis or trans, Edman chemistry produces both
15.1	PTC-Met	Stable, partial conversion product of ATZ-Met
15.2	Ser derivative	Breakdown product of serine due to Edman Chemistry
16.5	PTC-Val	Stable, partial conversion product of ATZ-Val
16.7	Arg[b]	Stable but may not be completely extracted from sequencer
16.8	Cys-3-nitro-2-pyridinesulfenyl (Npys)	Sensitive to DTT. Converts to Cys with successive cycles
17.4	Asp-O-allyl (OAl)	Partially stable, converts to Asp
17.7	Tyr	Stable
18.8	Ser derivative	Breakdown product of serine due to Edman Chemistry

Table 2 (continued)
Relative Elution Positions of PTH-Amino Acids on HPLC using REZ Cycle[a]

Retention time (min)	PTH-amino acid	Notes
19.1	Ser-allyloxycarbonyl (Aloc)	Partially stable, converts to Ser derivatives
19.9	PTC-Lys	Stable, partial conversion product of ATZ-Lys
20.7	PTC-Leu	Stable, partial conversion product of ATZ-Leu
20.7	PTC-Ile	Stable, partial conversion product of ATZ-Ile
20.8	Tyr-dimethoxyphosphoryl (OP(OMe)$_2$)	Rapidly converted to Tyr(P) in first few cycles
21.1	Pro	Stable
21.9	Met	Usually stable but susceptible to oxidation
22.3	Val	Stable
22.6	Arg-diallyloxycarbonyl (Aloc)$_2$ derivative	Product of conversion of Arg (Aloc)$_2$, may be Arg (Aloc)
23.3	Thr-allyloxycarbonyl (Aloc)	Stable
23.9	Diphenylthiourea (DPTU)	Edman chemistry by-product
23.9	Arg-4-toluenesulfonyl (Tos)	Stable
24.1	Lys-allyloxycarbonyl (Aloc)	Stable
24.7	His-2,4-dinitrophenyl (Dnp)	Stable
25.1	Glu-O-allyl (OAl)	Stable, slight conversion to Glu derivative
25.3	Trp	Typically only partially recovered at inconsistent levels
25.8	Asp-O-*tert*-butyl (OtBu)	Very unstable, converts to Asp
26.1	Phe	Stable
26.1	Trp-N^{in}-formyl (CHO)	20–40% recovered as PTH-Trp
26.2	Phe-pNitrophenyl	Stable
26.2	Cys-allyl (Al)	Stable
26.7	Ile	Stable
27.1	Lys (ptc)	Stable, both amino groups are derivatized with PITC
27.4	Leu	Stable

Table 2 (continued)
Relative Elution Positions of PTH-Amino Acids on HPLC using REZ Cycle[a]

Retention time (min)	PTH-amino acid	Notes
27.5	Norleucine	Stable
27.8	His-*tert*-butoxymethyl (Bum)	Converted to PTH-His with successive cycles
28.2	Ser-benzyl (Bzl)	Partially recovered, remainder converted to Ser derivatives
28.5	Cys-allyloxycarbonyl (Aloc)	Stable
28.9	Arg-4-methoxy-2,3, 6-trimethylbenzenesulfonyl (Mtr)	Partially recovered, converted to PTH-Arg with successive cycles
29.2	Arg-mesitylene-2-sulfonyl (Mts)	Typically 1–5% recovered as PTH-Arg
29.5	Cys-tert-butyl (*t*Bu)	Very unstable, converts to Cys
30.0	Asp-O-benzyl (OBzl)	~ 5–20% recovered as PTH-Asp
30.1	Arg-diallyloxycarbonyl(Aloc)$_2$	Partially stable
30.5	Trp derivative	One of multiple Trp degradation products
30.6	His-3-benzyloxymethyl (3-Bom)	Stable
30.6	Thr-benzyl (Bzl)	Partially stable, products include PTH-Thr and Thr derivatives
31.2	Tyr-allyl (Al)	Stable
31.3	Glu-O-benzyl (OBzl)	~ 10–40% recovered as PTH-Glu
31.7	Tyr-*tert*-butyl (*t*Bu)	Very unstable, converts to Tyr
32.1	Cys-4-methoxybenzyl (Mob)	Partially stable, products include Ser derivatives
32.5	Thr-*tert*-butyl (tBu)	Very unstable, converts to Thr
33.4	Asp-O-cyclohexyl (OcHex)	~ 5–20% recovered as PTH-Asp
33.6	Lys-chlorobenzyloxycarbonyl (ClZ)	Partially stable, some recovered as PTH-Lys (ptc)
33.6	Lys-2-chlorbenzyloxycarbonyl (2ClZ)	Stable, perhaps 1–2% recovered as PTH-Lys (ptc)
33.8	Thr-benzyl (Bzl) derivative	Produced from Thr-benzyl (Bzl)
33.9	Lys-2,4-dinitrophenyl (Dnp)	Stable
34.0	Trp derivative	One of multiple Trp degradation products

Table 2 (continued)
Relative Elution Positions of PTH-Amino Acids on HPLC using REZ Cycle[a]

Retention time (min)	PTH-amino acid	Notes
34.3	Hydroxyproline-4-benzyl (4-Bzl)	Unstable, converts to Hypro
34.3	Glu-O-cyclohexyl (OcHex)	~ 10–40% recovered as PTH-Glu
34.6	Trp derivative	One of multiple Trp degradation products
34.9	Tyr-2,6-dichlorobenzyl (2,6-diClBzl)	Stable, perhaps 1–5% recovered as PTH-Tyr
35.1	Cys-4-methylbenzyl (Meb)	Partially stable, products include Ser derivatives
35.4	Hydroxyproline-4-benzyl (4-Bzl)	Unstable, converts to Hypro
37.8	Lys-9-fluorenylmethyloxycarbonyl (Fmoc)	Converted to PTH-Lys(ptc) with successive cycles
38.6	Tyr-2-bromobenzyloxycarbonyl (2-BrZ)	Stable, perhaps 1–5% recovered as PTH-Tyr
38.7	Glu-O-9-fluorenylmethyl (OFm)	~ 10–40% recovered as PTH-Glu
38.9	Trp derivative	One of multiple Trp degradation products

Side Chain Protected Amino Acids Completely or Nearly Completely Unstable to Edman Chemistry:

His-allyloxycarbonyl (Aloc)	Converts to His
His-tosyl (Tos)	Converts to His
His-benzyloxycarbonyl (Z)	Converts to His
Asp/Glu -O-tert-butyl (OtBu)	Converts to Asp/Glu
Ser/Thr/Tyr/Hypro-tert-butyl (tBu)	Converts to Ser/Thr/Tyr/Hypro
Lys/His -tert-butyloxycarbonyl (Boc)	Converts to Lys/His
His/Cys/Asn/Gln-triphenylmethyl (Trt)	Converts to His/Cys/Asn/Gln
Arg-2,2,5,7,8-pentamethyl-chroman-6-sulfonyl (Pmc)	Converts to Arg
Arg-2,2,4,6,7-pentamethyldihydrobenzofuran-5-sulfonyl (Pbf)	Converts to Arg
Asn/Gln/Cys -2,4,6-trimethoxybenzyl (Tmob)	Converts to Asn/Gln/Cys

[a]Elution positions are relative and may vary somewhat depending on system and column. Some of these derivatives were originally done with different conditions and are now represented on a typical resin bound sequencing cycle (ABI "REZ" cycle). Table composed from material taken from references **10–15**.

[b]Position is variable depending on ionic strength and elution buffer composition.

Table 3
Programs for Normal Sequencing and Resin-Bound Peptide Sequencing on an Applied Biosystems Model 477 Sequencer[a]

	Normal			Resin-bound peptides		
	Reaction cartridge	Conversion flask	HPLC gradient	Reaction cartridge	Conversion flask	HPLC gradient
Cycle 1	BEGIN-1	BEGIN-1	NORMAL-1	BGN REZ-1	BGN REZ-1	REZ-1
Cycle 2 to n	NORMAL-1	NORMAL-1	NORMAL-1	REZ-1	REZ-1	REZ-1
Temperature	48°C	64°C	55°C[b]	53°C	64°C	55°C[b]
HPLC gradient:[c]	Time		%B	Time		%B
	0		12	0		12
	0.4		12	0.4		12
	18.0		38	38.0		63
	25.0		38	39.0		90
	25.1		90	41.0		90
	28.1		90			
Data collection time		29 min			42 min	

[a]As supplied by the manufacturer.
[b]Column temperature.
[c]Model 477 Sodium Acetate System: Reservoir A: 8.0 mL of 3 M sodium acetate, pH 3.8, and 2.6 mL 3 M sodium acetate, pH 4.6, to 1 L of 3.5% (v/v) tetrahydrofuran in water. Reservoir B: 100% acetonitrile, 500 nmol DMPTU. Flow rate : 0.21 mL/min.
Model 477 Premix System: Reservoir A: 20 mL Premix added to 1 L 3.5% (v/v) tetrahydrofuran in water (A3).
Reservoir B: 100% acetonitrile, 500 nmol DMPTU. Flow rate : 0.21 mL/min.

Table 4
Programs for Normal Sequencing on Applied Biosystems Model 49x and 49x cLC Sequencer

	Model 49x[a]			Model 49x cLC[b]		
	Reaction cartridge	Conversion flask	HPLC gradient	Reaction cartridge	Conversion flask	HPLC gradient
Default	Cart Pulsed-liquid	Flask Normal	Fast Normal I	Cart PL 6mmGFFcLC	Flask Normal cLC	Normal I cLC
Cycle 1	None	Flask Prep Pump	PrepPump	None	Prepare Pump cLC	Prepare Pump cLC
Cycle 2	None	Flask Blank	Fast Normal I	None	Flask Blank cLC	Normal I cLC
Cycle 3	Cart Begin	Flask Standard	Fast Normal I	Cart Begin cLC	Flask Standard cLC	Normal I cLC
Cycle 4	Cleavage cycle and number of repetitions specified in start-up menu. (This cycle is the first PTH-amino acid.)			Cleavage cycle and number of repetitions specified in start-up menu. (This cycle is the first PTH-amino acid.)		
Temperature	48°C	64°C	55°C[c]	48°C	64°C	55°C[c]
HPLC Gradient:	Time	%B	Flow (µL/min)	Time	%B	Flow (µL/min)
	0	2	325	0	10	40
	0.2	4	325	0.4	12	40
	0.4	18	325	4.0	22	40
	18.0	44	325	22.0	47	40
	18.5	44	325	22.6	90	40
	19.5	90	325	23.5	90	40
	20.0	90	325	29.0	90	60
	22.5	90	325			
	23.0	50	10			
Data collection time	22.0 min			28.0 min		

[a]Model 49x Premix System: Reservoir A: 20 mL Premix added to 1 L 3.5% (v/v) tetrahydrofuran in water (A3). Reservoir B: 100 % acetonitrile.
[b]Model 49x cLC Premix System: Reservoir A: 10 mL Premix added to 1, 450 mL bottle of 3.5% (v/v) tetrahydrofuran in water, 500 µL 1% acetone, 25 µL TFA. Reservoir B: 12% isopropanol in acetonitrile
[c]Column temperature.

261

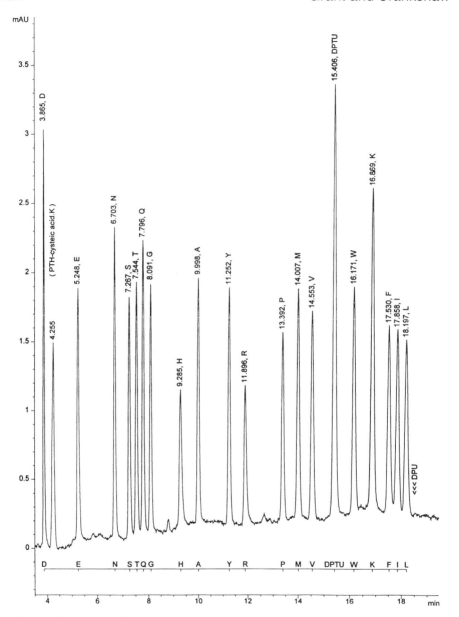

Fig. 4. Elution profile of standard PTH-amino acids on Hewlett Packard PTH-amino acid analyzers. The one-letter code designates standard PTH-amino acids and elution times are indicated in minutes. Gradient conditions are described in **Table 5**. The elution times of common PTH-cysteine derivatives are 5 min for S-carboxymethyl PTH-cysteine (CM-Cys), 8.6 min for S-carboxamidomethyl PTH-cysteine (CAM-Cys), 9 min for S-β-propionamide PTH-cysteine (PAM-Cys), and 15.7 min for S-pyridylethyl PTH-cysteine (PE-Cys).

3. Methods

3.1. Applied Biosystems Sequencers

3.1.1. Premix Solvent System

1. With standard columns and solvent components provided by the manufacturer, the only optimization necessary in this system is for the elution position of the protonatable amino acids, PTH-arginine, PTH-histidine, and others such as PTH-pyridylethyl cysteine. This is accomplished by adding a volume of the premix buffer to a solution of 3.5 % (v/v) tetrahydrofuran. The premix buffer contains 17.5 mM sodium acetate, 33 mM acetic acid, and 10 mM hexanesulfonic acid sodium salt, an ion-pairing reagent. The interaction of the positively charged amino acid side chains with exposed silanol groups on the column effect their elution position and peak shape. The purpose of the ion pairing reagent is to saturate these groups to produce stable elution positions and sharper peaks for the charged amino acids. The amount of premix added to accomplish this is empirical. The manufacturer suggests 15 mL of premix buffer per bottle of Solvent A (1 L) *(3)*, but this will vary with the column lot and its age. The preferred outcome is for PTH-histidine to elute before PTH-alanine and for PTH-arginine to elute before PTH-tyrosine and for them to be at least base line resolved from their adjacent peaks. Similarly, PTH-pyridylethyl cysteine should elute before PTH-proline. To maintain this state as the HPLC column ages, it is usually necessary to add more premix to buffer A (*see* **Note 3**). The manufacturer's recommended gradient elution programs are listed in **Tables 3** and **4**. Representative buffer recipes are presented in the table footnotes.

2. In this system, PTH-tryptophan may co-elute with diphenylurea (DPU). To resolve these two peaks, ABI suggests changing solvent B to 88% acetonitrile/12% isopropanol (v/v) as indicated for the cLC sequencer (*see* **Fig. 3**). This tends to move PTH-tryptophan out in front of DPU. However, experience has shown that the resolution often decreases as the column ages.

3. Tryptophan is also susceptible to oxidation either during sample purification or during sequencing, with PTH-kynurenine (*see* **Table 1**) as the primary product detected. Another possible source of tryptophan oxidation during sequencing is the presence of peroxides in the tetrahydrofuran solution. This can also lead to reduced signal from PTH-lysine. Replacing reservoir A with fresh solution should prevent this.

3.1.2. Sodium Acetate Solvent System

This system is used only on the sequencers prior to the 49x series (i.e., Model 477). PTH-amino acid separation with this solvent system can be optimized by adjusting the pH and ionic strength. The pH is adjusted by changing the ratio of the two sodium acetate solutions. The ionic strength can be varied by adjusting the total amount of sodium acetate added. As long as the ratio of the two sodium acetate solutions remains constant, the pH will remain constant as the ionic strength changes. In addition, some degree of optimization can also be accomplished by adjusting gradient parameters or column oven temperature. A summary of the effect of these operations on PTH-amino acid elution position is presented in **Fig. 5**. The starting amounts of the sodium acetate buffers to be added to a bottle of solution A vary somewhat with the column used. An example of starting concentrations are presented in the footnote to **Table 3**. Fine tuning of elution positions is then carried out as described earlier. Additional fine tuning usually has to be performed as the column ages. The manufacturers recommended gradient elution programs for the Model 477 are listed in **Table 3**.

3.2. Hewlett-Packard Sequencer

According to the manufacturer, no optimization is necessary for standard sequencing. This has been confirmed in practice. The three solvent system, which utilizes an ion pairing reagent in solution A, appears to consistently produce excellent resolution of the standard PTH-amino acids. The specific compositions of the solvents, however, is proprietary so that they must be purchased from the manufacturer if this system is to be used according to manufacturer's specifications. The manufacturers recommended gradient elution program is listed in **Table 5**.

4. Notes

1. All modern sequencers use on-line reverse-phase HPLC as the standard method for identifying the resulting PTH-amino acids. Although the chromatographic separation is similar on all three, there are differences, so care should be exercised in comparing PTH-amino acid elution positions from one to another. In doing so, use the positions of the common amino acids as a guide. Similarly, buffer and gradient adjustments that are productive on one system may not necessarily apply to another system.

2. The elution positions of many of the modified PTH-amino acids presented in this chapter were taken from a compilation undertaken by the Association of Biomolecular Resource Facilities *(4)* in an attempt to consolidate information of this type for easy reference. It should be noted that an exhaustive review of the literature was not conducted in this effort. Rather, individuals were asked to submit any information they had in this regard for inclusion in that booklet. Thus, please note that independent verification of the information presented there was not performed. In many instances the entries come from personal observations of the contributors and there are no published references to provide rigorous documentation. When literature references were provided they were included and can be found in **ref.** *(4)*. To the best of our knowledge, all of the information presented in regard to modified PTH-amino acids is correct but should be used only as a guide. Additional supporting analyses should be employed to verify the identity of any unknown residue.

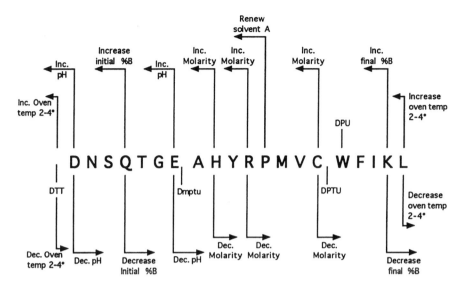

Fig. 5. Separation guidelines for Applied Biosystems PTH-amino acid analyzers. This table was developed with the Model 477 using the sodium acetate buffer system. (Kindly provided by Applied Biosystems.)

3. Many researchers have observed in the Applied Biosystems Premix system that as the column ages the amount of premix must be increased to maintain the separation and eventually baseline separation may no longer be possible with this strategy. An article by Chestnut *(5)*, with contributions by L. Packman, C. Beach, and J. Hempel, deals specifically with this problem and suggests that at the point where higher levels of premix no longer produce adequate resolution (35 mL is recommended as the upper limit) the premix should be reduced back to the original 15 mL. When this is done, the elution positions of PTH-histidine and PTH-arginine now follow those of PTH-alanine and PTH-tyrosine, respectively. A problem with this strategy is that PTH-pyridylethyl cysteine will elute with the DPTU peak. A solution to this is to employ an alternate cysteine modifying reagent such as 3-bromopropylamine *(6,7)* or N-isopropyliodoacetamide *(8)* to produce PTH-aminopropyl cysteine or PTH-carboxamidopropyl cysteine, respectively. Since neither of these adducts are protonatable, their elution position will not vary and they can both be baseline resolved with the standard system (*see* **Fig. 1** and **Table 1**).

4. Automated Edman degradation can be used to monitor the integrity of the peptide chain in solid-phase peptide synthesis prior to side chain deprotection and cleavage from the resin *(4,9–11)*. This procedure is often referred to as preview analysis because deletion sequences are picked up as a preview of a PTH-amino acid in a cycle preceding its expected occurrence. Since peptides sequenced "on resin" have not yet been subjected to side chain deprotection procedures, those

Table 5
Gradient Program for Hewlett-Packard PTH-Amino Acid Elution[a]

Time (min)	% B	% C	Flow (mL/min)
0.1	0	0	0.3
16.00	100	0	0.3
18.50	100	0	0.3
18.60	0	100	0.3
20.00	0	100	0.3
20.01	0	100	0.5
24.00	0	100	0.5
24.10	0	0	0.5
27.50	0	0	0.5
27.51	0	0	0.3

[a]Reservoir A: Acetonitrile/Water (15/85, v/v), pH 5.2, 0.1% "ion pairing reagent." Reservoir B: 2-propanol/ Water (31/69, v/v), 0.2% buffer. Reservoir C: Acetonitrile. Column temperature 40°C.

PTH-amino acids whose side chains still contain protecting groups will not elute at the normal PTH-amino acid positions. Generally the side chain protected PTH-amino acids elute later than their deprotected counterparts and often later than the latest eluting deprotected PTH-amino acid. Thus, for detection of side chain protected PTH-amino acids the gradient run times need to be extended. An example of an extended gradient for this purpose is the one used with the REZ cycle of Applied Biosystems sequencers and is listed in **Table 3**. Note that previous analysis works best with peptides synthesized using "Boc" chemistry, some side chain protecting groups used with "Fmoc" chemistry are easily removed during Edinen degradation.

5. The methyl lysines (mono-, di-, and tri-) have proven to be particularly problematic in establishing their elution position. Different investigators have shown them eluting in basically two places. The majority show them eluting in a wide area between alanine and DPTU and also after leucine. In some instances, an investigator will indicate both positions, and in others, only one of the two positions. In an effort to resolve this discrepancy, we obtained samples of the mono-, di-, and tri-methyl lysines and ran them on a standard Applied Biosystems 477A sequencer simply by loading an aliquot into the reaction vessel and running a sequencer cycle. Both mono- and dimethyl lysine show both early and late peaks, while trimethyl lysine shows only the early peak. The elution order of the early peaks is mono- before di- before tri-, with a fairly limited range between Ala and Met. The late eluting peaks tend to co-elute just after leu (and nleu). So, what is the explanation? Without chemical proof we can only speculate, but we offer the following possibility. Mono- and dimethyl lysine are alkyl amines that may be capable of becoming protonated and assuming a positive charge. Trimethyl lysine is a quaternary amine that is always positively charged. As such, the charge

should cause them to run relatively early in the chromatogram and, like His and Arg, their elution position will probably be very sensitive to ionic strength. Hence, varying ionic strength from different systems may explain the wide variance reported in the elution positions of the early peaks. The late peaks may be due to a portion of the mono- and dimethyl lysine side chains reacting with PITC in a manner similar to that of lysine, since they retain a free pair of electrons on the nitrogen that can participate in nucleophilic attack on the PITC. In our hands, this appears to be a major reaction for monomethyl lysine and a minor reaction for dimethyl lysine, but others have reported variable ratios. This variability may be cycle dependent. Trimethyl lysine does not possess an unbonded pair of electrons and thus would not be expected to react with PITC at this position. Hence, we do not see a late eluting peak.

Again, it must be stressed that this is only a hypothesis and particular care must be taken in interpreting your results. However, the general behavior of the methylated lysines, whatever the reason, is well documented and should aid in their identification.

Acknowledgments

The authors gratefully acknowledge the members of the Association of Biomolecular Resource Facilities (ABRF) who contributed information on the elution positions of unusual PTH-amino acids for a monograph originally published by the ABRF *(4)*. A list of contributors can be found in the monograph. We also thank David Dupont of Applied Biosystems and Ben Madden of the Mayo Foundation for providing information about their systems.

References

1. Grant, G. A. and Crankshaw, M. W. (1997) Identification of PTH-Amino Acids by High Performance Liquid Chromatography, in *Protein Sequencing Protocols* (Smith, B. J., ed.), Humana Press, Totowa, NJ, pp. 197–215.
2. Hewlett Packard (1993) Technical Note TN 93-4.
3. Applied Biosystems User Bulletin, Number 43 (1992).
4. Crankshaw, M. W. and Grant G. A. (1993) *Identification of Modified Amino Acids in Protein Sequence Analysis*. Association of Biomolecular Resource Facilities, Bethesda, MD.
5. Chestnut, W. (1995) Tips for Optimizing the Separation of PTH-Amino Acids During Protein Sequencing. *Assoc. Biomolec. Resource Facilities Newslett.* **6(1)**, 8–10.
6. Jue, R. A. and Hale, J. E. (1993) Identification of cysteine residues alkylated with 3-bromopropylamine by protein sequence analysis. *Anal. Biochem.* **210**, 39–44.
7. Jue, R. A. and Hale J. E. (1994) Alkylation of cysteine residues with 3-bromopropylamine: identification by protein sequencing and quantitation by amino acid analysis, in *Techniques in Protein Chemistry V* (Crabb, J., ed.), Academic Press, San Diego, CA, pp. 179–188.

8. Krutzsch, H. C. and Inman, J. K. (1993) N-Isopropyliodoacetamide in the reduction and alkylation of proteins: use in microsequence analysis. *Anal. Biochem.* **209**, 109–116.
9. Grant, G. A. (1992) Evaluation of the finished product, in *Synthetic Peptides: A User's Guide* (Grant, G. A., ed.), W. H. Freeman, New York, pp. 185–258.
10. Kent, S. B. H., Riemen, M., LeDoux, M., and Merrifield, R. B. (1982) A study of the Edman degradation in the assesment of the purity of synthetic peptides, in *Methods of Protein Sequence Analysis* (Elzinga, M., ed.), Humana Press, Totowa, NJ, pp. 205–213.
11. Kochersperger, M. L., Blacher, R., Kelly, P., Pierce, L., and Hawke, D. H. (1989) Sequencing of peptides on solid phase supports. *Am. Biotech. Lab.* **7**, 26–37.
12. McCormick, D. J., Madden, B. J., and Ryan, R. J. Identification of Side-Chain protected L-Phenylthiohydantoins on Cyano HPLS Columns: An Application to Gase Phase Microsequencing of Peptides Synthesized on Solid-Phase Supports, in *Proteins, Structure and Function* (L'Italien, J. J., ed.), Plenum Press, New York and London, 1987, p. 403.
13. Fields, C. G., VanDrisse, V. L., and Fields, G. B. (1993) Edman degradation sequence analysis of resin-bound peptides synthesized by 9-fluorenylmethoxycarbonyl chemistry. *Peptide Research* **6**, 39–47.
14. Applied Biosystems User Bulletin No. 13, pp. 1–18 (1985).
15. Fields, C. G., Loffet, A., Kates, S. A., and Fields, G. B. (1992) The development of high-performance liquid chromatographic analysis of allyl and allyloxy-carbonyl side-chain-protected phenylthiohydantoin amino acids. *Anal. Biochem.* **203**, 245–251.

22

Protein Sequencer Maintenance and Troubleshooting

Bryan Dunbar

1. Introduction

Modern automated sequencers, both gas-phase and pulsed-liquid forms, are normally efficient and reliable; this chapter is intended for use when they are not. The author's own experiences are almost wholly concerned with ABI (Applied Biosystems) instruments; however, the principles described will apply to most other manufacturers' sequencers. This chapter will address the various functions individually, in an attempt to identify typical faults and the best approach to remedy them.

Different sample supports require particular and individually-optimized cycles to achieve optimum performance, i.e., PVDF (polyvinylidenedifluoride) or traditional glass-fiber filters. Even when precisely optimized, individual settings may drift, and once every 2 or 3 wk the operator should sit and watch one complete cycle to check crucial delivery and drying steps, and so on, so that any anomalies may be rectified.

Routine, preventative maintenance is very worthwhile, e.g., changing HPLC guard columns every 4–6 mo, cleaning instrument air filters every 2–3 mo depending on how dusty your laboratory gets; replacing the vent trap 50% phosphoric acid (which should contain a little Methyl Red as a pH indicator) either when it starts to lose the red color or at a 1-yr interval (whichever comes sooner); replacement of the injection pathway components, i.e., the needle transfer line and the restrictor/overflow line every 3–4 mo, replacement of the HPLC pump seals (*see* **Note 1**).

The operator should become familiar with the typical or "normal" performance of his/her own instrument so that any potential problems can be spotted early on, e.g. the normal profile of the standard phenylthiohydantoin

From: *Methods in Molecular Biology, vol. 211: Protein Sequencing Protocols, 2nd ed.*
Edited by: B. J. Smith © Humana Press Inc., Totowa, NJ

amino acid (PTH-a.a.) separation. The basis of this chemistry is described in the **Note 2**. Even fully optimized programs can generate problems, however. **Table 1** outlines some typical problems, with reference to **Subheadings 3.** and **4.** for remedies. **Subheading 3.0** describes various checks and procedures for optimisation of performance, while **Subheading 4.0** provides further detail.

2. Materials

2.1. Spare Reagents and Solvents

Adequate stocks of sample supports, reagents, solvents, cartridge filters, reaction cartridge seals and other consumables should be maintained to keep up with consumption; ensures that the storage of each complies with the recommendations of the manufacturer (*see* **Note 3**). Keeping careful records of the installation of new batches of reagents is essential, so that a sudden deterioration in performance may be linked to the introduction of a new batch. Sequencer-grade chemicals should be used at all times (*see* **Note 4**).

2.2. Spare Parts

Instrument manufacturers will advise on a recommended list of spares necessary to cope with most common breakdowns and with routine maintenance. Such a list might include the following for ABI instruments:

1. Bottle-cap assembly.
2. Bottle seals.
3. Injection transfer needle lines.
4. Injector restrictor/overflow lines.
5. Conversion-flask assembly.
6. Cartridge inlet and outlet lines.
7. Guard columns for high-performance liquid chromatography (HPLC).
8. Analytical PTH-HPLC column.
9. Deuterium lamp for HPLC (*see* **Note 5**).
10. Heat sink compound.
11. Printer/integrator/chart recorder supplies, e.g., paper, pens, ink cartridges, and so on.

2.3. 50% Nitric Acid

This is used for cleaning the conversion flask (*see* **Subheading 3.4.**). One volume of conc. nitric acid should be added to 1 volume of dist. water, slowly, with stirring, in a fume cupboard. Suitable protective clothing should be worn.

2.4. Sonicating Waterbath

A sonicating water bath: for the thorough cleaning of many sequencer components with 50% nitric acid (*see* **Subheading 2.3.**).

Table 1
A Troubleshooting Guide for Protein Sequencers

1. Blank chromatograms: possible failure of solvent and reagent delivery (*see* **Subheading 3.1.**) or of PTH aa transfer (*see* **Subheadings 3.4.–3.6.**, and **Note 6**).
2. Erratic retention times on the PTH aa chromatograms (*see* **Subheading 3.6.** and **Note 7**).
3. New peak appearing in every PTH aa chromatogram, close to proline (*see* **Note 4**).
4. Compacting of the hydrophobic region of the PTH aa separation; most noticeable on the PTH aa standard; can also be accompanied by peak broadening (*see* **Note 8**).
5. Increased carryover or "lag" from cycle to cycle (*see* **Subheading 3.2.** and **Note 9**).
6. Abnormally high levels of DPTU and DPU (*see* **Subheadings 3.2.** and **3.3.** and **Note 10**).
7. Excessive background peaks on PTH aa chromatograms, starting within the early stages of a run (*see* **Subheading 3.2.** and **Note 11**).
8. Low repetitive yield, i.e., less than 90–92% (*see* **Subheading 3.2.** and **Note 12**).
9. Low yield of PTH-ser or thr, or generally low yield due to over- or underdrying in conversion flask (*see* **Subheading 3.4.**).
10. Inability to reach the C-terminus of peptides (*see* **Note 13**).
11. Artifact peaks from the addition of DTT (*see* **Note 14**).
12. Sloping baseline at the start of the PTH separation and abysmal resolution of the early-eluting amino acids (*see* **Note 15**).
13. A very concave "smiling baseline" on PTH chromatograms (*see* **Note 16**).
14. High pressure indicated on analyser display; usually first noticed as a high pressure shutdown during a solvent purge (*see* **Note 17**).
15. Huge, off-scale peaks on the PTH aa chromatograms (*see* **Note 18**).

3. Methods

Care should be taken when checking out any sequencer systems, as hazardous voltages and chemicals under gas pressure may be present. The instrument manual and safety data sheets should be consulted and protective clothing worn (laboratory coat, gloves, and face mask).

3.1. Check for Failure of Delivery of Any Reagent or Solvent, and Suggested Remedies

Nondelivery of any reagent or solvent, is of course, a major problem; when affecting all deliveries, the cause is almost certainly related to gas pressure: either low pressure or an empty cylinder; an empty reserve cylinder; or failure of the automatic changeover to a full reserve. With 49X series sequencers, an insufficient argon supply is indicated by messages on the computer monitor and in the event log. In addition to driving chemical deliveries, the gas supply on ABI instruments is used for generating vacuum for the vacuum-assisted valve blocks. Failure of this system leads to erratic chemical deliveries and sub-optimal performance.

1. Check that the supply of argon is sufficient and at the recommended pressure (see instrument manual). Replace gas cylinders as necessary. If the reserve argon supply is adequate, then check out the automatic changeover system and repair it if necessary. The main argon cylinder regulator should be set 5–10 psi higher than the reserve to ensure that it has priority of delivery after it has been replaced.
2. If the argon supply was satisfactory, then check for leaks in the supply lines and fittings (*see* **Note 19**).

3.2. Check for Failure of Delivery of One Reagent or Solvent Only, and Suggested Remedies

1. Check that there is sufficient reagent or solvent in the reservoir bottle; replace as required. Exhaustion of vapour-delivery R2 is not obvious to the eye, but will be by sequencer performance (*see* **Note 20**).
2. If available, select the appropriate programmed bottle leak test(s). Alternatively, check that the reagent/solvent reservoir bottle can be pressurized as follows; with the vent valve closed, pressurize the appropriate bottle and observe the bubbling of gas into the liquid. This bubbling should stop after after a few minutes; failure to do so will indicate a leak at the cap assembly. With reference to the instrument manual for identification of the cap assembly components, replace any seals and ensure that the ratchet mechanism is not worn, which would prevent correct sealing of the bottle.

   For reagents delivered in the vapor-phase, the gas-bubbling test is inappropriate, so replacing cap assembly seals and checking delivery as outlined in **Note 20**, is the best course of action.
3. Check the "Angar" solenoid valve in the relevant valve block (using the manual to correctly identify it). Proceed as follows: A malfunctioning vent or pressurise "Angar" solenoid valve is simply checked; when the valve is actuated and lightly touched with one finger; a working valve can be felt to snap open, whereas no movement will be felt with a faulty valve. The only remedy, and sometimes the only final confirmation of a "dead" valve, is to replace it and try a repeat delivery. Particular care should be taken to ensure that the new "Angar" valve is of precisely the correct type, e.g., trifluoroacetic acid (TFA) cleavage reagent for example, requires a special valve type.
4. Blockage of an individual delivery line is unlikely, but possible after a period of inoperation with liquid left in the line. If no other checks have revealed a problem, try clearing the obstruction by increasing the pressure on the reservoir gas regulator (up to a maximum of 5 psi). If this fails, then replacing the line may be the only option (*see* **Note 21**).

3.3. Checking the Reaction Cartridge

Given that delivery of the reagents/solvents is satisfactory, the Edman chemistry should proceed normally; if not then proceed as follows:

1. Check that the reaction cartridge is tightly sealed; leak testing of the cell prior to commencing a run should be taken seriously. It can be easy to fall into a "it never usually leaks, so it will be O.K." way of thinking, but if oxygen gets into the cell, then it will not be a satisfactory run. For sequencers without a programmed leak test, a simple but reliable way of checking for a leaky cell is to pressurize it fully (to about 5 psi), and back off the regulator. Lay a piece of white sticky tape onto the regulator dial, exactly along the top edge of the needle, with your eye at the same level. Wait for 2 min, then check to see if the needle's edge has disappeared underneath the white tape; it is very easy to spot any change. ABI suggest a 0.1 psi drop over a 5-min period; but if you can see little or no change over 2 min, then it is a rapid indication that the cell is very firmly sealed. If not, take the cell apart and look carefully at the Zitex seal. An illuminated circular magnifier is ideal for this and many other sequencer-related applications, including watching critical steps in the "Blott" cell. You may see a compressed, but uneven opaque circle, in the Zitex seal indicating incomplete tightening, or even a small cut, which is often seen when using Polybrene on a glass-fiber disc. This occurs when a crystal of Polybrene forms just off the edge of the G-F disc, and pierces the Zitex seal when the cell is tightened. This problem can be avoided by applying the Polybrene in two small aliquots, thus avoiding saturation of the disc, and subsequent formation of crystals on the edge of the glass surface (*see* **Note 22**).
2. Check that the reaction temperature is set correctly for the programmed cycle in use, and if necessary check the cartridge temperature with a contact thermometer. Failure to generate the correct temperature will probably require the attention of a service engineer.
3. Care and cleaning of the reaction cell (*see* **Note 12**): check for physical damage and blockage. Replace if necessary. To clean, as a remedial or precautionary measure, sonicate in 50% nitric acid (*see* **Subheading 2.3.**) for 30 min. Wash thoroughly in water, then methanol, and then dry and replace it in the sequencer.

3.4. Checking the Conversion Flask; ATZ to PTH Amino Acid

1. Check the conversion temperature (as in **Subheading 3.3.**, **step 2**).
2. Slow transfer or none at all, from the transfer flask can result from various causes. From a clean, fresh flask with its headspace pressure regulator set at 3.5 psi, the last of the flask contents should disappear into the pick-up line within 7–8 s; if it does not then there is something amiss, perhaps downstream in the needle or restrictor lines. The well-known problem of the pick-up line "silting up" with what is thought to be silica from the action of R4 (25% TFA in water) on the flask surface, or material gradually extracted from the glass-fiber filter in the reaction cartridge, has been addressed in a specific ABI user bulletin (*1*).

This bulletin contains suggestions for a clean-up procedure with KOH (potassium hydroxide) and so on, which is quite a hassle.

On 47X series instruments, it is simpler to pull the line through the white cap for about 0.5 cm or so, and snip off the affected part, making sure that the line reaches almost to the apex of the flask cone. This can be done twice, sometimes three times, before a new flask becomes necessary. I have been advised that this may not be so easy with the most recent design of flask from ABI; however, recent experience has shown that it will work. On 49X series instruments, the pick-up line may be removed completely before trimming and adjustment or replacement.

The white deposit on the inner surface of the flask can be removed with a 30-min sonication in 50% aqueous nitric acid.

Remember to coat (not plaster) only the base of the flask, with fresh heat-sink compound when it is reinstalled (47X only).

There is one direct effect of this pick-up line problem: the vital flask drying steps will be altered. The first step to be checked occurs just before the R4 delivery, which when properly optimized, should result in a tiny spot of liquid being left at the very base of the flask at the end of the step. Over-drying at this stage will destroy labile residues such as serine and threonine. The second step is after the pause for the conversion process and is also influenced by line blockage; the flask must be dry 40–60 s (47X) or 300 s (49X) before the end of the step to ensure that all the TFA is swept from the flask and associated tubing. Under-drying will cause the PTH (phenylthiohydantoin) amino acids to go to waste during the following "Clear injector to waste step." Any residual conversion TFA will be injected onto the column, causing a sloping baseline and loss of resolution of the early PTH amino acid peaks.

A final possible cause of conversion flask problems is that the seal for the flask vial can leak, usually as a result of overtightening or incorrect seating; the only remedy is replacement.

3.5. Check of PTH-Amino Acid Delivery to the On-Line HPLC

Given that the conversion process and sample drying steps are satisfactory; failure to observe the expected chromatograms require the following checks:

1. Check the delivery of sample solution to injection loop as follows; with the injector in the "load" position, fill the conversion flask with the normal measured aliquots of S4. When delivered to the injector, the flask should empty promptly and a drop of liquid should form at the end of the restrictor/overflow line within 1 s of the calibrated transfer time (optimized as described in the instrument manual). This system is one that needs constant monitoring. It takes very little to block or just partially occlude the 0.007 inch ID injector/overflow line. One of the author's 477A sequencers runs almost exclusively on blotted samples or samples immobilised onto "ProSorb" devices. The inherent risk is that microfine particulate material from the PVDF eventually builds up in the injection pathway, resulting in failed injection steps. ABI has a users' bulletin

(1), regarding this problem, but we have found a simple preventative protocol to be very successful.

With the X2 bottle position (on the 477A) filled with Sequencer grade methanol, a simple Wash Program is performed before the start of each new PVDF run. Solvents S1 and S3 are only used to wash each new PVDF sample in the BLOTT cartridge, while the conversion flask is filled with methanol from reservoir X2.

This is subsequently transferred through to the injector then ultimately to waste. The routine is performed twice. A cycle listing of "INJ-WASH" for this procedure is given in **Table 2**.

While this wash procedure minimizes any blocking problems in the restrictor line, it is strongly recommended that both the needle transfer line and the restrictor line be replaced (as part of a preventative maintenance plan) every 3–4 mo.

3.6. Checking PTH Analyzer Performance

Abnormal chromatography of samples requires the following checks:

1. Check that sufficient HPLC solvents are present (*see* **Note 23**).
2. The amount of Premix ion-pair reagent added to solvent A is precisely calculated to give optimum positioning of his, arg and pyridylethyl-cysteine. If these peaks drift out out of position, more or less Premix will need to be added (*see* **Note 24**). Chapter 21 contains further information on the analysis of PTH amino acids by HPLC.
3. Check that the column oven is set correctly, and that the selected temperature is being maintained.
4. HPLC columns which have been in use for some time will start to show signs of deterioration (*see* **Note 25** for further comments).
5. Check the deuterium lamp; if the wavelength selector is set to the visible end of the spectrum, e.g., 560 nm, then a clear, bright green disc should be seen at the flow cell window; if not check the lamp energy reading (*see* **Note 5**).
6. Failure of the HPLC pumps gives rise to various symptoms (*see* **Note 26**); trouble with the dynamic mixer is not so easy to establish unless failure is complete (i.e., the motor coupling has stopped rotating). It may be necessary to call a service engineer.
7. If peaks of non-PTH amino acid material appear, check that the conversion function is proceeding normally (*see* **Subheading 3.3.**). Check that the injector loop size being used is compatible with the S4 being used (*see* **Note 27**).
8. If no peaks are seen, not even the "system" peaks such as DPTU, look for failure of the Edman chemistry and check the transfer step to the injector loop.

4. Notes

1. Syringe pump systems driven by microprocessor-controlled stepper motors are reliable and remarkably trouble-free. Apart from very light lubrication of the drive cogs twice a year, these pumps need relatively little attention. An obvious source of potential problems are the pump seals, and the author has to admit to a

Table 2
A Listing of the Program "INJ WASH" for the Conversion Flask
and Injector Line of the ABI 477A Sequencer

Step	Function	Fxn#	Value	Elapsed Time
REACTION CYCLE "INJ-WASH" (Model 477A)				
1	Prep S1	13	3	0 min 3 s
2	Deliver S1	14	40	0 min 43 s
3	Pause	33	120	2 min 43 s
4	Prep S3	19	3	2 min 46 s
5	Deliver S3	20	40	3 min 26 s
6	Pause	33	120	5 min 26 s
7	Argon Dry	29	120	7 min 26 s
CONVERSION CYCLE "INJ-WASH" (Model 477A)				
1	Prep X2	7	3	0 min 3 s
2	Deliver X2	8	20	0 min 23 s
3	Argon Dry	22	4	0 min 27 s
4	Load Injector	15	180	3 min 27 s
5	Prep S4	10	3	3 min 30 s
6	Deliver S4	11	25	3 min 55 s
7	Argon Dry	22	4	3 min 59 s
8	Load Injector	15	180	6 min 59 s
9	Pause	25	10	7 min 9 s

certain lack of preventative maintenance in that direction (*see* **Note 7**). Pump seals will last for 3 to even 9 mo, depending mostly on the amount of use, and to a certain extent, on how carefully they were installed. In 90% of cases, failure is immediate; within 2 or 3 cycles retention times will become erratic, and replacement strictly according to the manufacturer's instructions, is the only cure. It must of course, be recommended that a maintenance programme of replacement is carried out before trouble strikes, i.e., change the seals every 6 mo.

2. The elegant original chemistry established by Pehr Edman in 1950 (manual) (*2*), and updated in 1967 (automated) (*3*), has not changed to any great extent; only the methodology of applying it has. This section will outline chemical problems frequently encountered, their diagnosis and rectification.

   REAGENTS:

   Coupling: R1 - 5% PITC in heptane (*see* **Note 28**).
   Coupling: R2 - TMA, DIPEA, methyl n-piperidine (*see* **Note 29**).
   Cleavage: R3 - Conc. Trifluoroacetic acid (*see* **Note 30**).
   Conversion: R4 - 25% triflouroacetic acid in water (*see* **Note 31**).
   Calibration: R5 - Standard PTH aa Mixture (*see* **Note 32**).

SOLVENTS:

The automated gas-phase sequencer uses a set of solvents similar, but not identical, to that of the original spinning-cup instrument *(3)*. The purpose of the solvents is to remove excess reagents that could adversely affect later stages of the degradation cycle or form by-products that might interfere with the PTH analysis

Solvent Wash: S1 - Heptane (*see* **Note 33**).

Solvent Wash: S2 - Ethyl acetate (*see* **Note 34**).

ATZ extraction: S3 - Butyl Chloride, Ethyl Acetate/Heptane 1:1 v/v, ethyl acetate or butyl chloride and ethyl acetate in combination (*see* **Note 35**).

PTH transfer: S4 - 10% or 20% Acetonitrile in water (*see* **Notes 17** and **36**).

3. There was a recent suggestion that storage of R1 at –20°C might be beneficial with regard to stability. The author and several other established sequencing persons have been keeping PITC at room temperature for the last 15 yr and as long as storage times are not excessive, i.e., no reagent lying on the shelf for more than 3–4 mo, then no problems seem to occur.

Indeed, freezer storage might create more potential problems, such as moisture pick-up. This will happen if the bottle is not tightly sealed when placed in the freezer, or if not allowed to come to room temperature before being opened. This will result in a poor repetitive yield and elevated levels of DPTU.

4. Attempts have been made to economise by substituting HPLC-grade chemicals for some of the solvents. Although this option might be satisfactory for sequencing at the high picomole or even nanomole level, it is unsuitable at the low and subpicomole level, and as such is hardly worth the trouble.

5. It is a fact of life that deuterium lamps start "dying" from the instant they are powered up. Currently supplied lamps from ABI seem to come on-line with an energy level of around 3.1–3.3, and will probably start to give a troublesome baseline when showing an energy level of about 2.6–2.7. What is perhaps not so widely known is that standby lamps, tucked away in your drawer, do not have an infinite shelf life. Indeed, after a storage period of 6 mo or more, the power-up energy level of such a stored lamp will be greatly reduced, perhaps to as low as 2.8. It is a good trick not to order a standby lamp until around 3 mo after you install the previous one; if the running lamp goes before that 3-mo period, then you will have been very unlucky! Your ABI service desk can almost certainly ship you a replacement within 24 h, if the worst happens.

When that fatal morning comes, and it will, when the overnight run has been killed off due to a blown lamp, or more annoyingly, there has been a power interruption and the lamp power factor displays some arbitrary number, such as –0.100, then pause for thought. Before resetting the 120A analyzer by the classical method of switching it off then on again after waiting for 1 min, it is very advisable to remove the deuterium lamp and carefully test it on a separate 110-volt supply. If a 120A or 130A instrument is powered-up with a blown lamp installed, then 9 times out of 10, the lamp power supply control board will blow; this is inconvenient and expensive, unless you have a skilled electronics engineer at hand.

6. The symptoms of injection pathway blockage are as follows; blank chromatograms and empty fraction collector tubes indicate a blocked needle line. Blank chromatograms and overfull fraction collector tubes indicate a blocked restrictor/overflow line. Replacement is the only sure option, flushing through the offending line and refitting it is only a short term ploy.

   I have not previously mentioned the fraction collector as a separate item. Many users (who do not collect fractions for radiosequencing) have actually severed the return line and let the effluent drain into a suitable vessel. However it is worth bearing in mind the value of this simple diagnostic trick. It is also possible that the injector has not switched from the "load" to "inject" position - check the gas supply which should be set to 60–80 psi; if this is satisfactory, then it may be necessary to rebuild the injector (see manual).

7. It is apparent that many users are now adding dithiothreitol (DTT) to solvents S2 and S3, (*see* **Notes 34** and **35**), giving rise to a new "system" peak; the DTT adduct of phenylisothiocyanate (PITC), which elutes too close for comfort to proline with 5% THF, hence the introduction of the lower concentration (3.5%) of solvent A.

8. The major symptom of an ageing column can be seen at the hydrophobic end of the elution profile, i.e., the compacting of the phe, ile, lys, and leu peaks; also some peak broadening may be apparent. Other indications may occur, such as a reduction in yield, dropping to as low as 60% of that expected, seen first of all on the standard injections. There have been some tricks reported concerning the regeneration of fading columns, i.e. the injection of several 100–200 µL aliquots of dimethylsulphoxide (DMSO) but none will substitute for replacement.

9. Excessive carryover can be the result of depleted R2 reagent (*see* **Note 9**); or nondelivery of PITC; this can be checked visually. The liquid delivery of TFA can obviously be checked visually, and the vapor delivery can be monitored by the previously mentioned pH paper method (*see* **Note 20**). It must be noted, however, that as the level of liquid gradually drops in the reagent bottle when using the gas-phase method, the pick-up line will eventually be too far away from the surface of the liquid to ensure an adequate delivery. This point is usually reached when two-thirds of the liquid has been used. An increase in bottle pressure is probably undesirable. Better to discard the remainder and open a fresh bottle.

10. Abnormally high levels of DPTU can result from a myriad of causes and an elevated level can give rise to a large diphenylurea (DPU) peak, making tryptophan assignments very difficult.

    These are some of the factors involved:
    a. Leaks in the reaction cell: the normal leak test before each run will spell that out.
    b. Inadequate solvent extraction following coupling: check the flow rates.
    c. R1 contamination by water: rare, but possible if the bottle has been in the freezer, perhaps with a slightly loose cap. New PITC is the only option.

    DPU is formed mostly from oxidative desulfurization of DPTU, and is normally present at about 5–10% of the DPTU level. Elevated DPU can be the result of the sequencer lying idle for several days, allowing infiltration of oxygen into the fluid pathways. The DPU level usually normalizes after the sequencer has been run for about 10 cycles. In the original THF/acetonitrile PTH analysis solvent system, DPU elutes directly on top of tryptophan, and can hide low amounts of that residue. There are protocols available that claim to resolve the two components *(4,5)*.

Applied Biosystems have introduced an alternative gas-phase coupling reagent; N-methylpiperidine *(6)*, which will not generate any DMPTU, as DMA will not be present. Another alternative R2 reagent (which is the personal preference of the author), is di-isopropylethylamine (DIPEA), which is made up as a 5% solution in 85% methanol/10% ultra high-quality (UHQ) grade water. This reagent can be left on the instrument for 5–6 wk, which usually translates as 400–500 cycles, with no obvious signs of lack of performance. Both of these new coupling reagents will generate a new artifact, the PITC adduct of methanol, which elutes slightly after DPTU, and may or may not resolve from it, depending on the status of your column. N-methylpiperidine will also generate the product of PITC and piperidine, piperidine-phenylthiourea (PPTU), which elutes just before DPTU.

As the water component of these alternative coupling reagents is mostly replaced by methanol, the observed DPTU level drops markedly; in the DIPEA example to approx 25–40% of that previously seen.

As previously mentioned, DPU is derived from DPTU and it will also diminish, and in many cases it will disappear from view completely. In this welcome situation, assignments of tryptophan become much more reliable, even at the subpicomole level. R2 delivery is normally monitored by adjusting the gas-pressure regulator to the recommended setting, placing the cartridge inlet line in a small beaker of methanol, and observing the bubble rate. This procedure is described in many instrument manuals.

11. The optimized delivery of the cleavage reagent is crucial: too little and cleavage will be incomplete, giving excess lag or carryover, too much and a high background will be seen on the chromatograms, owing to acid hydrolysis of the sample. Depending on the particular sample being sequenced, this can be a horrendous problem (*see* also **Note 30**).

12. Carefully check the instrument's cycle from start to finish. It may well be that the optimized settings have slipped somewhere. If not look for the possibility of a faulty reagent or solvent; it is usually a reagent, rarely a solvent. Replacing chemicals one by one is tedious but necessary. One short cut is that your customer support specialist may have recently heard of a similar problem to yours; this may be worth a telephone call. In the author's laboratory recently, a few sequencing runs showing a lower repetitive yield than normal were traced back to the bottle of R4, which was installed on the instrument just 6 d before. Most sequencing chemicals carry a stability storage limit of 1 yr after the date of purchase and this particular bottle was 11 mo into that period. Replacing it with a new batch solved the problem. There have also been one or two faulty batches shipped from the manufacturers, but the symptoms are the same in both cases: a low repetitive yield and possibly, strange peaks appearing in the chromatograms. Such peaks will only disappear when a new batch of R4 is installed and flushed well through the system.

13. It is possible to overdo the S2 wash, in an attempt to reduce the DPTU levels, and cause extractive loss or "wash-out" of some small peptides. In the 477A sequencer, the delivery should normally be 1.0–1.5 mL. over a time period of around 400 s. Newer instruments use a "pulsed-liquid" delivery of solvent (either ethyl acetate or a combination of ethyl acetate and butyl chloride) to reduce solvent consumption and overall cycle time without apparently reducing washing efficiency *(7)* (*see* **Note 34**).

14. A new chemistry artifact will be generated by the addition of DTT. The DTT/ PITC adduct elutes just prior to proline and may co-elute with pyridylethyl-cysteine, which is the product of the reduction and alkylation of cysteine, using 4-vinyl-pyridine *(8)* If this situation is a problem to the user, then modification to the gradient should help. Another artifact peak may be observed close to isoleucine, if:
    a. too much DTT is added to the S2;
    b. the S2+DTT has been left unused for too long;
    c. the stock of DTT, which should have been kept at minus –20°C is old. Splitting the DTT into aliquots, immediately on receipt, will help minimize degradation due to repeated freeze-thawing.

15. This problem may be due to underdrying of R4 (*see* **Subheading 3.4.**, **step 2**), or the wrong concentration of S4 for the size of the PTH analyzer injection loop (*see* **Note 27**).

16. This can have several causes. Pumps: acid wash procedure (see PTH analyzer manual); dirty flow cell: clean and replace windows; worn pump seals: replace; or possibly an ageing column: again an acid wash procedure will be found in the manual.

17. This is probably due to a blockage in the static mixer/guard column; sometimes sonication in acetonitrile will do the trick when no spare is available, but replacement is the best option.

18. This is a classic symptom of a bubble in the detector flow cell; check the cell visually (*see* **Subheading 3.6.**, **step 5**) and assuming manual control of the pumping system, run it at 350mL/min. and at 80%B for 15–30 min. This will clear away the bubble.

19. A wash bottle with soapy water and a little glycerol, filled to just below the bottom of the inlet line, so that it delivers mostly foam (after having been shaken vigorously), is as good a leak detector as any.

20. This reagent is normally replaced at a time recommended by the manufacturer, even when liquid still remains in the bottle. The delivery step can be monitored by holding a moistened strip of multi-range pH indicator paper close to the end of the cartridge inlet line: the color change should be strong and virtually instantaneous.

21. If a reagent/solvent reservoir is inadvertently left to run dry, the resulting slug of gas drawn up into the line will 8 times out of 10, stop delivery after the bottle has been replaced. If increasing the regulator pressure does not clear this, then attaching a 5 mL. syringe filled with a little of the relevant chemical to the inlet line with a suitable clean piece of tubing, actuating the delivery function and gently

pressing the syringe, should clear the line. Once a new bottle of reagent/solvent has been installed, the system should be thoroughly flushed through.

22. The author at one time experienced problems with batches of discs that were oversized; and gave persistent problems with the piercing of Zitex seals. The only short-term remedy was to reduce the diameter of these discs with a new, very clean, and very sharp cork borer. The use of glass-fiber discs is gradually being superseded by one means or another, thereby eliminating problems related to their use.

23. Solvent A for the model 120A PTH analyser was originally 5% THF (tetrahydrofuran) in water with pH 3.8 or pH 4.6 3 M acetate added; the relative amounts of each were varied to optimize the elution position of various PTH amino acids.

More recently, the THF concentration was lowered to 3.5% and a "Premix Buffer Concentrate" containing the ion-pairing agent hexanesulfonate was introduced as a substitute for the two acetate buffers. Solvent B is either neat acetonitrile with DMPTU added as an oxidant scavenger or 12% isopropanol in acetonitrile *(4)* (*see* **Note 10**), which may be used without DMPTU. In either case, at high sensitivity the baseline rise will may be inconvenient for easy data interpretation. Adding a compensator to solvent A has been suggested *(4)*, which in this case was 1 mL/L of a 1% acetone/water solution. The author's preference however, is to make buffer A 1 μM with respect to high-purity tryptophan *(9)*. This method not only gives a flatter baseline at high sensitivity, but the tryptophan seems to act as some sort of scavenger, giving a noticeably less noisy baseline.

24. As PTH columns age, more and more ion-pair reagent needs to be added (as much as 30 mL/L). Some users have attempted reducing the ion pair reagent ratio back to the usual starting point of 15 mL/L, which should move his after ala, and arg after tyr, thereby giving the column a new lease of life. The author has tried this with two ageing columns but without much success; apparently this technique gives the best results with recent batch numbers.

25. PTH HPLC columns will last for 2 mo or 12 mo. You pay your money and you take your chance, depending on what you do to them. ABI PTH columns have always been batch-dependent with regard to lifespan; and this is probably the case for all manufacturers' columns. Individual users will favor certain C-18 columns. The author has been fairly conservative in sticking with ABI (better the devil you know?), but one thing is certain: a fully capped stationary phase is essential.

26. Pump flow rates can be checked by assuming manual control of the "A" and "B" pumps in turn. Selecting a flow rate of 500 µL/min and collecting the effluent at the column inlet line will give a direct pointer as to the performance of each pump. Erratic flow rates indicate the need to change pump seals, which must be done strictly according to the instrument manual.

27. Many users have decided to change the proportion of the flask content, which is transferred to the PTH analyzer. However, if the size of the injector loop is doubled, then the resulting large volume of 20% aqueous acetonitrile will have a

ruinous effect on the separation of the early eluting PTH-amino acids, such as ser, thr, gln, and gly. A reduction in acetonitrile concentration to 10% will completely cure the problem, and has no adverse effect on the dissolution of any of the PTH-amino acids (5).

28. A 5% solution of PTH in heptane has been in use as this reagent for some time. This reagent was originally introduced as a 15% solution, the concentration later decreased to minimise the levels of both DMPTU and DPTU.

   PITC is a notoriously difficult reagent to purify, requiring vacuum distillation and subfractionation in a lengthy process. If you are unfortunate enough to be the recipient of a bad batch of R1, it will virtually kill the reaction, with the symptoms being new artifact peaks on the chromatograms and very poor or even no sequence. The delivery of R1 is normally adjusted so that it is seen to saturate the piece of PVDF or glass-fiber disc installed in the reaction cell. Obviously, inadequate or inconsistent R1 delivery will result in poor coupling efficiency.

29. The original Beckman spinning-cup sequencer utilised two coupling buffers, Quadrol, for proteins; and THEED (N,N,N',N'- tetrakis (2-hydroxyethyl) ethylenediamine, for peptides; but with the ABI 470A gas-phase sequencer came aqueous trimethylamine (TMA), initially at a concentration of 25%, later reduced to 12.5% (to minimize DMPTU levels), but just as foul-smelling. This coupling reagent gives rise to a few normal reaction by-products, the first of which is DMPTU. This is generated from the dimethylamine (DMA) present in TMA, reacting with R1 (PITC), during the coupling reaction. This by-product peak usually elutes just after glutamic acid in most PTH separations (*see* Chapter 22), at a more or less constant level of about 5–10 picomoles, and most users think no more of it. However, the acrylamide derivative of cysteine, PAM-CYS (cysteine-S-b-propionamide), elutes directly underneath DMPTU (10), and is extremely difficult to resolve; but some ways out of this problem will be mentioned later in this note.

   DPTU results from a slightly complex situation. Water present in R2, reacts with PITC to form H_2S(hydrogen sulphide), CO_2 (carbon dioxide), and aniline. The aniline then reacts with remaining PITC to form DPTU, which is normally present at a constant level on an individual instrument at amounts reported to vary from 10–100 picomoles. In the author's early days of gas-phase sequencing, it was once remarked that the DPTU peak was pleasingly small until it was noticed that the TMA reagent was virtually depleted, which was also the reason for the low yield of sample being sequenced that day. Another symptom of depletion is increased "lag" or carryover.

30. The R5 reservoir is normally used to introduce the standard mixture of PTH amino acids to the conversion flask for subsequent PTH analysis. As such, the formulation of this reagent is entirely in the user's hands, and abnormalities in the standard PTH chromatograms can usually be traced back to poor storage, careless formulation, or occasionally a bad batch of PTH standard supplied by the manufacturers.

31. Anhydrous TFA is in use almost exclusively, the term "anhydrous" being a vital one. If for any reason the TFA picks up any water at all, then it will be virtually useless. With the ability these days to uncap a reagent bottle, then smartly attach it to the instrument, the risk of any type of contamination is minimal. Users who still prefer to use gas-phase R3, will have to continue refilling a 200-mL reservoir, as there is insufficient surface area in the supplied 40-mL bottle. This refilling should be carefully and swiftly done in the dry atmosphere of a suitable fume hood.

    There are two methods of R3 delivery, gas-phase, with a typical step time of around 700 s, or pulsed-liquid delivery using either a full or partial fill of the volumetric loop, with a cleavage time of around 300 s. With the latter delivery mode, when using the ABI "Blott" cell, it is sufficient to have a partial delivery (2 s) followed by a precise argon-dry step (usually performed twice), to leave the PVDF membrane thoroughly saturated, but not bathed in liquid. Immediately after this step the cell is pressurized, thus forcing the TFA into the PVDF matrix. This procedure gives what appears to be the most efficient cleavage procedure currently in use *(7)*.

32. The reagent used for postinstrument conversion with the original Beckman sequencer was 200 µL 1 *M* HCL, with 10% ethanethiol used as a very pungent reducing agent, which was intended to help the recovery of labile residues such as ser, thr, and trp. This reagent was placed in the 5-mL fraction collector tubes, the resulting ATZ(anilinothiazolinone amino acids were delivered in butyl chloride, and the tubes were subsequently processed. The butyl chloride was blown off in a stream of nitrogen until only the bottom acid layer remained; the samples were heated for usually 30–60 min, dried down in a vacuum desiccator before the final pellet was reconstituted, and analyzed by HPLC.

    The 470A ABI instrument was equipped with on-line automatic conversion, and there was a choice of conversion reagents, 1 *M* HCl in methanol, or 25% TFA in water. Methanolic-HCl conversion generates primarily methyl-asp and methyl-glu, which were not well-resolved in some HPLC PTH separation systems. There are little or no common problems associated with TFA as R4, apart from deterioration after over-long storage.

    The manufacturer's lot code system should be familiar to the sequencing operator, so that a close eye may be kept on the age of individual batches of chemicals. R4 delivery is set up ensuring the correct filling of a volumetric loop, in pulsed-liquid mode, or by a short delivery step directly to the flask, followed by a short puff of argon. A visual check can be used to observe the level of reagent present.

33. In 470A, 475A, and 477A sequencers, the first postcoupling solvent wash, (heptane) is primarily intended to remove excess PITC, in order to avoid additional thiourea formation. The flow-rate should be monitored so that approx 0.5–0.8 mL is delivered in this step.

    S1 is also usually delivered to the mid-point of the reaction cell immediately before extraction of the ATZ amino acid, in order to dilute any residual R3. Undiluted residual R3 may partially elute the sample if a glass-fiber filter is being used.

34. The main solvent wash in 470A, 475A, and 477A sequencers, and the only postcoupling wash in the other 47X series sequencers is ethyl acetate. In 49X series instruments, a combination of butyl chloride and ethyl acetate is used. The purpose is to wash out excess PITC and associated by-products, mainly thioureas and ureas, which could interfere with PTH analysis. It also removes any residual R2 reagent, which might form salts during the delivery of R3.

In this technique, S2 is delivered to the cell until the meniscus is about 1 cm. into the cell exit line. After a pause of 20 s the solvent is blown out with argon for 15 s. This process is repeated three times with a final argon dry of 100 s. This feature is incorporated in the author's Fastblot cycle, and has been found to be highly successful. Another option with regard to S2 is the addition of DTT (7), at the concentration of 1–1.5 mg per 100 mL. This serves to improve, especially when sequencing at high sensitivity, the recoveries of ser, thr, trp, and Pam-cys.

The reagent must be prepared meticulously: 100 mL of S2 is purged with argon for 10 min on the instrument, then DTT, of the highest purity available, is added with brief stirring; the S2 is replaced on the instrument, and once more purged with argon for 10 min. This preparation should be kept on the sequencer for 1 wk only, then discarded.

35. In 470A, 475A, and 477A sequencers, butyl chloride is used to extract the cleaved ATZ-amino acid from the reaction cell. This step in the cycle should be carefully optimized to obtain the most efficient transfer from the cell to the conversion flask. This is done by adjusting the S3 pressure regulator, so that when S3 is delivered immediately after the S1 mid-point step (see S1 section), the slug of combined S1/S3 reaches the flask 15–20 s after leaving the mid-point of the cell. If this time is too short or too long, the pressure regulator should be adjusted accordingly.

There has been a drift away from the use of butyl chloride as the main extraction solvent, towards a 1:1 mixture of heptane/ethyl acetate, which has been shown to be a more efficient ATZ-amino acid extractant, originally for only PVDF-bound samples (8). In the author's hands at least, this mixture also performs extremely well with samples spotted onto conventional glass-fiber filters. DTT may also be added to this version of S3, following the protocol mentioned in the S2 section. Ethyl acetate alone (471A, 473A, and 476A) or a combination of ethyl acetate and butyl chloride (49X series) can also be used for ATZ-amino acid extraction.

36. The reaction cell is at its most vulnerable when out of the cell-heater assembly; i.e., when changing samples. If you drop it on a hard floor in your laboratory, it could be an expensive slip. If the edges of either half of the cell become chipped, then it may only appear unsightly; but if the sealing ring on the top half or either of the outer sealing surfaces become damaged, then replacement may be the only remedy.

The BLOTT cartridge has a machined slot in the top half for holding polyvinylidene difluoride (PVDF) samples; extreme care must be taken when pushing the sample down into the base of this slot.

A very thin strip of teflon is ideal, but the scalpel blade that has just been used to cut out the sample is not. If the sample slot becomes damaged in any way, it may affect the critical fluid dynamics of the Blott cell, which need to be maintained precisely in order to achieve efficient control over the delicate wetting and drying steps in PVDF cycles. All types of cartridges will benefit from regular cleaning; it not only keeps the glass surfaces in their optimum condition for wetting, but will remove the build-up of any stain from blots.

In the author's experience a dark deposit of unknown origin was once seen, along a 3-mm length of the inlet channel on the top half of the cell. This turned out to be debris from the main gas cylinder that had found its way through the system, through the 'E' valve block and had accumulated in the cell. It was later found that the particulate filter unit on the main gas line had been wrongly rerouted when moving the sequencer. The 'E' valve block gave up about 6 wk later, as the vast majority of the debris had been deposited there, effectively restricting the flow of chemicals. The cell was completely cleaned by sonication in 50% aqueous nitric acid; but the valve block had to be replaced.

37. At the end of the conversion step, the PTH amino acid must be dissolved prior to being transferred to an on-line PTH analyzer. S4, normally 20% aqueous acetonitrile, is delivered in two measured aliquots and then "paused" for dissolution. It is strongly recommended that several short bursts of flask bubbling, of about 5 s each, be carried out at regular intervals, to maximize solubilization.

Acknowledgments

My thanks go to Professor J. E. Fothergill of my own Department, and to Dr. Philip Jackson of Applied Biosystems for their assistance during the preparation of this article.

References

1. Applied Biosystems User Bulletin No. 30 (revised) Feb. 1988. Troubleshooting and flush procedures for the transfer line between the models 470A/477A and the model 120A.
2. Edman, P. (1950) Method for determination of the amino acid sequence in peptides. *Acta Chem. Scand.* **4**, 283–293.
3. Edman, P. and Begg, G. (1967) A protein sequenator. *Euro. J. Biochem.* **1**, 80–91.
4. Applied Biosystems User Bulletin No. 59. Dec. 1993 Solvent B2; improved separation for the identification of PTH-Trp.
5. Muller-Michel, T. and Bohlen, P. (1990) A method for the unambiguous identification of tryptophan in automated protein protein sequence analysis. *Anal. Biochem.* **191**, 169–173.
6. Applied Biosystems User Bulletin No. 60. Sept 1994 Using N-methylpiperidine as a coupling base for Edman sequencing.
7. Totty, N. (1992) Accelerated high sensitivity microsequencing of proteins and peptides using a miniature reaction cartridge. *Protein Sci.* **1**, 1215–1224.

8. Amons, R. (1987) Vapor-phase modification of sulfhydryl groups in proteins. *FEBS Lett.* **212**, 68–72.

9. Tempst, P. and Riviere, L. (1989) Examination of automated polypeptide sequencing using standard phenylisothiocyanate reagent and subpicomole high-performance liquid chromatographic analysis. *Anal. Biochem.* **183**, 290–300.

10. Brune, D. C. (1992) Alkylation of cysteine with acrylamide for protein sequencing analysis. *Anal. Biochem.* **207**, 285–290.

11. Speicher, D. W. and Reim, D. F. (1992) Microsequence analysis of electroblotted proteins. *Anal. Biochem.* **207**, 19–23.

23

N-Terminal Protein Sequencing
for Special Applications

Philip J. Jackson

1. Introduction

This chapter is aimed at enabling operators of protein microsequencers to design protocols and chemistry cycles for special N-terminal sequencing applications. The information provided is, as far as possible, intended to be generic and therefore appropriate for all types of instruments currently operating in laboratories worldwide. Specific details of microsequencer functions and cycle editing are of course available from the individual manufacturers. It is also assumed that the reader is conversant with Edman degradation chemistry and sequence data interpretation (*1*).

Many of the applications described are nonroutine for the majority of microsequencing laboratories. These include the use of Edman degradation for locating radiolabeled amino acid residues, the most common being [^{32}P]-phosphoryl-serine, -threonine, and -tyrosine. For these projects, the sequence should already be known because the normal practice is to extract and collect the ATZs for scintillation counting rather than continue the chemistry through conversion and phenylthiohydantoin (PTH) analysis (*2,3*). It is also important to maximize the radioactive signal by labeling at as high a specific activity as possible and sequencing in a timely manner to minimize the effect of a short half-life. Because of the usual progressive decrease in S:N ratio with successive degradation cycles, protein samples are typically fragmented to give peptides of 5–15 residues. These are then separated by reverse-phase high-performance liquid chromatography (HPLC) and the radioactive peptides sequenced following identification by on-line scintillation detection, scintillation counting or autoradiography of fractions applied to a strip of poly(vinylidine difluoride) (PVDF) membrane (*3*).

From: *Methods in Molecular Biology, vol. 211: Protein Sequencing Protocols, 2nd ed.*
Edited by: B. J. Smith © Humana Press Inc., Totowa, NJ

Other applications generally regarded as nonroutine include the identification of glycosylation sites. When normal sequencing protocols are used, only O-linked glycosylation sites where the modified serine and threonine residues possess monosaccharide moieties are sufficiently soluble in the ATZ extraction solvent(s) to give identifiable peaks on PTH analysis *(4)*. N-linked or larger O-linked glycosylation sites give rise to blank PTH analysis chromatograms and therefore may be analyzed by comparing sequence data before and after deglycosylation (*see* Chapter 30). An alternative sequencing protocol that generates detectable PTH derivatives for both N- and O-glycosylated amino acid residues is described here. As in the case of phosphorylation site identification, proteins should be cleaved into peptide fragments before sequencing. Additionally, this protocol should only be used to characterize glycoproteins at a superficial level because the information provided by the elution positions of the glyco-PTHs is limited *(5,6)*.

Applications that are probably encountered more frequently are sequencing proline containing samples, synthetic peptides and samples immobilized on PVDF membranes. It is well known that proline residues are inefficiently released as the ATZ derivative during the normal Edman cleavage reaction, giving 30–40% lag with a consequent decrease in readable sequence. Incorporation of modified cycles with more vigorous cleavage conditions in known proline positions enables an extended sequence. This strategy is useful for unknown samples where there is the opportunity for a second sequence analysis to provide more data to specify oligonucleotide probes for a cDNA library or perform a sequence database search. Incidentally, the cleavage conditions used by proline-specific cycles are not appropriate for an entire sequencing run because trifluoroacetic acid (TFA)-driven side-reactions such as serine β-elimination and peptide bond acidolysis would be increased.

Synthetic peptides are sometimes presented while still bound to resin beads, either to verify the sequence before cleavage from the resin or identify peptides that react in a screening assay. Those synthesized by Fmoc chemistry are best cleaved in a small volume of TFA and then applied to a sequencing support as per normal samples in solution. Those synthesized using tBoc chemistry are covered in this chapter because they can be sequenced while still attached to resin beads. A modified sequencing protocol is required to facilitate sample handling and ensure efficient cleavage and ATZ extraction.

In most laboratories, sequencing samples on PVDF membranes, for example after one-dimensional (1D) or two-dimensional (2D) polyacrylamide gel electrophoresis (PAGE) (*see* Chapters 2–4) is now considered to be completely routine and, in fact, often the only type of sample received by microsequencing facilities. All of the sequencer manufacturers have therefore provided tested and supported cycles for sequencing on PVDF and sequencer operators generally find them to be perfectly applicable and robust. The design of a PVDF-specific cycle described in this chapter is aimed at operators who wish to experiment with sequencing parameters to reduce cycle time and possibly enhance repetitive yield *(7,8)*.

Many of the protocols described in this chapter include suggested test samples for use in verifying and optimizing the sequencer's performance when running these special applications. Methods for radiolabeling test samples are also included to enable appropriate pilot experiments to be done. It is certainly advisable to do pilot experiments before committing real samples for sequencing on modified instruments. It is also advisable to exercise care when using modified chemistries and cycles that have not been tested and provided by the sequencer manufacturers. It is especially important to ensure that any hardware changes are thoroughly checked for potential chemical leakage.

2. Materials

In all protocols the majority of sequencer reagents and solvents used are the normal ones specified and supplied by the instrument manufacturers. Only additional or modified chemicals, together with suggested test proteins or peptides are specified below. All reagents and solvents should be sequencer or HPLC grade, as appropriate.

2.1. Proline-Containing Samples

1. Sample immobilization media, as appropriate for the real sample. For example a precycled Biobrene Plus™-coated glass fiber filter, ProSorb™ membrane (Applied Biosystems), protein or peptide support disk (Beckman Coulter), sample column (Agilent, formerly Hewlett Packard), or PVDF membrane after transfer from SDS-PAGE (all instruments).
2. Bovine α-chymotrypsinogen A or soybean trypsin inhibitor (e.g., Sigma C 4879 or T 9003, respectively) dissolved to 0.5–1.0 mg/mL in 0.1% TFA (v/v) in water (*see* **Note 1**).
3. 0.1% (v/v) TFA in water.

2.2. Samples Adsorbed onto PVDF Membrane

1. The instrument manufacturer's standard test protein, e.g., bovine β-lactoglobulin, serum albumin or α-lactalbumin.
2. Test protein (5–20 pmol) electro-transferred, after electrophoresis in a polyacrylamide gel, onto a sequencing grade PVDF membrane, e.g. ProBlott™ (Applied Biosystems), Immobilon PSQ™ (Millipore), or PVM™ (Pall Gelman) and stained with Coomassie Blue or Amido Black (*see* Chapter 4). Alternatively, the test protein may be adsorbed onto PVDF in a ProSorb cartridge (Applied Biosystems) according to the protocol supplied with the product.
3. Methanol.

4. 0.1% (v/v) triethylamine (e.g., Aldrich, 99+%) in methanol.
5. 50% (v/v) methanol in water.

2.3. Synthetic Peptides (tBoc) on Resin Beads

1. Test sample (*see* **Note 2**).
2. Sequencer grade PVDF membrane (*see* **Subheading 2.2.**) cut into either a 5 mm square or 5-mm diameter circle.
3. Fine forceps (2 pairs).
4. Glass microscope slide, cleaned with methanol or ethanol.
5. Optional: illuminated magnifying lens.
6. Metal micro-spatula.

2.4. Radiolabeled Peptides

2.4.1. Preparation of a ^{14}C-Radiolabeled Test Peptide

1. Test sample, e.g., frog atrial natriuretic peptide (Sigma A 0929) for radiolabeling (*see* **Note 3**).
2. Denaturing buffer: 6 M guanidine hydrochloride, 0.1 M Tris-HCl, pH 8.5.
3. Source of argon or oxygen-free nitrogen.
4. Dithiothreitol (DTT): 40 mg/mL in denaturing buffer, freshly prepared.
5. Iodo[2-^{14}C]acetic acid (e.g. Amersham Pharmacia Biotech CFA269). Obtain one vial containing 50 µCi of solid.
6. Iodoacetic acid: 500 mM in water, freshly prepared with the pH adjusted to 8.5 with NaOH.
7. HPLC system with a C8 reverse-phase column (30 × 2.1 mm).
8. 0.1% (v/v) TFA in water.
9. 0.085% (v/v) TFA in acetonitrile.
10. Vacuum centrifuge.

2.4.2. Preparation of an $[^{125}I]$-Iodotyrosine Test Peptide

1. Test sample, e.g., acyl carrier protein fragment 65–74 (e.g., Sigma A 6700 or Bachem H-8810) for radiolabeling (*see* **Note 4**).
2. Pierce IODO-GEN® pre-coated tube.
3. Iodination buffer: 25 mM Tris-HCl, pH 7.5, 0.4 M NaCl.
4. Sodium [^{125}I]-iodide (e.g., Amersham Pharmacia Biotech IM530).
5. Scavenging buffer: 25 mM Tris-HCl, pH 7.5, 0.4 M NaCl, 10 mg/mL tyrosine (*see* **Note 5**).
6. HPLC system with a C8 reverse-phase column (30 × 2.1 mm).
7. 0.1% (v/v) TFA in water.
8. 0.085% (v/v) TFA in acetonitrile.
9. Vacuum centrifuge.

2.4.3. Determination of Radiolabeled Residues in [³H]-, [¹²C]-, [³⁵S]-, and [¹²⁵I]-Peptides

1. Sample immobilization medium, as appropriate for the real sample.
2. Fraction collector and appropriate scintillation counter vials.
3. 0.5 mm ID PTFE tubing and appropriate fittings for connection between the reaction cartridge output valve block and the fraction collector.
4. Scintillation fluid suitable for nonaqueous samples (except [¹²⁵I]-peptides) (*see* **Note 6**).
5. Liquid scintillation or gamma counter.

2.4.4. Preparation of a Test [³²P]-Phosphopeptide

1. Test sample, e.g., Kemptide (Sigma K1127, Bachem M-1510 or Calbiochem 05-23-4900) (*see* **Note 7**).
2. Phosphorylation buffer: 40 mM Tris-HCl, pH 7.4, 20 mM magnesium acetate, 500 µM adenosine 5'-triphosphate (ATP, magnesium salt). Add the ATP just before use.
3. Bovine cAMP dependent protein kinase (protein kinase A), catalytic subunit (e.g., Sigma P 2645 or Calbiochem 539486). Dissolve this in a minimal volume of phosphorylation buffer immediately before use.
4. Adenosine 5'-[γ-³²P]triphosphate in aqueous solution. (e.g., Amersham Pharmacia Biotech PB170).
5. Heating block at 30°C.
6. HPLC system with a C8 reverse-phase column (30 × 2.1 mm).
7. 0.1% (v/v) TFA in water.
8. 0.085% (v/v) TFA in acetonitrile.
9. Vacuum centrifuge.

2.4.5. Determination of Radiolabeled Residues in a [³²P] Phosphopeptide

1. Methanol or 90% (v/v) methanol in water (*see* **Note 8**).
2. Fraction collector and appropriate scintillation vials.
3. 0.5 mm ID PTFE tubing and appropriate fittings for connection between the reaction cartridge output valve block and the fraction collector.
4. Sequelon-AA sample immobilization kit (Applied Biosystems, GEN920033).
5. 100 mL beaker.
6. Heating block at 55°C.
7. 50% (v/v) methanol in water.
8. Methanol.
9. Scintillation fluid suitable for nonaqueous or aqueous samples (*see* **Note 6**).

2.5. Glycosylated Peptides

1. Sequelon-AA sample immobilization kit (Applied Biosystems, GEN920033).
2. Human glycophorin A and chicken ovomucoid (e.g., Sigma G 9511 and T 2100 respectively) (*see* **Note 9**).
3. 20% (v/v) acetonitrile in water.

4. 0.2 *M* TFA in water.
5. Vacuum centrifuge.
6. PTH analysis HPLC solvents: Solvent A: 2 m*M* formic acid in water or 5 m*M* formic acid in water titrated to pH 4.0 with triethylamine (*see* **Note 10**). Solvent B: acetonitrile.
7. Optional: 90% (v/v) methanol in water.

3. Methods
3.1. Proline-Containing Samples

1. Edit the sequencer's Edman degradation cycle as appropriate for the particular sample immobilization medium, according to one of the following options:
   a. For liquid TFA cleavage, extend the cleavage wait or pause time by 5 min (*see* **Table 1**, **step 12**).
   b. For gas-phase TFA cleavage, extend the TFA vapor delivery time by 5 min (*see* **Table 1**, **step 12**).
   c. For either liquid or gas phase TFA cleavage, insert a step that sets the reaction cartridge temperature to 5°C higher than the normal cleavage temperature. To allow time for the temperature to be reached, this step should be inserted just before the postcoupling extraction (*see* **Table 1**, **step 9**). Insert another step to return the reaction cartridge temperature to normal between the postcleavage argon or nitrogen delivery and ATZ extraction steps (*see* **Table 1**, **steps 13** and **14**).
2. Dilute either α-chymotrypsinogen A or soybean trypsin inhibitor in 0.1% (v/v) TFA in water to 1.0 pmol/μL just before use.
3. Load 10–20 pmol of the test protein onto a sequencing support according to the manufacturer's protocol.
4. Enter the sequencer's run program, specifying the proline-specific cycles for either residues 4 and 8 for α-chymotrypsinogen A or residue 9 for trypsin inhibitor. Normal cycles should be specified for the nonproline residues.
5. Use the data from the α-chymotrypsinogen A or trypsin inhibitor test run to verify that the lag through the proline residues is significantly reduced from the usual 30–40% to less than 10%. To optimize the proline-specific cycle, adjust the cleavage time and temperature as necessary to minimize as far as possible both proline lag and the increase in background.

3.2. Samples Adsorbed onto PVDF Membrane

1. Excise the stained test protein band and de-stain it in 1 mL of 0.1% triethylamine in methanol (*see* **Note 12**).
2. Rinse the band in methanol and, while still wet (to minimize problems with static electricity), insert it into the sequencer reaction cartridge.
3. Optional: for instruments that deliver coupling base as vapor, install 50% methanol onto a spare reaction cartridge bottle position (*see* **Note 13**).
4. Edit the normal sequencing cycle as follows:

Table 1
A Generic N-Terminal Sequencer Reaction Cycle

Step number	Chemical delivery	Purpose
1	Coupling base (tertiary amine)	De-protonate N-terminal α-amino group
2	Phenylisothiocyanate (PITC)	Coupling (Edman) reagent
3	Coupling base	Maintain high pH for coupling
4	PITC	Coupling steps are
5	Coupling base	repeated to maximize
6	PITC	efficiency
7	Coupling base	
8	Argon or nitrogen	Evaporate excess base
9	Solvent through to waste	Extract excess PITC and reaction by-products
10	Argon or nitrogen	Evaporate excess solvent
11	Trifluoroacetic acid (TFA)	Cleavage reagent (*see* **Note 11**)
12	Pause or TFA vapor	(*see* **Note 11**)
13	Argon or nitrogen	Evaporate excess TFA
14	Solvent through to the conversion flask	Extract and transfer the ATZ derivative to the conversion flask
15	Argon or nitrogen	Evaporate excess solvent

a. If using 50% methanol as an additional chemical insert a step, before the pre-coupling base delivery (*see* **Table 1**, **step 1**), which loads 5–10 mL into the fluid pathway upstream of the reaction cartridge. Follow this step with a 5–10 sec argon or nitrogen delivery to drive the aqueous methanol onto the sample. Both of these steps should be optimized to just saturate the PVDF membrane(s) without leaving it bathed in excess liquid.

b. Adjust the number of PITC and coupling base deliveries to two (*see* **Table 1**, **steps 2–7**). Ensure that a single delivery of phenylisothiocyanate (PITC) and coupling base (if formulated for liquid delivery) is sufficient to completely wet the PVDF membrane(s). Adjust the total coupling time to two-thirds of that specified in the normal cycle by editing the coupling base vapor delivery or pause steps, as appropriate.

c. If necessary, substitute the existing postcoupling extraction routine (*see* **Table 1**, **step 9**) with three pulsed solvent extractions. Each solvent delivery should completely saturate the PVDF membrane(s) and just enter the tube immediately downstream of the reaction cartridge.

5. Depending on the type of instrument, the TFA cleavage can be configured according to the following options:

   a. If only TFA vapor delivery is possible, no adjustment to the manufacturer's delivery time or temperature is necessary.
   b. If only liquid TFA delivery is possible, observe the extent of PVDF wetting by a single delivery. By the beginning of the cleavage pause time (*see* **Table 1**, **step 12**) the PVDF membrane should be moist with TFA over its entire area but not bathed in excess liquid. If the TFA has not completely spread through the PVDF, edit the cycle to deliver a second aliquot of TFA. If there is excess TFA, extend the gas delivery which drives the liquid TFA into the reaction cartridge so that this excess is evaporated. This extended gas delivery may also be needed when two deliveries of TFA are incorporated into a cycle (*see* **Note 14**).
   c. If both vapor and liquid TFA delivery alternatives are possible, it is worthwhile comparing sequencing efficiency with a number of test samples. Compare repetitive yields, lag (especially through proline), serine recoveries, and rates of background increase.

3.3. Synthetic Peptides (tBoc) on Resin Beads

1. Manipulating the PVDF membrane square or circle with the fine forceps, fold it in half and apply pressure to make a clearly defined crease.
2. Open the membrane out and place 1–10 resin beads within the crease. The bead(s) can be manipulated in a drop of water or methanol on the microscope slide, with the aid of an illuminated magnifying lens.
3. Refold the PVDF and seal the open edges by applying pressure with the straight end of the spatula.
4. Insert the PVDF envelope into the sequencer reaction cartridge (*see* **Note 15**).
5. Edit the normal sequencing cycle as follows:

   a. For liquid TFA cleavage (*see* **Table 1**, **step 11**), insert an additional TFA delivery. Minimise the gas deliveries driving the TFA into the reaction cartridge so that the PVDF envelope is fully saturated with TFA to increase the penetration into the resin bead(s). With instruments that can only deliver TFA as a vapor, try doubling the cleavage time if excessive lag is a problem.
   b. Omit the post-cleavage gas delivery (*see* **Table 1**, **step 13**) to retain TFA in contact with the sample and enhance subsequent ATZ extraction.

3.4. Radiolabeled Peptides

3.4.1. Preparation of a ^{14}C-Radiolabeled Test Peptide

1. Dissolve the test peptide (25–50 µg) in 40 µL of denaturing buffer in a 1.5-mL Eppendorf tube.
2. Add 5 µL of DTT solution, purge the tube with inert gas, seal, mix, and incubate at room temperature for 10 min.

3. Dissolve the iodo[2-^{14}C]acetic acid in 10 µL of denaturing buffer to and transfer it to the reduced test peptide solution. **Care:** Manipulation of a radiolabeled reagent should be confined to a fume-hood and disposable protective gloves worn.

4. Purge the tube with inert gas, seal, mix, and incubate in the dark at 37°C for 30 min.

5. Add 20 µL of nonradiolabeled iodoacetic acid in 5 µL portions, mixing after each addition. Purge with inert gas, seal, and incubate in the dark at 37°C for 30 min.

6. De-salt the ^{14}C-labeled peptide by RP-HPLC as soon as possible after derivatization. Do not store it in the presence of iodoacetic acid. **Care:** Take special precautions to minimize radioactive contamination of the HPLC equipment. Run blank gradients through the column, monitoring the eluate by scintillation counting, until the radioactivity level is at background.

3.4.2. Preparation of an [^{125}I]-Iodotyrosine Test Peptide

1. Add 1.0 mL of iodination buffer to an IODO-GEN pre-coated tube to wet the inside surface and empty the tube.

2. Add 100 µL of iodination buffer directly to the bottom of the tube without allowing any to drain down the side.

3. Add 0.1–1.0 mCi of sodium [^{125}I]-iodide and incubate at room temperature for 6 min, with gentle swirling every 30 s. **Caution:** Radioiodinated material should be manipulated behind a lead screen in a fume hood. Disposable protective gloves should be worn.

4. Transfer the activated radioiodination reagent to a 1.5-mL Eppendorf tube containing 10–20 nmol of test peptide in 100 µL of iodination buffer. Incubate at room temperature for 6–9 min with mixing every 30 s.

5. Add 50 µL of scavenging buffer.

6. De-salt the ^{125}I-labeled peptide by RP-HPLC. **Caution:** Take special precautions to minimize radioactive contamination of the HPLC equipment. Run blank gradients through the column, monitoring the eluate by gamma counting, until the radioactivity level is at background.

3.4.3. Determination of Radiolabeled Residues in [^{3}H]-, [^{12}C]-, [^{35}S]-, and [^{125}I]-Peptides

1. Make the following adaptations to the sequencer hardware and cycle programming:

   a. Fit the 0.5 mm ID PTFE tubing to an appropriate port on the reaction cartridge output valve block. If a special port has not been designated by the manufacturer, use the pathway leading to the conversion flask. Ensure that the pathway connecting the reaction cartridge to waste is still functional. Connect the downstream end of the tubing to the fraction collector, which should be placed as close as possible to the sequencer to minimize the length of tubing needed.

   b. Connect the external device terminals on the sequencer to the appropriate terminals on the fraction collector. Where these are absent or incompatible, the fraction collector may be programmed according to the sequencer cycle time.

c. Edit the sequencer's reaction cartridge cycle so that the ATZ transfer to the conversion flask is substituted by a transfer to the fraction collector. There should be no need to increase the volume of extraction solvent unless there is excessive carry-over of radioactivity. However, the gas delivery times may need lengthening to drive the solvent along the tube to the fraction collector. Add a step to operate the external device relay at the end of the cycle, as appropriate. Retain the "Begin" (cleaning and first coupling) cycle at the start of the run but eliminate all programming for the conversion flask and PTH analysis HPLC.

2. Immobilize the radiolabeled test peptide onto an appropriate sequencing support, matching the amount and/or radioactivity as appropriate for the real sample. If necessary decrease the specific radioactivity of the test peptide by the addition of nonlabeled material.

3. After sequencing, add scintillation fluid (*see* **Note 6**) to the ATZ fractions (except in the case of ^{125}I) and determine their radioactivity in a scintillation or gamma counter, as appropriate.

3.4.4. Preparation of a Test [^{32}P]-Phosphopeptide

1. Dissolve the protein kinase substrate peptide (5–10 µg) in 50 µL of phosphorylation buffer in a 1.5-mL Eppendorf tube.

2. Add 10-50 µCi of [γ-^{32}P]ATP and 100 units of protein kinase, then incubate at 30°C for 3 h. **Care:** ^{32}P-radiolabeled material should be manipulated behind an acrylic screen in a fume hood. Disposable protective gloves should be worn.

3. De-salt the [^{32}P]-phosphopeptide by RP-HPLC (*see* Chapter 1). **Care:** Take special precautions to minimize radioactive contamination of the HPLC equipment. Run blank gradients through the column, monitoring the eluate by either Cerenkov or scintillation counting, until the radioactivity level is at background.

3.4.5. Determination of Radiolabeled Residues
in a [^{32}P]-Phosphopeptide (see also Chapters 24 and 31)

1. Make hardware changes to the protein sequencer as described in **Subheading 3.4.3.**

2. Substitute the reaction cartridge solvent with methanol or the ATZ transfer solvent with 90% methanol in water (*see* **Note 8**).

3. Test the timing of the methanol delivery to the reaction cartridge and, if necessary, increase the regulator pressure (*see* **Note 16**).

4. Edit the sequencer's reaction cartridge cycle as follows:

a. Substitute the ATZ transfer to the conversion flask (*see* **Table 1**, **step 14**) by a transfer to the fraction collector. Two pulses of solvent separated by a 20 s pause should be enough for efficient ^{32}P extraction *(3)*.

b. If necessary, increase the gas delivery time to drive the solvent along the tube to the fraction collector.

c. Add a step to operate the external device relay at the end of the cycle, as appropriate.

    d. Retain the "Begin" (cleaning and first coupling) cycle at the start of the run but eliminate all programming for the conversion flask and PTH analysis HPLC.

5. Immobilize the [^{32}P]-phosphorylated test peptide onto the Sequelon AA covalent attachment membrane according to the following adapted protocol (*see* **Note 17**):

    a. Place the Mylar sheet supplied with the Sequelon AA kit onto the heating block at 55°C.

    b. Place a Sequelon AA membrane onto the Mylar sheet and apply the test peptide dissolved in 0.1% TFA/acetonitrile, as it eluted from a reverse-phase column. Match the amount and/or radioactivity as appropriate for the real sample, if necessary making multiple 3 µL applications. Allow the membrane to dry between applications, covered with the upturned beaker.

    c. Transfer the Mylar sheet, membrane and upturned beaker to a refrigerator at 4°C.

    d. Dissolve EDC in coupling buffer (both supplied in the Sequelon AA kit) at 0.2 mg/µL and apply 5 µL of this solution to the membrane. Incubate the membrane, covered in the refrigerator for 15 min.

    e. Vortex the membrane, first in 1.0 mL of 50% (v/v) methanol in water then 1.0 mL of neat methanol. Allow to dry, covered, at 55°C.

6. After sequencing, add scintillation fluid (*see* **Note 6**) to the ATZ fractions and determine their radioactivity in a liquid scintillation counter.

3.5. Glycosylated Peptides

1. Dissolve the test glycoprotein in 20% (v/v) acetonitrile in water to 20–30 nmol/mL.
2. To 20 µL in a 1.5-mL Eppendorf, add an equal volume of 0.2 *M* TFA in water, mix, and incubate at 80°C for 1 h (*see* **Note 18**).
3. Add 360 µL of 20% (v/v) acetonitrile in water and concentrate the glycopeptide solution to about 10 µL in a vacuum centrifuge.
4. Add 90 µL of 20% (v/v) acetonitrile in water and re-concentrate to about 10 µL.
5. Immobilize 50–100 pmol on a Sequelon AA membrane as described in **Subheading 3.4.5.**
6. Make the following adaptations to the sequencer:

    a. Where the sequencer uses a separate ATZ transfer solvent, replace it with 90% (v/v) methanol in water and, if necessary increase its delivery pressure (*see* **Note 16**).

    b. Where liquid TFA is available for cleavage and only a single reaction cartridge solvent is used, delete the postcleavage gas delivery (*see* **Table 1**, **step 13**) to maintain the presence of TFA during the ATZ extraction and transfer.

    c. Install the alternative PTH analysis HPLC solvents, retaining the normal gradient profile. Although the separation of the usual PTH derivatives may not be ideal, an 8-min gap between the injection artefact peaks and PTH-Asp provides evidence of glycosylation sites in the sequence.

4. Notes

1. The N-terminal 15 residues of these test proteins are:

   Bovine α-chymotrypsinogen A: CGVPA IQPVL SGLSR.
   Soybean trypsin inhibitor: DFVLD NEGPL ENGGT.

   They have been selected because of the proximity of proline to their N-termini, unlike the manufacturers' standard test proteins. Additionally, the use of bovine α-chymotrypsinogen A enables fine-tuning of the proline-specific cleavage conditions by monitoring the serine recovery in cycles 11 and 14.

2. Resin-bound tBoc synthetic peptide test samples should be obtainable from the same laboratory as the real sample(s). It is advisable to not only to check the sequencing performance of the modified cycles but also practice manipulating resin beads. This is particularly important when the sample consists of a single resin bead. When analyzing a sample of bulk resin, assume that a typical peptide loading is 25–100 pmol per bead. Therefore no more than 10 beads should be placed in the sequencer reaction cartridge.

3. The sequence of frog atrial natriuretic peptide is:

   SSDCF GSRID RIGAQ SGMGC GRRF.

   After sequencing the reduced and S-[^{14}C]-carboxymethylated peptide, radioactivity should be released in cycles 4 and 20 (indicated in bold).

4. The sequence of acyl carrier protein fragment 65-74 is:

   VQAAI DYING.

   After sequencing the [^{125}I]-iodinated peptide, radioactivity should be released in cycle 7 (indicated in bold).

5. This buffer is a saturated solution with respect to tyrosine therefore some crystals will remain undissolved.

6. The choice of scintillation fluid depends on whether standard non-polar solvent is used for radiolabeled ATZ extraction or neat or aqueous methanol in the case of [^{32}P]phosphoryl ATZs (*see* **Note 8**). If quenching is a problem the fractions can be dried in a vacuum centrifuge and non-aqueous scintillation fluid used.

7. The sequence of the protein kinase substrate Kemptide is:

   LRRAS LG.

   After sequencing the [^{32}P]-phosphorylated peptide, radioactivity should be released in cycle 5 (indicated in bold).

8. For instruments that use a single reaction cartridge solvent, efficient extractions at both postcoupling (*see* **Table 1, step 9**) and [^{32}P]-phosphoryl ATZ transfer steps are ensured by using neat methanol. Otherwise, 90% (v/v) methanol in water can be used for ATZ transfer to the fraction collector while retaining the normal postcoupling extraction solvent(s) (*3*).

9. The N-terminal 15 residues of these test glycoproteins are:

   Glycophorin A: S/LSTTG/E VAMHT STSSS.
   Ovomucoid: AEVDC SRFPN ATDKE.

Glycosylation sites are indicated in bold. In glycophorin A residues 1 and 5 are blood group variants.

10. The formic acid and triethylamine formate versions of solvent A are described in **refs.** *(5)* and *(6)*, respectively. It is worth testing both alternatives to determine which gives the better PTH elution profile for a particular column and chromatography system. A column already used for normal PTH analysis may not be acceptable for the modified separation because of conditioning of the stationary phase with the normal buffer and ion-pairing agent. If the performance of a used column is not satisfactory, replace it with a new one.

11. The cleavage reagent TFA can be delivered either as a measured pulse of liquid or as vapor carried in a stream of argon or nitrogen. In the former case ("pulsed-liquid" sequencing) step 11 would be the liquid TFA delivery and step 12 a pause while the cleavage reaction takes place. On some instruments it is possible to substitute this pause step with a reaction cartridge pressurization function thereby decreasing the cleavage time *(8)*. In "gas-phase" sequencing steps 11 and 12 would be combined as a TFA vapor delivery for the duration of the cleavage.

12. Complete extraction of Coomassie Blue or Amido Black stain gives a cleaner chemical background on PTH analysis and is therefore a benefit at high sensitivity. Amino acid analysis has shown this procedure not to extract any protein from the PVDF membrane (L. C. Packman, personal communication).

13. Delivery of a small volume of 50% aqueous methanol to wet a PVDF-immobilized sample before N-methylpiperidine coupling base delivery decreases lag by enhancing coupling efficiency.

14. The extent of PVDF saturation by liquid TFA is easier to observe when the membrane is left stained for cycle testing (c.f. **Note 12**). Delivery of sufficient TFA is indicated by a change of the normal blue or blue-black to pale beige. For optimum performance, it may be useful to design specific TFA/gas delivery regimes according to the number of PVDF pieces in the reaction cartridge *(9)*.

15. The idea of enclosing resin beads in a PVDF envelope instead of placing them on a glass fiber filter for sequencing was supplied by C. Ioannou (personal communication).

16. Methanol (neat and 90% [v/v] in water) is more viscous than the normal nonpolar sequencer solvents. Increasing the delivery pressure not only keeps the normal timing of the ATZ extraction and transfer but also prevents the delivery from sticking.

17. The Sequelon AA membrane may need trimming with a sharp scalpel blade to fit into the sequencer's reaction cartridge. The protocol described for covalent attachment is adapted from the manufacturer's protocol according to **refs.** *(3)* and *(6)*.

18. Treatment with TFA at 80°C is to desialylate the glycoprotein. Otherwise, terminal sialic acids would form covalent bonds during subsequent attachment to Sequelon AA membrane, immobilising the ATZ-glycoamino acids.

Acknowledgments

I wish to thank Bryan Dunbar (University of Aberdeen), Nick Morrice (University of Dundee), Len Packman (University of Cambridge), and Nick Totty (ICRF, London) for invaluable discussions.

References

1. Jackson, P. J. and Bayne, S. J. (1998) Quality control of protein primary structure by automated sequencing and mass spectrometry, in *Bioseparation and Bioprocessing: A Handbook* (Subramanian, G., ed.), Wiley-VCH, Weinheim, Germany, pp. 291–323.
2. Wettenhall, R. E. H., Aebersold, R. H., and Hood, L. E. (1991) Solid-phase sequencing of ^{32}P-labeled phosphopeptides at picomole and subpicomole levels. *Methods Enzymol.* **201**, 186–199.
3. Casamayor, A., Morrice, N. A., and Alessi, D. R. (1999) Phosphorylation of ser241 is essential for the activity of 3-phosphoinositide-dependent protein kinase 1: Identification of 5 sites of phosphorylation in vivo. *Biochem. J.* **342**, 287–292.
4. Gerken, T. A., Owens, C. L., and Pasumarthy, M. (1997) Determination of the site-specific O-glycosylation pattern of the porcine submaxillary mucin tandem repeat glycopeptide. *J. Biol. Chem.* **272**, 9709–9719.
5. Gooley, A. A., Packer, N. H., Pisano, A., Redmond, J. W., Williams, K. L., Jones, A., et al. (1995) Characterisation of individual N- and O-linked glycosylation sites using Edman degradation, in *Techniques in Protein Chemistry VI* (Crabb, J. W., ed.), Academic Press, San Diego, pp. 83–90.
6. Pisano, A., Packer, N. H., Redmond, J. W., Williams, K. L., and Gooley, A. A. (1995) Identification and characterization of glycosylated phenylthiohydantoin amino acids, in Methods in Protein Structure Analysis (Atassi, M. Z. and Appella, E., eds.), Plenum Press, New York, NY, pp. 69–80.
7. Henzel, W. J., Tropea, J., and Dupont, D. (1999) Protein identification using 20-minute Edman cycles and sequence mixture analysis. *Anal. Biochem.* **267**, 148–160.
8. Totty, N. F., Waterfield, M. D., and Hsuan, J. J. (1992) Accelerated high-sensitivity microsequencing of proteins and peptides using a miniature reaction cartridge. *Protein Sci.* **1**, 1215–1224.
9. Dunbar, B. and Wilson, S. B. (1994) A buffer exchange procedure giving enhanced resolution to polyacrylamide gels prerun fro protein sequencing. *Anal. Biochem.* **216**, 227–228.

24

Identification of Phosphorylation Sites by Edman Degradation

Laurey Steinke and Richard G. Cook

1. Introduction

Identification of specific sites of phosphorylation (Ser, Thr, Tyr) in proteins is important for understanding protein-protein interactions, intracellular signaling pathways, and regulation of cell growth and differentiation. There are several approaches used to determine specific sites of phosphorylation. If one is working with a known protein, educated guesses can be made based on sequence motifs as to which sites might be phosphorylated. These residues can then be mutated to nonphosphorylatable residues and the resulting protein analyzed for loss of a phosphorylated (^{32}P-labeled) peptide following digestion and two-dimensional (2D) thin-layer chromatography. This method is quite labor-intensive and can fail to give conclusive results if the protein is phosphorylated on nonconventional sites. A second approach, which has been widely employed in recent years, is to use mass spectrometry to identify phosphorylation sites following enzymatic digestion of the protein (*1,2*). A third strategy is to use Edman degradation and radioactive counting to identify ^{32}P-labeled sites in proteins following enzymatic digestion and peptide separation by high-performance liquid chromatography (HPLC). Over the years, various methods based on Edman degradation for sequencing peptides and identifying the site of phosphorylation have been developed. Most of these suffer from an inability to identify the amino acids present in the peptide while concomitantly monitoring the release of radioactivity. Also problematic is the fact that phosphorylated residues are poorly extracted from the sequencing support and/or do not produce stable PTH-derivatives for identification. Using standard Edman degradation procedures, phosphoserine and phosphothreonine break down and result

From: *Methods in Molecular Biology, vol. 211: Protein Sequencing Protocols, 2nd ed.*
Edited by: B. J. Smith © Humana Press Inc., Totowa, NJ

in production of inorganic phosphate, which is not soluble in the normal transfer solvents used; phosphotyrosine is stable but also poorly extracted. Thus the radioactivity remains associated with the filter during sequencing. If solvents that will remove the inorganic phosphate/residues to the conversion flask are used, the peptide is also removed from the support.

Methods that allow sequencing a peptide of known sequence and identification of the area phosphorylated have also been developed, but they are rather unwieldy and require removing a piece of the support after each cycle during sequencing for residue extraction *(3)*. Wettenhall et al. *(4)* describe a method for sequencing of ^{32}P-labeled peptides by modifying a glass-fiber filter such that the peptide can be covalently attached. They describe the modification of the S3 transfer agent to 90% methanol containing phosphate buffer, along with modifications to S2. Shannon and Fox *(5)* describe a method for attachment of a radioactively labeled peptide to a commercially available Sequelon arylamine membrane (Sequelon-AA), and modifications made to the sequencer reagent positions to allow collection of ATZ amino acids in neat TFA with no chromatographic analysis. In the method described here, the radiolabeled peptide is covalently attached to a Sequelon-AA membrane, and the transfer solvent is changed to one that will wash the inorganic phosphate/residue from the membrane into the flask. Following the conversion, a portion of the sample is routed to a fraction collector for subsequent counting and the remainder is sent to the HPLC for residue identification *(see also* Chapters 23 and 31).

2. Materials

2.1. Chemicals and Reagents

1. Sequelon-AA Reagent Kit (Applied Biosystems, Foster City, CA, Cat. no. GEN920033) *(see* **Note 1**).
2. Methanol (Applied Biosystems, Foster City, CA; Fisher Scientifics, Pittsburgh, PA; Pierce, Rockford, IL; or Aldrich, Milwaukee, WI).
3. Phosphoric Acid, ACS reagent (Sigma, St. Louis, MO).
4. Acetonitrile (ACN)(Applied Biosystems or Fisher).
5. Trifluoroacetic Acid (TFA) (Applied Biosystems or Pierce).
6. HPLC-grade (Fisher Scientifics) or Milli-Q water (water prepared by using a Milli-Q UV plus water purification system on reverse osmosis water).

2.2. Laboratory Equipment

1. Multi-Block Heater (Lab-line Instruments, Melrose Park, IL, or Pierce).
2. Analytical balance (Fisher Scientifics).
3. Glass or plastic scintillation vials *(see* **Note 2**).
4. Radiation survey meter.
5. Liquid scintillation counter.

2.3. Protein Sequencing

1. Gas/Liquid Phase Automated Protein Sequencer using Edman degradation (*see* **Note 3**) with standard reagents for the instrument and equipped with a Blott Cartridge.

2.4. Preparing the Reagents

1. S3 Reagent for sequencer: 90% methanol/0.015% phosphoric acid (90:10: 0.015 v/v) 90 mL methanol, 10 mL water, 15 µL phosphoric acid.
2. X2 reagent for sequencer: fill bottle of appropriate size with Milli-Q water.
3. Carbodiimide Solution: according to the instructions on the Sequelon kit, weigh out 1 mg water soluble carbodiimide, add 100 µL of the provided coupling buffer and vortex. Must be absolutely fresh, prepare when indicated in Methods.
4. Methanol (and H_2O) wash: place approx 3 mL of methanol (or H_2O) into a glass scintillation vial (*see* **Note 2**).
5. Acetonitrile wash: TFA/ACN/H_2O (10:50:40 v/v) 0.3 mL TFA, 1.5 mL ACN, 1.2 mL H_2O. Place into a glass scintillation vial (*see* **Note 2**).

3. Methods
3.1. Cycle Modifications for the Sequencer

1. Start with the conventional Blott Reaction cycle for the 477, using the manufacturer's modifications if you are using N-methyl piperidine instead of TMA (*see* **Note 4** for other sequencers).
2. Cleavage: Change the R3 (TFA) load time in step 26 to 8–9 s. Leave the "Argon Dry" time in step 27 at 6 s. Change the "Pause" time in step 28 to 330 s to allow complete cleavage.
3. Extraction: Remove the "Argon Dry" after the cleavage step. Reduce the "Prep Transfer" in what is now step 33 to 5 s. Heptane (S1) is delivered to the midpoint of the cartridge block (5–6 s) followed by a 10-s pause prior to each transfer.
4. Transfer: The S3 position on the sequencer is changed from butyl chloride to 0.015% phosphoric acid, 90% methanol. The programming for the sequencer is changed so that three transfer steps of 30 s each using the phosphoric acid/methanol are used. All "Pause" steps are 10 s, all "Transfer with Argon" steps are 24 s. The sequence should look like this:

    34 Deliver S1
    35 Pause
    36 Transfer with S3
    37 Pause
    38 Transfer with Argon
    39 Deliver S1
    40 Pause
    41 Transfer with S3
    42 Pause
    43 Transfer with Argon
    44 Deliver S1
    45 Pause

46 Transfer with S3
47 Pause
48 Transfer with Argon
49 End Transfer

At this point return to the remainder of the manufacturer's recommended Blott cycle.

5. Conversion Cycle: For solubilization of the PTH-derivative, S4 20% acetonitrile,) is loaded 4 times for a total volume of 300 µL. 50 µL is injected onto the attached ABI 120 HPLC, while the remainder of the sample is collected into the attached fraction collector (*see* **Note 5**).

6. Wash Cycle: Develop a new cycle for your sequencer to wash the injector and the flask after the sequencing cycles are completed.
Deliver X2 (H_2O), 60 s
Argon Dry, 15 s
Clear Inj to waste, 70 s.
Repeat these steps twice, then repeat twice again using S4 (20% ACN/water) instead of X2. Finally,
Deliver S4,
Load Injector,
Inject,
Clear Inj to FC,
FC advance.

3.2. Digestion of Protein and Identification of Peptides of Interest

1. Phosphorylated protein is excised from an SDS gel and peptide fragments are generated by in-gel digestion with trypsin, Lys-C or other enzymes *(6)*. Fragments are separated on a Vydac C18 column (2.1 mm inner diameter [ID] × 250 mm) using a Hewlett-Packard 1090 HPLC with diode array detector. Digestion, separation, and detection can be accomplished by the routine method of the laboratory.

2. Collect peptide fractions into 1.5-mL Eppendorf tubes with screw caps using a peak-detecting fraction collector, or collect using the routine method of the laboratory.

3. After the run is completed, number and cap the tubes, place them inside your usual scintillation vials. Count for 1 min each by Cherenkov radiation. (Windows on the scintillation counter fully open. Do not use scintillation fluid.) Select peaks associated with radioactivity. Choose these to subject to N-terminal sequencing (*see* **Note 6**).

3.3. ³²P-Labeled Peptide Attachment to Sequelon-AA

1. Place a 4 × 4 inch square Mylar sheet from the Sequelon kit, or a clean glass plate, onto the heating block, and preheat to 55°C. Use a fresh piece of Mylar each time. Place the Sequelon-AA membrane on the sheet/plate with clean forceps.

2. Add the peptide to the membrane using 5-µL aliquots, allowing the membrane to dry (5 min) before applying additional aliquots.

3. Count the empty tube to see if the majority of the peptide was applied to the filter.
4. If the majority of the counts remain in the tube, transfer the membrane to a glass plate (if you are using the Mylar sheet that comes with the kit). Wash the tube with 15 µL of neat TFA and load the TFA onto the membrane in one aliquot. This usually rescues material left in the tube.
5. Allow the membrane to dry.
6. Weigh out 1 mg of the carbodiimide supplied with the kit, add 100 µL of the coupling buffer (also supplied with the kit), and vortex. Remove the sheet/plate from the heating block.
7. Add 6 µL of the carbodiimide solution to the membrane. Allow the membrane to dry (20 min).
8. Place the membrane into a vial with approx 3 mL of methanol and rinse for 5 min.
9. Decant the methanol.
10. Add approx 3 mL of Milli-Q water into the vial and rinse for 5 min. Decant the water. Repeat **steps 8–10**.
11. Add the 3 mL of acetonitrile wash solution (10% TFA, 50% ACN, 40% H_2O) to the membrane, and rinse for 5 min.
12. Decant the solvent and allow the membrane to dry on a clean glass plate.
13. Cut the membrane into 6 pieces, such that it fits into the sequencer reaction cartridge (Blott).
14. Place all the pieces of the membrane into the Blott cartridge.

3.4. Sequencer Setup

1. Change S3 to 90% methanol/10% H_2O with 0.015% phosphoric acid. Change X2 to water.
2. Load Blott cartridge and run leak test as usual. Setup method using the Beginblott cycle and standard to start, the modified cycles for the run, and the flask wash cycle to finish.
3. Make sure the fraction collector is full of clean tubes (*see* **Note 7**).
4. Use your standard protocols to check the sequencer and run the sample.

3.5. Data Analysis

1. Unload the tubes from the fraction collector and transfer either the entire tube or the liquid into standard scintillation vials (we use plastic). Add scintillation fluid and count for 5 min using the normal window for ^{32}P. The first and last tubes are blanks and do not contain amino acid residues. The chromatogram from the first amino acid residue on the sequencer will have several artifact peaks present, the largest being near serine and alanine. These decrease in size over the next several cycles.
2. Read the sequence of the peptide from the chromatograms by standard procedures. Phosphoserine and phosphothreonine do not appear on the chromatograms, but the radioactivity will be present in the fractions collected. If the nonphosphorylated peptide has eluted with the phosphorylated peptide you may see the amino acid.

4. Notes

1. The box will still say Millipore, but Millipore doesn't sell the membranes. It is supplied by Applied Biosystems. The part number on the box is correct. A Mylar sheet comes with the kit, but a glass plate worked better in our hands. The plate is necessary if washing the peptide containing tube with neat TFA is necessary. The Mylar sheet will melt messily.

2. 15-mL polypropylene conical tubes may be substituted for the glass scintillation vials.

3. The cycle changes described are for an ABI 477. This type of sequencer is easy to use for this application since it has an integral fraction collector. Steinke modified the restrictor line on the attached 120 HPLC by replacing it with larger internal diameter tubing to minimize clogging with the phosphate salts. Any change in the restrictor line necessitates changing and optimizing the load injector time. Cook did not find this necessary, nor does he use a wash cycle. This may be due to the use of 10% acetonitrile for solubilization (*see* **Note 5**). The 477 sequencer is no longer sold by Applied Biosystems, and will not be supported by its service personnel much longer. The cycle modifications should be applicable to an HT Procise instrument, equipped with an external fraction collector, but care should be taken to avoid clogging the valve blocks. The fraction collector on the Procise is plumbed directly from the cartridge, so the transfers from the cartridge would need to be split, with half sent to the fraction collector, and half sent to the flask. The modifications should not be attempted on a Procise cLC.

4. If using the Applied Biosystems Procise, start with the cycle "Cart PL PVDF Protein." The various transfers would be inserted between steps 68 (Del S3 cart, sensor) and 79 (transfer complete) of the cycle. The transfers would need to be split, alternating between delivery to the external fraction collector, and delivery to the flask. Since the Procise does not generally use heptane, you will need to add it back to the machine in the S1 position. Please note that this has not been tried experimentally in either of the authors' laboratories.

5. 10% Acetonitrile can also be used for solubilization, as can injector loops of various sizes such as 100 μL.

6. Best results have been obtained on peptides which have greater than 1000 cpm by Cherenkov counting (no scintillation fluid, windows on the scintillation counter wide open). Acceptable results have been obtained with counts as low as 200 cpm. To obtain an acceptable number of counts in the isolated peptide fragments, it is best to start with a slice of gel from sodium dodecyl sulfate polyacrylamide gel electrophoresis (SDS-PAGE) with between 100,000 and 1,000,000 counts present, depending on the number of phosphorylation sites. There seems to be a fair amount of ^{32}P nonspecifically and noncovalently associated with proteins in SDS gels, so that the sum of the radioactivity found in fractions collected from the HPLC will not come close to the total radioactivity found in the original gel slice.

7. One possibility for fraction collecting tubes is Waters limited volume inserts. They fit snugly into the fraction collector of the 477, so that collections are not

missed due to tube wobble. They come in glass and plastic. The number for the plastic part is: WAT72030.

8. Returning the sequencer to normal, nonradioactive function may take about half a day on a Procise. Always run a standard protein as the first sample after reverting.

References

1. Zhang, X., Herring, C. J., Romano, P. R., Szczepanowska, J., Brzeska, H., Hinnebusch, A. G., and Qin, J. (1998) Identification of phosphorylation sites in proteins separated by polyacrylamide gel electrophoresis. *Anal. Chem.* **70**, 2050–2059.
2. Neubauer, G. and Mann, M. (1999) Mapping of phosphorylation sites of gel-isolated proteins by nanoelectrospray tandem mass spectrometry: potentials and limitations. *Anal. Chem.* **71**, 235–242.
3. Roach, P. J. and Wang, Y. (1991) Identification of phosphorylation sites in peptides using a microsequencer. *Methods Enzymol.* **201**, 200–206.
4. Wettenhall, R. E. H., Aebersold, R. H., and Hood L. E. (1991) Solid-phase sequencing of [32]-P labeled phosphopeptides at picomole and subpicomole levels. *Methods Enzymol.* **201**, 186–199.
5. Shannon, J. D. and Fox, J. W. (1995) Identification of phosphorylation sites by edman degradation. *Tech. Protein Chem.* **VI**, 117–123.
6. Rosenfeld, J., Capedevielle, J., Guillemot, J. C., and Ferrara, P. (1992) In-gel digestion of proteins for internal sequence analysis after one-or two-dimensional gel electrophoresis. *Anal. Biochem.* **203**, 173–179.

25

Validation of Protein Sequencing in a Regulated Laboratory

Philip J. Jackson and Neil Dodsworth

1. Introduction

A variety of protein analytical techniques are required in the process of discovery, development, and production of protein or peptide pharmaceuticals. Included among these is protein sequencing (both N- and C-terminal). Sequence analysis may be used to confirm that the quality of the protein product is to the level required by the regulatory authorities, comparable with reference samples, and consistent from batch to batch (*see* **Notes 1** and **2**). The approach to protein sequencing in the circumstances of protein pharmaceutical production is more regulated than that in the research environment. Clearly, this is to ensure for reasons of product safety that the results from analysis are reliable and consistent. The details of the requirements of the regulatory authorities may vary from time to time, area to area, and product to product, so the reader is directed to the appropriate authorities for further guidance: see websites for the Food and Drug Administration (FDA) *(1)*, Medicines Control Agency (MCA) *(2)*, Center for Biologics Evaluation and Research (CBER) *(3)* and the United States Pharmacopeia (USP) *(4)*. Another useful source is F-D-C Reports, which provides information on developments affecting regulation of healthcare products *(5)*.

Manufacturers of protein and peptide pharmaceuticals will most probably have adopted ISO9000 and certainly both Good Manufacturing Practice (GMP) and Good Laboratory Practice (GLP). ISO9000 is voluntary and allows for some flexibility in terms of which aspects are put in place. It is focused on system rather than product quality, the aim being to provide consumer satisfaction by the refinement of business practices. While there is considerable overlap between ISO9000 and GMP, the latter is more detailed, legally binding, and aimed at consumer protection. GMP is that part of Quality Assurance

From: *Methods in Molecular Biology, vol. 211: Protein Sequencing Protocols, 2nd ed.*
Edited by: B. J. Smith © Humana Press Inc., Totowa, NJ

that ensures that products are manufactured consistently and controlled to the quality standards appropriate to their intended use, as required by the marketing authorization or product specification. GMP is concerned with both production and Quality Control, which in turn includes all testing and release procedures. The principles of operating to GLP in a regulated analytical laboratory are the same as those for GMP except that they apply to supporting nonclinical laboratory studies as opposed to production for clinical applications.

The regulatory authorities require controlled systems within laboratories, with GMP/GLP specifying that all records of methods and analyses are made in an approved fashion for absolute traceability. These records should include for example, the type of instrument used, Equipment Qualification (EQ; *see* **Note 3**), identifying serial numbers, reagent lot numbers and expiry dates, maintenance log, event log (an instrument-generated record of analysis start and stop times, and unusual occurrences such as power failures), standard operating procedures (SOPs), certification of operator training, method validation protocols, and raw data (signed and dated by the operator and another person). When an analysis fails to meet acceptance criteria, an investigation should be carried out and documented: it is not permissible to repeat the analysis until the results fall within specification. Guidance on GMP and GLP is available on the MCA website *(2)*.

Validation is the process of establishing, in accordance with the principles of GMP, that any method used for the analysis and release of a product has performance characteristics that meet the requirements for the intended analytical applications. Performance characteristics are expressed in terms of, for example, accuracy, specificity, precision, and robustness. The limits of the method are determined by investigating these parameters and setting acceptance criteria defined in a validation protocol.

This chapter aims to provide an outline of the approach to protein sequencing in a regulated laboratory: the impact of Good Laboratory Practice (GLP) and Good Manufacturing Practice (GMP), instrument and software performance, and protein sequencer system validation, but principally the latter. There are more protocols described than most regulated laboratories employ: A typical laboratory will use only a selection of the protocols and there is considerable variation between different laboratories. The reader is advised to adopt a customized approach to validation according to the specific purpose of the protein product testing.

2. Materials

2.1. Protein Sequencing System

A protein-sequencing system comprises the unit that holds the sample(s) and delivers the reagents and solvents for the coupling, cleavage, and conversion reactions. Also included is the PTH analysis high-performance liquid chromatography (HPLC) system, consisting of the solvent pump, C18 reverse-phase column (and column heater) and spectrophotometric detector.

2.2. Data Acquisition and Analysis Software

The purpose of this software is to enable a computer to receive input from the PTH analysis HPLC detector via an analog-to-digital converter. Chromatograms can then be displayed on the screen or printed, with phenylthiohydontoin (PTH) amino acids identified and quantitated by comparison with the retention times and peak areas/heights of standards. Software specifically designed for protein sequencing interprets the data for an entire run to provide a sequence call together with calculations for the initial and repetitive yields. Software-based sequence interpretation should always be verified by a trained operator.

Software developed for the latest generation of sequencers incorporates an event log to record dates and times of instrument operation, power failures, reagent replacement, and operator intervention, for example.

Documentary evidence that the software was developed in accordance with internationally recognized guidelines should be obtained from the manufacturer as part of the system validation process. This may be in the form of either a certificate supplied by the manufacturer or a questionnaire provided by the purchaser's Quality Control department for a representative of the manufacturer to complete.

2.3. Chemicals

Material Safety Data Sheets (MSDSs) for each type of chemical used and Certificates of Analysis (C of As) for specific batches of chemicals should be obtained from the manufacturer. Although these documents are not particularly required for system validation, they are required for running a protein sequencer in a regulated laboratory. Reagents and solvents should be stored as recommended by the manufacturer and used within the specified expiry dates. Chemicals that are installed on the protein sequencer should be used within the recommended on-instrument time periods (*see* **Note 4**).

2.3.1. PTH Amino Acid Standards

These are as specified by the instrument manufacturer. For installation on the sequencer, dilute the stock solution(s) so that the amount injected onto the PTH analyzer is the same as that specified for the Operational Qualification (OQ; *see* **Note 3**).

2.3.2. Standard Proteins (see **Note 5**)

1. Bovine β-lactoglobulin (Applied Biosystems 400979).
2. Horse apomyoglobin (Sigma A 8673).

3. Bovine serum albumin (Calbiochem 126609).
4. Bovine α-lactalbumin (Fluka 61289).
5. Bovine α-chymotrypsinogen A (Sigma C 4879).
6. Soybean trypsin inhibitor (Sigma T 9003).
7. 0.1% (v/v) TFA in water to dissolve all standard proteins except β-lactoglobulin.
8. 0.1% (v/v) TFA, 20% (v/v) acetonitrile in water to dissolve β-lactoglobulin.

2.3.3. Reference Proteins

These should be fully characterized samples of the protein products for which sequence analysis is being validated. A suitable solvent or diluent (in the case of liquid-formulated proteins) should be used. It may also be appropriate to obtain a protein that has a potential modification of a protein product for Selectivity and Specificity testing (*see* **Subheading 3.2.4.** and **Note 6**).

2.3.4. Sequencer Reagents and Solvents

These are as specified by the manufacturer for the instrument model and sample support (*see* **Subheading 2.3.5.** and **Note 7**).

2.3.5. Sample Supports

These are as specified by the sequencer manufacturer, as appropriate for the formulations of the protein products to be analyzed (*see* **Note 8**).

3. Methods

3.1. PTH Amino Acid Identification and Quantitation

3.1.1. Precision (see **Note 9**)

1. Optimize the PTH separation, establishing a suitable gradient profile and solvent composition for the installed reverse-phase chromatography column.
2. Program the sequencer to process and inject the normal amount of PTH amino acid standards (e.g., 10 pmol) for 10 repetitions.
3. Run this protocol each day for a total of 5 consecutive days.
4. Calculate the intra-day and inter-day mean, SD and % RSD for each PTH retention time and peak height/area.

3.1.2. Accuracy, Linearity, Range and Limits of Detection and Quantitation (see **Notes 10–14**)

1. Verify the PTH separation as above.
2. Program the sequencer to perform 10 consecutive injections of PTH amino acid standards for each amount at 1, 5, 10, 50, 100, and 500 pmol.
3. For accuracy, calculate the mean peak height/area for each PTH amino acid amount and express this as a percentage of the true amount (*see* **Note 10**).
4. For linearity, calculate the linear regression equation and correlation coefficient for each PTH amino acid.

5. For range, record the minimum and maximum amounts of each PTH that have acceptable precision, accuracy, and linearity.
6. For limit of detection either:
   a. Visually evaluate the data to determine the minimum amounts of each PTH that can be reliably detected, or
   b. Calculate [3.3 σ]/S where σ is the SD of y-intercepts of regression lines and S is the slope of the linear regression equation.
7. For limit of quantitation either:
   a. Visually evaluate the data to determine the minimum amounts of each PTH that can be reliably quantified, or
   b. Calculate [10 σ]/S where σ is the SD of y-intercepts of regression lines and S is the slope of the linear regression equation.

3.1.3. Ruggedness (see **Note 15**)

1. Replace the PTH standards with those from a different manufacturing batch.
2. Program the sequencer to process and inject the normal amount of PTH standards (e.g., 10 pmol) for 10 repetitions.
3. Calculate the mean, SD and % RSD for each PTH retention time and peak height/area.
4. Compare these results with those of **Subheading 3.1.1.** to determine whether the change has any effect.
5. Replace the HPLC solvents with those from a different manufacturing batch.
6. Repeat **steps 2–4**.
7. Compare these results with those of **Subheading 3.1.1.** to determine whether the change has any effect.

3.1.4. Robustness (see **Note 16**)

1. Program the sequencer to process and inject the normal amount of PTH standards (e.g., 10 pmol) for 10 repetitions for each of three column heater temperature settings: normal, 2°C above and 2°C below normal.
   Compare the chromatograms to determine whether the PTH amino acids are identified and quantitated at the incorrect temperature settings.
2. Program the sequencer to process and inject the normal amount of PTH standards (e.g., 10 pmol) for 10 repetitions for each of three detector wavelength settings: normal (usually 269 nm), 5 nm above and 5 nm below normal.
3. Compare the chromatograms to determine whether the PTH amino acids are identified and quantitated at the incorrect wavelength settings.

3.2. Edman Chemistry

3.2.1. Precision (see **Note 9**)

1. Load either the manufacturer's standard protein at the level specified for the OQ (*see* **Note 3**) or a reference protein at an appropriate level (*see* **Note 1**), where specified in a PQ (*see* **Note 3**), onto suitable sample supports for each reaction cartridge.

2. Program the sequencer to run each reaction cartridge in turn for the number of cycles specified for the OQ or PQ.
3. Repeat the analyses a further 5×.
4. Calculate initial and repetitive yields for each run according to the OQ or PQ protocols and then the intra- and inter-cartridge means, SDs and %RSDs.

3.2.2. Accuracy, Linearity, Range and Limits of Detection and Quantitation (see **Notes 10–14**)

1. Load 1.0 pmol of either the manufacturer's standard protein or a reference protein onto a suitable sample support and install this in one of the reaction cartridges.
2. Run the sequence for the number of cycles specified in the OQ or PQ.
3. Repeat the test with 5, 10, 100, 500, and 1000 pmol of the protein in the same reaction cartridge
4. Sequence each amount for 6 repetitions in the same reaction cartridge.
5. Calculate the initial yield for each sequence analysis.
6. For accuracy, calculate the mean initial yield for each amount of protein loaded and express this as a percentage of the true amount (*see* **Note 10**).
7. For linearity, calculate the linear regression equation and correlation coefficient for initial yield vs amount of protein loaded.
8. For range, record the minimum and maximum amounts of protein that have acceptable precision, accuracy, and linearity.
9. For limit of detection either:
   a. Visually evaluate the data to determine the minimum amount of protein that can be reliably detected, or
   b. Calculate [3.3 σ]/S where σ is the SD of y-intercepts of regression lines and S is the slope of the linear regression equation.
10. For limit of quantitation either:
    a. Visually evaluate the data to determine the minimum amount of protein that can be reliably quantified, or
    b. Calculate [10 σ]/S where σ is the SD of y-intercepts of regression lines and S is the slope of the linear regression equation.

3.2.3. Ruggedness (see **Note 15**)

1. Using only the written SOP, allow a second operator to run the OQ or PQ protocol 6 times on one of the reaction cartridges.
2. Verify that the means, SDs and %RSDs of the initial and repetitive yields are comparable to those of the intra-cartridge values in the precision determination (*see* **Subheading 3.2.1.**).
3. Run the OQ or PQ protocol on 3 different batches of standard or reference protein.
4. Verify that the OQ or PQ specifications are achieved.
5. Run the OQ or PQ protocol with the following modifications, verifying that the specifications are achieved after each:
   a. Change all chemicals, including HPLC solvents, to those of different manufacturing batches.

b. Change the HPLC column (the solvent A composition and gradient profile may need re-optimizing).

3.2.4. Robustness (see **Note 16**)

1. Run the OQ or PQ Protocol with the following modifications:
2. Increase and decrease the column temperature by 2°C from the normal setting.
3. Increase and decrease the standard or reference protein loading volume by 25% while keeping the concentration the same.
4. Increase and decrease the standard or reference protein concentration by 25% while keeping the volume the same.
5. Determine whether the change has any effect and verify that the specifications are achieved after each.

3.2.5. Selectivity and Specificity (see **Note 6**)

3.2.5.1. IDENTIFICATION

1. Dissolve the standard proteins in the solvents indicated in **Subheading 2.3.2.** at the appropriate concentration for the sample support and a loading of 25–50 pmol.
2. Sequence the proteins to 15 amino acid residues.
3. Verify that the correct sequences, as cited in the databases, are generated.

3.2.5.2. DISCRIMINATION

1. Dissolve or dilute, as appropriate, the buffer or medium in which the protein product is normally formulated.
2. Apply this 'blank' solution to the sample support and subject to sequencing as per the protein product.
3. Verify that the PTH analysis chromatograms contain no evidence of spurious sequences or peaks that would interfere with sequence identification.
4. Prepare a series of samples containing a reference protein at its normal concentration for sequencing spiked with a 1, 2, 5, and 10% relative molar concentration of a second protein (either one of the standard proteins or a known potential contaminant of the protein product).
5. Sequence each sample 3 times.
6. Verify that the reference protein sequence is identifiable in the presence of the second protein and determine any effect on the initial and repetitive yields.

3.3. System Suitability Testing (see **Note 16**)

1. Immediately before sequence analysis of a test batch of the protein product, run either the OQ or PQ protocol or both.
2. Confirm the following parameters before proceeding with the test analysis:
   a. The standard or reference protein sequence is identifiable.
   b. The minimum repetitive yield is 2 SDs from the combined intra- and intra-cartridge mean determined in the precision test (*see* **Subheading 3.2.1.**).

    c. The PTH amino acid peak heights/areas in the standards cycle are within 2 SDs of the mean determined in the precision test (*see* **Subheading 3.1.1.** and **Note 17**).

    d. The PTH amino acid retention times in all cycles are within 2 SDs of the mean determined in the precision test (*see* **Subheading 3.1.1.** and **Note 17**).

3. Perform the sequence runs for the protein product test batch or batches.

4. Verify that these test analyses are within specification according to the criteria used in **step 2** (*see* **Note 18**).

5. Optional: repeat the OQ/PQ protocols as per **step 1** to verify that the sequencing system is still performing within specification after the test analyses.

4. Notes

1. When used for Quality Control, nanomolar amounts of sample are typically available for protein sequencing. The methods chosen may not be the same as those used for research purposes: it is preferable to have a robust method, even at the expense of sensitivity or economy of sample. Also, sample loading should be sufficient to bring any impurities (*see* **Note 2**) that must be determined as part of the analysis within the limits of detection and quantitation (*see* **Notes 12** and **13**).

2. Protein sequence analysis can provide qualitative and semi-quantitative information on:

    a. Confirmation of identity;

    b. Degradation of N-terminus;

    c. Background above that of reference sample or additional identifiable sequences, indicative of impurities and/or proteolysis at internal sites;

    d. Deamidation of Asn and Gln, seen as coincidental appearance of Asp or Glu respectively (but beware of deamidation as a side-reaction of Edman chemistry);

    e. Sites of modification, e.g., glycosylation;

    f. Presence of unusual residues or artifacts.

    g. Approximation of initial yield and agreement with sample size (as estimated by other means such as amino acid analysis or A_{280}): significant disagreement indicates error in estimation of sample amount, or blockage (modification) of sequencing chemistry. Initial yield estimates may also be useful in approximating relative amounts of more than one polypeptide present, such as H and L chains in an immunoglobulin preparation.

    If additional techniques are employed, such as Cys alkylation to positively identify it in the sequence (*see* Chapter 27), these must be integrated into the validation.

3. Equipment Qualification (EQ) consists of the following four stages, as defined in **ref.** *(6)*:

    (I) Design Qualification (DQ) specifies the functional and operational specifications of the instrument together with the decisions made in selecting the manufacturer.

    (II) Installation Qualification (IQ) provides evidence that the instrument is received as specified in the DQ and that it is properly installed in an environment suitable for its operation.

(III) Operational Qualification (OQ) is the process of verifying that the instrument functions according to its specifications. In the case of a protein sequencer, the OQ would be the analysis of the manufacturer's standard protein to demonstrate that all amino acid residues in 15–20 cycles are identifiable and the repetitive yield specification is achieved.

(IV) Performance Qualification (PQ) is the process of verifying that the instrument functions according to specifications for its routine operation. If the user is satisfied that the manufacturer's specification testing is appropriate, the OQ and PQ may be incorporated into a single process at installation. Alternatively, the PQ may be the analysis of a user-specified protein, normally a fully characterized reference sample of protein product for which sequence analysis is being validated.

4. Once installed on the sequencer, chemicals may deteriorate more rapidly than when they are stored unopened at low temperature and/or in the dark. Follow the manufacturer's guidelines in determining suitable on-instrument expiry dates, taking the laboratory environment into account.

5. The N-terminal 15 residues of these standard proteins are:

Bovine b-lactoglobulin: LIVTQ TMKGL DIQKV.
Horse apomyoglobin: GLSDG EWQQV LNVWG.
Bovine serum albumin: DTHKS EIAHR FKDLG.
Bovine a-lactalbumin: EQLTK CEVFR ELKDL.
Bovine a-chymotrypsinogen A: CGVPA IQPVL SGLSR.
Soybean trypsin inhibitor: DFVLD NEGPL ENGGT.

These are suggested standard proteins that are used to demonstrate the sequencer's selectivity and specificity (*see* **Note 6**). Any other high purity proteins or peptides that are not N-terminally blocked would be suitable. Proteins such as bovine a-lactalbumin and a-chymotrypsinogen A, with Cys within the first 15 residues should be reduced and alkylated (*see* Chapter 27) before sequencing.

6. Selectivity and specificity are the ability of the analysis to uniquely determine the sample in the presence of components that may be expected to be in the sample matrix and to differentiate the sample from compounds of similar character or structure.

7. When specifying sequencer reagents, solvents and sample supports in the Standard Operating Procedure (SOP), allow for the possibility that the manufacturer may modify these products from time to time. Incorporating flexibility into the SOP may eliminate the need for complete re-validation following any modification. For example, a reference to a reagent might read: "N-methylpiperidine in methanol/water (R2), Applied Biosystems part no. 401535 or equivalent."

8. The appropriate sample support may depend on the protein product formulation. If excipients which interfere with Edman chemistry are present, use a sample support that allows desalting during protein immobilization.

9. Precision is the degree of agreement between individual test results when an analysis is applied to multiple sampling of a homogenous sample.

10. Accuracy is the closeness between the mean value determined over a range and the true value determined experimentally expressed as a percentage. For PTH amino acid standards the true values may be inferred from the nominal amounts in the vials supplied by the manufacturer. For proteins the true values may be determined by amino acid analysis. This test, both for PTH standards and Edman chemistry, is often not performed for protein sequencing because it is not an absolutely quantitative analysis. Also, the initial yield is always less than the true amount.

11. Linearity is the ability of an analysis to provide results that are proportional to the amount of the sample within a given range.

12. Range is the interval between the minimum and maximum levels of sample amount that have acceptable precision, accuracy and linearity.

13. Limit of detection is the lowest amount of sample that can be distinguished from the background.

14. Limit of quantitation is the lowest amount of sample that retains acceptable precision and accuracy.

15. Ruggedness is the degree of precision as determined under a variety of conditions that are found within the normal variability of the analysis.

16. Robustness is the measure of the capacity of an analytical method to remain unaffected by small but deliberate variations in method parameters and provides an indication of reliability during normal usage.

17. System suitability testing is done on the basis of specified performance parameters that are evaluated before sample analysis. This testing verifies that the analysis meets the appropriate performance specifications determined by the validation.

18. Some laboratories only use the PTH-Ala peak for system suitability testing. Other laboratories that use all PTHs reduce the stringency with respect to PTH-His and -Arg because of their higher degree of variability.

19. If a protein product batch release test fails to meet the documented specifications, the analysis is referred to as Out-of-Specification (OOS). In this situation, it is not acceptable to simply repeat the sequencing. The reason for the problem has to be investigated and an OOS Report generated.

Acknowledgments

We wish to thank Bryan Smith (Celltech Chiroscience, Slough, UK), Gordon Garde (Schering-Plough, Brinny, Ireland), and Benne Parten (Applied Biosystems, Foster City, CA) for providing valuable input.

References

1. http://www.fda.gov
2. http://www.open.gov.uk/mca/home.htm
3. http://www.fda/cber/reading.htm
4. http://www.usp.org
5. http://www.fdcreports.com
6. Bedson, P. and Sargent, M. (1996) The development and application of guidance on equipment qualification of analytical instruments. *Accred. Qual. Assur.* **1**, 265–274.

26

Automated C-Terminal Protein Sequence Analysis Using the Alkylated-Thiohydantoin Method

David R. Dupont, Sylvia W. Yuen, and Kenneth S. Graham

1. Introduction

The alkylated-thiohydantoin (ATH) method for C-terminal sequence analysis, which exploits some unique properties of the thiohydantoin (TH) ring system, was first described in 1992 (*1*). The ATH method incorporates alkylation of the C-terminal TH with an aralkylbromide reagent. The resulting ATH is a better leaving group than the unmodified TH, eliminating the need for strongly acidic or strongly nucleophilic cleavage conditions and minimizing ring opening of the TH and hydrolysis of the protein. The ATH is readily cleaved under mildly acidic conditions by nucleophilic displacement with thiocyanate ion, [NCS]⁻, conditions that simultaneously form the penultimate (C-1) TH. This is a key advantage of the ATH method over other methods for C-terminal sequence analysis. Combination of ATH-cleavage and TH-derivatization into one step eliminates the need for activation of the C-terminal amino acid before each cycle of sequencing. Therefore, any new C-termini arising from fragmentation of peptide bonds in the protein can not form TH. This prevents detection of amino acid background resulting from internal cleavage of the protein. In other methods for C-terminal sequence analysis, internal cleavage followed by repeated C-terminal activation can produce a rising TH-background, making accurate sequence calling difficult. In the ATH method, each alkylation of the proteinyl-TH, followed by cleavage and derivatization with [NCS]⁻, comprises one cycle of sequencing. Additional advantages of the ATH method are the introduction of a chromophore or a fluorophore and improved separation and detection of the resulting ATH-amino acids.

From: *Methods in Molecular Biology, vol. 211: Protein Sequencing Protocols, 2nd ed.*
Edited by: B. J. Smith © Humana Press Inc., Totowa, NJ

The ATH method sequences through and detects derivatives for 19 of the 20 genetically coded amino acids. Proline residues are not detected and stop sequencing progress. The ATH method includes modification of amino acids with reactive side-chains in order to improve the ability to sequence through and detect these residues. Initial and repetitive yields are amino acid and sequence specific and so vary significantly from sample to sample. The cycle time is approximately 65 min. Analysis of horse heart apomyoglobin verifies the method; one nanomole, applied to polyvinylidene difluoride (PVDF) membrane and loaded into the reaction cartridge, must provide at least five residues of readable sequence.

The ATH method is compatible with conventional methods of protein sample preparation. Samples prepared by polyacrylamide gel electrophoresis (PAGE) followed by electrotransfer to a PVDF membrane can be directly analyzed. Any protein stain which is compatible with N-terminal (Edman) sequence analysis will not interfere with C-terminal analysis using the ATH method. Samples can be directly applied to PVDF membrane for analysis; samples can also be covalently attached to media compatible with Edman sequencing reagents and solvents. Recently, we reviewed the ATH method *(2)* and several groups have evaluated the method, reporting successful C-terminal analysis of samples following Edman sequence analysis *(3,4)*.

2. Materials

1. The Procise C Sequencing System is used to run the ATH method. The system includes the Procise C Sequencer, the Model 140C Pump (two 10 mL syringes) and a variable wavelength UV detector set to 260 nm. The ATH-amino acid derivatives are separated on a PTC 5 μm, C18 2.1 × 200 mm cartridge-type column.
2. Heating block set to 55-60°C.
3. 1.5 mL microcentrifuge tubes.

2.1. Reduction and Alkylation of Protein Sample

1. 6 *M* guanidine hydrochloride, 0.3 *M* Tris-HCl, pH 8.5.
2. 10 m*M* dithiothreitol (DTT) in water. Prepare fresh as needed.
3. 30% (w/v) acrylamide in water. **Warning:** Acrylamide is highly toxic and an irritant. Use appropriate precautions when handling.

2.2. Application to PVDF Membrane

1. ProSorb Sample Preparation Cartridges.
2. Methanol.
3. 0.1% (v/v) trifluoroacetic acid (TFA) in water.

2.3. Derivatization of Amino Groups

1. 3% (v/v) diisopropylethylamine in acetonitrile (DIEA).
2. 3% (v/v) phenylisocyanate in heptane (PIC).

2.4. Automated C-Terminal Sequence Analysis

1. Reaction cartridge wash solvent: ethyl acetate.
2. Sample reconstitution solvent (for HPLC): 20% (v/v) acetonitrile in water.
3. Flask wash solvent: acetonitrile.
4. HPLC solvents:
   Solvent A: 3.5% (v/v) tetrahydrofuran (THF) in water (Solvent A3), 75 mM sodium acetate (3 M concentrate, pH 5.5, from Applied Biosystems), 2 mM tetrabutylammonium acetate (electrochemical grade). Add approx 5 mL of 1% (v/v) acetone in water per liter of Solvent A to flatten baseline.
   Solvent B: 18% (v/v) THF in acetonitrile.
5. t-Boc-protected ATH-AA reference standard containing 19 of the 20 genetically coded amino acids (proline excluded) in acetonitrile.

2.4.1. C-Terminal Activation

2.4.1.1. ACETIC ANHYDRIDE ACTIVATION

1. 10% (v/v) acetic anhydride/10% (v/v) lutidine in acetonitrile.
2. 5% (w/v) tetrabutylammonium isothiocyanate in acetonitrile.
3. Trifluoroacetic acid (TFA).

2.4.1.2. DIPHENYLPHOSPHOROISOTHIOCYANATIDATE (DPP-ITC) ACTIVATION

1. 5% (v/v) DPP-ITC in toluene/acetonitrile (1:1).
2. 2% (v/v) DIEA in heptane.
3. Pyridine (neat).

2.4.2. Asp and Glu Amidation

1. 10% (v/v) acetic anhydride/10% (v/v) lutidine in acetonitrile.
2. 5% (w/v) piperidine thiocyanate in acetonitrile.

2.4.3. Ser and Thr Acylation

1. 10% (v/v) acetic anhydride/10% (v/v) lutidine in acetonitrile.
2. 15% (v/v) N-methylimidazole in acetonitrile.

2.4.4. Alkylation, Cleavage, and Thiohydantoin Formation

1. 2% (v/v) DIEA in heptane.
2. 2.5% (w/v) 2-(bromomethyl)naphthalene in acetonitrile.
3. 5% (w/v) tetrabutylammonium isothiocyanate in acetonitrile.
4. Trifluoroacetic acid (TFA).

3. Methods

3.1. Reduction and Alkylation of the Protein Sample (see Notes 1,2)

This procedure for Cys alkylation is adapted from Brune *(5)*.

1. Dissolve or dilute the protein in a solution of 6 M Gu•HCl, 0.3 M Tris-HCl, pH 8.5. The final concentration should be at least 4M Gu•HCl, and the final volume less than 500 μL.

2. Add sufficient 10 mM DTT solution to make the ratio of DTT to the number of protein disulfide bonds 100:1.
3. Incubate the solution for 30 min at 60°C.
4. Add sufficient 30% (w/v) acrylamide in water to make the molar ratio of acrylamide to DTT 100:1.
5. Incubate the solution for 1 h at 60°C, or overnight at room temperature.

3.2. Application of the Sample to PVDF Membrane (see Note 3)

The protein is applied to PVDF membrane using a ProSorb device as follows:

1. Assemble the insert and filter.
2. Wet the PVDF membrane at the bottom of the insert with 15 µL of methanol. Press the insert snug against the filter so that the methanol is drawn through the membrane.
3. Add 50 µL of 0.1% (v/v) TFA in water to the insert. Allow all the fluid to be drawn through the membrane.
4. Add the sample solution to the insert. Allow all the fluid to be drawn through the membrane.
5. If the sample contains salts (e.g., Gu•HCl), wash with 150 µL of 0.1% (v/v) TFA in water. Allow all the fluid to be drawn through the membrane.
6. Remove the insert from the holder and dry for 10 min on a 60°C heating block.

3.3. Derivatization of Amino Groups (see Notes 4–6)

This derivatization converts the ε-amino group of Lys (and the N-terminal amino group) to phenylurea. Before starting this derivatization, the sample should be placed in a 60°C heating block for 5–10 min.

1. Apply a 3 µL aliquot of 3% (v/v) PIC in heptane directly to the PVDF membrane.
2. Apply a 3 µL aliquot of 3% (v/v) DIEA in acetonitrile.
3. Incubate for 5 min at 60°C.

3.4. Automated C-Terminal Sequence Analysis

Figure 1 displays the chromatographic separation of 20 picomoles of the ATH-amino acid standard mixture that includes 19 of the 20 genetically coded amino acids (proline excluded) and the major sequencing by-product, naphthylmethylthiocyanate (NMTC). The *t*-Boc-protected ATH reference standards are stable in acetonitrile, but decompose within 24 h after removal of the *t*-Boc protecting group. Therefore, the *t*-Boc group is removed in the flask during an automated cycle, immediately prior to HPLC analysis (*see* **Note 7**).

Figure 2 compares the steps of a typical thiohydantoin chemistry method to the ATH method. Note that the ATH method requires activation of the C-terminal carboxyl group only once at the start of the chemistry, the repetitive chemistry alternates between TH and ATH.

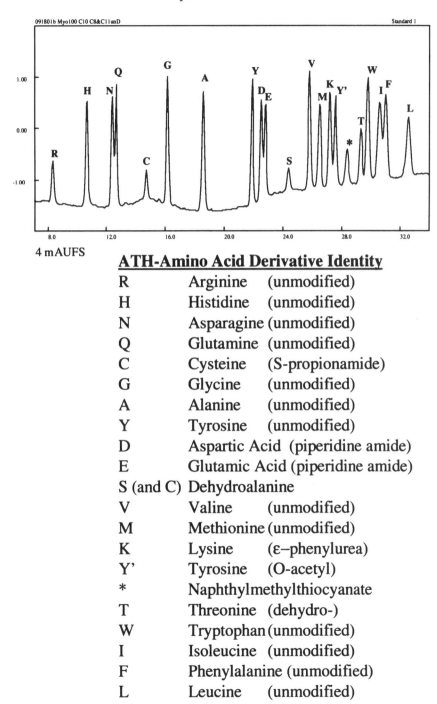

ATH-Amino Acid Derivative Identity

R	Arginine	(unmodified)
H	Histidine	(unmodified)
N	Asparagine	(unmodified)
Q	Glutamine	(unmodified)
C	Cysteine	(S-propionamide)
G	Glycine	(unmodified)
A	Alanine	(unmodified)
Y	Tyrosine	(unmodified)
D	Aspartic Acid	(piperidine amide)
E	Glutamic Acid	(piperidine amide)
S (and C)	Dehydroalanine	
V	Valine	(unmodified)
M	Methionine	(unmodified)
K	Lysine	(ε–phenylurea)
Y'	Tyrosine	(O-acetyl)
*	Naphthylmethylthiocyanate	
T	Threonine	(dehydro-)
W	Tryptophan	(unmodified)
I	Isoleucine	(unmodified)
F	Phenylalanine	(unmodified)
L	Leucine	(unmodified)

Fig. 1. HPLC separation of a standard mixture of 19 ATH- amino acids (20 picomoles).

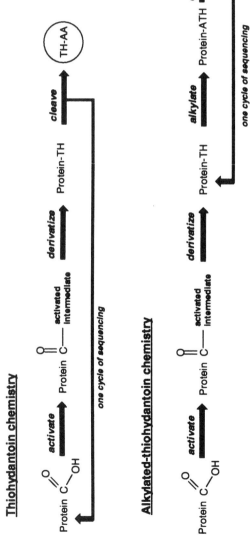

Fig. 2. Comparison of the steps of a typical thiohydantoin (TH) method and the alkylated-thiohydantoin (ATH) method.

4. Notes

4.1. Reduction and Alkylation of the Protein Sample

1. Unmodified Cys residues are detected as dehydroalanine-ATH, and therefore, cannot be distinguished from Ser residues.
2. A number of reagents were tested for the alkylation of Cys and evaluated for chemical stability and chromatographic positioning of the Cys derivative formed. The reagents investigated include: 2-(bromomethyl)naphthalene, iodoacetamide, 3-bromopropylamine, 4-vinylpyridine, maleimide, N-methyl maleimide, N-ethyl maleimide, and N-phenyl maleimide *(6)*. The derivatives of Cys formed by most of these reagents were unstable to the ATH sequencing conditions, resulting in the detection of dehydroalanine-ATH as the primary product. Others could not be detected during HPLC analysis, either due to coelution with other ATH derivatives or chemical artifacts, or because the derivatives were too hydrophobic and eluted in the column wash.

4.2. Application to PVDF Membrane

3. The ATH method is compatible with conventional methods of protein sample preparation. The reagents, solvents and conditions are compatible with PVDF membrane as the sequencing support. Samples can be prepared by PAGE followed by electroblotting to PVDF membrane. Any stains compatible with Edman sequencing will not interfere in C-terminal sequencing. Samples can be directly applied to PVDF membrane for analysis; better results are obtained using PVDF membrane with a pore size of 0.2 µm or smaller. Samples can also be covalently attached to membranes (e.g., Sequelon-DITC) or DITC-glass beads or DITC-glass fiber filters.

4.3. Derivatization of Amino Groups

4. Extended treatment of a protein with PIC may amidate carboxyl groups, including the C-terminus, blocking subsequent sequencing *(7)*.
5. In the absence of the PIC-treatment, the ε-amino group of Lys is acetylated during the initial C-terminal activation with acetic anhydride. The resulting acetylated ATH-Lys derivative is chemically stable, but is difficult to resolve chromatographically from one of the chemistry artifact peaks. In order to form a Lys derivative with better chromatographic characteristics, samples are treated with PIC after immobilization on PVDF membrane.
6. PIC reacts with the hydroxyl group of Tyr to form an arylcarbamate, and thereby minimizes acetylation of Tyr during exposure to acetic anhydride.

4.4. Automated C-Terminal Sequencing

7. Ser-ATH and Thr-ATH, whether derived from the benzyl-protected *t*-Boc-ATH-amino acid standards or from sequencing, are detected chromatographically as their dehydro-analogs. There are two isomers of Thr-ATH that can be detected as distinct chromatographic peaks. Using the current chromatography conditions, one of the isomer peaks co-elutes with the NMTC peak.

Table 1
Effect of Activation Chemistry on C-Terminal Amino Acid Yield - pmol Detected / 1 nmol Loaded

Amino acid	Acetic anhydride activation	DPPITC activation
Glutamic acid	21	80
Aspartic acid	75	92
Leucine	217	613
Isoleucine	232	644
Alanine	245	654
Histidine	47	530
Arginine	56	578
Glycine	106	424
Phenylalanine	314	605
Lysine	125	455
Serine	30	54
Threonine	39	53
Asparagine	51	80
Glutamine	131	461
Methionine	135	424
Cysteine	23	120
Tyrosine	89	80
Tryptophan	238	556
Valine	185	408

4.4.1. C-Terminal Activation

8. Activation of the free carboxylic acid group at the C-terminus occurs only once; at the start of the ATH sequencing method in order to form the first proteinyl-TH. We compared the results of C-terminal activation and TH formation using acetic anhydride activation and DPP-ITC activation. **Table 1** compares the effects of activation chemistry on the recoveries of 19 of 20 amino acids at the C-terminal position. Sets of synthetic peptides were covalently attached to modified-PVDF membrane and submitted for C-terminal sequencing using the ATH method with initial activation using acetic anhydride or DPP-ITC. The theoretical loading of each amino acid is one nanomole and the penultimate residue is glutamine. The recoveries of nearly all amino acids are significantly improved using DPP-ITC activation. Using acetic anhydride activation, Tyr is detected in acetylated and unacetylated forms; only unacetylated Tyr is detected using DPP-ITC activation and the yield is slightly lower than the combined yield from the acetic anhydride activation. It is unclear why the Tyr yield is lower using DPP-ITC activation; it is likely that a small amount forms a diphenylphosphoryl derivative. For these experiments the peptides were not reduced and alkylated so both Cys and Ser are

detected as dehydroalanine-ATH; Thr is also detected as its dehydro-analog. The recovery of Cys, Ser, and Thr derivatives is rather low because the methylimidazole-catalyzed acylation does not improve recovery of Ser, Thr or unmodified Cys in the C-terminal position. No derivative is detected for Asp and Glu if the amidation cycle is deleted. The low yield of Asn is unexpected and appears to be an artifact of the peptide set rather than an indication of difficulty in forming a TH. For the acetic anhydride activation, the yields of His and Arg are the sum of multiple peaks resulting from acetylated and unacetylated derivatives. Clearly, the recoveries of His and Arg are significantly improved in the absence of acetic anhydride.

4.4.1.1. ACETIC ANHYDRIDE ACTIVATION

9. A mixture of acetic anhydride and lutidine is used for carboxyl group activation. Activation of the carboxyl group of a peptide or protein with acetic anhydride under basic conditions results in the formation of an oxazolone. In a subsequent step, the oxazolone reacts with [NCS]⁻ under acidic conditions to form TH. The proteinyl-TH is stable in the presence of the reagents used for TH formation. The C-terminal oxazolone can react with excess activating reagent under conditions that are strongly basic. O-acetylation of an oxazolone, the first step of the Dakin-West reaction, is promoted in the presence of a strong base, such as triethylamine, or an acylation catalyst, such as pyridine or DMAP. Because the O-acetylated-oxazolone reacts very slowly (relative to the unmodified oxazolone) to form a TH, the sequencing initial yield is reduced if the first step, or any subsequent step, of the Dakin-West reaction occurs. Therefore, lutidine, a weak base and poor acylation catalyst, is used during activation with acetic anhydride.

4.4.1.2. DPP-ITC ACTIVATION

10. An alternative approach to activating the C-terminus of a protein or peptide is to use a single reagent that incorporates the ability to activate the C-terminus and form the first TH. One such reagent, DPP-ITC, is proposed to form a TH without proceeding through an oxazolone intermediate *(8)*. The best results were obtained using a cycle in which a delivery of DPP-ITC (5% [v/v] in 1:1 toluene: acetonitrile [v/v]) was made to the reaction cartridge, followed by a delivery of pyridine vapor for 60 s. Valves at the inlet and outlet of the reaction cartridge were then closed and the reaction proceeded at 60°C for 450 s. This sequence of steps was repeated 5 times during the cycle, with an ethyl acetate wash between each coupling.

4.4.2. Asp and Glu Amidation

11. Amidation of Asp and Glu improves the ability to detect these residues. The carboxylic acid groups of Asp and Glu residues can be selectively amidated, without amidating the C-terminal carboxyl group. During the initial activation, the side-chain carboxyl groups of Asp and Glu react to form mixed anhydrides while the C-terminal amino acid is derivatized to a TH. Formation of the C-terminal TH protects the C-terminus from modification during the subsequent amidation.

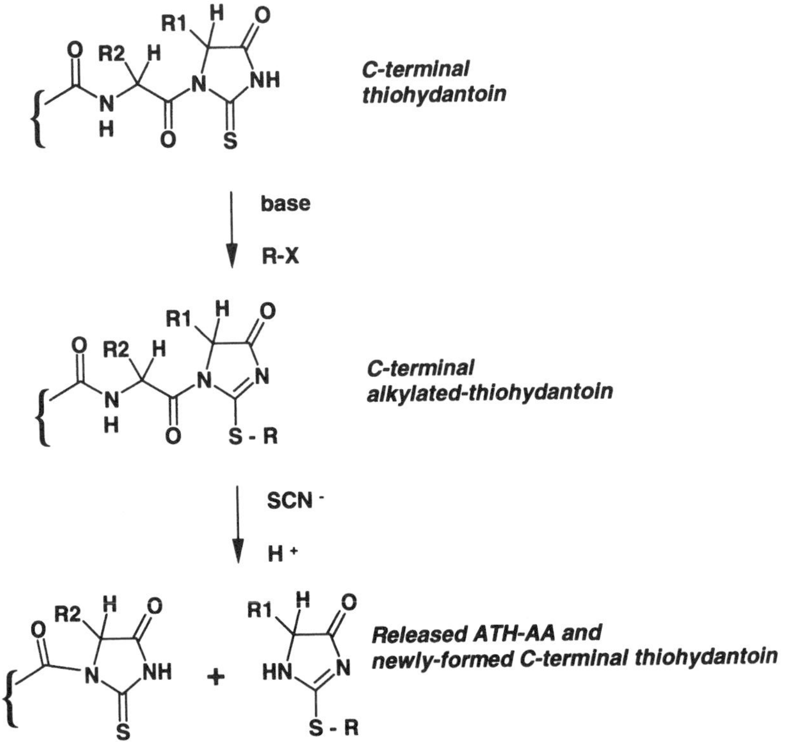

Fig. 3. The repetitive steps of the ATH method: thiohydantoin alkylation followed by cleavage and thiohydantoin formation.

The side chain carboxyls of Asp and Glu are amidated by piperidine thiocyanate in the presence of lutidine. The derivatized C-terminus will not react with piperidine thiocyanate under these mildly basic conditions. However, in our investigations, we found that substituting piperidine for piperidine thiocyanate did cause amidation of the C-terminus.

4.4.3. Ser and Thr Acylation

12. An unmodified Ser or Thr hydroxyl adjacent to an ATH may promote displacement of the alkylated sulfur (*1*). A desulfurized-TH can not be cleaved from the protein, and sequencing stops at the residue that precedes the Ser or Thr (from the C-terminus). Forming an acylated or "capped" hydroxyl group by reaction with acetic anhydride minimizes desulfurization. To ensure complete modification of Ser and Thr hydroxyl groups, an acylation catalyst, such as N-methylimidazole or DMAP, is used along with acetic anhydride. A proteinyl-ATH is stable in the presence of N-methylimidazole and acetic anhydride; therefore, the acylation catalyst is introduced after alkylation of the first TH.

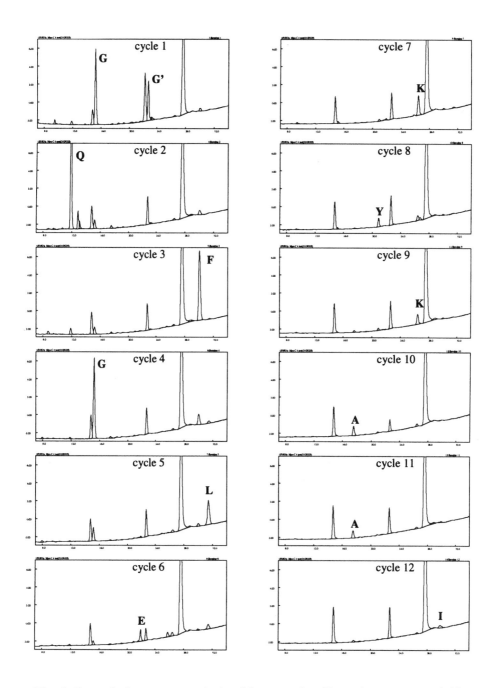

Fig. 4. C-terminal sequence analysis of 1 nanomole of horse heart apomyoglobin, acetic anhydride activation. Chromatogram scale is 10 mAUFS.

13. Under the conditions to modify Ser and Thr hydroxyl groups, acetic anhydride also reacts with Arg, His, Tyr, and Lys (in the absence of the PIC pre-treatment).

4.4.4. Alkylation, Cleavage, and Thiohydantoin Formation

14. After the initial activation and conversion of the C-terminus of a protein into a TH, three deliveries of 2-(bromomethyl)naphthalene and DIEA to the reaction cartridge, separated by 10 minute pauses for reaction result in the S-alkylation of the C-terminal TH. After washing with ethyl acetate, the protein is treated with tetrabutylammonium thiocyanate and TFA vapor. The quaternary ammonium salt of [NCS]⁻ was selected because the amine counter-ion is non-nucleophilic and sufficiently soluble in organic solvents. The [NCS]⁻ generated under acidic conditions cleaves the ATH and forms a new TH at the C-terminus. The repetitive steps of the ATH method are shown in **Fig. 3**; alkylation of the C-terminal TH followed by cleavage of the C-terminal ATH-amino acid and simultaneous TH-derivatization of the penultimate amino acid. The results of C-terminal sequence analysis of 1 nanomole of horse heart apomyoglobin are shown in **Fig. 4**. This sample was run using acetic anhydride activation and acylation of Ser and Thr hydroxyls. Using these conditions, C-terminal Gly is detected in both unacetylated (G) and acetylated (G') forms.

Acknowledgments

We thank Dr. Richard L. Noble of Applied Biosystems for the sets of synthetic peptides used to compare activation chemistries.

References

1. Boyd, V. L., Bozzini, M., Zon, G., Noble, R. L., and Mattaliano, R. J. (1992) Sequencing of proteins and peptides from the carboxy terminus. *Anal. Biochem.* **206**, 344–352.

2. Dupont, D. R., Bozzini, M., and Boyd, V. L. (2000) The Alkylated Thiohydantoin Method for C-terminal sequence analysis, in *Proteomics in Functional Genomics, Protein Structure Analysis* (Jolles, P. and Jornvall, H., eds.), Birkhäuser, Verlag, Basel, pp. 119–131.

3. Bergman, T., Cederlund, E., and Jornvall, H. (2001) Chemical C-terminal protein sequence analysis: improved sensitivity, length of degradation, proline passage, and combination with Edman degradation. *Anal. Biochem.* **290(1)**, 74–82.

4. Samyn, B., Hardeman, K., Van der Eycken, J., and Van Beeumen, J. (2000) Applicability of the alkylation chemistry for chemical C-terminal protein sequence analysis. *Anal. Chem.* **72**, 1389–1399.

5. Brune, D. C. (1992) Alkylation of cysteine with acrylamide for protein sequence analysis. *Anal. Biochem.* **207**, 285–290.

6. Nguyen, D. N., Becker, G. W., Riggin, R. M., Boyd, V. L., Bozzini, M., Yuan, P.-M., and Loudon, G. M. Alkylation of Cysteine: Application to Automated C-terminal Sequencing of Proteins, presented at the Ninth Symposium of the Protein Society, Boston, MA, 1995.

7. Blagbrough, I. S., Mackenzie, N. E., Ortiz, C., and Scott, A. I. (1986) The condensation reaction between isocyanates and carboxylic acids. A practical synthesis of substituted amides and anilides. *Tetrahed. Lett.* **27**, 1251–1254.
8. Bailey, J. M., Nikfarjam, F., Shenoy, N. R. and Shively, J. E. (1992) Automated carboxy-terminal sequence analysis of peptides and proteins using diphenyl phosphoroisothiocyanatidate. *Protein Sci.* **1**, 1622–1633.

27

Chemical Modifications of Proteins as an Aid to Sequence Analysis

Alex F. Carne

1. Introduction

The most widely used applications for chemical modification are in primary structure analysis and in the identification of essential groups involved in the binding and catalytic sites of proteins. Chemical modifications, using an ever increasing number of cross-linkers, are involved in both protein "active center" identification *(1)*, subunit cross-linking, protein-protein interaction, and many more (see http://www. piercenet. com/Technical/Docs/08/apps.pdf), and also in mapping the positions of disulfide linkage patterns in peptides containing tightly clustered cysteines *(2)*. The methods discussed here are those used frequently in primary structure analysis.

The introduction of the gas-phase *(3)* and pulsed-liquid automated sequencers *(4)* with the ability, and the necessity, to sequence proteins only available in the low picomole range, has made it essential to modify some of the more well-known chemical modifications methods to suit the smaller amounts of polypeptide. Also, the increasing use of mass spectrometry has led to specific modifications designed to facilitate the interpretation of data, e.g., methyl esterification of free carboxyl groups.

Cysteine presents a problem both in amino acid and sequence analysis. It is subject to auto-oxidation leading to a variety of products such as mixed disulfides *(5)* and is involved in the formation of protein disulfide bridges. The presence of these bridges within a polypeptide chain leads to difficulties in peptide isolation after enzymic digestion and in sequence determination. Reduction, normally after denaturation to expose the sulfhydryl groups, is followed by alkylation. Dithiothreitol (DTT) is the most widely used reductant but the use of tributylphosphine (TBP) and tris (2-carboxyethyl)phosphine

From: *Methods in Molecular Biology, vol. 211: Protein Sequencing Protocols, 2nd ed.*
Edited by: B. J. Smith © Humana Press Inc., Totowa, NJ

(TCEP) *(6)* are gaining in popularity because they are much less pungent than DTT. Also, TCEP can be used at acid pH, allowing reduction in 0.1% trifluoroacetic acid (TFA) prior to high-performance liquid chromatography (HPLC) analysis for disulfide bond assignment. The structures of DTT and TCEP are shown in **Figs. 1** and **2**. The mechanism of disulfide reduction by thiols is an exchange of the thiolate anion (XS^-) as shown in Reactions [1] and [2].

$$XS^- + RSSR \rightarrow RSSX + RS^- \qquad [1]$$

$$XS^- + RSSX \rightarrow XSSX + RS^- \qquad [2]$$

TCEP, in aqueous solutions, reacts only with disulfides, in a stoichiometric and irreversible manner, as there is no free thiol in this reductant. The mechanism is shown according to Reaction [3].

$$(CH_2CH_2COOH)_3P: + RSSR \rightarrow (CH_2CH_2COOH)_3P = O + 2RSH \quad [3]$$

The reduction can be carried out with TCEP and iodoacetamide though the reaction is slower due to the formation of a complex between these two reagents. However, using TCEP at a concentration of 0.1 m*M* gives 80–90% labeling efficiency *(6)*.

Reaction of cysteine with iodoacetic acid yields a stable carboxymethyl derivative *(see* **Fig. 3**) and also introduces a charged group that tends to enhance the solubility of the polypeptide in aqueous buffers at alkaline pH. This makes the protein more susceptible to digestion by proteases such as trypsin, chymotrypsin, and so on. A similar reaction with iodoacetamide is used to produce the carboxamidomethyl-cys derivative *(see* **Fig. 4**), both for mass mapping *(7)* and to alkylate cysteine residues after isoelectric focusing during two-dimensional (2D) gel electrophoresis where protein solubility and reduction of point streaking is an important factor for satisfactory analysis *(8)*. The use of iodoacetamide mixed with its ^{14}C isotope as the alkylating agent enables a radioactive label to be introduced into the polypeptide chain. A method will also be described for alkylation of proteins electroblotted onto polyvinylidene (PVDF) membrane *(9)*.

Other alkylating agents include N-isopropyliodoacetamide (NIPIA) *(10)* and 6-iodoacetamidofluorescein (6-IAF) *(11)*. The latter reagent introduces a fluorophor and can be used to prelabel proteins prior to polyacrylamide gel electrophoresis (PAGE) *(12)*. Phosphoserine residues can be modified via the β-elimination of phosphate followed by addition of ethanedithiol (EDT) onto the resulting double bond thus introducing a new sulfydryl group into the dehydroalanyl residue *(13)* which can then be alkylated with 6-IAF. Labeled bands can be excised for internal sequence analysis *(see* Chapters 5 and 6). The reaction mechanisms of NIPIA and the fluorescein derivatives are shown in **Figs. 5–7**. NIPIA is now one of the most popular reagents for sequence use since the phenylthiohydantoin (PTH-*N*-isopropylcarboxyamidomethyl-cys)

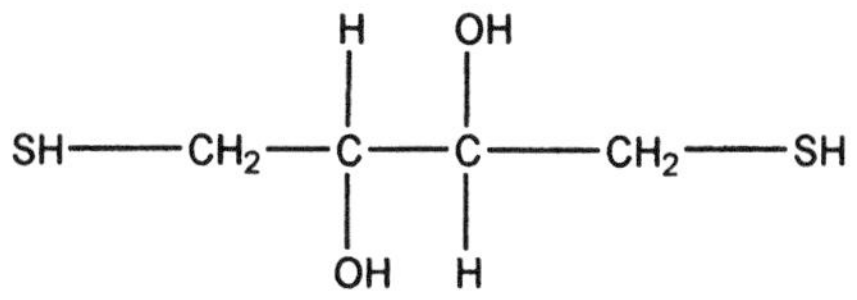

Fig. 1. Dithiothreitol (DTT).

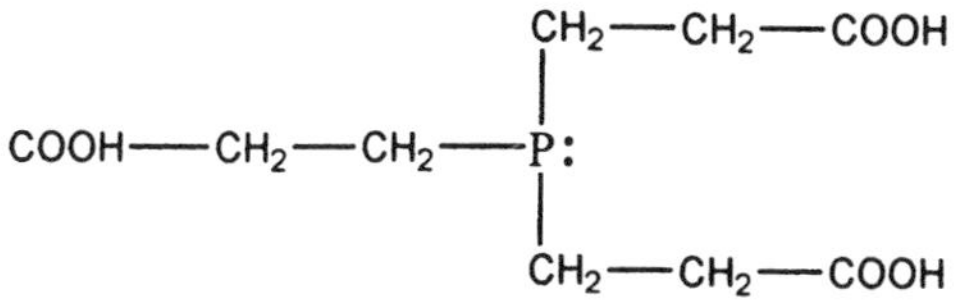

Fig. 2. Tris(2-carboxyethyl)phosphine (TCEP).

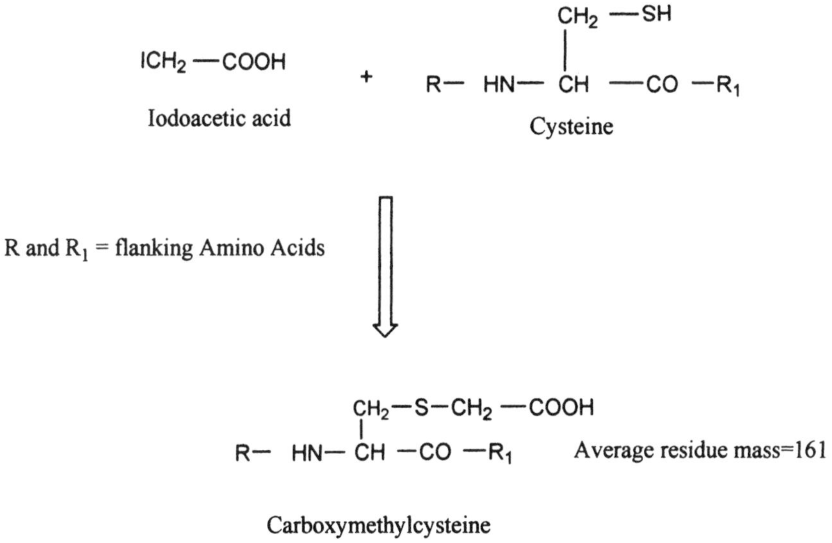
Iodoacetic acid
Cysteine
R and R1 = flanking Amino Acids
Average residue mass=161
Carboxymethylcysteine

Fig. 3. Carboxymethylation of Cysteine using Iodoacetic acid.

derivative elutes in the gap between tyrosine and proline during PTH amino acid analysis. In all cases, the addition of excess reductant quenches the reaction, by reacting with free alkylating agent. Also, for C-terminal sequencing, reductive alkylation is a necessary step to differentiate cysteine from serine in the thiohydantoin chemistry (*see* Chapter 26) used by Applied Biosystems ***(14)***.

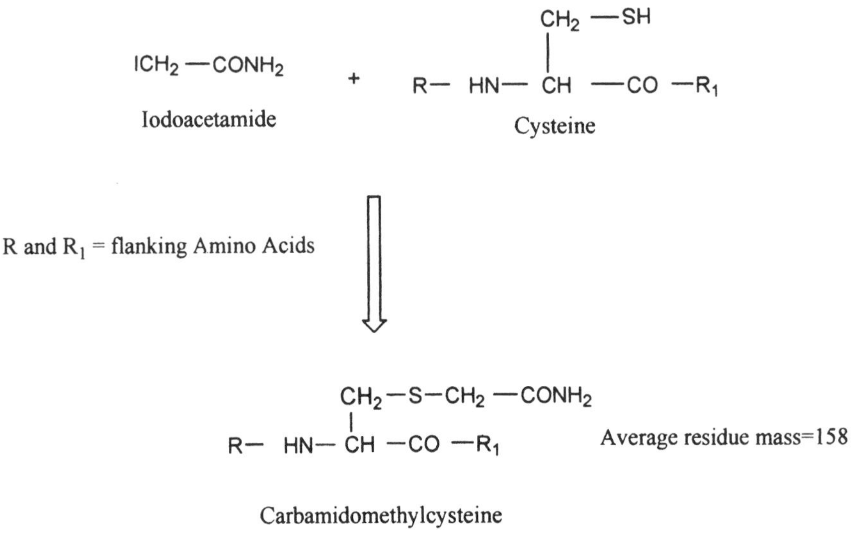

Fig. 4. Alkylation of Cysteine using Iodoacetamide.

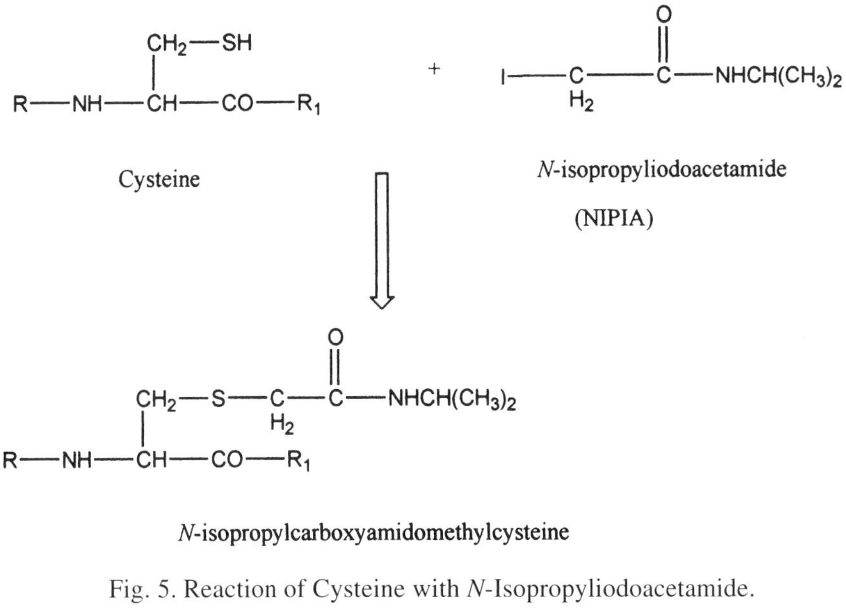

N-isopropylcarboxyamidomethylcysteine

Fig. 5. Reaction of Cysteine with N-Isopropyliodoacetamide.

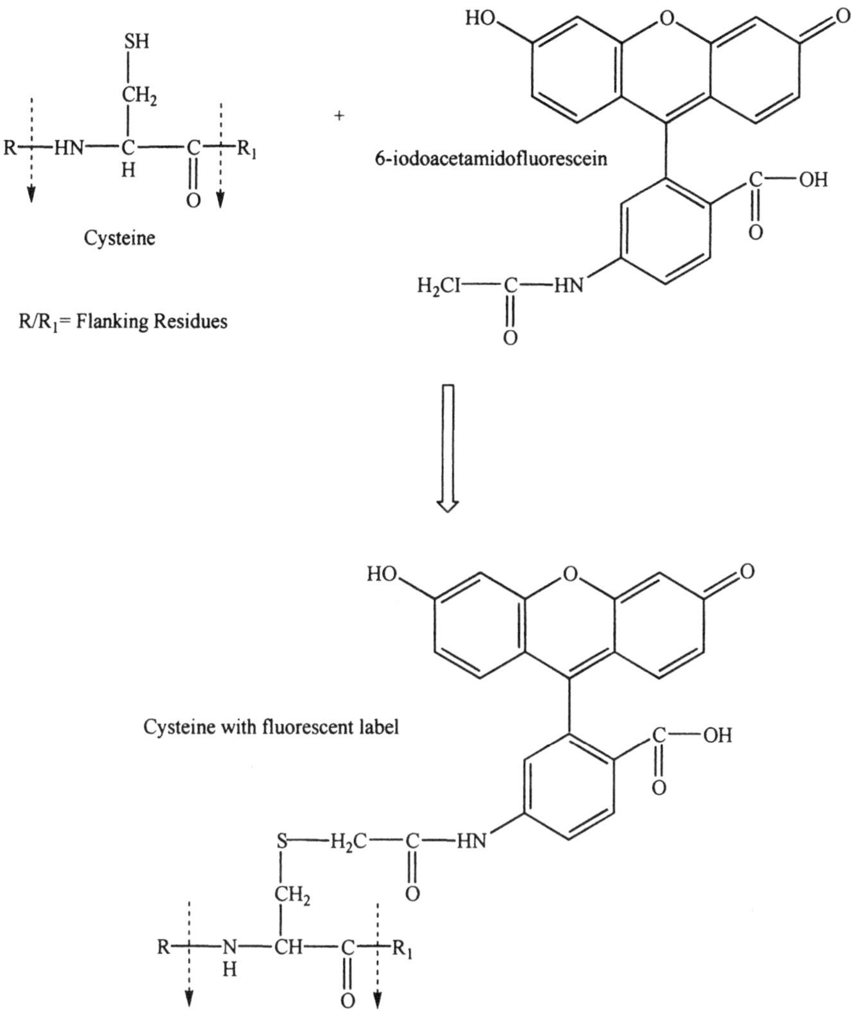

Fig. 6. Reaction of 6-IAF with Cysteine.

Pyridylethylation by 4-vinylpyridine is an alternative method of cystine/cysteine modification for both sequence analysis and for those amino acid analyzers that utilize phenylisothiocyanate as an amino terminal derivatizing agent (*see* **Fig. 8**). The reaction can be carried out in a vapor phase where the sample has been transferred to an inert support, e.g., glass-fiber *(15)* or PVDF *(16)*. The vapor-phase reaction circumvents the need to remove excess reagents by microbore reverse-phase high performance liquid chromatography (RP-HPLC), and hence partial loss of sample on the column. The loading of the sample onto

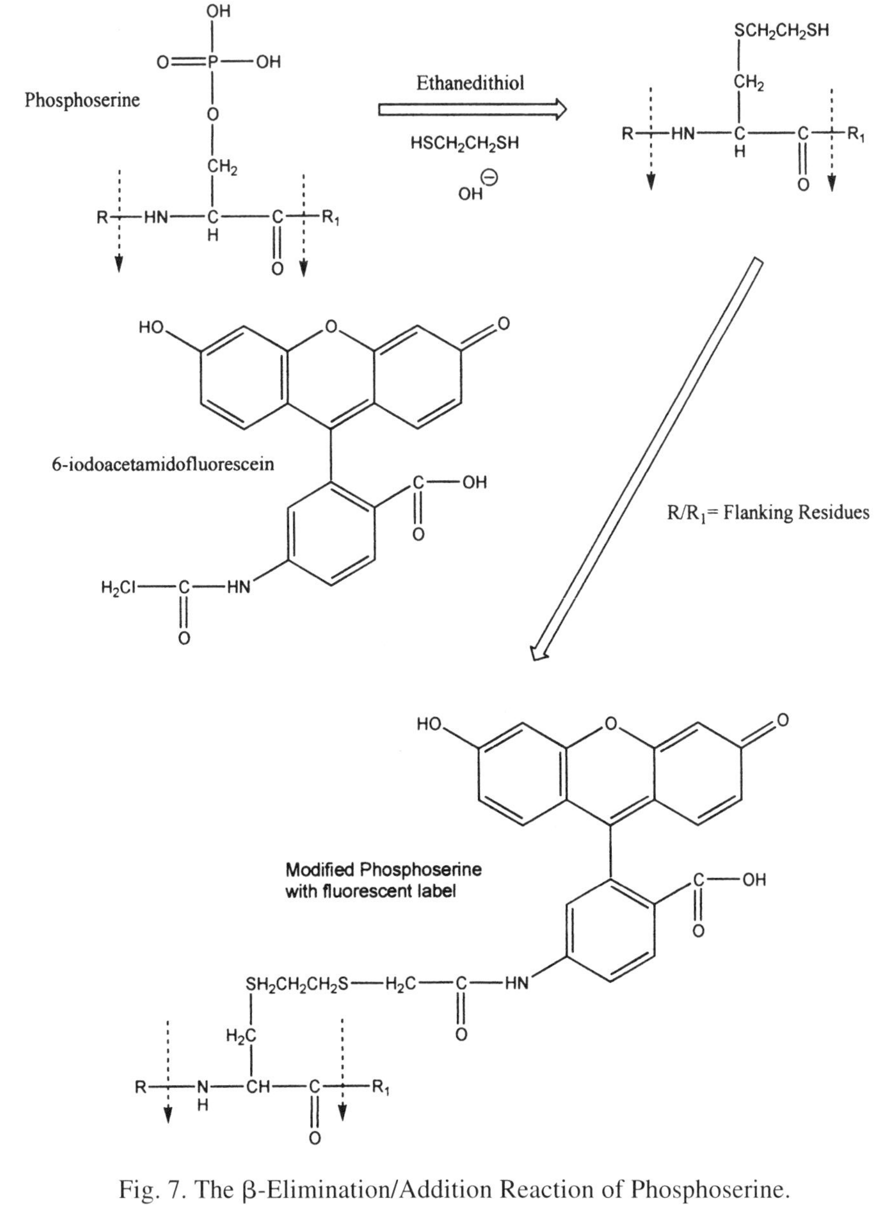

Fig. 7. The β-Elimination/Addition Reaction of Phosphoserine.

a PVDF membrane of a sample preparation cartridge, as described by Sheer *(17)*, allows the removal of all contaminants with minimal sample loss. The membrane then serves as an inert support for subsequent N-terminal sequence analysis. The first commercial cartridges were termed ProSpin™ *(18)* as the

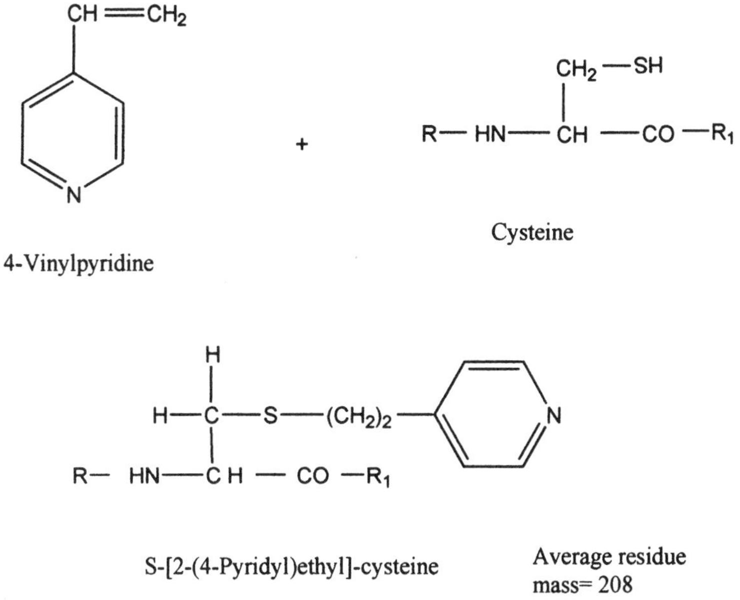

CH=CH₂ (4-Vinylpyridine)

+

CH_2—SH
R—HN—CH—CO—R_1
Cysteine

S-[2-(4-Pyridyl)ethyl]-cysteine

Average residue mass= 208

Fig. 8. Pyridylethylation of Cysteine.

sample was spun onto PVDF membrane. These have now been superceded by the ProSorb™ *(19)* cartridge, which utilizes an absorbent filter to draw the sample and wash solutions through the sample reservoir membrane. Unlike the ProSpin™, this device does not require centrifugation thus reducing the time necessary for sample binding.

The reaction can also be carried out in the liquid phase, the excess reagents being removed by RP-HPLC, and cysteine containing peptides from protein digests identified by their absorbance at 254 nm *(20)*.

An alternative method for sequence analysis is the reaction of cysteine with acrylamide, under mild alkaline conditions, to yield the thioether derivative, Cys-S-propionamide (Cys-S-Pam). This compound is stable during Edman degradation and its phenylthiohydantoin derivative, PTH-Cys-Pam, can be separated from other PTH-amino acids by HPLC. This was first noticed with polypeptides blotted onto PVDF membranes after acrylamide gel electrophoresis, the cysteine (s) reacting with residual unpolymerized acrylamide to produce the Cys-Pam modification *(21)*. The method can also be applied to samples spotted onto glass-fiber disks *(22)*. The formation of PTH-Cys-Pam is shown in **Fig. 9**.

For quantitative amino acid analysis, involving cystine or cysteine residues, carboxymethylation is not ideal since it is difficult to assess the completeness

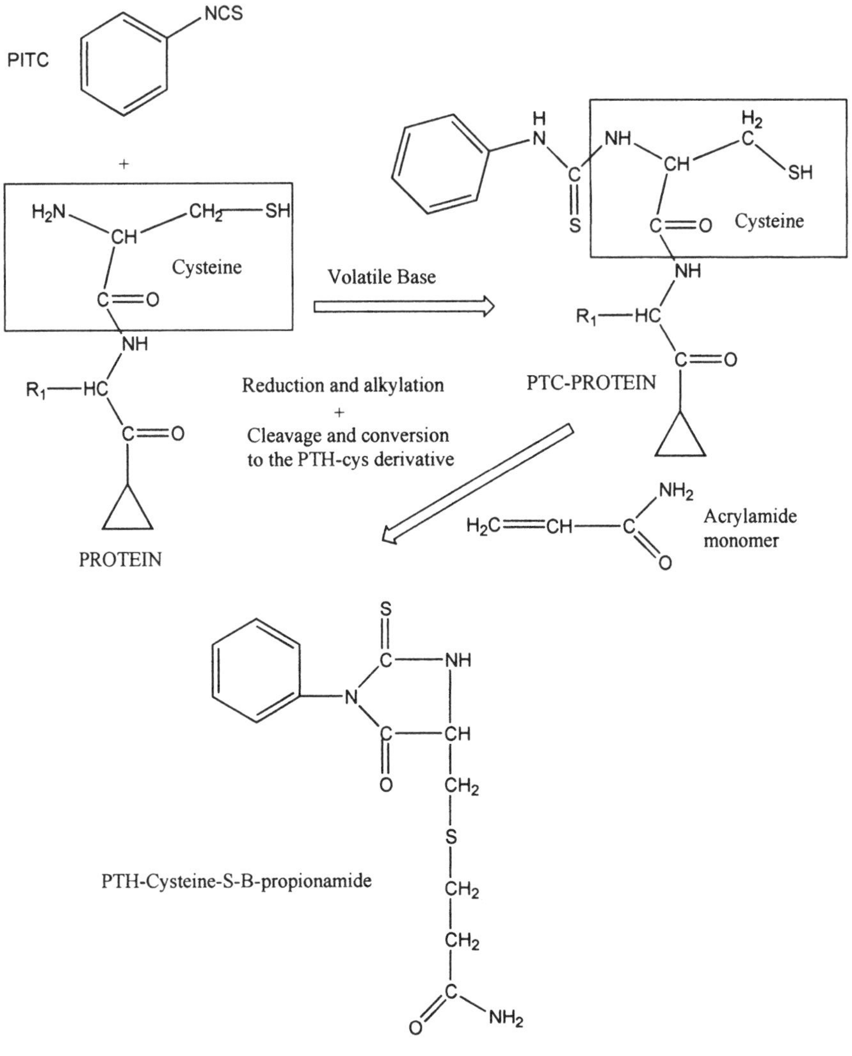

Fig. 9. Alkylation by Acrylamide.

of the reaction. A better method is to oxidize the S-S or SH groups using performic acid *(5)*, hydrolyze the polypeptide, and identify the cystine/cysteine as cysteic acid (*see* **Fig. 10**). Performic acid oxidation also improves protease digestion by removal of disulfide bonds. It is often used prior to digestion of phosphoproteins to help eliminate partial cleavages and simplify further analysis. If performic acid oxidized peptides are sequenced, the methionine can be identified as its oxidation product, methionine sulfone.

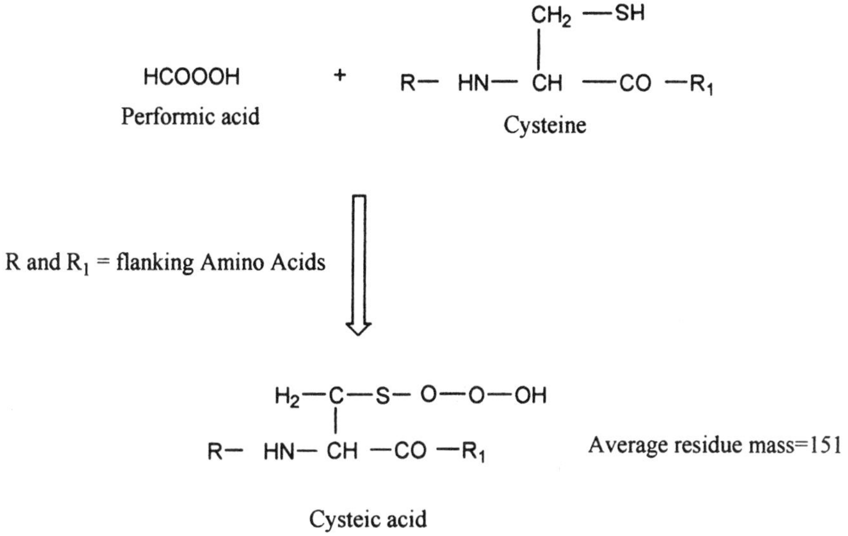

$$CH_2 - SH$$
$$|$$
$$HCOOOH \quad + \quad R- HN- CH -CO -R_1$$

Performic acid   Cysteine

R and R$_1$ = flanking Amino Acids

$$H_2-C-S- O-O-OH$$
$$|$$
$$R- HN- CH -CO -R_1 \qquad \text{Average residue mass=151}$$

Cysteic acid

Fig. 10. Performic acid oxidation of Cysteine.

A further useful modification is the succinylation of the ϵ-NH$_2$ amino side-chain of lysine (*see* **Fig. 11**). This is an irreversible reaction and hinders tryptic attack. Because cleavage cannot occur at lysine residues, larger peptides are generated with arginine at the carboxy-terminus, which may be advantageous for sequence analysis. As with carboxymethylation, the use of ^{14}C-succinic anhydride allows the incorporation of a radiolabel, and also proteins modified by dicarboxylic anhydrides are generally more soluble at pH 8.0, and the polypeptide chains are unfolded allowing ready digestion with proteases. Substance P (M$_r$ 1347.6), containing two ϵ-NH$_2$ groups, has been succinylated and the sample subjected to RP-HPLC. Two major peaks were present and on further analysis by Matrix-Assisted Laser Desorption Ionization (MALDI) mass spectrometry were shown to be the singly and doubly modified polypeptides (AFC, unpublished results). Therefore care must be taken to ensure full succinylation with low sample amounts of larger molecular weight species.

2. Materials (*see* Note 1)

2.1. Reductive Alkylation with Iodoacetic Acid/Iodoacetamide (see Note 2)

1. Denaturing buffer: 6 *M* guanidinium chloride, 0.1 *M* Tris-HCl, pH 8.5.
2. Source of oxygen-free nitrogen.
3. Dithiothreitol (DTT-Amresco, Biotechnology grade): 4 m*M* in HPLC water. This solution should be freshly made.

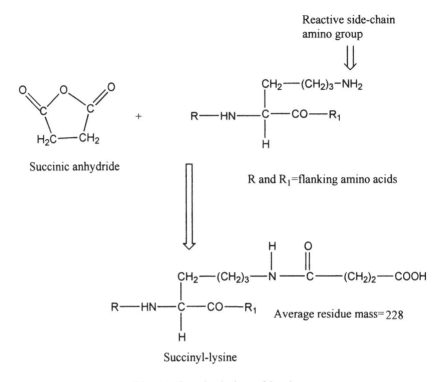

Fig. 11. Succinylation of Lysine.

4. Iodoacetic acid (Pierce)/Iodoacetamide (Calbiochem): 50 mM in HPLC water, pH adjusted to 8.5 with sodium hydroxide (*see* **Note 3**). Both reagents are light-sensitive and it is preferable to make up a fresh solution each time, though the solution can be stored in the dark at –20°C (*see* **Notes 4–6**). Discard the solution if it becomes discolored.
5. Ammonium bicarbonate (50 mM) in water (HPLC grade), pH approx 8.0 (*see* **Note 7**).
6. Microdialysis equipment (as supplied by Pierce: http://www.piercenet.com/dialysis/mini.html).
7. HPLC system with a Brownlee RP-300 "cartridge" type column (30 × 2.1 mm).
8. 0.1% (v/v) TFA (Applied Biosytems Sequencer grade) in water (HPLC grade).
9. Acetonitrile (HPLC, Far UV grade).

2.1.1. Reduction and Alkylation on PVDF

1. 45mg/mL polybrene in 50% acetonitrile (*see* **Note 8**).
2. 0.4 M N-ethylmorpholine in 40% acetonitrile. N-EM (Pierce HPLC Sequanal Grade).
3. Tri-*n*-butylphosphine (Sigma): 2% in acetonitrile.
4. 20 mM iodoacetamide.

2.2. Alkylation with N-Isopropyliodoactamide

1. Stock solution of NIPIA:Dissolve in methanol and then dilute with water to 100 m*M*. The final methanol concentration should be 20–30%. Prepare by dissolving 2.3 mg of NIPIA (Molecular Probes) in 20 μL of methanol and dilute to 100 μL with 80 μL of water. The NIPIA solution is stable for at least 2–3 h and solid NIPIA has been stored at 4°C for over 2 mo with no observed decomposition.
2. Tri-*n*-butylphosphine: 0.1 *M* in 2-propanol. This solution should be freshly made.
3. Denaturing buffer: 6 *M* guanidinium chloride, 0.1 *M* Tris-HCl, pH 8.5.
4. 10% (v/v) TFA (Applied Biosytems Sequencer grade) in water (HPLC grade).

2.3. Alkylation with 6-Iodoacetamidofluorescein (6-IAF)

2.3.1. Cysteine Residues

1. 0.1 *M* Tris-HCl, 1.0% sodium dodecyl sulfate (SDS), pH 8.0 (for dissolution of the protein).
2. 10 m*M* DTT (Amresco) in HPLC-grade water.
3. 100 m*M* stock solution of 6-IAF (*see* **Note 9**). 6-IAF is available from Molecular Probes (http://www.probes.com) as the compound I-15.
4. Gel Electrophoresis equipment.

2.3.2. Phosphoserine Residues (see **Note 10**)

1. Performic acid oxidized protein (*see* **Subheadings 2.6.** and **3.6.**).
2. 0.1% TFA.
3. ProSorb™ Cartridge (Applied Biosystems).
4. β-elimination/addition mixture:
    a. 100 μL H$_2$O.
    b. 30 μL of 5 *N* NaOH.
    c. 3 μL of 250 m*M* EDTA (Serva).
    d. 170 μL of ethanol.
    e. 7 μL of ethanedithiol (Aldrich).

    This mixture should be freshly made prior to use.
5. Phosphate buffer: 100 m*M* Na$_2$HPO$_4$, pH 9.0, 2.5 m*M* EDTA, 1. 0 m*M* DTT.
6. 0.2 *M* Ammonium bicarbonate (NH$_4$HCO$_3$).

2.4. Pyridylethylation

2.4.1. Vapor-Phase Derivitization

1. 4-Vinylpyridine (Sigma): Store at –20°C (*see* **Note 11**).
2. Tri-*n*-butylphosphine (Sigma): Store at –20°C.
3. Pyridine (BDH, Aristar): Store in the dark at 4°C.
4. n-heptane (Applied Biosystems, Sequencer-grade).
5. Ethyl acetate (Applied Biosystems, Sequencer-grade).
6. TFA-washed glass-fiber disks (Applied Biosystems).

7. PVDF membrane, e.g., Immobilon™ (Millipore), ProBlott™ (Applied Biosystems).
8. Pyrex glass reaction vessel (*see* **Fig. 12**). These tubes can be custom made by any company offering a glass-making service.
9. Vacuum source: A diaphragm or rotary pump can be used.
10. Glass-blowing torch with gas and oxygen flame.

2.4.2. Liquid-Phase Derivatization

1. 4-Vinylpyridine: Store at –20°C (*see* **Note 11**).
2. N-Ethylmorpholine (Pierce).
3. Acetic acid (Fluka, puriss).
4. Dithiothreitol (DTT) (Amresco): 50 mM in 0.2 M N-ethylmorpholine acetate buffer (NEM acetate), pH 8.0. The NEM acetate buffer is prepared by adding 230 µL of N-ethyl-morpholine to approx 8.0 mL of HPLC water. Acetic acid is added to give a pH of 8.0 and the volume adjusted to a total of 10 mL. This reducing buffer must be freshly prepared when required.
5. HPLC with variable wavelength detector and suitable column.
6. 0. 1% TFA (v/v) in water (HPLC-grade reagent).
7. Denaturing buffer: 8 M urea (Biorad, Electrophoresis grade) in 0.2 M NEM acetate buffer.
8. Acetonitrile (HPLC, far UV grade).
9. ProSorb™ Cartridge. For N-terminal sequencing only.
10. 20% methanol (v/v) in water (HPLC-grade reagents). Used for washing the PVDF membrane of the cartridge prior to sequence analysis.

2.5. Alkylation with Acrylamide

1. Tri-*n*-butylphosphine (Sigma): 0.1 M in 2-propanol. This solution should be freshly made.
2. Acrylamide (Biorad, electrophoresis grade): 0.5 M in 85% methanol/10% water/ 5% N'N-diisopropylethylamine (Applied Biosystems), (v/v/v) (*see* **Note 12**). Care: Acrylamide is a neurotoxin. Dispensing should be confined to a fume-hood and protective gloves worn. Store acrylamide solutions at 4°C.
3. Ethyl acetate (Applied Biosystems, Sequencer-grade).
4. n-heptane (Applied Biosystems, Sequencer-grade).
5. TFA-washed glass-fiber disks (Applied Biosystems). Use of an Applied Biosystems Protein Sequencer or similar (*see* **Note 13**).

2.6. Performic Acid Oxidation

1. Hydrogen peroxide (Aldrich, 30% w/v). Care: Strong oxidizing agent.
2. Formic acid (Aldrich, 88% w/v).
3. Hydrobromic acid (Aldrich, 48% w/v). Care: this acid gives off a caustic, irritating vapor and its use should be confined to a fume hood.

2.7. Succinylation

1. Guanidinium chloride: 6 M in distilled water.
2. Succinic anhydride (+ ^{14}C succinic anhydride if required). Both reagents from Sigma.

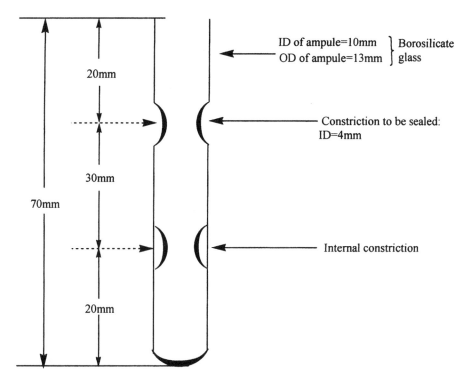

Fig. 12. Ampule for vapor-phase pyridylethylation.

3. Sodium hydroxide: 1 *M* in distilled water.
4. Micro-pH probe.
5. Ammonium bicarbonate: 50 m*M* in distilled water, pH approx 8.0 (*see* **Note 7**).
6. Microdialysis or HPLC equipment.

3. Methods
3.1. Reductive Carboxymethylation

1. Dissolve the protein (10–20 µg) in 50 µL of denaturing buffer in a 1.5-mL Eppendorf tube.
2. Gently blow N$_2$ over the top of the solution for 2–3 min, cap tube and leave at room temperature for 5 min.
3. Add an equal volume of DTT solution to a concentration of 2 m*M*.
4. Blow N$_2$ over the solution, cap the tube, and incubate for 60 min at room temperature.
5. Wrap the Eppendorf tube in aluminum foil before the addition of iodoacetic acid (*see* **Note 4**).
6. 40 µL of iodoacetic acid solution is added dropwise using a micropipet. During this addition the solution should be stirred (*see* **Note 14**).

7. N_2 is again blown over the surface of the solution and the reaction tube sealed.
8. Incubate in the dark at 37°C for 30 min. The reaction is carried out in the dark to prevent the formation of iodine, from the iodoacetate, thus preventing the iodination of tyrosine residues.
9. Removal of excess reagents. This may be performed in one of two ways:
    a. Microdialysis: The solution is injected into a dialysis cassette (Pierce). The cassette is then placed into the dialysis buffer, usually 50 mM ammonium bicarbonate. Alternatively, for very small sample volumes, mini-dialysis units are available (see http://piercenet. com/dialysis/mini. html). Dialysis is then carried out for about 24 h (*see* **Note 15**).
    b. HPLC: The protein can be separated from the reaction by-products by RP-HPLC. To minimize product loss, a 30 × 2.1 mm microbore "cartridge"column is employed. For proteins, a C4 or C8 matrix with a 10 micron particle diameter and a 300 angstrom pore size can be used, (e.g., Aquapore octyl RP-300). The column is equilibrated in 0. 1% aqueous TFA and the sample is injected onto the column and eluted using an acetonitrile gradient (1%/min, flowrate 200 μL/min). The excess reagents and reaction by-products all appear in the breakthrough, i.e., they are not retained by the column packing material.

3.1.1. Alkylation on PVDF Membrane

1. Place a piece of PVDF on a glass-fiber sequencing disc in the reaction cartridge.
2. Wet the PVDF with the minimum amount of the polybrene solution.
3. Air dry and add 10 μL of 0.4 M N-ethylmorpholine in 40% acetonitrile and 10 μL of 2% tributylphosphine.
4. Seal with parafilm and incubate for 1 h at 55°C.
5. Add 20 μL of 20 mM iodoacetamide.
6. Seal with parafilm and incubate for 30 min at room temperature.
7. Wash the PVDF with distilled water, dry, and place the membrane pieces into the sequencer Blott™ cartridge.

3.2. N-Isopropyliodoacetamide

1. Dissolve the protein, preferably 200 pmol or less, in 50 μL of denaturing buffer.
2. Make the solution 0.1% in tributylphosphine.
3. Incubate for 2 h at room temperature.
4. Add 30% excess of NIPIA over the total thiol concentration.
5. Incubate for 30 min at 37°C in the dark.
6. Terminate the reaction by HPLC, acidification with 10% TFA to give a final acid concentration of 0.1% or by freezing.

3.3. 6-Iodoacetamidofluorescein

3.3.1. Cysteine Residues

1. Dissolve the protein (0.1–0.25 μg) in 0.1 M Tris-HCl containing 1.0% SDS pH 8.0.
2. Reduce the disulfide bonds by the addition of 10 mM DTT for 1 h in the dark.

3. Add the alkylating agent to 20 mM and incubate in the dark at room temperature for 1 h.
4. Load the protein onto an appropriate % SDS-acrylamide gel.
5. Visualize under UV and excise bands of interest for in-gel digestion if required (*see* Chapters 5 and 6).

3.3.2. Phosphoserine Residues *(see* **Note 10***)*

1. Performic acid oxidize the protein (0.5–1 µg) as in **Subheading 3.6.** and dissolve the sample in 100 µL 0.1% TFA.
2. Load the protein solution onto the ProSorb cartridge. Use the absorbent filter to draw excess liquid through the PVDF leaving the protein bound to the membrane.
3. Remove the filter from the holder and add 50 µL of the β-elimination/reaction mix to the cartridge.
4. Incubate at 55°C in the dark for 90 min.
5. Remove the reaction mixture using the filter and wash the cartridge membrane five times with 100 µL aliquots of the phosphate buffer, changing the filter when necessary, (filter capacity = 750 µL).
6. Add 10 µL of the 6-IAF stock solution to 40 µL of the phosphate buffer (20 mM 6-IAF) and add to the cartridge.
7. Incubate at room temperature for 1–2 h.
8. Repeat wash **step 5**.
9. Wash with the NH$_4$HCO$_3$ buffer (100 µL, 5X).
10. Punch out the membrane ready for further analysis, either direct N-terminal sequencing or on-membrane digestion (*see* Chapters 5 and 6).

3.4. Pyridylethylation

3.4.1. Vapor-Phase *(see* **Note 16***)*

1. 1–10 µg protein dissolved in a minimum volume of distilled water or other appropriate solvent is spotted in small aliquots onto a glass-fiber disk. If PVDF is used, the membrane needs to be wetted with methanol before the application of the sample.
2. The sample is then allowed to dry on the matrix.
3. The lower chamber of the glass ampule (*see* **Fig. 12**) is carefully filled with a freshly prepared mixture of:
   a. 100 µL H$_2$O.
   b. 100 µL pyridine.
   c. 20 µL 4-vinylpyridine.
   d. 20 µL tributylphosphine.
4. The disk/PVDF is then placed in the central part of the ampule.
5. The ampule is evacuated and sealed at the top constriction using a gas/oxygen flame.
6. Allow the reaction to proceed at 60°C for 2 h.
7. Score and break open the ampule and remove the disk.

8. Wash three times with 1.0 mL of each of the following:
   a. n-heptane.
   b. n-heptane:ethyl acetate (2:1 v/v).
   c. Ethyl acetate.
9. Air dry the disk/PVDF, which is now ready for further analysis, e.g., N-terminal sequencing.

3.4.2. Liquid-Phase (see **Note 17**)

1. Dissolve the protein at an approximate concentration of 1 mg/mL in 8 M urea/N-ethylmorpholine acetate buffer.
2. Add DTT in NEM acetate buffer to achieve a final DTT concentration of 2.5 mM.
3. Leave at 25°C for 2h.
4. Add 2 µL of 4-vinylpyridine/100 µL of protein solution.
5. Allow to react for 90 min at 25°C.
6. Desalt the S-pyridylethylated protein by RP-HPLC using the method described in **Subheading 3.1., step 9**.
7. Remove the solvent *in vacuo*, if the HPLC method is used, and redissolve in the required buffer.
8. The sample is now ready for sequencing directly or enzymic digestion and peptide mapping (*see* **Note 18**).
9. It is possible to remove the reaction by-products by binding the polypeptide onto PVDF membrane using a ProSorb cartridge as described by the supplier (*19*).
   Wait until all the solution has gone through and then add 50 µL of 20% methanol and allow the liquid to be drawn through. Remove the membrane and rinse with 0.1% TFA followed by water to remove any contaminants. The membrane is allowed to dry and can then be used for sequence or amino-acid analysis (*see* **Notes 19–21**). Alternatively, change the absorbent filter and wash *in situ* before removing the PVDF membrane.

3.5. Alkylation with Acrylamide

1. Dissolve 1–10 µg protein in a minimum of distilled water or other appropriate solvent e.g., aqueous acetonitrile (50–80% v/v) containing 0.1% TFA (v/v).
2. Spot the required amount of protein onto a glass-fiber disk, allow to dry, and fit into the sequencer cartridge.
3. Subject the sample to one coupling step of the Edman degradation to prevent alkylation of the N-terminus (*see* **Note 22** and **Fig. 9**).
4. Remove excess phenylisothiocyanate with three washes of n-heptane and ethyl acetate, 30 s for each solvent. Dry the disk under N_2 for 60 s.
5. Add 15 µL of tri-*n*-butylphosphine reagent to the disk and, using the cartridge drying function, dry under a stream of N_2 for 30 s. This removes most of the propanol.
6. Add 15 µL of the acrylamide reagent, close the reaction cartridge, and allow to react for 20 min at 48°C.

7. Using the sequencer, dry the sample under N_2 for 60 s and then wash three times with 30 s deliveries of ethyl acetate.
8. The sample can now be sequenced and the Cys-Pam phenylthiohydantoin derivative(s) identified (*see* **Note 23**).

3.6. Performic Acid Oxidation

1. Add 100 µL of hydrogen peroxide to 900 µL of formic acid and allow to stand at room temperature for 1 h. This produces performic acid (HCOOOH).
2. Cool the performic acid on ice to approx 0°C.
3. Dissolve the protein in 50 µL of performic acid (about 200–500 µg/mL), in a precooled tube.
4. Keep at 0°C for 4 h (*see* **Note 24**).
5. Add 7.5 µL of cold HBr to neutralize the performic acid. Care must be taken at this stage since toxic bromine will be liberated, albeit a small amount.
6. Dry the sample thoroughly *in vacuo* over NaOH to remove bromine and formic acid.
7. The sample can then be hydrolyzed for amino acid analysis or digested and the peptides used for sequence analysis (*see* **Notes 25–29**).

3.7. Succinylation (see Notes 30–32)

1. Dissolve the protein or protein derivative at a concentration of 10–50 µg in 200 µL of 6 *M* guanidinium chloride and using a micro-pH probe adjust the pH to 9.0 with the sodium hydroxide solution.
2. Crush 1–2 crystals (1–2 mg) of solid succinic anhydride and add to the stirred solution over a period of about 15 min, while maintaining the pH of the solution at 9.0 with 1 *M* NaOH (*see* **Note 33**).
3. Excess reagents are removed by microdialysis or by HPLC (*see* **Note 34**).

4. Notes

1. Reagents for all the methods should be of the highest quality available, and though it is not possible to list every vendor, sources for the most critical reagents are included in the Materials Section. HPLC-grade water (BDH) is used for all methods.
2. At pH 8.0, IAA will react with SH groups very rapidly taking only 15–20 min for completion. 4-VP is much slower and usually takes 2–4 h. IAA reacts with methionine at a nearly constant rate over the range of pH. If the reaction is not quenched with excess thiol after 20 min, modification of methionine can occur.
3. If iodoacetamide is used it is not necessary to neutralize with sodium hydroxide.
4. The iodoacetic acid used must be colorless; any iodine present, revealed by a yellow color, causes rapid oxidation of thiol groups preventing alkylation and possibly modifying tyrosine residues. Iodoacetamides in solution undergo rapid photo-decomposition to unreactive products. Minimize exposure to light prior to carrying out the reaction and totally exclude light during the reaction.
5. If good-quality iodoacetic acid is not available it may be recrystallized from hexane.

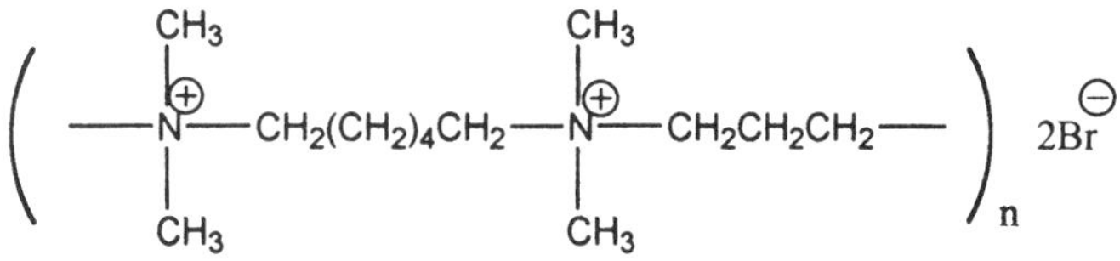

$$\left(\begin{array}{c} CH_3 \\ | \\ -N^{\oplus}-CH_2(CH_2)_4CH_2-N^{\oplus}-CH_2CH_2CH_2- \\ | \\ CH_3 \end{array} \begin{array}{c} CH_3 \\ | \\ | \\ CH_3 \end{array} \right)_n \quad 2Br^{\ominus}$$

Fig. 13. Chemical Structure of Polybrene (Hexadimethrine bromide).

6. The method described for reductive alkylation does not include the use of iodo[2-^{14}C] acetic acid. The radiolabeled iodoacetic acid should be diluted before use with carrier iodoacetic acid to the required final specific activity according to the requirements of the worker. With this modification the method is otherwise as described by Walker et al. *(23)*.

7. Ammonium bicarbonate (50 mM) gives a pH of approx 8.0. No pH adjustment is necessary.

8. Polybrene is the polymeric quaternary amine chemically known as hexadimethrine bromide (*see* **Fig. 13**) and is used to prevent "wash-out" of protein during sequencing. It is available from Applied Biosystems as "Biobrene™".

9. Long-term storage of aqueous stock solutions is not recommended. Use within 24 h.

10. Pretreatment of unknown samples with performic acid, to oxidize sulfhydryls, is necessary to avoid false positives from labeled cysteines.

11. If the 4-vinylpyridine is dark in color, it should be discarded as the coloration is due to air oxidation.

12. Alkylation is carried out on glass-fiber disks according to the method of Brune *(21)* except that a modified acrylamide solution is used. After application of the sample to the disk, all the steps require the use of an automated sequencer. The published method utilizes 0.5 M acrylamide in 2.75% triethylamine, but the protocol described uses N'N-diisopropylethylamine (DIPEA) as the volatile base with similar results. With triethylamine, the approximate concentration needs to be determined by titration, which is unnecessary with the DIPEA.

13. The sequencer used in my own laboratory is an Applied Biosystems Model 477A, though the method can be used with most commercially available instruments.

14. Continuous stirring during the addition of the iodoacetic acid is necessary to keep the pH constant throughout the solution especially when dealing with small volumes. This is not critical when using iodoacetamide.

15. Dialysis tubing should be avoided since the losses involved when handling small amounts of protein can be high. The microdialysis system which uses a sheet of dialysis membrane, is preferable, but losses can still occur. If the reader has access to an HPLC system, then I would recommend this as the method of choice.

16. Pyridylethylation may be carried out in the vapor-phase according to the method of Amons *(15)* except that modifications were made to the glass ampule and also PVDF membrane can be used as an alternative to the glass-fiber disks.

17. This is a modified version of the method used by Fullmer for chick calcium-binding protein *(20)*.

18. The liquid phase method of pyridylethylation is used to introduce a reporter group into cysteine containing peptides from digests of derivatized proteins and facilitates detection of these peptides at 254 nm *(20)*.

19. The brownish color left on a ProSorb membrane after binding is normal.

20. Pyridylethylation on PVDF membrane enables the sample to be hydrolyzed since S-B- (4-Pyridylethyl) cysteine [PE-cysteine] is acid stable *(24)*.

21. The use of a ProSorb cartridge enables better yields of pyridylethylated material with subsequent removal of excess reagents. For small quantities of protein (<100 pmol), this would be the preferred method for direct sequence analysis. However, if it is necessary to generate peptides, either by chemical or enzymic cleavage methods, it is preferable to have the protein in solution and I would therefore advise the HPLC method for removal of excess reagents.

22. It is necessary to react the sample with phenylisothiocyanate (PITC), i.e., the coupling step of the Edman degradation, prior to alkylation to prevent any reaction of the amino-terminus with the acrylamide which would prevent sequence analysis.

23. PTH Cys-Pam elutes just after PTH glutamic acid with our gradient and using the ABI buffer system *(25)* and an on-line PTH analyzer.

24. During performic acid oxidation the temperature of all the reactants must be at or near 0°C to minimize any side-reactions such as the oxidation of phenolic groups and the hydroxyl groups of serine and threonine.

25. Oxidation of methionine occurs producing methionine sulfone, which can be quantified by amino acid analysis.

26. The addition of HBr causes bromination of tyrosine residues producing a mixture of mono and di-substituted bromotyrosines. These derivatives have lower color values than tyrosine and quantitation by ninhydrin is generally not good after acid hydrolysis.

27. As several residues are modified, this method is best used in conjunction with amino acid analysis. It is not recommended to use performic acid oxidized samples for sequence analysis.

28. For the estimation of S-S or S-H groups by amino acid analysis the most efficient method is still performic acid oxidation.

29. Instruments using RP-HPLC for the analysis of derivitized amino acids may need gradient modification to separate cysteic acid from aspartate.

30. Succinylation: This is best carried out in proteins that have been carboxymethylated to prevent S-S or SH groups from reacting. This gives complete selective modification.

31. Succinylation is a better method than maleylation or citraconylation, for the modification of lysine residues, as it avoids problems arising from the still reactive ethylenic groups. However, if there is a need for the removal of the blocking groups, for subsequent tryptic cleavage at lysine residues, maleylation or citraconylation (*see* **Fig. 14**) should be used since the groups introduced by citraconic or maleic anhydride can be removed by treatment with dilute acid *(26)*.

32. Both clostripain, (EC 3.4.22.8) and endoproteinase Arg-C, (EC 3.4.21.40) can be used for cleavage at arginine residues instead of succinylation followed by trypsin

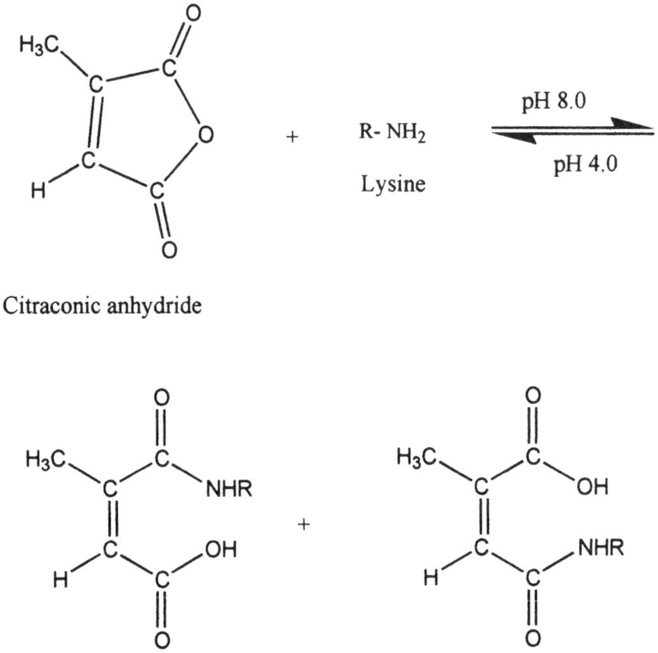

Citraconic anhydride

Two isomeric e-N-acyl derivatives of lysine are produced,differing in pKa.

Fig. 14. The reversible reaction of Citraconic anhydride.

digestion. However, clostripain may cleave at other sites *(27)* and endoproteinase Arg-C *(28)*, although highly specific for the C-terminal side of arginine residues, has been shown to cleave at other sites *(29,30)*. Thus succinylation is still of use to the protein chemist trying to achieve specific cleavages coupled with the property of enhanced solubility.

33. The addition of small amounts of succinic anhydride over a 15-min time period enables the worker to control the pH so that it can be maintained at or near pH 9.0.
34. For the removal of excess reagents the same criteria apply as for carboxymethylation.

References

1. Pfleiderer, G. (1985) Chemical modification of proteins, in *Modern Methods in Protein Chemistry*, Review Articles, vol. 2 (Tschesche, H. ed.), Walter de Gruyter, Berlin, pp. 207–259.
2. Gray, W. R. (1993) Disulfide structures of highly bridged peptides:a new strategy for analysis. *Protein Sci.* **2**, 1732–1748.
3. Hewick, R. M., Hunkapillar, M. W., Hood, L. E., and Dreyer, W. J. (1981) A gas-liquid solid phase peptide and protein sequenator. *J. Biol. Chem.* **256**, 7990–7997.

4. Applied Biosystems (1992) New multimode sequencer permits range of sequencing conditions. *Appl. Biosyst. Reporter* **17**, 6–7.
5. Glazer, A. N., Delange, R. J., and Sigman, D. S. (1975) Chemical characterization of proteins and their derivatives, in *Chemical Modification of Proteins* (Work, T. S. and Work, E., eds.), North Holland, American Elsevier, Amsterdam, pp. 21–24.
6. Getz, E. B., Xiao, M., Chakrabarty, T., Cooke, R., and Selvin, P. R. (1999) A comparison between the sulfhydryl reductants tris (2-carboxyethyl) phosphine and dithiothreitol for use in protein chemistry. *Anal. Biochem.* **273**, 73–80.
7. Pappin, D. J. C., Hojrup, P., and Bleasby, A. J. (1993) Rapid identification of proteins by peptide-mass fingerprinting. *Curr. Biol.* **3**, 327–332.
8. Yan, J. X, Sanchez, J. C, Rouge, V, Williams, K. L, and Hochstrasser, D, F. (1999) Modified immobilized pH gradient gel strip equilibration procedure in SWISS-2DPAGE protocols. *Electrophoresis* **20**, 723–726.
9. Atherton, D., Fernandez, J., and Mische, S. M. (1993) Identification of cysteine residues at the 10-pmol level by carboxamidomethylation of protein bound to sequencer membrane supports. *Anal. Biochem.* **212**, 98–105.
10. Krutzsch, H. C. and Inman, J. K. (1993) *N*-Isopropyliodoacetamide in the reduction and alkylation of proteins: use in microsequence analysis. *Anal. Biochem.* **209**, 109–116.
11. Martelli, A. M., Billi, A. M., De Marchis, C., Manzoli, L., and Cocco, L. (1990) Probing the sulfhydryl groups of nuclear matrix proteins with 6-Iodoacetamidofluorescein. *Cell Biol. Int. Rep.* **14**, 409–418.
12. Hsi, K-L., O'Neill, S. A., Dupont, D. R., and Yuan, P-M. (1998) Visualization of proteins by modification of lysines, cysteines, and phosphorylated serines facilitates sample preparation for microsequencing. *Anal. Biochem.* **258**, 38–47.
13. Fadden, P. and Haystead, T. A. (1995) Quantitative and selective fluorophore labeling of phosphoserine on peptides and proteins: characterization at the attomole level by capillary electrophoresis and laser-induced fluorescence. *Anal. Biochem.* **225**, 81–88.
14. Samyn, B., Hardeman, K., Van der Eycken J., and Van Beeumen, J. (2000) Applicability of the alkylation chemistry for chemical C-terminal protein sequence analysis. *Anal. Chem.* **72**, 1389–1399.
15. Amons, R. (1987) Vapor phase modification of sulfhydryl groups in proteins. *FEBS Lett.* **122**, 68–72.
16. Kao, M. C. C. and Chung, M. C. M. (1993) Pyridylethylation of Cysteine Residues in Proteins. *Anal. Biochem.* **215**, 82–85.
17. Sheer, D. (1990) Sample Centrifugation onto membranes for sequencing. *Anal. Biochem.* **187**, 76–83.
18. Applied Biosystems (1991) Efficient recovery of pyridylethylated CPP for high sensitivity protein sequencing. Application Note #2.
19. Applied Biosystems (1996) Prosorb Sample Preparation Cartridge. User Bulletin No. 64.
20. Fullmer, C. S. (1984) Identification of cysteine containing peptides in protein digests by high performance chromatography. *Anal. Biochem.* **142**, 336–339.

21. Brune, D. C. (1992) Alkylation of cysteine with acrylamide for protein sequence analysis. *Anal. Biochem.* **207**, 285–290.

22. Applied Biosystems (1995) Derivatization of cysteine residues for protein sequencing. User Bulletin Number 61.

23. Walker, J. E., Carne, A. F., Runswick. M. J., Bridgen, J., and Harris, J. I. (1980) D-Glyceraldehyde-3-phosphate dehydrogenase. Complete amino-acid sequence of the enzyme from *Bacillus stearothermophilus. Eur. J. Biochem.* **108**, 549–565.

24. Tous, G. I., Fausnaugh, J. L., Akinyosoye, O., Hackland, H., Winter-Cash, P., Vitorica, F. J., and Stein, S. (1989) Amino acid analysis on polyvinylidene difluoride membranes. *Anal. Biochem.* **179**, 50–55.

25. Applied Biosystems (1992) Premix buffer for protein sequencers. User Bulletin #43.

26. Dixon, H. B. and Perham, R. N. (1968) Reversible blocking of amino groups with citraconic anhydride. *Biochem. J.* **109**, 312–314.

27. Mitchell, W. M. and Harrington, W. F. (1970) Clostripain, in *Methods in Enzymology, vol. XIX Proteolytic Enzymes* (Perlman, G. E. and Lorand, L., eds.), Academic, New York, pp. 635–642.

28. Levy, M., Fishman, L., and Schenkein, I. (1970) Mouse submaxillary gland proteases, in *Methods in Enzymology, vol. XIX, Proteolytic Enzymes* (Perlman, G. E. and Lorand, L., eds.), Academic, New York, pp. 672–681.

29. Thomas, K. A. and Bradshaw, R. A. (1981) γ-Subunit of mouse submaxillary gland 7S nerve growth factor: an endopeptidase of the serine family, in *Methods in Enzymology, vol. 80, Proteolytic Enzymes*, part C (Lorand, L., ed.), Academic, New York, pp. 609–620.

30. Krueger, R. J., Hobbs, T. R., Mihal, K. A., Tehrani, J., and Zeece, M. G. (1991) Analysis of endoproteinase Arg-C action on adrenocorticotrophic hormone by capillary electrophoresis and reversed-phase high-performance liquid chromatography. *J. Chromatogr.* **543**, 451–461.

28

Deblocking of N-Terminally Modified Proteins

Hisashi Hirano and Roza M. Kamp

1. Introduction

A small amount of proteins separated by one-dimensional (1D) or two-dimensional (2D) polyacrylamide gel electrophoresis (PAGE) and transferred from the gel onto a polyvinylidene difluoride (PVDF) membrane by Western blotting can be sequenced directly by a gas-phase protein sequencer *(1)*. This Western blotting/sequencing technique has come to be widely used in protein sequence analysis. However, even if proteins are successfully separated by PAGE and then transferred onto a PVDF membrane, N-terminal amino acid sequences of the proteins with a blocking group at the N-terminus cannot be determined by Edman degradation in the sequencer. Many proteins are N-terminally blocked *(2,3)*. Brown and Roberts *(4)* found over 50% of soluble mammalian proteins to be N-terminally blocked. Thus, a simple and rapid technique for obtaining N-terminal sequence information on blocked proteins should be developed.

Most techniques proposed so far for this purpose are applicable to obtain only the internal sequence of proteins. A new technique for deblocking has been developed, and the N-terminal sequencing of proteins with an acetyl, formyl, or pyroglutamyl group at the N-terminus, electroblotted from a polyacrylamide gel onto a PVDF membrane *(5)*, can be carried out.

Acetyl modification is the most prevalent in the blocking groups identified so far *(5)*, and the formyl and pyroglutamyl groups are also frequently detected (*see* **Note 1**). In this chapter, techniques for deblocking of the N-terminally modified proteins separated by PAGE and subsequent gas-phase sequencing are described (*see* **Note 2**). Techniques for PAGE and protein electroblotting of proteins are presented elsewhere *(6,7)* (*see* **Note 3** and Chapters 2–4).

From: *Methods in Molecular Biology, vol. 211: Protein Sequencing Protocols, 2nd ed.*
Edited by: B. J. Smith © Humana Press Inc., Totowa, NJ

2. Materials

2.1. Deblocking of N-Formylated Proteins

1. Hydrochloric acid (0.1 M).
2. Acetonitrile (Sequencing grade).

2.2. Deblocking of Proteins with N-Terminal Pyroglutamic Acid

1. Acetonitrile (Sequencing grade).
2. Polyvinylpyrrolidone (PVP)-40 (0.5% w/v) in acetic acid (100 mM).
3. Phosphate buffer (0.1 M, pH 8.0) containing dithiothreitol (DTT) (5 mM) and ethylenediaminetetraacetic acid (EDTA) (10 mM).
4. Pyroglutamyl peptidase (Takara Shuzo, Japan).

2.3. Deblocking of N-Acetylated Proteins

2.3.1. Deblocking with Trifluoroacetic Acid (TFA)

1. Trifluoroacetic acid (TFA) (Sequencing grade).
2. Nitrogen gas.

2.3.2. Deblocking with TFA in Methanol

1. TFA (Sequencing grade).
2. Methanol (Analytical grade).

2.3.3. Deblocking with Acylamino Acid Releasing Enzyme

1. Acetonitrile (Sequencing grade).
2. PVP-40 (0.5% w/v) in acetic acid (100 mM).
3. Trypsin (Promega, USA, modified by reductive methylation).
4. Ammonium bicarbonate buffer (0.1 M, pH 8.0).
5. Pyridine (50% v/v).
6. Phenylisothiocyanate (Sequencing grade).
7. Formic acid.
8. Hydrogen peroxide (30% w/v).
9. Phosphate buffer (0.2 M, pH 7.2).
10. DTT (1 mM).
11. Acylamino acid-releasing enzyme (AARE) (Takara Shuzo).
12. Nitrogen gas.

2.4. Deblocking of Proteins with N-Acyl Groups by PyrococcusAminopeptidase

1. Ethylmorpholine buffer (50 mM, pH 8.0) containing CoCl$_2$ (0.1%, v/v).
2. *Pyrococcus furiosus* aminopeptidase (Takara Shuzo).

3. Methods

3.1. Deblocking of N-Formylated Proteins

1. Separate the N-terminally blocked protein (<100 pmol) by SDS-PAGE or 2D-PAGE and electroblot it on a PVDF membrane.
2. Cut out the portion of the PVDF membrane carrying the protein band or spot.
3. Wet the membrane with a small amount of acetonitrile and soak it in 200–500 µL 0.6 M HCl at 25°C for 24 h.
4. Wash the PVDF membrane adequately with deionized water.
5. Dry the membrane *in vacuo* and apply it to a gas-phase sequencer.

3.2. Deblocking of Proteins with N-Terminal Pyroglutamic Acid

1. Separate the N-terminally blocked protein (<100 pmol) by SDS-PAGE or 2D-PAGE and electroblot it on a PVDF membrane.
2. Cut out the portion of the PVDF membrane carrying the protein band or spot.
3. Wet the membrane with a small amount of acetonitrile and soak it in 200 µL 0.5% (w/v) PVP-40 in 100 mM acetic acid at 37°C for 30 min to block the unbound protein-binding sites on the membrane.
4. Wash the membrane with 5 mL of deionized water at least 10 times.
5. Soak the membrane in 100 µL 0.1 M phosphate buffer, pH 8.0, containing 5 mM DTT and 10 mM EDTA.
6. Add pyroglutamyl peptidase (5 µg, enzyme:substrate, 1:1–1:10 w/w).
7. Incubate the reaction solution at 30°C for 24 h.
8. Wash the membrane with deionized water.
9. Dry the membrane *in vacuo* and apply it to a gas-phase sequencer (*see* **Note 4**).

3.3. Deblocking of N-Acetylated Proteins

3.3.1. Deblocking with TFA

1. Separate the N-terminally blocked protein (<200 pmol) by SDS-PAGE or 2D-PAGE and electroblot it on a PVDF membrane.
2. Cut out the portion of the PVDF membrane carrying the protein band or spot and place the membrane in an Eppendorf tube.
3. Place the Eppendorf tube in a vial (2.6 × 6 cm) and then add 100–300 µL TFA.
4. Purge the vial with nitrogen gas for 30 s, seal with a stopper, and incubate at 60°C for 30 min (*see* **Note 5**).
5. Dry the membrane *in vacuo* and apply to a gas-phase sequencer (*see* **Notes 6** and **7**).

3.3.2. Deblocking with TFA in Methanol

1. Incubate the N-terminally blocked protein (<100 pmol) on glass fiber filters or PVDF membranes with TFA in methanol (1:1 v/v) at 47°C for 2–3 d (*8*).
2. Dry the sample and apply it to a gas-phase sequencer.

3.3.3. Deblocking with Acylamino Acid Releasing Enzyme

1. Separate the N-terminally blocked protein (<100 pmol) by SDS-PAGE or 2D-PAGE and electroblot it onto a PVDF membrane (*see* Chapters 2–4).
2. Cut out the portion of the PVDF membrane carrying the protein band or spot.
3. Wet the membrane with a small amount of acetonitrile and pretreat the membrane with 200 µL 0.5% (w/v) PVP-40 in 100 m*M* acetic acid at 37°C for 30 min.
4. After thorough washing with deionized water, digest the protein on the PVDF membrane with 5–10 µg trypsin (enzyme:substrate, 1:1–1:10, w/w) in 100 µL 0.1 *M* ammonium bicarbonate buffer, pH 8.0, containing 10% (v/v) acetonitrile, at 37°C for 24 h with shaking (*see* **Notes 8** and **9**).
5. Pool the digestion buffer containing tryptic peptides in an Eppendorf tube.
6. Wash the membrane by vortexing with 100 µL deionized water and the washing solution with the digestion buffer.
7. Evaporate the digestion mixture to dryness *in vacuo* and add 100 µL 50% (v/v) pyridine and 10 µL phenylisothiocyanate to react with the free, but not blocked N-terminal amino acids of tryptic peptides (*see* **Note 10**).
8. Flush with nitrogen gas, vortex the reaction solution and centrifuge at 3000*g* for 1 min.
9. Discard the resultant supernatant containing reaction byproducts and excess reagent. Repeat this washing procedure three times and evaporate the sample to dryness *in vacuo*.
10. Prepare performic acid solution by mixing 9 mL formic acid and 1 mL 30% hydrogen peroxide and keep the mixture at room temperature for 1 h. Add 100 µL performic acid to the dried sample and, after mixing, place the tube on ice (1 h) to convert the N-terminal phenylthiocarbamyl groups of the peptide to phenylcarbamyl groups by oxidation.
11. Evaporate the sample solution to dryness *in vacuo*, resuspend in deionized water and again dry *in vacuo*.
12. Resuspend the sample in 100 µL 0.2 *M* phosphate buffer, pH 7.2, containing 1 m*M* DTT.
13. Dissolve 50 mU AARE in 50 µL of the same buffer and add to the mixture. Incubate at 37°C for 12 h to remove the N-acetylated amino acid.
14. Apply the sample solution to a Polybrene-coated glass fiber filter mounted into the upper glass block of the reaction chamber of a gas-phase sequencer (*see* **Notes 11–13**).

3.4. Deblocking of Proteins with N-Acyl Groups by Pyrococcus Aminopeptidase

An aminopeptidase from the archaeon *Pyrococcus furiosus* can be applied to the deblocking, followed by internal sequence analysis of the N-terminally blocked proteins *(9–11)*. This enzyme can cleave all peptide bonds of protein sequentially from the N-terminus, except the peptide bond at the N-terminal side of proline. Therefore, the N-terminally blocked protein can be truncated from the N-terminus to one residue before proline after *Pyrococcus* aminopeptidase digestion. It is possible to sequence the generated polypeptide with a gas-phase sequencer (*see* **Note 14**).

1. Cut out the portion of the PVDF membrane carrying the protein (<100 pmol) band or spot (*see* **Note 15**).
2. Wash the membrane with 5 mL deionized water.
3. Immerse the membrane in 1 mL 70% (v/v) acetic acid and sonicate for 1/2 h.
4. Remove the acetic acid solution and evaporate the solution to dryness.
5. Resuspend the sample in 50 μL 50 m*M* ethylmorpholine buffer, pH 8.0, containing 0.1% (w/v) $CoCl_2$.
6. Add the *Pyrococcus* aminopeptidase solution. Incubate the sample solution at 50°C for from 4 h for small peptides to 2 d for large proteins (*see* **Note 16**). The optimized enzyme:substrate molar ratio is from 1:100 for small peptides to 1:1 for large proteins.
7. Apply the sample solution to a gas-phase sequencer (*see* **Note 17**).

4. Notes

1. If the N-terminal peptide can be purified from a protein, the N-terminal blocking group and the sequence of the peptide may be determined by mass spectrometry.
2. Deblocking of proteins electroblotted on a PVDF membrane has several advantages:

   a. Proteins purified by PAGE can be efficiently deblocked and sequenced.
   b. Chemical and enzymatic deblocking can be easily performed without desalting, since after electroblotting on a PVDF membrane, the proteins become almost completely salt-free.
   c. Proteins can be deblocked at picomole levels.
   d. Sequential deblocking as described in **Note 18** is possible.

3. Proteins are often posttranslationally modified and N-terminal blockage is a well-known posttranslational modification. Proteins are N-terminally blocked not only in vivo but also in vitro. It is possible to prevent artificial in vitro blocking generated during protein extraction, PAGE, and Western blotting. The use of highly pure reagents, addition of thioglycolic acid as a free-radical scavenger during the extraction, electrophoresis and electroblotting buffers, or pre-electrophoresis for removing the free-radicals from the gel may help to prevent in vitro blocking *(12,13)* (*see* Chapters 2 and 3). However, if proteins are blocked in vivo, a chemical or enzymatic deblocking procedure such as that described in this chapter is required to determine the N-terminal sequence.
4. Miyatake et al. *(14)* treated the protein-blotted PVDF membrane with anhydrous hydrazine vapor at 20°C for 8 h to deblock proteins with a pyroglutamic acid at the N-terminus. The N-terminal pyroglutamic acid of the protein was converted to Glu-γ-hydrazide which then could undergo Edman degradation. But this often

causes partial modification of asparagine and glutamine to their hydrazides and the conversion of arginine residue to ornithine. Milder hydrazinolysis at –5°C for 8 h may be useful to deblock the N-formylated residue *(14)*.

5. Wellner et al. *(15)* indicated that an *N–O* acyl shift may be involved in the removal of acetyl group of N-terminal acetylserine and acetylthreonine. The advantage of this method is that deblocking is easy and rapid, although overall sequencing yields obtained by this procedure are usually low (~5%).

6. When the membrane is exposed to TFA vapor for longer than 2 h, yields of the PTH-amino acids increase, but at the same time, so do those of the reaction byproducts, which prevent identification of PTH-amino acids. Reaction byproducts may be generated primarily through cleavage of polypeptide with TFA *(16)*.

7. Proteins with either acetylserine or acetylthreonine can be deblocked by TFA treatment but not proteins with other acetylamino acids. About 36 and 4% of acetylated proteins carry acetylserine and acetylthreonine, respectively, as N-terminal amino acids. Treatment with TFA vapor should thus deblock about 40% of acetylated proteins and the remaining may be deblocked with AARE.

8. Pyroglutamated proteins with relatively high molecular mass such as hen egg riboflavin-binding protein (34 kDa) and *Geotrichum candidum* lipase (59 kDa) can be deblocked on a PVDF membrane by direct treatment with pyroglutamyl peptidase. AARE, in contrast, does not directly digest proteins of high molecular mass. As described by Tsunasawa et al. *(17)*, AARE specifically cleaves the X-Y bond of RCO-X-Y type peptides shorter than 40 residues (R: alkyl group, X/Y: L-amino acid). Thus, prior to AARE digestion, proteins on the membrane should be digested with a protease such as trypsin. Short peptides produced are subsequently released from the membrane. In this protease digestion, 10% acetonitrile is included in the digestion buffer to reduce hydrophobic interaction between peptide and PVDF membrane *(18)*. In this organic solvent, peptides are more efficiently eluted from the membrane. The acetylated amino acid of the N-terminal peptide extracted is selectively removed by digestion with AARE.

9. If the N-terminal peptide obtained by protease digestion has more than 10 residues, its extraction from the PVDF membrane would be difficult. In such a case, a second digestion with another protease should be performed on the same membrane.

10. Krishna et al. *(19)* reported an alternative method in which after fragmentation, peptides with a free amino group at the N-terminus are succinylated and the N-terminal blocked peptide is then deblocked with AARE.

11. If butyl chloride (reagent S3 in ABI protein sequencers) is delivered for 30 s prior to Edman degradation in the gas-phase sequencer to recover the N-terminal blocked amino acid released by AARE, the blocking group and N-terminal amino acid can be identified by mass spectrometry.

12. In the sequencing of cytochrome *c*, an N-acetylated protein, overall initial yields from successive steps including electrophoresis, electroblotting, deblocking, and sequencing ranged from 23–25%. The efficiency of deblocking and sequencing described here depends primarily on that of tryptic digestion and subsequent pep-

tide elution from the PVDF membrane. Shaking the digestion solution containing the PVDF membrane during tryptic digestion should facilitate elution of the tryptic peptides.

13. The N-terminal myristoyl group of the blocked proteins can be removed when, instead of AARE, peptide N-fatty acylase (Wako Pure Chemicals, Japan) is used as described in **Subheading 3.3.**

14. The *Pyrococcus* aminopeptidase removes the acetylated amino acid slowly, and after cleavage of the first blocked amino acid, the speed of protein degradation rapidly increases. This makes impossible to remove only the N-terminal blocking group with the *Pyrococcus* aminopeptidase. Therefore, the blocked N-terminal region is truncated and the internal sequence from a residue before proline can be determined by Edman degradation. Since the direct on-membrane digestion of proteins with the *Pyrococcus* aminopeptidase is not easy, it is prerequisite to elute proteins from the membranes prior to the digestion.

15. It is recommended to use the Immobilon-P membrane or teflon membrane (GoreTex). These membranes allow more efficient elution of proteins than the Immobilon-P^{SQ}, Immobilon-CD, Hyperbond, Fluorotrans, ProBlott, Transblot, and polypropylene membranes.

16. If longer time and higher temperature (60–80°C) were used, the deacetylation is more efficient, but high background level significantly decreases the sensitivity and make identification of amino acid impossible. The treatment with the *Pyrococcus* aminopeptidase under optimized conditions results in sufficient deblocking with initial yields up to 50%. However, the initial yields in the sequencer for some large proteins are below 5%.

17. When the sequence of the truncated polypeptide is analyzed, the *Pyrococcus* aminopeptidase itself is simultaneously sequenced. The N-terminal sequence of the aminopeptidase is MVDYELLKKVVEAPGV. The digest has the N-terminal sequence of Xaa-Pro-. If the sample on a glass fiber filter is treated with o-phthalaldehyde (OPA) after the first cycle of Edman degradation, the N-terminus of the *Pyrococcus* aminopeptidase is specifically blocked. However, the digest is not blocked, since the N-terminal residue is proline. Therefore, the sequence of only the digest can be determined from the second cycle *(9)*.

18. The deblocking techniques may be used in combination to make possible sequential deblocking and sequencing of unknown proteins immmobilized on PVDF membranes (*see* **Fig. 1**).

    a. A protein sample is transferred from the PAGE gel onto a PVDF membrane by Western blotting.

    b. The membrane carrying the protein is directly subjected to gas-phase sequencing.

    c. If sequencing (2–3 cycles) fails at step a, the membrane is removed from the sequencer, and treated with TFA vapor at 60°C for 30 min to remove the acetyl groups of acetylserine and acetylthreonine. If the protein is N-terminally blocked by acetylserine or acetylthreonine, the acetyl group is removed by this procedure and sequencing from the N-terminus becomes possible.

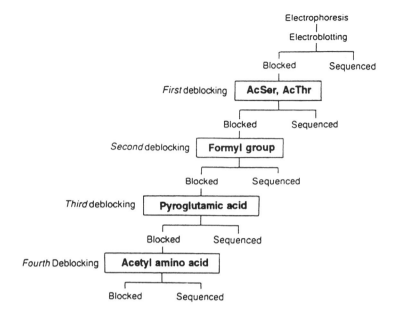

Fig. 1. Strategy of sequential deblocking for N-terminally blocked proteins electroblotted on a PVDF membrane *(6)*.

d. If sequencing fails at step b, the membrane is incubated in 0.6 *M* HCl at 25°C for 24 h to remove formyl groups, and then subjected to sequence analysis again.

e. If sequencing fails at step c, the sample is subjected to on-membrane pyroglutamyl peptidase digestion to remove N-terminal pyroglutamic acid and sequencing is again conducted.

f. Finally, if sequencing fails at step (d), deblocking with AARE is performed to remove acetylamino acids that are not removed in step (b) and sequencing is then attempted again. If sequencing still fails, different methods must be used.

References

1. Matsudaira, P. (1987) Sequence of picomole quantities of proteins electroblotted onto polyvinylidene difluoride membranes. *J. Biol. Chem.* **262**, 10,035–10,038.
2. Aitken, A. (1990) *Identification of Protein Consensus Sequences.* Ellis Horwood Ltd., Chichester, West Sussex, UK.
3. Tsunasawa, S. and Sakiyama, F. (1992) Amino-terminal acetylation of proteins, in *The Posttranslational Modification of Proteins* (Tuboi, S., Taniguchi, N., and Katsumura, N., eds.), Japan Scientific Societies, Tokyo, pp. 113–121.
4. Brown, J. and Roberts, W. (1976) Evidence that approximately eighty percent of the soluble proteins from Ehrlich ascites cells are N-acetylated. *J. Biol. Chem.* **251**, 1009–1014.

5. Tsunasawa, S. and Hirano, H. (1993) Deblocking and subsequent microsequence analysis of N-terminally blocked proteins immobilized on PVDF membrane, in *Methods in Protein Sequence Analysis* (Imahori, K. and Sakiyama, F., eds.), Plenum, New York, pp. 43–51.

6. Hirano, H. and Watanabe, T. (1990) Microsequencing of proteins electrotransferred onto immobilizing matrices from polyacrylamide gel electrophoresis: application to an insoluble protein. *Electrophoresis* **11**, 573–580.

7. Dunbar, B. S. (1994) *Protein Blotting, a practical approach*. IRL Press, Oxford University Press, Oxford, p. 242.

8. Gheorghe, M. T., Jörnvall, H., and Bergman, T. (1997) Optimized alcoholytic deacetylation of N-acetyl-blocked polypeptides for subsequent Edman degradation. *Anal. Biochem.* **254**, 119–125.

9. Tsunasawa, S. (2000) Deblocking aminopeptidase from *Pyrococcus furiosus* and its application for proteomics. *Protein, Nucleic Acid Enzyme* **45**, 186–192.

10. Kamp, R. M., Tsunasawa, S., and Hirano, H. (1998) Application of new deblocking aminopeptidase from *Pyrococcus furiosus* for microsequence analysis of blocked proteins. *J. Protein Chem.* **17**, 512–513.

11. Kamp, R. M. and Hirano, H. (1999) Enzymatic and chemical deblocking of N-terminally modified proteins, in *Proteome and Protein Analysis* (Kamp, R. M., Kyriakidis, D., and Choli-Papadopoulou, T. eds.), Springer, Berlin, pp. 303–319.

12. Moos, M., Jr., Nguyen, N. Y., and Liu, T.-Y. (1988) Reproducible high yield sequencing of proteins electrophoretically separated and transferred to an inert support. *J. Biol. Chem.* **263**, 6005–6008.

13. Ploug, M., Stoffer, B., and Jensen, A. L. (1992) In situ alkylation of cysteine residues in a hydrophobic membrane protein immobilized on polyvinylidene difluoride membranes by electroblotting prior to microsequence and amino acid analysis. *Electrophoresis* **13**, 148–153.

14. Miyatake, N., Kamo, M., Satake, K., Uchiyama, Y., and Tsugita, A. (1992) Removal of N-terminal formyl groups and deblocking of pyrrolidone carboxylic acid of proteins with anhydrous hydrazine vapor. *Eur. J. Biochem.* **212**, 785–789.

15. Wellner, D., Panneerselvam, C., and Horecker, B. L. (1990) Sequencing of peptides and proteins with blocked N-terminal amino acids: N-acetylserine or N-acetylthreonine. *Proc. Natl. Acad. Sci. USA* **87**, 1947–1949.

16. Hulmes, J. D. and Pan, Y.-C. (1991) Selective cleavage of polypeptides with trifluoroacetic acid: applications for microsequencing. *Anal. Biochem.* **197**, 368–376.

17. Tsunasawa, S., Takakura, H., and Sakiyama, F. (1990) Microsequence analysis of N-acetylated proteins. *J. Protein Chem.* **9**, 265–266.

18. Aebersold, R. H., Leavitt, J., Saavedra, R. A., Hood, L. E., and Kent, S. B. H. (1987) Internal amino acid sequence analysis of proteins separated by one- or two-dimensional gel electrophoresis after in situ protease digestion on nitrocellulose. *Proc. Natl. Acad. Sci. USA* **84**, 6970–6974.

19. Krishna, R. G., Chin, C., and Wold, F. (1991) N-terminal sequence analysis of N-acetylated proteins after unblocking with N-acylaminoacyl peptide hydrolase. *Anal. Biochem.* **199**, 45–50.

29

Deblocking of Proteins Containing N-Terminal Pyroglutamic Acid

Jacek Mozdzanowski

1. Introduction

Pyroglutamic acid (pGlu) is the product of cyclization of N-terminal glutamine and is one of the common modifications found at the N-termini of proteins. This modification is typical for immunoglobulins, especially IgGs, since they contain glutamine at the amino terminus of the heavy chain. This chemical modification, in addition to other modifications of the N-terminus results in a "blocked" protein which cannot be sequenced using Edman degradation. Many deblocking procedures have been reported to remove common modifications at the N-terminal end of proteins for the sequence determination by Edman chemistry (1). Deblocking of N-terminal pGlu is usually achieved by the use of either chemical methods (2,3) or by the use of pyroglutamate aminopeptidase (PGAP) (1,4,5). Removal of pGlu from peptides by PGAP with high yield has been described (1,6–10) and recommended as a standard procedure (1). However, when this protocol is applied to large proteins such as IgGs, it results in relatively low yield (<40%) of the deblocked protein. The most commonly cited procedure of Podell and Abraham (11) for removal of pGlu from proteins is also recommended in the handbooks (1,12). However, this procedure is not consistent with the results, which demonstrated that incubation at 4°C did not yield any deblocking (13). Other existing protocols use PGAP for digestion in solution or for proteins blotted onto PVDF membranes and result in variable yields of deblocked protein (2,11,12,15,16). The need to confirm cDNA coded protein sequence and to determine the purity of recombinant therapeutic proteins by reliable N-terminal sequencing requires highly efficient procedure that can remove N-terminal pGlu without protein fragmen-

From: *Methods in Molecular Biology, vol. 211: Protein Sequencing Protocols, 2nd ed.*
Edited by: B. J. Smith © Humana Press Inc., Totowa, NJ

tation *(13)*. This chapter describes such a procedure utilizing pyroglutamate aminopeptidase for deblocking of proteins in solution.

2. Materials
2.1. Reduction and Carboxymethylation of Proteins

1. Dithiothreitol (DTT).
2. 8 *M* guanidine buffer: 8 *M* guanidine, 0.35 *M* Tris-HCl, pH 8.5 (this buffer may be stored at room temperature for approx 1 yr).
3. Iodoacetic acid, sodium salt.

2.2. Desalting/Buffer Exchange

1. Sephadex G-25 column (NAP™-5, Pharmacia Biotech, gel-bed dimensions 0.9 × 2.8 cm).
2. Phosphate buffer: 0.1 *M* sodium phosphate, 2 m*M* EDTA, pH = 8.0. It is highly recommended that this buffer be prepared fresh for each experiment.

2.3. Digestion with Pyroglutamate Aminopeptidase

1. Glycerol.
2. 1.0 *M* DTT solution in phosphate buffer (*see* **Subheading 2.2.**, phosphate buffer). Prepare fresh for each experiment.
3. Pyroglutamate aminopeptidase (Takara Shuzo Co. Ltd., Code No. 7334).

3. Methods
3.1. Reduction and Carboxymethylation of Protein

1. Dissolve 0.5 mg of protein (e.g., m IgG) in 1.0 mL of guanidine buffer (*see* **Note 1**).
2. Add 8.5 mg DTT and dissolve it the reaction mixture.
3. Blanket the reaction mixture with argon.
4. Incubate at 60°C for 90 min.
5. Cool the reaction mixture down to room temperature.
6. Add 28 mg of solid sodium iodoacetate and dissolve in the reaction mixture (*see* **Note 2**).
7. Incubate at room temperature for 45 min in the dark (wrapped with aluminum foil) (*see* **Note 3**).
8. Add approx 5 mg DTT and dissolve in the reaction mixture (*see* **Note 4**).

3.2. Desalting/Buffer Exchange

1. Equilibrate Sephadex G-25 (NAP-5) column with approx 10 mL of 0.1 *M* phosphate buffer (0.1 *M* phosphate, 2 m*M* EDTA, pH 8.0) (*see* **Note 5**).
2. Load 0.3 mL of the reaction mixture containing reduced/carboxymethylated protein to the column (*see* **Note 6**).
3. Wait until there is no solution left above the top frit of the NAP-5 column. Discard the eluate (*see* **Note 7**).

4. Apply 0.55 mL of phosphate buffer (0.1 *M* phosphate, 2 m*M* EDTA, pH 8.0) to the column (*see* **Note 6**).
5. Wait until there is no solution left above the top frit of the NAP-5 column. Discard the eluate.
6. Place a clean vial under the column and apply 0.4 mL of phosphate buffer to the column.
7. Wait until there is no solution left above the top frit of the NAP-5 column. Collect the eluate (*see* **Note 8**).

3.3. Digestion with Pyroglutamate Aminopeptidase

1. Add 20 µL of glycerol to 0.4 mL of desalted reaction mixture (see desalting/buffer exchange, **Subheading 3.2.**).
2. Add 5 µL of 1.0 *M* DTT solution in phosphate buffer (0.1 *M* phosphate, 2 m*M* EDTA, pH 8.0) to the reaction mixture (*see* **Note 9**).
3. Dissolve pyrogutamate aminopeptidase in 100 µL of phosphate buffer (0.1 *M* phosphate, 2 m*M* EDTA, pH 8.0) (*see* **Note 10**).
4. Add 10–20 µL of pyroglutamate aminopeptidase solution to the reaction mixture (*see* **Note 11**).
5. Blanket the reaction mixture with argon.
6. Incubate at 37°C for 24 h (*see* **Note 12**).
7. Sequence the sample (*see* **Note 13**).

4. Notes

1. Typically proteins are available as a solution. In most cases the protein solution may be mixed with the guanindine buffer to produce a final protein concentration of 0.5 mg/mL. Dilution of the guanidine buffer from 8 *M* to 4 *M* does not affect the reduction/carboxymethylation process. Proteins in dilute solutions may require concentration. The amount of protein needed for sequencing is much lower than 0.5 mg. This procedure may be modified for use with limited amounts of protein.
2. The amount of sodium iodoacetate has to be in stoichiometric excess of DTT in the reaction mixture. If the amount of DTT used in **step 2** is greater than 8.5 mg, 28 mg of sodium iodoacetate may be insufficient for carboxymethylation. Sodium iodoacetate decomposes with time producing free iodide. Free iodide may result in iodination of tyrosines. Sodium iodoacetate which is pale yellow or yellow should not be used. Exposure to light increases the degradation of sodium iodoacetate.
3. Suggested incubation time for both the reduction and carboxymethylation steps is to assure a complete reaction. With many proteins a complete reaction may be achieved in a shorter time.
4. DTT is added to neutralize the unreacted iodoacetate. The exact amount of DTT added at this step is not critical.
5. Excess phosphate buffer (~10 mL) will provide very good equilibration of the NAP-5 column. The exact volume is not critical and larger volumes used for equilibration will not affect the procedure. If a desalting column of different vol-

ume is used, the volume used for equilibration should be equal to at least 3 column volumes.

6. All the volumes are given for Sephadex G-25 NAP-5 columns from Pharmacia. Other desalting columns may be used after the elution volume for a protein is established. The elution volumes may vary for different proteins. Low molecular-weight proteins or peptides cannot be processed using the described desalting protocol. The volumes given in this protocol are based on elution of IgGs.

7. The buffer flow through the column will stop as soon as there is no more liquid above the top frit. There is no danger of air getting inside the resin bed.

8. The method was not optimized for quantitative recovery of protein. This volume may require modifications for different proteins. If quantitative recovery is desired, the exact elution profile established by any suitable method (e.g., absorbance at 280 nm) will be very helpful.

9. Pyroglutamate aminopeptidase is a thiol protease and the presence of DTT is essential for the activity of the enzyme.

10. Takara sells pyroglutamate aminopeptidase in lyophilized form, which is stable for several months under the proper storage conditions. After the enzyme is dissolved, it should be stored frozen (–70°C) and used within 1 mo.

11. The amount of PGAP given here was optimized for IgGs. Larger amount of PGAP will improve deblocking yield. Commercially available enzyme is highly purified and the use of larger amount will have minimal effect on sequencing background.

12. Pyroglutamate aminopeptidase from *Pyrococcus furiosus* (Takara) can be used at higher temperatures. Higher incubation temperature reduces the time necessary for digestion but often increases protein degradation and results in higher background in sequencing. In some cases incubation at 37°C for 2 or 3 d may be beneficial for the deblocking yield.

13. This procedure was used for the deblocking of several IgGs. There is no reason to anticipate problems when deblocking other proteins. In general the expected deblocking yield is between 80–100%.

References

1. Fowler, E., Moyer, M., Krishna, R. G., Chin, C. C. Q., and Wold, F. (1996) Removal of N-terminal blocking groups from proteins, in *Current Protocols in Protein Science* (Colligan, J. E., Dunn, B. M., Ploegh, H. L., Speicker, D. W., Wingfield, P. T., eds.), Wiley, New York, pp. 11.7.1–11.7.4.

2. Miyatake, N., Kamo, M., Satake, K., and Tsugita, A. (1992) Hydrazine deblocking method for formylated and pyroglutamyl residues from N-terminus of protein. *J. Protein. Chem.* **11**, 383–383.

3. Hashimoto, T., Saito, S., Ohki, K., and Sakura, N. (1996) Highly selective cleavage of pyroglutamyl-peptide bond in concentrated hydrochloric acid. *Chem. Pharm. Bull.* **44**, 877–879.

4. Hashimoto, T., Ohki, K., and Sakura, N. (1996) Hydrolytic cleavage of pyroglutamyl-peptide bond. III. a highly selective cleavage in 70% methanesulfonic acid. *Chem. Pharm. Bull.* **44**, 2033–2036.

5. Saito, S., Ohki, K., Sakura, N., and Hashimoto, T. (1996) Hydrolytic cleavage of pyroglutamyl-peptide bond. II. Effects of amino acid residue neighboring the pGlu moiety. *Biol. Pharm. Bull.* **19**, 768–770.

6. Awade, A. C., Cleuziat, P. H., Gonzales, T. H., and Robert-Baudouy, J. (1994) Pyrrolidone carboxyl peptidase (Pcp): An enzyme that removes pyroglutamic acid (pGlu) from pGlu-peptides and pGlu-proteins. *Proteins Struct. Funct. Genet.* **20**, 34–51.

7. Gade, G., Hilbich, C., Beyreuther, K., and Rinehart, K. L. (1988) Sequence analysis of two neuropeptides of the AKH/RPCH-family from the Lubber grasshopper, *Romalea microptera. Peptides* **9**, 681–688

8. Fisher, W. H. and Park, M. (1992) Sequence analysis of pyroglutamyl peptides: microscale removal of pyroglutamate residues. *J. Protein Chem.* **11**, 366–366.

9. Powell, J. F. F., Fisher, W. H., Park, M., Craig, A. G., Rivier, J. E., White, S. A., et al. (1995), Primary structure of solitary form of gonadotropin-releasing hormone (GnRH) in cichlid pituitary; three forms of GnRH in brain of cichlid and pumpkinseed *Fish. Regul. Pept.* **57**, 43–53.

10. Kim, J. and Kim, K. (1995) The use of FAB mass spetrometry and pyroglutamate aminopeptidase digestion for the structure determination of pyroglutamate in modified peptides. *Biochem. Mol. Biol. Int.* **35**, 803–811.

11. Podell, D. N. and Abraham, G. N. (1978) A technique for the removal of pyroglutamic acid from the amino terminus of proteins using calf liver pyroglutamate amino peptidase. *Biochem. Biophys. Res. Commun.* **81**, 176–185.

12. Sweeney, P. J. and Walker, J. M. (1993) Aminopeptidases, in *Methods in Molecular Biology* vol. 16, (Burrell, M. M., ed.), Humana Press, Totowa, NJ, pp. 319–329.

13. Mozdzanowski, J., Bongers, J., and Anumula, K. (1998) High-yield deblocking of amino termini of recombinant immunoglobulins with pyroglutamate aminopeptidase. *Anal. Biochem.* **260**, 183–187.

14. Abraham, G. N. and Podell, D. N. (1981) Pyroglutamic acid. Non-metabolic formation, function in proteins and peptides, and characteristics of the enzymes effecting its removal. *Mol. Cell. Biochem.* **38**, 181–190.

15. Lu, H. S., Clogston, C. L., Wypych, J., Fausset, P. R., Lauren, S., Mandiaz, E. A., et al. (1991) Amino acid sequence and post-translational modification of stem cell fctor isolated from buffalo rat liver cell-conditioned medium. *J. Biol. Chem.* **266**, 8102–8107.

16. Hirano, H., Komatsu, S., Kajiwara, H., Takagi, Y., and Tsunasawa, S. (1993) Microsequence analysis of the N-terminally blocked proteins immobilized on polyvinylidene difluoride membrane by western blotting. *Electrophoresis* **14**, 839–846.

30

Identification of Sites of Glycosylation

David J. Harvey

1. Introduction

Glycoproteins typically contain three types of glycans (*1*), the so-called N-linked glycans which are attached via an amide bond to asparagine in a Asn-Xxx-Ser(Thr) motif where Xxx can be any amino acid except proline; the O-linked glycans that are attached to serine or threonine, and glycosyl-phosphatidylinositol lipid anchors attached to the carboxy-terminus of some proteins. Glycoproteins containing N-linked glycans typically possess from 1–20 glycosylation sites that may or may not be occupied, usually with a range of carbohydrate structures. Each individual glycoprotein is known as a "glycoform". All of these N-linked carbohydrate structures contain a trimannosyl-chitobiose [Manα1→3(Manα1→6)Manβ1→4GlcNAcβ1→ 4GlcNAc] pentasaccharide core with one or more glycan chains (antennae) attached to each of the nonreducing mannose residues (*see* **Fig. 1**). Glycans containing only mannose in the antennae are termed "high-mannose," those with galactose and GlcNAc in both antennae are termed "complex" and glycans with both mannose and GlcNAc on different antennae are known as "hybrid" glycans. Fucose, sialic acid, other monosaccharides and sulphate are frequently also present. O-linked glycans are usually smaller, lack a common core structure and are usually found in groups on adjacent or closely spaced amino acids. Several structural types are recognised (*see* **Fig. 1**). Unlike the case of N-linked glycans, there is no consensus sequence of amino acids directing O-linked glycosylation. Thus, in addition to the identification of which amino acids are glycosylated, it is also necessary to identify the extent of glycosylation (site occupancy) and the type of glycan attached to each site.

Several methods are available to enable the glycosylation site and occupancy to be determined. The most important methods depend on:

From: *Methods in Molecular Biology, vol. 211: Protein Sequencing Protocols, 2nd ed.*
Edited by: B. J. Smith © Humana Press Inc., Totowa, NJ

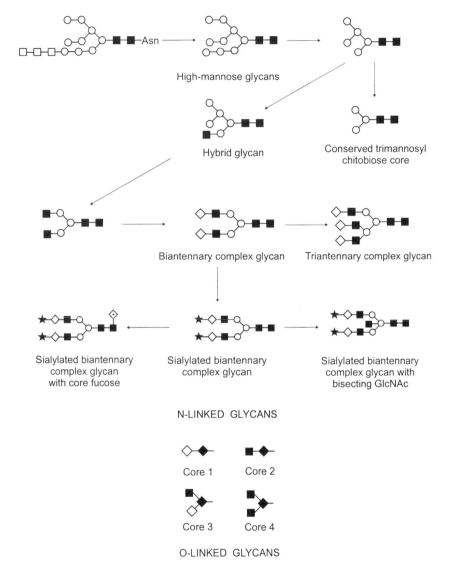

High-mannose glycans

Hybrid glycan

Conserved trimannosyl
chitobiose core

Biantennary complex glycan   Triantennary complex glycan

Sialylated biantennary   Sialylated biantennary   Sialylated biantennary
complex glycan   complex glycan   complex glycan with
with core fucose   bisecting GlcNAc

N-LINKED GLYCANS

Core 1   Core 2

Core 3   Core 4

O-LINKED GLYCANS

Fig. 1. Top, Simplified biosynthetic pathway for N-linked glycans showing the main structural types. Bottom, The four common core structures for O-linked glycans. ○, mannose; ◇, galactose; □, glucose; ■, GlcNAc; ◆, GlcNAc; ◇, fucose; *, sialic acid.

a. Amino acid analysis,
b. mass determination of the (glyco)peptides before and after deglycosylation, and
c. the use of endoglycosidases and stable isotope labeling to leave a "marker" at the site of glycosylation.

The presence of both N- and O-linked glycans can be revealed by classical Edman sequencing techniques as glycosylated amino acids are generally not recovered to a significant extent, unless they contain a single monosaccharide, resulting in "blanks" in the analysis. Potential, but unoccupied glycosylation sites, yield the amino acid as normal. Although fully occupied and unoccupied sites can be detected, the method is unreliable at revealing the extent of occupancy of partially occupied sites, particularly when occupancy is low.

With small glycoproteins containing a single N-linked glycosylation site, such as ribonuclease B (15 kDa) with five mass-different glycoforms *(2)*, a matrix-assisted laser desorption/ionization (MALDI) or electrospray (ESI) mass spectrum will provide a profile containing a peak from unglycosylated protein and a peak from each of the glycoforms *(3,4)*. Thus, site occupancy is immediately apparent. If the mass of the protein is known from its sequence, the mass of each glycan, and hence its composition, can be determined by difference. Masses of the common constituent monosaccharide residues present in N-linked glycans is given in **Table 1**. The method does not, however, reveal the presence of isomeric glycans or give any further information on their structure. This information must be obtained from further experiments involving, for example, specific exoglycosidase digestion (*see* **Note 1**) or mass spectrometric fragmentation. With larger glycoproteins (currently above about 35 kDa *[4,5]*) and those containing many glycans, the resolution of current mass spectrometers is usually insufficient to resolve the individual glycoforms and the compound should then be treated in the same way as glycoproteins containing several occupied N-linked sites.

Mass spectra of glycans containing two or more occupied N-linked sites are usually too complex to allow resolution of individual glycoforms and site analysis is performed by cleavage of the protein chain, usually with trypsin, into a mixture of peptides and glycopeptides. These compounds can be separated or isolated by high-performance liquid chromatography (HPLC) prior to examination by mass spectrometry and, providing that the proteolysis cleaves the protein between each glycosylation site, the occupancy and glycan type at each site can be determined. Because the masses of the resulting glycopeptides are usually relatively small, there is no problem with insufficient instrumental resolution.

One of the main problems with this approach is the mass spectrometric identification of glycopeptides in the presence of peptides as the signals produced by the glycopeptides are frequently much weaker than those from their unglycosylated counterparts. The most widely accepted technique for resolving this problem, particularly for complex mixtures, is to the use liquid chromatographyl mass spectrometry (LC/MS) with an ESI ion source and the "orifice stepping" technique devised by Huddleston and Carr *(6)*. This technique makes use of the ready fragmentation of the glycan portion of the mol-

Table 1
Residue Masses of Common Monosaccharides

Oligosaccharide	Residue formula	Residue mass[a]
Deoxy-pentose	$C_5H_8O_3$	116.047
		116.117
Pentose	$C_5H_8O_4$	132.042
		132.116
Deoxy-hexose	$C_6H_{10}O_4$	146.058
		146.143
Hexose	$C_6H_{10}O_5$	162.053
		162.142
Hexosamine	$C_6H_{11}NO_4$	161.069
		161.158
HexNAc	$C_8H_{13}NO_5$	203.079
		203.179
Hexuronic-Acid	$C_6H_8O_6$	176.032
		176.126
KDO	$C_8H_{12}O_7$	220.058
		220.195
N-Acetyl-neuraminic acid	$C_{11}H_{17}NO_8$	291.095
		291.258
N-glycoyl-neuraminic acid	$C_{11}H_{17}NO_9$	307.090
		307.257

[a] Top figure = monoisotopic mass (based on C = 12.000000, H = 1.007825, N = 14.003074, O = 15.994915), Lower figure = average mass (based on C = 12.011, H = 1.00794, N = 14.0067, O = 15.9994.

ecule at relatively high orifice potentials to give such species as [hexose + H]$^+$ (*m/z* 163), [HexNAc + H]$^+$ (*m/z* 204), [sialic acid + H]$^+$ (*m/z* 292) and [(hexose-HexNAc) + H]$^+$ (*m/z* 366). The mass spectrometer is scanned repeatedly throughout the elution of the peptides and glycopeptides from the column. The orifice potential is high at the low mass end of the scan but low for the remainder in order to record molecular ions. At the end of the experiment, a comparison of the single-ion plots of the glycan-containing ions with the total ion plot, immediately identifies those peaks produced by glycopeptides.

When glycopeptides can be detected in, for example a MALDI spectrum, considerable information can be obtained by deglycosylation with an endoglycosidase (*see* **Note 2**) such as peptide-N-glycosidase F, (PNGase-F) (*7*) followed by re-examination of the peptide mixture to detect both missing peaks (deglycosylated peptides) and the appearance of new ones. Again, the mass difference between the two reveals the mass of the glycan. This method, however, can be rather imprecise due to the fact that some glycopeptides might be absent from the original spectrum and, secondly, with multiply glycosylated proteins, coincidences in mass between different combinations of putative glycans and peptides may lead to erroneous conclusions. A method for deglycosylation of glycoproteins from within sodium dodecyl sulfate polyacrylamide gel electrophoresis (SDS-PAGE) gels has been described *(8)* leaving the protein available for subsequent tryptic digestion.

In addition to enzymatic release, several chemical-based methods are also available and are generally preferred for the release of O-linked glycans as general purpose O-glycanases are not available. The most common method for releasing O-linked glycans is by β-elimination in the presence of dilute alkali (*see* **Note 3**). The reaction is usually performed in the presence of a reducing agent to prevent degradation of the released glycans by "peeling" reactions and is relatively specific for glycans O-linked to serine and threonine. N-linked glycans and glycans linked to hydroxyproline, hydroxylysine or tyrosine are usually resistant. Under nonreducing conditions, serine and threonine are converted into the dihydro-amino acids, 2-aminopropenic and 2-amino-2-butenoic acid, respectively, which again can be readily detected in the peptide chain by mass spectrometry. If strong alkali (e.g., 1 M NaOH) and high temperatures (100°C) are used, N-linked glycans can also be released.

A more precise mass spectrometric technique for revealing glycan location but not glycan type utilises the fact that PNGase-F and PNGase-A cleave N-linked glycans at the glycan-Asn bond leaving aspartic acid at the occupied site. The mass of this amino acid differs from that of asparagine by one unit and can be readily detected by mass spectrometry of derived peptides. If the amino acid sequence of the protein is known, substitution of aspartic acid for asparagine reveals the glycosylation site. If the amino acid sequence is not known, the generated aspartic acid residues can be readily distinguished from native residues if the deglycosylation is performed in the presence of 50% ^{18}O-labeled water *(9,10)*. Mass spectrometric examination of the resulting peptides enables the formerly conjugated aspartic acid residues to be identified by their incorporation of the ^{18}O isotope, which provides doublets spaced by 2 Da in the resulting spectra.

GPI Anchors contain a heterogeneous carbohydrate core and a membrane-binding lipid moiety linked to the protein through an ethylamine-phosphate group *(11)*. Hydrofluoric acid can be used to cleave the GPI anchor from the carboxy-terminal amino acid by hydrolysis of the phosphate linkage *(12)*. The resulting peptide-ethanolamine conjugate can be detected by MALDI mass spectrometry using the same matrix (usually α-cyano-4-hydroxycinnamic acid) as used for the analysis of unsubstituted peptides.

2. Materials

2.1. Glycosidases (see Note 2)

2.1.1. N-Glycanase (see **Note 4**)

1. N-Glycanase (peptide-N-glycosidase F, PNGase-F): from *Flavobacterium meningosepticum* (EC 3.5.1.52) (Glyko Inc., Novato, CA; Calbiochem, Boehringer Mannheim, Lewes, UK). Also available as a recombinant preparation (from *Escherichia coli*) from Glyko Inc. (*see* **Note 5**). Store at –20°C. Resuspend lyophilized enzyme in water to a concentration of 200 U/mL. Store at –20°C. Prepare a working solution by diluting 1 in 3 to 50 U/mL.
2. Digest buffer (nondenaturing conditions): 20 m*M* ammonium or sodium bicarbonate, pH 8.0, 50% (v:v) methanol (*see* **Note 6**).

2.1.2. Endoglycosidase H (Endo-H) (see **Note 7**)

1. Endoglycosidase H from *Streptomyces plicatus* (EC 3.2.1.96), recombinant in *E. coli* (Glyko Inc., Calbiochem or Boehringer Mannheim). Endoglycosidase H is supplied as a lyophilysed solid. Resuspend in water to a concentration of 1 U/mL. Store at –20°C.
2. Digest buffer: 50 m*M* sodium citrate/phosphate, pH 5.0.

2.1.3. O-Glycosidase (see **Note 8**)

1. O-Glycosidase from *Streptococcus pneumoniae* (EC 3.2.1.97) (Glyko Inc. or Calbiochem). Store at –20°C. The enzyme is supplied in solution in 25% (v:v) glycerol and BSA (100 mg/mL).
2. Digest buffer: 50 m*M* sodium citrate/phosphate, pH 6.0.

2.2. Other Materials

1. Freshly prepared solution of 0.1 *M* NaOH in 2 *M* NaBH$_4$.
2. 48% Hydrofluoric acid solution (obtainable from Merck Ltd., Lutterworth, Leicestershire, UK) (*see* **Note 9**).
3. Solution of 10 mg of sinapinic acid (Aldrich Chemical Co., Poole, Dorset, UK) in acetonitrile (0.3 mL) and water (0.7 mL) containing 0.1% TFA.

3. Methods

3.1. Deglycosylation of Glycopeptides

3.1.1. PNGase-F Digestion

3.1.1.1. GLYCOPROTEINS (*SEE* NOTE 10)

1. Denature the glycoprotein as follows: Freeze dry the protein from water containing 0.1% TFA. Add 10 µL denaturing buffer (made from 50 µL of 5 *M* Tris/HCl, pH 8.0, 10 µL 10 *M* dithiothreitol (DTT), 25 µL 10% SDS, 415 µL distilled water (*see Note 11*) and heat to 100°C for 5 min.

2. Cool and add 5 µL of 5% NP-40 (*see* **Note 12**). Divide sample into two 7 µL aliquots, incubate one overnight at 37°C with 7 µL glycerol-free PNGase-F (200–250 U/mL) and the other with 7 µL of the denaturing buffer as a control. (*see* **Notes 13** and **14**).
3. Store at –20°C to stop the reaction.
4. Samples can be purified for MALDI analysis with C18 ZipTips™ (Millipore Corporation, Bedford, MA) according to the directions supplied by the manufacturers.

3.1.1.2. GLYCOPEPTIDES

1. Add 1 µL of digest buffer and 4 µL of PNGase F (200 mU) to the peptide solution (5 µL) in 10% methanol/90% water.
2. Incubate at 37°C overnight.
3. Store at –20°C to stop the reaction.
4. Samples can be purified for MALDI analysis with C18 ZipTips™ according to the directions supplied by the manufacturers.

3.1.2. Endoglycosidase H Digestion

1. Add 0.1 µL of digest buffer and 0.1 µL of endo H (0.1 mU) to 0.8 µL of sample in 10% methanol/90% water (v:v).
2. Incubate at room temperature overnight.
3. Store at –20°C to stop the reaction.
4. Samples can be purified for MALDI analysis with C18 ZipTips™.

3.1.3. O-Glycosidase Digestion

1. Add 0.1 µL of digest buffer and 0.1 µL of O-glycanase (0.1 mU) to 0.8 µL of sample in 10% methanol/90% water (v:v).
2. Incubate at room temperature overnight.
3. Store at –20°C to stop the reaction.
4. Samples can be purified for MALDI analysis with C18 ZipTips™.

3.1.4. Removal of O-Glycans by β-Elimination (see **Note 13**)

1. Add sodium hydroxide to a solution of the glycoprotein (10–20 µg/µL) to give a pH of 10.0. Add an equal volume of a freshly prepared solution of NaOH in 2 M NaBH$_4$.
2. Incubate at 45°C for 16 h (overnight).
3. Reduce the pH to 6.0 by the addition of 50% acetic acid to the cooled solution.
4. Samples can be purified for MALDI analysis with C18 ZipTips.

3.1.5. GPI Cleavage with Hydrofluoric Acid (see **Note 15**)

1. Add 0.5 µL of 48% hydrofluoric acid solution to 0.5 µL of sample in 10% methanol/90% water (v:v).
2. Incubate at 0°C for 60 h.
3. Lyophilize to stop the reaction.

3.2. MALDI Mass Spectrometry

3.2.1. Proteins with Molecular Weights Above About 5 kDa

1. Place 0.5 μL of the sample solution on the MALDI target (e.g., Micromass TofSpec 2E) followed by 0.5 μL of the matrix solution. Allow to dry (*see* **Note 16**).
2. Place the target in the instrument and sum spectra from multiple laser shots until a satisfactory signal:noise ratio is obtained (*see* **Note 17**).

3.2.2. Proteins with Molecular Weights Below 5 kDa

The procedure is essentially the same except that the matrix is a-cyano-4-hydroxycinnamic acid (Aldrich Chemical Co.) (*see* **Note 18**).

3.2.3. Released Glycans

1. The best matrix for these compounds is 2,5-dihydroxybenzoic acid (DHB) or a mixture of DHB and 5% (w:w) 5-methoxy-2-hydroxybenzoic acid, a mixture known as "super DHB" *(14)*. Other matrices are discussed in **ref.** *(15)*.
2. After the spot has dried, redissolve in the minimum amount of ethanol and allow to recrystallize (*see* **Note 19**).
3. Obtain the MALDI spectrum as above.

4. Notes

1. Suitable enzymes are listed in several publications e.g., **refs.** *(16,17)*.
2. Store enzymes according to the manufacturers' recommendations - not all glycosidases should be stored frozen. In order to prolong the life of a glycosidase requiring storage at –20°C, once it has been resuspended in water, split the solution into small aliquots (e.g., for microdigest experiments use 2 μL per aliquot). Use one aliquot at a time or freeze-thaw a maximum of five times. Small amounts of bovine serum albumin (BSA) (e.g., 100 μg/mL) can be included in the digest buffer as a carrier protein to prevent absorption of enzymes and glycopeptides onto the side of the incubation tube. The glycosidases and BSA do not interfere with analysis by MALDI but may be detected by sequencing or by capillary electrophoresis.
3. β-Elimination is ineffective if the carboxyl group of serine or threonine is free.
4. A number of endoglycosidases (*see* **Table 2**) are available for releasing glycans from glycoproteins. Although some are relatively non-specific with respect to the glycans cleaved, others show various specificities that can yield (limited) structural information on the types of glycan present at various sites. PNGase-F *(7)* cleaves a wide range of N-linked carbohydrates (high-mannose, hybrid, bi-, tri- and tetra-antennary complex, sulphated, and phosphorylated high-mannose and sulphated biantennary from proteins and peptides by splitting the link between the glycan and asparagine to leave aspartic acid and a glycosylamine. The glycosylamine rapidly hydrolyses to the free glycan, particularly in the presence of dilute acid. However, PNGase-F will not cleave N-linked glycans carrying a fucose residue α1→3-linked to the reducing-terminal GlcNAc residue. In these situations, PNGase-A is effective *(18,19)*. The reaction with PNGase-F is par-

ticularly rapid (usually complete within 1–5 h) for isolated glycopeptides in non-denaturing buffers. However, the enzyme usually cannot access the cleavage site if the secondary or tertiary protein structure is intact and, thus, reduction and/or proteolytic cleavage is often performed prior to glycosidase digestion. Leaving the digests overnight has no detrimental effect on the deglycosylated peptide. Only large (tetra-antennary) glycans carrying additional N-acetyllactosamine units (e.g., those from human α1-acid glycoprotein [20]), which are bulky and which may cause steric hindrance, present any resistance. In these cases, it is frequently necessary to make a second addition of N-glycanase after 24 h and to extend the incubation period to 48 h.

5. If the glycans are to be examined by MALDI mass spectrometry, the glycerol should be removed by dialysis.

6. If the free glycans are required, ammonium-containing buffers should be avoided as their use promotes retention of the glycosylamine (21).

7. Endo-H selectively cleaves oligomannose and hybrid type structures but does not release complex N-linked glycans. It cleaves the bond linking the two GlcNAc residues of the chitobiose core, leaving GlcNAc attached to the protein. Endo-D, which also cleaves this bond, does not show the narrow spectrum of active shown by Endo-H.

8. O-Glycanase is relatively specific in that it cleaves Galβ1→3GalNAc from serine and threonine. The presence of sialic acid, fucose or additional GalNAc residues is sufficient to block the reaction. Pretreatment of the glycopeptide(protein) with a sialidase may be beneficial to the action of the O-glycanase (22).

9. Hydrofluoric acid is very toxic and corrosive. Do not allow skin contact. Handle in a fume cupboard, wearing gloves and safety glasses.

10. PNGase F works much more efficiently with glycopeptides and, thus, it is preferable to cleave the glycoprotein with trypsin or another suitable enzyme prior to PNGase digestion. However, this method can work for small glycoproteins.

11. Many laboratory sources of "pure" water, although relatively ion free, still contain organic material such as polyethylene glycol that can interfere with mass spectrometric analysis. Organic material can be removed by distillation from a dilute solution of sodium permanganate. Sodium rather than potassium permanganate should be used to avoid introducing potassium ions, which, under MALDI conditions, will give the $[M + K]^+$ ion in addition to the required $[M + Na]^+$ ion.

12. PNGase F is inactive in SDS, an ionic detergent, therefore, an excess of the non-ionic detergent NP-40 should be added.

13. To monitor the deglycosylation, analyse both sample and control by SDS-PAGE. Deglycosylation is indicated by the decrease in the apparent molecular weight of the enzyme-treated sample to the predicted mass of the de-N-glycosylated protein.

14. If the glycans are subsequently required for analysis, remove the protein by passing the digest through a Pro-Spin filter (P85100 Pro-Mem, Radleys/Life Sciences International) that has been pre-washed with water. Wash the filter with water three times, combining the washings with the filtered digest, and evaporate to dryness.

Table 2
Endoglycosidases Used to Cleave Glycans from Glycoproteins

Endoglycosidase	Common name	Source	EC Number	Cleavage site
Endo-N-acetylgalactosaminidase	O-Glycanase	Recombinant, *E. coli*	EC 3.2.1.97	Ser/Thr to Galβ1-3GalNAc
Endoglycosidase D	Endo D	*Streptococcus pneumoniae*	EC 3.2.1.96	GlcNAc to GlcNAc from $(Man)_5(GlcNAc)_2$
Endoglycosidase F1	Endo F1	*Chryseobacterium meningosepticum* Recombinant, *E. coli*	–	GlcNAc to GlcNAc from high-mannose and hybrid N-linked glycans
Endoglycosidase F2	Endo F2	*Chryseobacterium meningosepticum* Recombinant, *E. coli*	–	GlcNAc to GlcNAc from hybrid and biantennary, complex N-linked glycans
Endoglycosidase F3	Endo F3	*Chryseobacterium meningosepticum* Recombinant, *E. coli*	–	GlcNAc to GlcNAc from bi- and tri-antennary, complex N-linked glycans
Endoglycosidase H	Endo H	*Streptomyces plicatus*, Recombinant, *E. coli*	EC 3.2.1.96	GlcNAc to GlcNAc from high-mannose and hybrid N-linked glycans
N-Glycosidase A	PNGase A	Almond	EC 3.5.1.52	GlcNAc to Asn from all N-linked glycans, including those containing core 1-3fucose
N-Glycosidase F	PNGase F	*Chryseobacterium meningosepticum* Recombinant, *E. coli*	EC 3.5.1.52	Glc to Asn from all N-linked glycans except those containing core 1-3fucose

15. Hydrofluoric acid hydrolyses the phosphate linkage between the peptide and GPI-anchor moieties *(12)*. The peptide-ethanolamine product (an additional 44.1 units above the expected mass of the carboxy-terminal peptide) can be analyzed by MALDI mass spectrometry using the conditions outlined in **Subheadings 2.3.** and **3.3.**

16. It is important to ensure that the protein solution is as free from other compounds (buffer salts, etc.) as possible as the presence of these will reduce or abolish the signal. The maximum tolerated amounts of various commonly encountered contaminants are given in **ref.** *(23)*.

17. The laser power should be kept as low as possible in order to obtain the best resolution, particularly on older, linear time-of-flight (TOF) mass spectrometers not equipped with time-lag focusing (delayed extraction). The instrument should be calibrated with proteins of a similar mass, spanning the mass of the sample protein. For the most accurate measurement, the calibration compounds should be added to the sample solution at such a concentration as to give peaks of comparable intensity.

18. The laser power for obtaining spectra from α-cyano-4-hydroxycinnamic acid is usually considerably lower than that for sinapinic acid.

19. Aqueous solutions of DHB usually crystallize from the periphery of the target spot producing large crystals that project towards the centre. Much of the sample is deposited as an amorphous solid between the crystals. Re-dissolving in ethanol and allowing to recrystallize produces a more homogeneous target with a better incorporation of sample into the crystal lattice *(24)*.

References

1. Sturgeon, R. J. (1988) The glycoproteins and glycogen, in *Carbohydrate Chemistry* (Kennedy, J. F., ed.), Oxford University Press, Oxford, UK, pp. 263–302.

2. Fu, D., Chen, L., and O'Neill, R. A. (1994) A detailed structural characterization of ribonuclease B oligosaccharides by 1H NMR spectroscopy and mass spectrometry. *Carbohydrate Res.* **261**, 173–186.

3. Mock, K. K., Davy, M., and Cottrell, J. S. (1991) The analysis of underivatised oligosaccharides by matrix-assisted laser desorption mass spectrometry. *Biochem. Biophys. Res. Commun.* **177**, 644–651.

4. Bonfichi, R., Sottani, C., Colombo, L., Coutant, J. E., Riva, E., and Zanette, D. (1995) Preliminary investigation of glycosylated proteins by capillary electrophoresis and capillary electrophoresis/mass spectrometry using electrospray ionization and by matrix-assisted laser desorption ionization/time-of-flight mass spectrometry. *J. Mass Spectrom./Rapid Commun. Mass Spectrom.* S95–S106.

5. Tsarbopoulos, A., Bahr, U., Pramanik, B. N., and Karas, M. (1997) Glycoprotein analysis by delayed extraction and post-source decay MALDI-TOF-MS. *Int. J. Mass Spectrom. Ion Processes* **169/170**, 251–261.

6. Huddleston, M. J., Bean, M. F., and Carr, S. A. (1993) Collisional fragmentation of glycopeptides by electrospray ionization LC/MS and LC/MS/MS - methods for selective detection of glycopeptides in protein digests. *Anal. Chem.* **65**, 877–884.

7. Tarentino, A. L., Gómez, C. M., and Plummer, T. H., Jr. (1985) Deglycosylation of asparagine-linked glycans by peptide:N-glycosidase F. *Biochemistry* **24**, 4665–5671.

8. Küster, B., Wheeler, S. F., Hunter, A. P., Dwek, R. A., and Harvey, D. J. (1997) Sequencing of N-linked oligosaccharides directly from protein gels: In-gel deglycosylation followed by matrix-assisted laser desorption/ionization mass spectrometry and normal-phase high performance liquid chromatography. *Anal. Biochem.* **250**, 82–101.

9. Gonzalez, J., Takao, T., Hori, H., Besada, V., Rodriguez, R., Padron, G., and Shimonishi, Y. (1992) A method for determination of N-glycosylation sites in glycoproteins by collision-induced dissociation analysis in fast atom bombardment mass spectrometry: identification of the positions of carbohydrate-linked asparagine in recombinant α-amylase by treatment with peptide-N-glycosidase F in ¹⁸O-labelled water. *Anal. Biochem.* **205**, 151–158.

10. Küster, B. and Mann, M. (1999) ¹⁸O-labeling of N-glycosylation sites to improve the identification of gel-separated glycoproteins using peptide mass mapping and database searching. *Anal. Chem.* **71**, 1431-1440.

11. Ferguson, M. A. J. (1991) Lipid anchors on membrane proteins. *Curr. Opin. Struct. Biol.* **1**, 522–529.

12. Puoti, A. and Conzelmann, A. (1992) Structural characterisation of free glycolipids which are potential precursors for glycosylphosphatidylinositol anchors in mouse thymoma cell lines. *J. Biol. Chem.* **267**, 22673–22680.

13. Montreuil, J., Bouquelet, S., Debray, H., Fournet, B., Spik, G., and Strecker, G. (1986) Glycoproteins, in *Carbohydrate Analysis: A Practical Approach* (Chaplin, M. F., and Kennedy, J. F., eds.), IRL Press, Oxford, UK, pp. 143–204.

14. Karas, M., Ehring, H., Nordhoff, E., Stahl, B., Strupat, K., Hillenkamp, F., et al. (1993) Matrix-assisted laser desorption/ionization mass spectrometry with additives to 2,5-dihydroxybenzoic acid. *Org. Mass Spectrom.* **28**, 1476–1481.

15. Harvey, D. J. (1999) Matrix-assisted laser desorption/ionization mass spectrometry of carbohydrates. *Mass Spectrom. Rev.* **18**, 349–451.

16. Dwek, R. A., Edge, C. J., Harvey, D. J., Wormald, M. R., and Parekh, R. B. (1993) Analysis of glycoprotein-associated oligosaccharides. *Ann. Rev. Biochem.* **62**, 65–100.

17. Harvey, D. J., Kuster, B., Wheeler, S. F., Hunter, A. P., Bateman, R. H., and Dwek, R. A. (2000) Matrix-assisted laser desorption/ionization mass spectrometry of N-linked carbohydrates and related compounds, in *Mass Spectrometry in Biology and Medicine* (Burlingame, A. L., Carr, S. A., and Baldwin, M. A., eds.), Humana Press, Totowa, NJ, pp. 403–437.

18. Takahashi, N. (1977) Demonstration of a new amidase acting on glycopeptides. *Biochem. Biophys. Res. Commun.* **76**, 1194–1201.

19. Sugiyama, K., Ishihara, H., Tejima, S., and Takahashi, N. (1983) Demonstration of a new glycopeptidase from jack-bean meal acting on aspartylglucosylamine linkages. *Biochem. Biophys. Res. Commun.* **112**, 155–160.

20. Yoshima, H., Matsumoto, A., Mizuochi, T., Kawasaki, T., and Kobata, A. (1981) Comparative study of the carbohydrate moieties of rat and human plasma α1-acid glycoproteins. *J. Biol. Chem.* **256**, 8476–8484.

21. Küster, B., and Harvey, D. J. (1997) Ammonium-containing buffers should be avoided during enzymatic release of glycans from glycoproteins when followed by reducing terminal derivatization. *Glycobiology* **7**, vii–ix.

22. Sutton, C. W., O'Neill, J. A., and Cottrell, J. S. (1994) Site-specific characterization of glycoprotein carbohydrates by exoglycosidase digestion and laser desorption mass spectrometry. *Anal. Biochem.* **218**, 34–46.

23. Mock, K. K., Sutton, C. W., and Cottrell, J. S. (1992) Sample immobilization protocols for matrix-assisted laser-desorption mass spectrometry. *Rapid Commun. Mass Spectrom.* **6**, 233–238.

24. Harvey, D. J. (1993) Quantitative aspects of the matrix-assisted laser desorption mass spectrometry of complex oligosaccharides. *Rapid Commun. Mass Spectrom.* **7**, 614–619.

31

Analysis of Sites of Protein Phosphorylation

Alastair Aitken and Michele Learmonth

1. Introduction

In this chapter, methods will be reviewed for the identification of the type of modified amino acid residue (serine, threonine, or tyrosine) and position in the sequence of a phosphopeptide.

Reversible phosphorylation of proteins is now established as a major intracellular regulatory mechanism. Different protein kinases show widely preferred substrate specificities in their target proteins and knowledge of the amino acid sequence surrounding a site of phosphorylation is vitally important to establish which class of kinase is responsible for regulating the activity of the protein or enzyme being studied. The protein kinases that regulate the intracellular processes described earlier, are themselves subject to regulation in a wide variety of ways. Many kinases are regulated by second messengers and a number are under control of calcium ions. Other mechanisms for regulation of protein kinases include the presence of pseudosubstrate sequences in the regulatory domain. In addition, many protein kinases undergo autophosphorylation.

Analysis of modified peptides, where no ^{32}P or other radiolabel is present, may be particularly difficult. Even the identification of phosphorylation sites where ^{32}P-radioactivity is present may pose problems if one relies on phosphopeptide map analysis. Depending on exact type of column and gradient phosphorylated peptides will in general elute slightly ahead (1–2% acetonitrile) of their unphosphorylated analogs on reverse-phase high-performance liquid chromatography (RP-HPLC) *(1)* (but *see* **Note 1**).

The ability to quantitate phosphoamino acids at nanomolar levels is a great advance on qualitative techniques such as thin-layer electrophoresis and allows an estimation of the degree of phosphorylation of the protein to be made.

From: *Methods in Molecular Biology, vol. 211: Protein Sequencing Protocols, 2nd ed.*
Edited by: B. J. Smith © Humana Press Inc., Totowa, NJ

A particularly useful method utilizing pre-column derivatization with o-phthalaldehyde and separation on reverse-phase chromatography was described by Carlomango and co-workers *(2)* and is detailed in **Subheading 3.2.** This procedure can also separate the phosphoforms of a number of other amino acids *(see* **Note 2**).

The best recovery of phosphate from phosphoamino acids is achieved with solid phase sequencing *(3)*. In this chemistry the thiazolinone derivatives of the residues are removed at each cycle of Edman degradation along with a high percentage of the ^{32}P-phosphate, by washing with aqueous methanol *(see* **Note 3**). The sensitivity of the method is tens of cpm of ^{32}P by scintillation counting.

Phosphoseryl residues may be specifically modified to the stable adduct S-ethylcysteine *(4)* by β-elimination of the phosphate followed by addition of ethanethiol. A particularly useful method for this derivatization of phosphoserine is outlined in **Subheading 3.6.** This can detect picomolar levels of this phosphoamino acid.

2. Materials

2.1. Identification of Phosphoamino Acids by Thin Layer Electrophoresis

1. 6 *M* HCl.
2. 5 *M* KOH.
3. Hydrolysis tubes.
4. Vacuum desiccator.
5. Flat-bed thin-layer electrophoresis apparatus (e.g., Hunter Thin-Layer Electrophoresis Apparatus, CBS Scientific).
6. Plastic backed thin layer cellulose sheets (BDH, 20 cm × 20 cm).
7. pH 1.9 electrophoresis buffer: acetic acid: formic acid: water 4:1:45 (v/v/v).
8. pH 3.5 electrophoresis buffer: pyridine: acetic acid: water 1:10:89 (v/v).
9. Ninhydrin spray solution (0.2% Ninhydrin [w/v] in ethanol).
10. Microdialysis kit (e.g., "Slide-A-Lyzer" or "Micro DispoDialyzer" from Pierce-Warriner).
11. Ammonium bicarbonate.

2.2. HPLC Identification of Phosphoamino Acids

1. HPLC apparatus.
2. Reverse-phase chromatography column, Hamilton PRP-1 polymer-based reverse-phase column or a Spherisorb S5ODS-2 column.
3. Fluorimetric detector fitted with OPA filters.
4. Phosphoamino acid stock solutions: phosphoserine, phosphothreonine, and phosphotyrosine; 60 m*M* in water, store at –20°C; from Sigma.
5. o-Phthalaldehyde (OPA) stock solution: 10 mg o-phthalaldehyde; 14.9 mL methanol (HPLC-grade); 5.3 mL saturated boric acid solution, pH 9.5, and 25

μL ethanethiol. This solution is prepared daily and stored in the dark at room temperature.

6. HPLC buffers (*see* **Note 4**): High purity water; Phosphate buffer: 0.3 *M* sodium dihydrogen phosphate (NaH$_2$PO$_4$).

   a. Buffer A: 100 mL phosphate buffer; 25 mL tetrahydrofuran (THF); 1875 mL water.

   b. Buffer B: 45 mL phosphate buffer; 1100 mL acetonitrile; 855 mL water.

2.3. Gas (or Pulsed-Liquid) Phase Microsequencing of Phosphopeptides

1. For coupling of phosphopeptides to solid support *see* **Subheading 3.5.**

2.4. Manual Edman Degradation of Radiolabeled Phosphopeptides

1. Plastic-backed thin-layer cellulose sheets (Camlab/Machery-Nagel)
2. pH 1.9 electrophoresis buffer: acetic acid: formic acid: water 4:1:45 (v/v/v).

2.5. Solid-Phase Microsequencing of Phosphopeptides

1. Sequelon-AA kit (Applied Biosystems).
2. Peptide solvent (e.g., 20% acetonitrile/0.1% TFA).
3. Scintillation fluid (e.g., Universol ES, ICN).
4. Heating block.
5. Scintillation counter.

2.6. Modification of Phosphoserine to S-Ethylcysteine

1. Incubation mixture: 0.2 mL water, 0.2 mL dimethylsulphoxide, 100 μL ethanol, 65 μL 5 *M* sodium hydroxide, and 60 μL 10 *M* ethanethiol.
2. Glacial acetic acid.
3. Nitrogen gas source.

2.7. Selective Detection of Phosphopeptides on Mass Spectrometry

1. Mass spectrometer.
2. Reagents for modification to S-ethylcysteine (*see* **Subheading 2.6.**).

3. Methods

3.1. Identification of Phosphoamino Acids by Thin Layer Electrophoresis

1. ^{32}P-phosphopeptides are hydrolyzed for 1–4 h at 110°C. Optimal hydrolysis times for a particular phosphoamino acid are given (*see* **Note 5**). Phosphoserine/ phosphothreonine, 6 *M* HCl, 2 h and 4 h, respectively, at 110°C *(5)*. Phosphotyrosine, 6 *M* HCl, 1 h at 110°C or 5 *M* KOH, 30 min at 155°C *(6)*.
2. After desalting by microdialysis in a volatile buffer such as ammonium bicarbonate, the sample is transferred to a small hard glass tube precleaned by pyrolysis at

500°C (annealing furnace). For identification, tubes should be numbered with a diamond scriber. The volatile sample buffer should be removed in vacuo, either in a dessicator or a "Savant" or "Gyrovap" centrifugal evaporator to minimise loss or contamination.

3. Hydrolysis is best done (especially for large numbers of samples) in the vapor phase in a screw-top PTFE vial (e.g., Pierce "tuftainer") or in a small "dry-seal" vacuum dessicator that may be evacuated. These are available from several manufacturers including Waters-Milligen and Ciba-Corning. The numbered tubes are placed in the container with 2 mL of 6 *M* HCl and 0.5% (v/v) phenol. This is flushed with 99.998% argon (sequencer grade) for 2 min, closed and evacuated if possible. The container is heated in an oven or block heater either at 110°C for 24, 48 or 72 h or at 150–160°C for 1–4 h. For acid-labile amino acids, special hydrolysis conditions are given in **Subheading 3.1., step 1** (*see* **Note 6**).

4. After cooling, the hydrolysis tubes are quickly transferred to a Savant type centrifugal evaporator and evacuated (without initially turning on the rotation to prevent washing down traces of HCl into the tubes). The residues are redissolved in 50 µL of HPLC-grade water and centrifugally evaporated in the conventional manner. The recovery of phosphoamino acids varies with the primary structure around the phosphorylation site as well as the acid lability of the phosphoamino acid *(5)*. The order of lability of free phosphoamino acids is phosphotyrosine > phosphoserine > phosphothreonine.

5. The dried residue is redissolved in a few mL unlabelled phosphoserine (P-Ser), phosphothreonine (P-Thr), and phosphotyrosine (P-Tyr). This is spotted onto one corner of the thin-layer cellulose sheet. This is subjected to electrophoresis at pH 1.9 (acetic acid:formic acid:water, 4:1:45, v/v/v) in the first dimension for 20 min at 1.5 kV and after turning through 90°, electrophoresis at pH 3.5 (pyridine: acetic acid: water, 1:10:89, v/v/v) in the second dimension for 16 min at 1.3kV (*see* **Fig. 1**) on plastic backed thin-layer cellulose sheets.

6. The electrophoretogram is stained with ninhydrin, by spraying or dipping in the ethanolic solution and allowing to develop. Heating in a warm oven will hasten the process. Areas containing phosphoamino acids are revealed by this stain (and by autoradiography if desired) and they are subsequently cut out and quantitatively counted for radioactivity.

3.2. HPLC Identification of Phosphoamino Acids

Protein/peptide should be hydrolyzed using conditions optimal to the likely phosphoamino acid(s) present (*see* **Subheading 3.1.**). Stock solutions of phosphoamino acids should be prepared in water at 60 m*M* (stored at –20°C) and are used to calibrate the HPLC column.

1. Amino acid samples are mixed with an equal volume of fresh OPA solution and allowed to react for 3 min at room temperature prior to injection of 10–100 µL onto the HPLC column.

2. A Hamilton PRP-1 polymer-based reverse-phase column was used originally, but we have also obtained good results using a Spherisorb S5ODS-2 column.

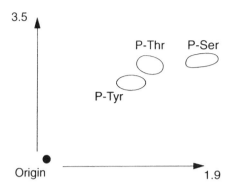

Fig. 1. Separation of phosphoamino acids by two-dimensional thin-layer chromatography. *See* **Subheading 3.1.**, **step 1** for optimum hydrolysis times for each
phosphoamino acid. Electrophoresis is carried out at pH 3.5 in the first dimension and
pH 1.9 in the second dimension. The electrophoretogram is stained with ninhydrin.
Phosphoamino acids are revealed by this stain (and by autoradiography if desired).
These areas are subsequently cut out and quantitatively counted for radioactivity.

3. Column equilibrated at room temperature with 88% Buffer A, 12% Buffer B and
   the system run isocratically at 1 mL/min flow rate. Typical elution times:
   phosphoserine 5.7 min; phosphotyrosine 12.3 min; phosphothreonine 16.2 min.
   Unphosphorylated serine, tyrosine, and threonine show much longer retention
   times and therefore do not interfere. Detection is by fluorescence using a fluorescence detector fitted with OPA filters. Column subjected to wash step (to 100%
   B) prior to re-equilibration for subsequent injections.

3.3. Gas (or Pulsed-liquid) Phase Microsequencing of Phosphopeptides

1. In the gas or pulsed liquid-phase microsequencer, with autoconversion of
   anilinothiazolinones and online HPLC identification of PTH derivatives, the
   remaining sample after injection of an aliquot into the microbore reversed-phase
   column is recovered at each cycle in a fraction collector (*see* **Note 7**).
2. The ^{32}P radioactivity that is removed by the butyl chloride extraction during
   Edman degradation of the radiolabeled phosphopeptide is measured by Cerenkov counting. Since this is a strong β emitter no scintillant is required and the
   measurement is nondestructive.
3. The ratio of dehydro-Ala or dehydro-α-ABA to Thr or Ser may be used to give a
   strong indication of the presence of a phosphorylated derivative at a particular
   cycle. If the residue is fully phosphorylated, only the DTT-adduct of dehydro-Ser
   is detected at that position, due to the sequencing chemistry, which results in
   β-elimination of phosphoserine. This is compared to the normal ratio of PTH-
   Ser/DTT-adduct of dehydro-Ser and that of the analogous derivatives of Thr that
   will be detected at the other sequencing cycles.

3.4. Manual Edman Degradation
of Radiolabeled Phosphopeptides

1. Start with the peptide or a subdigested fragment containing the suspected phosphoamino acid near the amino terminus (*see* **Note 8**).
2. Remove an aliquot after each cycle and dry down twice, redissolving in water after the first time.
3. Run the samples on thin-layer electrophoresis (microcrystalline cellulose sheets, 20 χ 20 cm) with pH 1.9 buffer for 1 h at 300V, Samples should be spotted onto a pencil line drawn halfway across the plate, a sample of the peptide that has not been subjected to Edman degradation should also be loaded.
4. Air-dry the plate and locate the position of radioactivity by autoradiography.
5. At the cycle where the radioactivity no longer moves to the cathode and is not associated with the peptide, but appears towards the anode (co-migrating with a sample of genuine inorganic phosphate); this is the position in the sequence of a phosphorylated residue.

3.5. Solid-Phase Microsequencing of Phosphopeptides
(see *Chapters 23 and 24*)

Below is given a suitable protocol to follow for the solid-phase microsequencing of a ^{32}P-labeled peptide on the Applied Biosystems gas or pulsed liquid-phase protein sequencers.

1. Dissolve radiolabeled peptide in 20 µL suitable solvent (aqueous acetonitrile, up to 30% with or without 0.1% TFA). Heat at 55°C and mix to aid dissolution.
2. Count tube in scintillation counter (Cerenkov program, use NO scintillant).
3. Place Sequelon-AA disk from kit onto a sheet of Mylar on a heating block at 55°C.
4. Pipet peptide sample onto disk. Allow disk to dry thoroughly. Count both disk and sample tube in scintillation counter (use NO scintillant).
5. Rinse tube 3 times with 20 µL solvent and pipet washings onto disk, drying between each application. If significant counts are left in the tube after washing, repeat this procedure.
6. Count disk in scintillation counter again.
7. Pipet into a test tube, 100 µL of coupling buffer and add 1 mg of the water soluble carbodiimide reagent, 1(-3-dimethylaminopropyl)-3-ethylcarbodiimide hydro-chloride (EDC) supplied with the kit.
8. Pipet 5 µL of the EDC solution onto the sample disk. Allow the reaction to proceed at room temperature for 20 min.
9. Wash disc twice with 0.5ml 0.1% TFA and once with 20% acetonitrile, to remove the low amounts of radioactive peptide that have not coupled covalently. Check this amount by combining washings and count in the scintillation counter.
10. Count this washed disk in scintillation counter without scintillant.
11. Replace S3 (butyl chloride) with 9:1 methanol: water (containing 2 mM sodium phosphate).

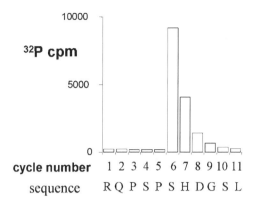

Fig. 2. Automated sequencer analysis on an Applied Biosystems 477 of a phosphopeptide labeled with ^{32}P orthophosphate. The radioactive protein was phosphorylated by casein kinase II and digested in-gel with chymotrypsin. The ^{32}P-labeled phosphopeptide recovered by microbore HPLC on a 1-mm Vydac C_{18} column. The material extracted from the sequencer by the aqueous methanol after each cycle of Edman degradation was measured by scintillation counting. The amount of ^{32}P radioactivity loaded onto the instrument (Cerenkov counting) indicated a typical recovery of 40–50% at cycle 6. The analysis of the sequence shows clearly that the other residues are not phosphorylated.

3.5.1. Sequencing

1. Place disk in the blot cartridge of the Applied Biosystems sequencer using 9:1 methanol: water (containing 2 m*M* aqueous phosphate) instead of butyl chloride (S3). Carry out standard Edman blot sequencing using an ATZ-collect program where all the ATZ derivatives are recovered into the fraction collector (*see* **Note 9**).
2. Count samples from the fraction collector in scintillant (for greater counting efficiency).
3. After sequencing, remove disk from sequencer cartridge, allow to dry and count in scintillation counter to check level of radioactivity remaining.
4. Efficiency of coupling and cpm/pmol for each residue can be calculated.
5. *See* **Fig. 2** for example of results obtained.

3.6. Modification of Phosphoserine to S-Ethylcysteine (see Note 10)

1. In a fume cupboard, the peptide is dissolved in a capped tube containing 50 µL incubation mixture in a capped Eppendorf tube.
2. The tube is flushed with nitrogen and incubated for 1 h at 50°C. After cooling, 10 µL of glacial acetic acid is added.
3. The derivatized peptide is either applied directly to the sequencer or concentrated by vacuum centrifugation (*see* **Note 11**).

4. Extended reaction time (18 h at 50°C) may be used since the β-elimination step of derivatization of a phophoserine adjacent to a proline residue is slow. This can be carried out with the use of acetonitrile instead of the normal solvents. This minimizes subsequent manipulations.

5. This procedure may be used for the selective isolation of phosphoseryl-peptides *(7)*. When the S-ethylcysteinyl-peptides are applied to a reverse-phase HPLC column (e.g., Vydac C_{18}) and eluted with linear gradients of water/acetonitrile in 0.1% trifluoroacetic acid (TFA), the derivatized peptides emerge on average 4–5% acetonitrile later than the native phosphopeptide. A derivatized peptide from a doubly phosphorylated species will elute correspondingly later than the singly derivatised species indicating the applicability of the method to multiple phosphoseryl-peptides. HPLC before and after derivatisation should produce highly purified peptides even from a very complex mixture since the elution position of all others will be unaffected.

3.7. Selective Detection of Phosphopeptides on Mass Spectrometry (see Note 12)

Phospho-Ser may be converted to S-ethylcysteine, which may be followed by ESMS as well as sequencing (*see* **Subheading 3.6.**).

Phosphate-specific fragment ions of 63Da (PO_2^-) and 79 Da (PO_3^-) are produced by collision-induced dissociation during negative ion LC-ESMS *(8)*. This technique of selective detection of post-translational modifications through collision -induced formation of low-mass fragment ions that serve as characteristic and selective markers for the modification of interest has been extended to identify other modifications such as glycosylation, sulphation and acylation (*see* **Note 10**).

Development in ESMS sources permit on-line microbore HPLC using matrices such as 10 mm Poros resins slurry-packed into columns less than 0.25 mm diameter. Polypeptides can be separated in formic or acetic acid (0.1%) on gradients of 5–75% acetonitrile in a few minutes at flow rates of around 10 μL min^{-1} *(9)*.

3.8. Ladder Sequencing by Mass Spectrometry

Ladder sequencing by mass spectrometry *(10)* involves the generation of a set of nested fragments of a polypeptide chain followed by analysis of the mass of each component (*see* **Note 13**). Each component in the ragged polypeptide mixture differs from the next by loss of a mass which is characteristic of the residue weight (which may involve a modified side chain). In this manner, the sequence of the polypeptide can be read from the masses obtained in MS.

The ladder of degraded peptides can be generated by Edman chemistry or by exopeptidase digestion from the N-and C-termini. This essentially a subtractive technique (one looks at the mass of the remaining fragment after each cycle). When a phosphoserine residue is encountered, a loss of 167.1 Da is observed in place of 87.1 for a serine residue. This technique therefore avoids one of the major problems of analysing post-translational modifications. Although the majority are stable during the Edman chemistry, some modifications such as O- or S- linked esters, including O-phosphate for example may be lost by β-elimination during the cleavage step to form the anilinothioazolidone or undergo O- or S- to N- acyl shifts which block further Edman degradation.

4. Notes

1. If the HPLC separation is combined with mass spectrometric characterization, the level of TFA required to produce sharp peaks and good resolution of peptide (circa 0.1% v/v) results in almost or complete suppression of signal. This does not permit true on-line HPLC-MS as the concentration of TFA in the eluted peptide must first be drastically reduced. However, the new "low TFA" reverse-phase HPLC columns from Vydac (300 Å pore size, 218MS54) are available in C_4 and two forms of C_{18} chemistries. They are also supplied in 1 mm diameter columns that are ideal for low levels of sample eluted in minimal volume. We have used as little as 0.005% TFA without major loss of resolution and have observed minimal signal loss. There may be a difference in selectivity compared to "classical" reverse-phase columns, for example we have observed phosphopeptides eluting approx 1% acetonitrile later than their unphosphorylated counterparts (the opposite to that conventionally seen). This is not a problem, but it is something of which one should be aware, and could be turned to advantage.

2. Identification by HPLC is also applicable to the identification of phosphoarginine and 1-phospho and 3-phosphohistidine. These residues may also account for the presence of phosphate in a protein. They are separated with baseline resolution in the following order: 1-phosphohistidine, phosphoserine, 3-phosphohistidine, phosphotyrosine, phosphothreonine, and phosphoarginine. Anion-exchange chromatography on HPLC with low-ionic-strength phosphate buffers has also been used in the analysis of phosphoserine, phosphothreonine, and phosphotyrosine *(12)*. This method for separation of phosphoamino acids includes post-column derivatization and fluorimetric detection. Capony and Demaille *(13)* have described a method of analyzing phosphoamino acids on a conventional amino acid analyzer using isocratic elution with TFA (10 m*M*). O-phthalaldehyde detection was used in both systems. A rapid isocratic technique for the separation and quantitation of phosphoamino acids, with their detection by "UV visualization," has been developed *(14)*. The visualization of these ions with o-phthalate yields a sensitive method with rapid analysis time. Using a column of Spherisorb 5-ODS2 and an eluant of 0.5 m*M* tetrabutylammonium hydroxide/phthalic acid, pH 6.3., all four ions including orthophosphate could be separated.

3. This strong wash is possible due to the covalent linkage of the peptide or protein to the derivatized PVDF membrane supports. The development of these allow

the attachment of proteins via side-chain lysine and N-terminus (Sequelon-DITC) or via side-chain carboxyl groups (Sequelon-AA). These were marketed in conjunction with an automated solid-phase protein sequencer by Millipore. The Sequelon-AA membranes are now used in the sequencing of phosphorylated peptides in the more conventional gas- or pulsed-liquid-phase protein sequencers marketed by Applied Biosystems.

4. All buffers should be prepared using freshly deionized water (Milli-Q), filtered before use through 0.45-µm filters and degassed.

5. The analysis of phosphoamino acids is frequently confined to qualitative or semi-quantitative determinations based on their separation by one- or two-dimensional thin-layer or paper electrophoresis (*11*), *see* **Fig. 1**. This is still an extremely useful technique since inexpensive equipment is required and the spots corresponding to phosphoamino acids may be excised and radioactivity (^{32}P) measured in a scintillation counter.

6. Protein or peptide samples must be substantially free of high salt concentrations, buffers, and detergents for good results in pre-column derivatization. Buffer salts in particular give rise to problems by keeping the pH too low for complete reaction of reagents with primary and secondary amines in the sample. Purity of all reagents and cleanliness of surfaces in contact with the sample is likewise essential. Heavy metal contamination leads to low recoveries of certain amino acids, especially that of lysine. Unless using vapor-phase hydrolysis as described, together with stringent cleanliness and minimal time elapse between derivatization and analysis, reproducible results below 20–50 pmol of each amino acid will prove difficult to obtain.

7. There is always considerable "tailing" of this peak of the "burst" of ^{32}P radioactivity that is detected into subsequent cycles and where closely spaced multiple phosphorylation sites occur the interpretation may be difficult (*15*). Very poor recoveries of phospho-serine, -threonine and -tyrosine derivatives are achieved in gas-phase (or pulsed-liquid) sequencers. The recovery of ^{32}P radioactivity may be as low as 0.05% to 5% since most of the phosphorylated derivative of the amino acid (in the case of phosphothreonine and phosphoserine) undergoes β elimination to the dehydro-derivative and inorganic phosphate. The latter product remains in the spinning cup or glass fibre disc, being poorly extracted with butyl chloride. Normal levels of PTH-phospho-Tyr are recovered. In contrast to the other phosphoamino acids this cannot undergo β-elimination but this derivative elutes very early, therefore a modified gradient may be required. For example, a triethylamine formate buffer (pH 4.0.), which can also be used to separate PTH glycosylated derivatives of Asn, Ser and Thr (*16*). If it is certain that only one type of phosphorylated amino acid is present in a peptide, (after partial acid hydrolysis and thin layer electrophoresis (*see* **Subheading 3.4.**) and amino acid analysis shows only one residue of that particular amino acid present, then the assignment of the site of phosphorylation may be made with more confidence. Otherwise it is strongly advisable to check the assignment of the site of phosphorylation by manual Edman degradation of the radiolabeled peptide.

8. A method has been described *(17)* that involves cutting the glass-fiber disc containing the phosphopeptide, into a number of pieces and removing one piece after each cycle in which phosphorylated residues may be expected. The residual peptide is extracted with 50% formic acid and subjected to reverse-phase HPLC. The radioactivity will coelute with the peptide until the phosphorylated amino acid(s) is/are reached, when the ^{32}P radioactivity (which has been converted to inorganic phosphate) will elute in the void volume.

9. This delivers the resulting ATZ-products and the ^{32}P phosphate directly to the fraction collector. It is advisable to check through each step and modify the sequencer programme to optimize the delivery times of the reagents and solvents. The aqueous methanol has a higher viscosity than butyl chloride and will require longer delivery time.

10. As well as direct sequencing of phosphopeptides, methods have also been developed that convert the phosphoamino acid to a more stable derivative. These are generally applicable to phosphoserine and to a lesser extent, phosphothreonine. Only one of these methods has proved particularly useful in practice when multiple phosphorylation sites are encountered (as is very common). The method developed by Annan et al. *(18)* involves β-elimination and addition of methylamine, while β-elimination followed by reduction with sodium borohydride to convert phosphoserine to alanine has also been used *(19)*. O-glycosylated serine residues can be similarly converted to alanine *(20)*. During these reactions the corresponding threonine derivatives are converted, with lower efficiency, to 2-aminobutanoic acid. Therefore independent confirmation of the presence of phosphate in a given peptide should be sought.

12. PTH-S-ethylcysteine elutes just before the contaminant diphenylthiourea (DPTU) on reverse-phase HPLC during microsequencing. Some DTT adduct of PTH-Ser is also seen, resulting from β-elimination during conversion of ATZ-S-ethylcysteine. Due to the alkaline conditions required for the conversion to S-ethyl-Cys, Asn, and Gln may be deamidated and identified as PTH-Asp or -Glu.

13. Since the recovery of serine, threonine, or tyrosine after complete acid hydrolysis of a peptide may be very low if one of these is present as a phosphoamino acid, mass spectrometric analysis is essential to confirm the exact amino acid composition and number of phosphorylated residues. If this is not available then subdigestion of a peptide containing multiple phosphorylatable peptides, with an appropriate proteinase, is advisable (*see* Chapter 5). Identification of either positive or negative ions may yield more information, depending on the mode of ionization and fragmentation of an individual peptide. Phosphopeptides may give better ions in the negative ion mode since they have a strong negative charge due to the phosphate group. The S-ethylcysteinyl- derivatives described earlier, provide excellent structural information on MS. However, phosphopeptides do run well on ES-MS and MALDI-TOF in positive ion mode. Particular problems may be associated with ES-MS of phosphopeptides, where high levels of Na$^+$ and K$^+$ adducts are regularly seen on species at charge states above that predicted by the theoretical number of basic groups (*see* **Fig. 3**). Electrospray mass spectrometers

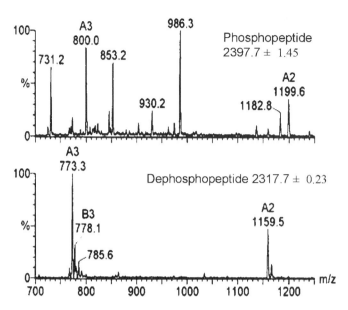

Fig. 3. Electrospray mass spectrometry of tryptic peptides from isoforms of 14-3-3 protein. The spectrum of the phosphopeptide is shown in the upper panel. The unmodified peptide (which elutes approx 1 min later on HPLC) from a parallel tryptic digestion of unphosphorylated 14-3-3 is reproduced in the lower panel. A2 and A3 are the doubly and triply charged forms of the peptides. The calculated masses of phosphorylated and unmodified peptides are 2397.7 and 2317.7 Da, respectively. The other masses are due to incompletely separated peptides from other 14-3-3 isoforms.

will give an accurate molecular weight up to 80–100,000 Da (and in favorable cases up to 150,000 Da has been claimed), (accuracy obtainable is >0.01%). Lower resolution obtained with laser desorption time-of-flight mass spectrometers (ca. 0.1% up to 20,000 Da). Sequence information is readily obtained using tandem mass spectrometry after collision induced dissociation *(21)*. Ion trap mass spectrometry technology (called "LCQ") is now well-established, which also permits sequence information to be readily obtained. Not only can MS-MS analysis be carried out but due to the high efficiency of each stage, further fragmentation of selected ions may be carried out to MS^n. The charge state of peptide ions is readily determined by a "zoom-scan" technique, which resolves the isotopic envelopes of multiply charged peptide ions. The instrument still allows accurate molecular weight determination to over 100000 Da at 0.01% mass accuracy. The recent development of Fourier transform ion cyclotron resonance mass spectrometry *(22)* in which the ions can be generated by a wide variety of techniques has very high resolution and sensitivity.

3. The technique of ladder sequencing has particular application in MALDI-TOF-MS, which has high sensitivity and greater ability to analyze mixtures. The

Edman chemistry is modified to generate the ladder by carrying out the coupling step with phenylisothiocyanate (PITC) in the presence of a small amount of phenylisocyanate (PIC), which acts as a chain-terminating agent. A development of this technique involves the addition of volatile trifluoroethylisothiocyanate (TFEITC) to the reaction tube to which a fresh aliquot of peptide is added after each cycle *(10)*. This avoids steps to remove excess reagent and by-products. This may be combined with subsequent modification of the terminal NH_2 group with quaternary ammonium alkyl N-hydroxysuccinimidyl (NHS) esters which allows increased sensitivity in MALDI-TOF down to 1 pmol. level. The use of exopeptidase digestion in ladder sequencing may be difficult as the rate of release of individual residues may vary greatly.

Acknowledgment

We wish to thank Steve Howell for the ES-MS analysis in **Fig. 3**.

References

1. Walsh, K. A. and Sasagawa, T. (1986) High performance liquid chromatography probes for post-translationally modified amino acids. *Methods Enzymol.* **106** , 22–29.
2. Carlomagno, L., Huebner, V. D., and Matthews, H. R. (1985) Rapid separation of phosphoamino acids including the phosphohistidines by isocratic high-performance liquid chromatography of the orthophthalaldehyde derivatives. *Anal. Biochem.* **149**, 344–348.
3. Coull, J. M., Dixon, J. D., Laursen, R. A., Koster, H., and Pappin, D. (1989) Development of membrane supports for the solid phase sequence analysis of proteins and peptides, in *Methods in Protein Sequence Analysis* (Wittmann-Liebold, B., ed.), Springer-Verlag, Berlin, pp. 69–78.
4. Meyer, H. E., Hoffmann-Posorske, E., Korte, H., and Heilmeyer, L. M. G., Jr.(1986) Sequence analysis of phosphoserine-containing peptides. Modification for picomolar sensitivity. *FEBS Lett.* **204**, 61–66.
5. Kemp, B. E. (1980) Relative alkali stability of some peptide o-phosphoserine and o-phosphothreonine esters. *FEBS Lett.* **110**, 308–312.
6. Martensen, T. M. (1982) Phosphotyrosine in proteins. Stability and quantification. *J. Biol. Chem.* **257**, 9848–9852.
7. Holmes, C. F. B. (1987) A new method for the selective isolation of phosphoserine-containing peptides. *FEBS Lett.* **215**, 21–24.
8. Bean, M. F ., Annan, R. S., Hemling, M. E., Mentzer, M. Huddleston, M. J., and Carr, S. A. (1995) LC-MS methods for selective detection of posttranslational modifications in proteins: glycosylation, phosphorylation, sulfation and acylation, in *Techniques in Protein Chemistry VI* (Crabb, J. W., ed.), Academic Press, San Diego, pp. 107–116.
9. Aitken, A., Patel, Y., Martin, H., Jones, D., Robinson, K., Madrazo, J., and Howell, S. (1994) Electrospray mass spectrometric analysis with on-line trapping of posttranslationally modified mammalian and avian brain 14-3-3 isoforms. *J. Prot. Chem.* **13**, 463–465.

10. Bartlet-Jones, M., Jeffery, W. A. Hansen, H. F., and Pappin, D. J. C. (1994) Peptide ladder sequencing by mass spectrometry using a novel, volatile degradation reagent. *Rapid Comm. Mass Spec.* **8**, 737–742.

11. Hunter, T. and Sefton, B. M. (1980) Transforming gene product of Rous sarcoma virus phosphorylates tyrosine. *Proc. Nat. Acad. Sci. USA* **77**, 1311–1315.

12. Yang, J. C., Fujitaki, J. M., and Smith, R. A. (1982) Separation of phosphohydroxyamino acids by high performance liquid chromatography. *Anal. Biochem.* **122**, 360–363.

13. Capony, J-P. and Demaille, J. G. (1983) A rapid microdetermination of phosphoserine, phosphothreonine and phosphotyrosine in proteins by automatic cation exchange on a conventional amino acid analyser. *Anal. Biochem.* **128**, 206–212.

14. Morrice, N. and Aitken, A. (1985) A simple and rapid method of quantitative analysis of phosphoamino acids by high performance liquid chromatography. *Anal. Biochem.* **148**, 207–212.

15. Amess,B., Manjarrez-Hernandez, H. A., Howell, S. A., Learmonth, M. P., and Aitken, A. (1992) Multisite phosphorylation of the 80kDa (MARCKS) protein kinase C substrate in C3H/10T1/2 fibroblasts. Quantitative analysis of individual sites by solid-phase microsequencing. *FEBS Lett.* **297**, 285–291.

16. Pisano, A., Packer, N. H., Redmond J. W., Williams, K. L., and Gooley, A. A. (1995) In *Methods in Protein Structure Analysis* (Atassi, M. Z. and Appella, E., eds.), Plenum Press, New York, p. 69.

17. Wang, Y. H., Fiol, C. J., DePaoli-Roach, A. A., Bell, A. W., Hermodson, M. A., and Roach, P. J. (1988) Identification of phosphorylation sites in peptides using a gas-phase sequencer. *Anal. Biochem.* **174**, 537–547.

18. Annan, W. D. , Manson, W., and Nimmo, J. A.(1982) The identification of phosphoseryl residues during the determination of amino acid sequence in phosphoproteins. *Anal. Biochem.* **121**, 62–68.

19. Richardson, W. S., Munksgaard, E. C., and Butler, W. T. (1978) Rat incisor phosphoprotein. The nature of the phosphate and quantitation of the phosphoserine. *J. Biol. Chem.* **253**, 8042–8046.

20. Downs, F., Peterson, C., Murty , V. L. N., and Pigman, W. (1977) Quantitation of the beta-elimination reaction as used on glycoproteins. *Int. J. Peptide Protein Res.* **10**, 315–322.

21. Hunt, D. F., Yates, J. R., Shabanowitz, J., Winston, S., and Hauer, C. R. (1986) Protein sequencing by tandem mass spectrometry. *Proc. Natl. Acad. Sci. USA* **83**, 6233–6237.

22. Hendrickson, C. L. and Emmett, M. R. (1999) Electrospray ionization Fourier transform ion cyclotron resonance mass spectrometry. *Ann. Rev. Phys. Chem.* **50**, 517–536.

32

Quantitation and Location of Disulfide Bonds in Proteins

Alastair Aitken and Michele Learmonth

1. Introduction

Sulfydryl and disulfide groups are of great structural, functional, and bio-logical importance in protein molecules. For example, the Cys sulfydryl is essential for the catalytic activity of some enzymes (e.g., thiol proteases) and the interconversion of Cys SH to cystine S-S is directly involved in the activity of protein disulfide isomerase. The conformation of many proteins is stabilized by the presence of disulfide bonds *(1)* and the formation of disulfide bonds is an important post-translational modification of secretory proteins.

Four methods will be covered in this chapter:

1. Quantification of disulfide (S-S) bonds and free thiol(S-H) groups using Ellman's reagent. A micro method using an enzyme-linked immunosorbent assay (ELISA) plate reader, suitable for reading a large number of chromatography column frac-tions, is also described..
2. Electrophoretic determination of the number of S-S bonds per protein molecule.
3. The use of high-performance liquid chromatography (HPLC) in the identifica-tion and purification of S-S bonded peptides; and
4. The use of mass spectrometry in the determination of S-S bonds.

The amount of sample required for determination of the number of S-S bonds depends on the sensitivity of the particular method.

Ellman's reagent 5,5'-dithiobis(2-nitrobenzoic acid) (DTNB) was first introduced in 1959 for the estimation of free thiol groups *(2)*. The procedure is based on the reaction of the thiol with DTNB to give the mixed disulfide and 2-nitro-5-thiobenzoic acid (TNB), which is quantified by the absorbance of the anion (TNB^{2-}) at 412 nm.

From: *Methods in Molecular Biology, vol. 211: Protein Sequencing Protocols, 2nd ed.*
Edited by: B. J. Smith © Humana Press Inc., Totowa, NJ

The reagent has been widely used for the quantitation of thiols in peptides and proteins. It has also been used to assay disulfides present after blocking any free thiols (e.g., by carboxymethylation) and reducing the disulfides prior to reaction with the reagent. It is also commonly used to check the efficiency of conjugation of sulfydryl containing peptides to carrier proteins in the production of antibodies.

Amino acid analysis quantifies the molar ratios of amino acids per mole of protein. This generally gives a nonintegral result, yet clearly there are integral numbers of the amino acids in each protein. A method was developed by Creighton *(1)* to count integral numbers of amino acid residues and it is particularly used in the determination of cysteine residues.

Creighton's method exploits the charge differences introduced by specific chemical modifications of cysteine. Cys residues may be reacted with iodoacetic acid, which introduces acidic carboxymethyl ($HO_2CCH_2S^-$) groups or with iodoacetamide, which introduces the neutral carboxyamidomethyl ($NH_2COCH_2S^-$) groups. A similar method was first used in the study of immunoglobulins by Feinstein *(3)*.

The reaction with either reagent is essentially irreversible, thereby producing a stable product for analysis.

Using varying iodoacetamide/iodoacetate ratios, the acidic and neutral agents will be compete for the available cysteines and a spectrum of protein molecules having $0,1,2,....n$ acidic carboxymethyl residues per molecule can be produced (where n is the number of cysteine residues in the protein). These species will have correspondingly $n, n-1, n-2....0$ neutral carboxyamidomethyl groups. These species may then be separated by electrophoresis, isoelectric focusing or by a combination of both *(1,3,4)*.

Creighton used a low pH discontinuous system *(1)*. Takahasi and Hirose recommend a high pH system *(5)*, while Stan-Lotter and Bragg *(6)* used the Laemmli electrophoresis system followed by isoelectric focusing. It may therefore be necessary to carry out preliminary experiments to find the best separation conditions for the protein under analysis.

In order to ensure that all thiol groups are chemically equivalent the reactions must be carried out in denaturing (in the presence of urea) and reducing (in the presence of dithiothreitol [DTT] conditions. The electrophoresis must be carried out with the unfolded protein in order that the modification has the same effect irrespective of where it is in the polypeptide chain.

The commonly used methods are given below.

The original method has been modified into a two-stage process to allow for the quantification of both sulfydryl groups and disulfide bonds (*see* **Note 1**) *(3,5)*. The principal of the method has also been adapted to count the number of lysine resides after progressive modification of the ε-amino groups with succinic anhydride, which converts this basic group to a carboxylic acid-containing moiety.

Classical techniques for determining disulfide bond patterns usually require the fragmentation of proteins into peptides under low pH conditions to prevent disulfide exchange. Proteases with active site thiols should be avoided (e.g., papain, bromelain). Pepsin or cyanogen bromide are particularly useful. Diagonal techniques to identify disulfide-linked peptides were developed by Brown and Hartley (*see* **Note 2**). A modern micromethod employing reverse-phase HPLC is described here.

Mass spectrometry, is playing a rapidly increasingly role in protein chemistry and sequencing (*7,8*). This is particularly useful in determining sites of co- and post-translational modification and application in locating disulfide bonds is no exception. Mass spectrometry can of course readily analyze peptide mixtures, therefore it is not always necessary to isolate the constituent peptides.

Like the diagonal methods outlined earlier, mass spectrometry can be used to identify peptides linked in pairs by -S-S-bonds most readily when cleavage methods are employed that minimize disulfide exchange (i.e., acid pH, *see* **Subheading 3.**). Partial acid hydrolysis, although nonspecific, has been successfully used in a number of instances. Combined with computer programs that will predict the cleavage position of any particular proteinase or chemical reagent, simple knowledge of the mass of the fragment will, in most instances, give unequivocal answers as to which segments of the polypeptide chain are disulfide-linked. If necessary, one cycle of Edman degradation can be carried out on the peptide mixture and the mass spectrometry analysis repeated. The shift in mass(es), which correlate with loss of specific residues, will confirm the assignment.

2. Materials

2.1. Ellman's Reagent Method

1. Reaction buffer: 0.1 *M* phosphate buffer, pH 8.0.
2. Denaturing buffer: 6 *M* guanidinium chloride, 0.1 *M* Na_2HPO_4, pH 8.0. (*see* **Note 3**).
3. Ellman's solution: 10 m*M* (4mg/mL) DTNB (Pierce) in 0.1 *M* phosphate buffer, pH 8.0. (*see* **Note 4**).
4. Dithiothreitol (DTT) solution: 4 m*M* in distilled water.
5. ELISA plate reader

2.2. Electrophoretic Method

2.2.1. Reaction Solutions

1. 1 *M* Tris-HCl, pH 8.0.
2. 0.1 *M* EDTA, pH 7.0.
3. 1 *M* Dithiothreitol (DTT, good quality, e.g., Calbiochem).

4. 8 *M* Urea (BDH, Aristar grade) (*see* **Note 3**).
5. Solution A: 0.25 *M* iodoacetamide, 0.25 *M* Tris-HCl, pH 8.0.
6. Solution B: 0.25 *M* iodoacetic acid, prepared in 0.25 *M* Tris-HCl, pH re-adjusted to pH 8.0 with 1 *M* KOH.

2.2.2. Solutions for Electrophoretic Analysis in the Low pH System (pH 4.0)

1. 30% acrylamide solution containing 30 g acrylamide (Extreme caution: NEU-ROTOXIN; work in fume hood) 0.8 g bisacrylamide made up to 100 mL with distilled water.
2. 10% acrylamide solution containing 10 g acrylamide (Extreme caution: NEU-ROTOXIN; work in fume hood) and 0.25 g bisacrylamide made up to 100 mL with distilled water.
3. Low pH buffer (8 times concentrated stock) for separating gel: 12.8 mL glacial acetic acid, 1 mL N,N,N',N',-tetramethylethylenediamine (TEMED); 1 *M* KOH (approx 35 mL) to pH 4.0, made up to 100 mL with distilled water.
4. Low pH buffer (8 times concentrated stock) for stacking gel: 4.3 mL glacial acetic acid, 0.46 mL TEMED, 1 *M* KOH to pH 5.0, made up to 100 mL with distilled water.
5. 4 mg riboflavin/100 mL water
6. Low pH buffer for electrode buffer: dissolve 14.2 g β-alanine in ~800 mL water then adjust with acetic acid to pH 4.0. made up to 1 L with distilled water.
7. Tracking dye solution (5 times concentrated stock); 20 mg methyl green; 5 mL water and 5 g glycerol.

2.2.3. Gel Solution Recipes for Low pH Electrophoresis (pH 4.0) (see **Note 5**)

1. 30 mL Separating gel (10% acrylamide, photo-polymerised with riboflavin). This is made up with: 10 mL 30% acrylamide stock, 4 mL pH 4 buffer stock, 3 mL riboflavin stock, 14.7 g urea, water to 30 mL.
2. 8 mL Stacking gel (2.5% acrylamide, photo-polymerised with riboflavin) is made up with: 2 mL 10% acrylamide stock, 1 mL pH 5.0 buffer stock, 1 mL riboflavin stock, 3.9 g urea, water (approx 1.2 mL) to 8 mL.

2.2.4. Electrophoresis Buffers for the High pH Separation (pH 8.9)

1. 30% acrylamide solution containing 30 g acrylamide, 0.8 g bisacrylamide (Extreme caution: NEUROTOXIN; work in fume hood) made up to 100 mL with distilled water.
2. 10% acrylamide solution containing 10 g acrylamide (Extreme caution: NEU-ROTOXIN; work in fume hood) and 0.25 g bisacrylamide made up to 100 mL with distilled water.
3. High pH buffer (4 times concentrated stock) for separating gel: 18.2 g Tris-HCl base (in approx 40 mL water), 0.23 mL TEMED, 1 *M* HCl to pH 8.9, made up to 100 mL with distilled water.

4. High pH buffer (4 times concentrated stock) for stacking gel: 5.7 g Tris-HCl base (in approx 40 mL water), 0.46 mL TEMED, 1 M H_3PO_4 to pH 6.9, made up to 100 mL with distilled water.
5. 4 mg riboflavin in 100 mL water
6. 10% ammonium persulphate solution (consisting of 0.1 g ammonium persulphate in 1 mL water).
7. High pH buffer for electrode buffer: 3 g Tris-HCl base, 14.4 g glycine, made up to 1 L with distilled water.
8. Tracking dye solution (5 times concentrated stock); 1 mL 0.1% bromophenol blue; 4 mL water and 5 g glycerol.

2.2.5. Gel Solution Recipes for High pH Electrophoresis (pH 8.9) (see *Note 5*)

1. 30 mL Separating gel (7.5% acrylamide, polymerised with ammonium persulphate). This is made up with 7.5 mL 30% acrylamide stock; 7.5 mL pH 8.9 buffer stock; 0.2 mL 10% ammonium persulphate (add immediately before casting); 14.7 g urea and water (approx 4 mL) to 30 mL Degas.
2. 8 mL Stacking gel (2.5% acrylamide, photo-polymerised with riboflavin) is made up with 2 mL 10% acrylamide stock; 1 mL pH 6.9 buffer stock; 1 mL riboflavin stock solution; 3.9 g urea and water (approx. 1.2 mL) to 8 mL. Degas.

2.3. Materials for HPLC Method

1. 1 M DTT (good quality, e.g., Calbiochem).
2. 100 mM Tris-HCl, pH 8.5.
3. 4-vinylpyridine.
4. 95% ethanol.
5. Isopropanol.
6. 1 M triethylamine-acetic acid, pH 10.0.
7. Tri-n-butyl-phosphine (1% in isopropanol).
8. HPLC system.
9. Vydac C_4, C_8, or C_{18} reverse phase HPLC columns (*see* **Note 6**).
10. Mass spectrometer.

3. Methods
3.1. Ellman's Reagent (2)
3.1.1. Analysis of Free Thiols

1. It may be necessary to expose thiols that may be buried in the interior of the protein. Dependent upon the protein under investigation therefore, the sample may be dissolved in Reaction buffer or Denaturing buffer. A solution of known concentration should be prepared with a reference mixture without protein.

   Sufficient protein should be used to ensure at least one thiol per protein molecule can be detected, in practice at least 2 nmol of protein (in 100 μL) is usually required.

2. Sample and reference cuvets containing 3 mL of the Reaction buffer of denaturing buffer should be prepared and should be read at 412 nm. The absorbance should be adjusted to zero (A_{buffer}).

3. Add 100 µL of Reaction buffer to the reference cuvet.

4. Add 100 µL of Ellman's solution to the sample cuvet. Record the absorbance (ADTNB).

5. Add 100 µL of protein solution to the reference cuvet.

6. Finally add 100 µL protein solution to the sample cuvette and, after thorough mixing, record the absorbance until there is no further increase. Record the final reading (A_{final}).

7. The concentration of thiols present may be calculated from the molar absorbance of the TNB anion using the following formula.

$\delta A_{412} = E_{412}TNB^{2-}[R\text{-}SH]_o$
Where $\delta A_{412} = A_{final} - (3.1/3.2)\,(A_{DTNB} - A_{buffer})$
$E_{412}TNB^{2-} = 1.415 \times 10^4\,cm^{-1}M^{-1}$ (*see* **Note 7**).
$E_{412}TNB^{2-}$ (guanHCl) $= 1.37 \times 10^4\,cm^{-1}M^{-1}$

3.1.2. Modification of Ellman Method for ELISA Plates

1. Require peptide at a concentration of approx 2 mg/mL.

2. Use Ellman's reagent (DTMB) at a concentration of 1 mg/mL.

3. To each well of a 96-well microtiter plate add: 200 µL conjugation buffer, 10 µL sample/blank, 20 µL Ellman's reagent.

4. Mix thoroughly by agitation on a plate shaker.

5. Measure optical density at 405 nm (which is a common wavelength available) on an ELISA plate reader.

3.1.3. Analysis of Disulfide Thiols

1. Sample should be carboxymethylated (*see* Chapter 27) or pyridylethylated (*see* Chapter 27) without prior reduction. This will derivatise any free thiols in the sample but will leave intact any disulfide bonds.

2. The sample (at least 2 nmol protein in 100 µL is usually required) should be dissolved in 6 *M* guanidinium HCl, 0.1 *M* Tris-HCl, pH 8.0, or denaturing buffer, under a nitrogen atmosphere.

3. Add freshly prepared DTT solution to give a final concentration of 10–100 m*M*. Carry out reduction for 1–2 h at room temperature.

4. Remove sample from excess DTT by dialysis for a few h each time with two changes of a few 100 mL of the Reaction buffer or Denaturing buffer (*see* **Subheading 2.1.**). Alternatively, gel-filtration into the same buffer may be carried out.

5. Carry out analysis of newly exposed disulfide thiols as described in **Subheading 3.1.**

3.2. Quantitation of Cysteine Residues by Electrophoresis

3.2.1. Reduction and Denaturation

1. To a 0.2 mg aliquot of lyophilized protein add 10 µL of each of the solutions containing 1 *M* Tris-HCl, pH 8.0, 0.1 *M* EDTA and 1 *M* DTT (*see* **Note 8**).

2. Add 1 mL of the 8 *M* urea solution (*see* **Note 3**).
3. Mix and incubate at 37°C for at least 30 min.

3.2.2. Reaction

1. Freshly prepare the following solutions using solutions A and B listed in the materials section.

   a. Mix 50 µL of solution A with 50 µL solution B (to give solution C).
   b. Mix 50 µL of solution A with 150 µL solution B (to give solution D).
   c. Mix 50 µL of solution A with 450 µL of solution B (to give solution E).

2. Label six Eppendorf tubes 1–6.
3. Add 10 µL of solutions A, B, C, D, and E to tubes 1–5. Reserve tube 6.
4. Add 40 µL of denatured, reduced protein solution prepared as given in **Subheading 3.1.** to each of tubes 1–5.
5. Gently mix each tube and leave at room temperature for 15 min. Thereafter store on ice.
6. After the 15 min incubation period, place 10 µL aliquots from each of tubes 1–5 into tube 6. Mix.

The samples are now ready for analysis (*see* **Notes 10** and **11**)

3.2.3. Electrophoretic Analysis

1. Aliquots of each sample (50 mL), labeled 1–6, mixed with 12 µL of appropriate tracking dye solution are loaded onto successive lanes of a polymerised high or low pH gel set up in a suitable slab gel electrophoresis apparatus.
2. Low pH buffer system. Electrophoresis is carried out towards the negative electrode, using a current of 5–20 mA for each gel, overnight at 8°C.
3. High pH buffer system *(10)*. Electrophoresis is carried out towards the positive electrode at 10–20 mA per gel (or 100–180V for 3–4 h).
4. Electrophoresis is stopped when the tracking dye reaches the bottom of the gel.
5. Proteins are visualized using conventional protein gel stains. The sensitivity of the method therefore depends on the stain. For example, the limits of detection *of each band* are approx 100 ng for Coomassie blue and 0.1 ng for silver staining. Remember that there may be 30 or more bands of protein, some which will be more faint than others.

3.3. Method for HPLC

1. Alkylate the protein (1–10 µg in 20–50 µL buffer) without reduction to prevent possible disulfide exchange by dissolving in 100 m*M* Tris-HCl, pH 8.5, and adding 1 µL of 4-vinylpyridine (*see* **Note 10**). Incubate for 1 h at room temperature and de-salt by HPLC or precipitate with 95% ice-cold ethanol followed by bench centrifugation. The pellet obtained after the latter treatment may be difficult to redissolve and may require addition of 10-fold concentrated acid (HCl, formic or acetic acid) before digestion at low pH. It may be sufficient to resuspend the

pellet with acid using a sonic bath if necessary, then commence the digestion. Vortex the suspension during the initial period until the solution clarifies.

2. Fragment the protein under conditions of low pH (*see* **Note 11**) and subject the peptides from half the digest to reverse-phase HPLC. Vydac C_4, C_8, or C_{18} columns give particularly good resolution depending on the size range of fragments produced (*see* **Note 6**). Typical separation conditions are: column equilibrated with 0.1% (v/v) aqueous TFA, elution with an acetonitrile/0.1% TFA gradient. A combination of different cleavages, both chemical and enzymatic may be required if peptide fragments of interest remain large after one digestion method.

3. To the other half of the digest (dried and re-suspended in 10 µL of isopropanol) add 5 µL of triethylamine-acetic acid (1 *M*) pH 10.0, 5 µL of tri-n-butyl-phosphine (1% in isopropanol) and 5 µL of 4-vinylpyridine. Incubate for 30 min at 37°C, and dry *in vacuo*, resuspending in 30 µL of isopropanol twice. This procedure cleaves the disulfide bonds and modifies the resultant -SH groups.

4. Run the reduced and alkylated sample on the same column, under identical conditions on reverse-phase HPLC. Cysteine-linked peptides are identified by the differences between elution of peaks from reduced and unreduced samples. Collection of the alkylated peptides (which can be identified by re-chromatography with detection at 254 nm) and a combination of sequence analysis and mass spectrometry (*see* **Subheading 3.4.**) will allow disulfide assignments to be made on low pmol and fmol quantities of peptides, respectively.

3.4. Method for the Identification of Disulfide Linkages by Mass Spectrometry

1. Peptides can be generated by any suitable proteolytic or chemical method that minimize disulfide exchange (i.e., acid pH; *see* Chapters 5 and 6). Partial acid hydrolysis, although nonspecific, has been successfully used in a number of instances. The peptides are then analyzed by a variety of mass spectrometry techniques *(7,8)*, the use of thiol and related compounds should be avoided for obvious reasons. Despite this, it is possible that disulfide bonds will be partially reduced during the analysis and peaks corresponding to the individual components of the disulfide linked peptides will be observed. Control samples with the aforementioned reagents are essential to avoid misleading results. The peptide mixture is incubated with reducing agents such as mercaptoethanol and DTT and re-analyzed as before. Control samples with the aforementioned reagents (employing the same matrix compounds) are essential to avoid misleading results due to additional matrix-derived peaks.

2. Peptides that were disulfide linked disappear from the spectrum and reappear at the appropriate positions for the individual components. For example, in the positive ion mode the mass (M) of disulfide-linked peptides (of individual masses A and B) will be detected as the pseudomolecular ion at $(M+H)^+$ and after reduction this will be replaced by two additional peaks for each disulfide bond in the polypeptide at masses $(A+H)^+$ and $(B+H)^+$. Remember that A+B equals M+2 since reduction of the disulfide bond will be accompanied by a consistent increase

in mass due to the conversion of cystine to two cysteine residues i.e., -S-S- -> -SH + HS- and peptides containing an intramolecular disulfide bond will appear at 2 a.m.u higher. Such peptides, if they are in the reduced state can normally be readily reoxidized to form intra-molecular disulfide bond by bubbling a stream of air through a solution of the peptide for a few minutes.

3. It may be possible to estimate the number of intramolecular disulfide bonds by mass spectrometry by measuring the mass of the intact polypeptide before and after reduction. Electrospray mass spectrometers will give accurate estimates of molecular weight above 100,000 Da but the increased mass of 2Da for each disulfide bond will in all probability be too small to obtain an accurate estimate for polypeptide of mass greater than ca 10,000 (accuracy obtainable is >0.01%) (*see* **Note 12**).

4. Notes

1. The method of Takahashi and Hirose *(5)* can be used to categorise the half-cystines in a native protein as:

   a. disulfide bonded,
   b. reactive sulfydryls, and
   c. nonreactive sulfydryls.

   In the first step, the protein sulfydryls are alkylated with iodoacetic acid in the presence and absence of 8M urea. In the second step the disulfide-bonded sulfydryls are fully reduced and reacted with iodoacetamide. The method described earlier is then used to give a ladder of half-cystines so that the number of introduced carboxymethyl groups can be quantified.

2. The diagonal method utilizes the change in electrophoretic mobility of peptides containing either a cysteine or cystine in a disulfide link *(9)*. The peptide is electrophoresed on paper or on a thin layer. The electrophoretogram is oxidized by performic acid vapor and rerun after turning the electrophoretogram through 90°. Peptides that are unaltered have the same mobility in both dimensions and therefore lie on a diagonal. After oxidation to cysteic acid, cysteine/cystine-containing peptides produce one or two spots that lie off the diagonal.

3. The use of guanidinium HCl is recommended rather than urea since the latter can readily degrade to form cyanates which will react with thiol (and amino) groups. It is possible to use urea if desirable for a particular protein but the highest grade of urea should always be used and solutions should be prepared immediately before use. Urea should be deionized immediately before use to remove cyanates, which react with amino and thiol groups. The method is to filter the urea solution through a mixed bed Dowex or Amberlite resin in a filter flask.

4. Unless newly purchased it is usually recommended to recrystallize DTNB from aqueous ethanol.

5. Electrophoresis in gels with higher or lower percentage acrylamide may be necessary, depending on the molecular weight of the particular protein. The recipes are given for large format gels (approx 13 cm × 12 cm and 1.5 mm spacers). For

"mini-gel" formats, 10 mL separating gel solutions and 4ml stacking gel are sufficient for ease of preparation and to leave some excess.

6. If the HPLC separation is combined with mass spectrometry, by on-line electrospray mass spectrometry (ES-MS) the level of TFA required to produce sharp peaks and good resolution of peptide (circa 0.1% v/v) results in almost or complete suppression of signal. This does not permit true on-line HPLC-MS as the concentration of TFA in the eluted peptide must first be drastically reduced. In this case it is recommended to use the new "low TFA," 218MS54, reverse phase HPLC columns from Vydac (300 Å pore size). They are available in C_4 and two forms of C_{18} chemistries and in 1 mm diameter columns. We have used as little as 0.005% TFA without major loss of resolution and have observed minimal signal loss. There may be a difference in selectivity compared to "classical" reverse phase columns, for example we have observed phosphopeptides eluting approx. 1% acetonitrile later than their unphosphorylated counterparts (the opposite to that conventionally seen). This is not a problem, but it is something of which one should be aware, and could be turned to advantage.

7. Standard protocols for use of Ellman's reagent often give $E_{412} TNB^{2-} = 1.36 \times 10^4 \ cm^{-1} \ M^{-1}$, a more recent examination of the chemistry of the reagent indicates that these are more suitable values.

8. If the protein is already in solution, the pH should be adjusted to 8.0 and the DTT and urea concentrations should be made up to 10 mM and 8 M respectively.

9. Additional ratios of iodoacetic acid to iodoacetamide may be necessary, depending on the molecular weight of the particular protein. Otherwise if more than about eight cysteines are present, a sufficiently intense band of each component may not be visible. Also, a greater ratio of iodoacetic acid is required if the acidic species are too faint (and vice versa).

10. The vinylpyridine or iodoacetic acid used must be colorless. A yellow colour in the latter indicates the presence of iodine, this will rapidly oxidize thiol groups, preventing alkylation and may also modify tyrosine residues. It is possible to recrystallize from hexane. Reductive alkylation may also be carried out using iodo-[^{14}C]-acetic acid or iodoacetamide (*see* **Chapter 26**). The radiolabeled material should be diluted to the desired specific activity before use with carrier iodoacetic acid or iodoacetamide to ensure an excess of this reagent over total thiol groups.

11. Fragmentation of proteins into peptides under low pH conditions to prevent disulfide exchange is important. Pepsin, Glu-C or cyanogen bromide are particularly useful (*see* Chapters 5 and 6). Typical conditions for pepsin are 25°C for 1–2 h at pH 2.0–3.0 (10 mM HCl or 5% acetic or formic acid) with an enzyme:substrate ratio of about 1:50. Endoproteinase Glu-C has a pH optimum at 4.0 as well as an optimum at pH 8.0. Digestion at the acid pH (typically 37°C overnight in ammonium acetate at pH 4.0 with an enzyme:substrate ratio of about 1:50) will also help minimize disulfide exchange. CNBr digestion in guanidinium HCl (6 M) / 0.1–0.2 M HCl may be more suitable acid medium due to the inherent redox potential of formic acid which is the most commonly used protein sol-

vent. When analyzing proteins, which contain multiple disulfide bonds it may be appropriate to carry out an initial chemical cleavage (CNBr is particularly useful) followed by a suitable proteolytic digestion. The initial acid chemical treatment will cause sufficient denaturation and unfolding as well as peptide bond cleavage to assist the complete digestion by the protease. If a protein has two adjacent cysteine residues this peptide bond will not be readily cleaved by specific endopeptidases. For example, this problem was overcome during mass spectrometric analysis of the disulfide bonds in insulin by using a combination of an acid proteinase (pepsin) and carboxypeptidase A as well as Edman degradation *(11)*.

12. Analysis can also be carried out by electrospray mass spectrometers (ES-MS) which will give an accurate molecular weight up to 80–100,000 Da (and in favorable cases up to 150,000 Da), the increased mass of 2Da for each disulfide bond will be too small to obtain an accurate estimate for polypeptide of mass greater than ca 10,000 (accuracy obtainable is >0.01%). There has been a recent marked increase in resolution obtained with both electrospray mass spectrometers and laser desorption time-of-flight mass spectrometers that could now permit a meaningful analysis. On the other hand, oxidation with performic acid will cause a mass increase of 48 Da for each cysteine and 49 Da for each half cystine residue. (Note that Met and Trp will also be oxidized, the former to a mass identical to that of Phe). Sequence information is readily obtained using triple quadrupole tandem mass spectrometry after collision-induced dissociation *(12)*. Ion trap mass spectrometry technology (called "LCQ") is now well-established which also permits sequence information to be readily obtained. Not only can MS-MS analysis be carried out but due to the high efficiency of each stage, further fragmentation of selected ions may be carried out to MS^n. The charge state of peptide ions is readily determined by a "zoom-scan" technique, which resolves the isotopic envelopes of multiply charged peptide ions. The instrument still allows accurate molecular weight determination to over 100000 Da at 0.01% mass accuracy. The recent development of Fourier transform ion cyclotron resonance mass spectrometry *(14)* in which the ions can be generated by a wide variety of techniques has very high resolution and sensitivity. Fourier transform mass spectrometers have ppm accuracy that is well within the necessary mass accuracy range even for large proteins.

References

1. Creighton, T. E. (1980) Counting integral numbers of amino acid residues per polypeptide chain. *Nature* **284**, 487–488.
2. Ellman, G. L. (1959) Tissue sulfhydryl groups. *Arch. Biochem. Biophys.* **82**, 70–77.
3. Feinstein, A. (1966) Use of charged thiol reagents in interpreting the electrophoretic patterns of immune globulin chains and fragments. *Nature* **210**, 135–137.
4. Anderson, W. L. and Wetlaufer, D. B. (1975) A new method for disulfide analysis of peptides. *Anal. Biochem.* **67**, 493–502.

5. Takahashi, N. and Hirose, M. (1990) Determination of sulfhydryl groups and disulfide bonds in a protein by polyacrylamide gel electrophoresis. *Anal. Biochem.* **188**, 359–365.

6. Stan-Lotter, H. and Bragg, P. (1985) Electrophoretic determination of sulfhydryl groups and its application to complex protein samples, in vitro protein synthesis mixtures and cross-linked proteins. *Biochem. Cell Biol.* **64**, 154–160.

7. Costello C. E. (1999) Bioanalytical applications of mass spectrometry. *Curr. Opin. Biotechnol.* **10**, 22–28.

8. Burlingame, A. L., Boyd, R. K., and Gaskell, S. J. (1998) Mass spectrometry. *Anal. Chem.* **70**, 647R–716R.

9. Davis, B. J. (1964) Disk electrophoresis II: Method and application to human serum proteins *Ann. NY Acad. Sci.* **21**, 404–427.

10. Brown, J. R. and Hartley, B. S. (1966) Location of disulfide bridges by diagonal paper electrophoresis. *Biochem. J.* **101**, 214–228.

11. Toren, P., Smith, D., Chance, R., and Hoffman, J. (1988). Determination of interchain crosslinkages in insulin B-chain dimers by fast atom bombardment mass spectrometry. *Anal. Biochem.* **169**, 287–299.

12. Sun, Y. and Smith, D. L. (1988) Identification of disulfide-containing peptides by performic acid oxidation and mass spectrometry. *Anal. Biochem.* **172**, 130–138.

13. Hunt, D. F., Yates, J. R., Shabanowitz, J., Winston, S., and Hauer, C. R. (1986) Protein sequencing by tandem mass spectrometry. *Proc. Natl. Acad. Sci. USA* **83**, 6233–6237.

14. Hendrickson, C. L and Emmett, M. R. (1999) Electrospray ionization Fourier transform ion cyclotron resonance mass spectrometry. *Annu. Rev. Phys. Chem.* **50**, 517–536.

33

Getting the Most from Your Protein Sequence

Richard R. Copley and Robert B. Russell

1. Introduction

The diversity and complexity of Bioinformatics tools currently available for protein sequence analysis can make it difficult to know where to begin when presented with a new sequence. In this chapter we assume that the reader has a protein sequence (full-length or partial) identified from mass spectrometry or translation of a putative gene and wishes to identify aspects of its structure and function via Bioinformatics. We go through a protocol outlining one approach that should give the most complete picture possible given the limits of available tools, and then provide a worked example to illustrate the procedures involved. The nature of this paper is such that we are unable to give complete details of all the methods discussed. We refer the reader to references and websites described in the text for more information.

2. Database Searching

A key part of any sequence analysis is the identification of homologous proteins (i.e., those sharing a common ancestor). Homologous sequences are likely to have similar (though not necessarily identical) functions. For this reason, database searches, which identify homologous sequences, are at the heart of all analyses of protein sequences. It is now possible to use a variety of search techniques to compare a protein sequence to virtually all major public sequence databases via the world wide web and thus easily place a protein within a wider context of homologs. However, there are a few things to be considered before and during these searches that we outline below.

The standard tool of database searching is the BLAST package (Basic Local Alignment Search Tool) *(1,2)*. This consists of programs for searching a protein against protein or nucleic acid databases (and also nucleic acids against

From: *Methods in Molecular Biology, vol. 211: Protein Sequencing Protocols, 2nd ed.*
Edited by: B. J. Smith © Humana Press Inc., Totowa, NJ

protein databases). Both are made easily accessible via the NCBI World Wide Web site (http://www.ncbi.nlm.nih.gov/BLAST/). Searching using this interface will provide a list of database hits. The advantage of using a resource such as the NCBI, is that the hits are linked to the primary literature describing the sequences. It is often an excellent way of getting a rapid handle on function.

It can often be helpful to use a number of tools for database searching. The SCANPS package at the EBI (*[3]* http://www.ebi.ac.uk/scanps/) provides an implementation of the Smith-Waterman algorithm, which under some circumstances can reveal a larger number of significant database hits than BLAST. The most sensitive tool easily available for web based searching is PSI-Blast (Position Specific Iterated Blast), again from the NCBI. This performs iterated database searches, constructing a position specific scoring matrix for the query sequence, based on the preceding round of database searching. This matrix is then used to score the following round of database searching, in an iterative fashion. The procedure continues until a set number of iterations have been performed, or no new hits have been found. A potential pitfall of this procedure is that the use of multi-domain proteins as queries can lead to the retrieval of sequences that may have quite different functions to the protein of interest. More thorough methods of database searching use Hidden Markov Models (HMM) (discussed later). However, effective use of these techniques can be difficult and time-consuming — under most circumstances, a PSI-Blast search will provide ample sensitivity for detecting remote homologs.

Database search methods do vary but, as they are quick to perform, it is not difficult to compare several methods. A unifying feature of modern search methods is that they provide statistical measures of the significance of the relationship between a query and database sequence. The statistics are based on the probability of finding a match similar to the one observed by chance in a given database. Although these statistics may not be directly comparable between different search algorithms, they are vastly superior to *ad hoc* rules of thumb based on criteria such as percentage identity. It is useful to report the statistical estimate of significance when reporting the results of database searches in the literature.

2.1. Choosing Which Database to Search

The choice of which database to search depends on what you want to know about your sequence. If you are interested in all homologs, it is best to search a so-called nonredundant (NR) database. Basically, this will include all sequences, which are not 100% identical to each other; even so, considerable redundancy remains in proteins with different lengths. The more sequences in a database, the less significant any given match will appear. Searching against smaller, better annotated databases (e.g., SWISS-PROT) can be useful to highlight hits where functional information is available. Similarly, finding if your sequence has a representative of known structure can be easier if only the sequences of known structures are searched. Finally, if one is interested in homologs in a particular organism, results are often easier to interpret if a database specific to that species is used.

2.2. Pre-processing of Protein Sequence Data and Intrinsic Features of Sequences

A number of protein features can produce misleading results during database searches. Searches operate on the assumption that the protein of interest has an average distribution of amino acids; if this is not the case, standard scoring systems can lead to results that may seem significant, even when two proteins are not related in a biologically meaningful sense. Long low-complexity regions, with only one or two different amino acid types, commonly create this effect. Other features, such as coiled-coils, trans-membrane (TM) domains, or cell-sorting signals can show sequence similarity and thus link proteins that are otherwise quite different. To counter this, the query sequences should be "filtered," usually before searching, but certainly before interpreting any detected sequence similarity as an indication of homology. A variety of programs to filter sequences exist. A protein sequence is fed into them, and the result is a new sequence with amino acids that form low complexity regions being replaced by an "X," which is neutral under standard scoring models.

2.2.1. Low-Complexity Regions

Low-complexity regions, for example extended runs of one or two types of amino acids, are often present in full-length proteins. They have a variety of roles, such as linkers, or possibly the formation of unusual regular structures (e.g., poly Gln [4]). Searching a database with low complexity regions present in the query can lead to very misleading results. To counter this, they can be removed using SEG (5). The XNU program can be used to filter out short overlapping repeats (e.g., a repetitive element of a few amino acids length). For large-scale automated analyses, it can be useful to use SEG to remove all regions of low complexity within the database, as well as the query sequence.

2.2.2. Coiled-Coil Regions

Coiled-coils are elongated helical structures that contain a highly specific heptad repeat (with positions in the heptad usually labeled 'a' through 'g') that conforms to a specific sequence pattern. Positions 'a' and 'd' are hydrophobic while the remaining positions generally show preferences for polar residues. The presence of coiled-coil regions within a sequence is usually associated with function. However, like low complexity regions, coiled-coils from proteins that are otherwise nonhomologous can be significantly similar during sequence database searches. Coiled-coils can be predicted quite accurately using COILS *(6)*, PAIRCOIL *(7)* or Multi-Coil *(8)* (*see* **Table 1** for details) and database searches often must be performed with these regions removed to avoid misleading results.

If a protein contains mostly coiled-coil regions, then clearly one should not remove coils before searching. However, matches found to any sequence in a database that are not along the whole length of the sequences should be treated cautiously. The identification of proteins that share a function with a coiled coil protein can be extremely difficult, and may be best approached with more sophisticated techniques (e.g., Pfam – *see* below).

2.2.3. Transmembrane Regions

Transmembrane (TM) segments function to localize a protein to the plasma membrane. They can occur either singly, or in bundles that criss-cross the membrane multiple times. Several methods have been developed to predict these regions (e.g., **refs.** *9,10,11*) (*see* **Table 1**). They do not always agree with each other, and can miss TM segments, so we recommend running many methods and building a consensus prediction. Like coiled-coil domains, TM segments may lead to otherwise non-homologous proteins detecting one another during a database search. However, this is usually only a major issue when very sensitive database search techniques (e.g., PSI-Blast) are being used. In general, TM regions should not be pre-processed as this would often mean that whole proteins (e.g., G-protein coupled receptors or ion-channels) would be removed.

2.2.4. Cellular Localization Signals

Transport of proteins from the cytosol to other parts of the cell after synthesis is generally accomplished by systems that recognise short localization signals. Export of proteins outside the cell is usually performed by way of a "signal peptide," which occurs at the N-terminus of the protein. This peptide is either cleaved off when a protein is freed to the extracellular environment, or left attached to anchor the protein to the plasma membrane. Because these peptides show good similarities in length, and distribution of amino acids around key regions, methods for predicting them have existed for many years, and are quite accurate *(12,13)*. Similarly, it is also possible to predict the presence of glycosylphosphatidylinositol (GPI) lipid-anchoring sites, which localize proteins to the extracellular membrane *(14)*.

Other transport signals include mitochondrial and chloroplast transit peptides and those that localise proteins to the nucleus *(15)*. Methods now exist that can predict mitochondrial or chloroplast signals accurately (mitoP, chloroP) *(16,17)*, but no method can currently predict nuclear localisation accurately.

Table 1
WWW Pages for Applications Referred to in the Text [a]

Sequence searches:

Blast	http://www.ncbi.nlm.nih.gov/BLAST
SCANPS	http://www.ebi.ac.uk/scanps
PSI-Blast	http://www.ncbi.nlm.nih.gov/BLAST
HMMsearch	http://hmmer.wustl.edu/

Pre-processing of
 sequence data:

SEG	ftp://ftp.ncbi.nlm.nih.gov/pub/seg/
coils	http://www.ch.embnet.org/software/COILS_form.html
paircoil	http://nightingale.lcs.mit.edu/cgi-bin/score

TM domains:

TMpred	http://www.ch.embnet.org/software/TMPRED_form.html
TMHMM	http://www.cbs.dtu.dk/services/TMHMM/
TMAP	http://www.mbb.ki.se/tmap/index.html
DAS	http://www.sbc.su.se/~miklos/DAS/
PredictProtein	http://dodo.cpmc.columbia.edu/predictprotein/

Cellular localization
 signals:

SignalP	http://www.cbs.dtu.dk/services/SignalP/
ChloroP	http://www.cbs.dtu.dk/services/ChloroP/
MitoP	http://www.cbs.dtu.dk/services/TargetP/
PSORT	http://psort.nibb.ac.jp/
GPI anchors	http://mendel.imp.univie.ac.at/gpi/

Analysis of proteomics
 data:

Protein Prospector	http://prospector.ucsf.edu
ExPASy Proteomics	http://www.expasy.ch/tools

Domain analysis:

PROSITE	http://www.expasy.ch/prosite/
SMART	http://smart.embl-heidelberg.de/
Pfam	http://www.sanger.ac.uk/Pfam/
COGS	http://www.ncbi.nlm.nih.gov/COG/
PRINTS	http://www.bioinf.man.ac.uk/dbbrowser/PRINTS/
BLOCKS	http://www.blocks.fhcrc.org/
CDD	http://www.ncbi.nlm.nih.gov/Structure/cdd/cdd.shtml
InterPro	http://www.ebi.ac.uk/interpro/

(continued)

Table 1 *(continued)*
WWW Pages for Applications Referred to in the Text

Sequence alignment:	
CLUSTAL	http://www-igbmc.u-strasbg.fr/BioInfo/ClustalX/Top.html
HMMer	http://hmmer.wustl.edu/
SAM	http://www.cse.ucsc.edu/research/compbio/sam.html
Jalview	http://circinus.ebi.ac.uk:6543/jalview/
Seaview	http://biom3.univ-lyon1.fr/software/seaview.html
Comparative modeling:	
SWISSMODEL	http://www.expasy.ch/swissmod/SWISS-MODEL.html
WHATIF	http://www.cmbi.kun.nl/whatif/
MODELLER	http://guitar.rockefeller.edu/modeller/modeller.html
Secondary Structure Prediction:	
PSI-pred	http://insulin.brunel.ac.uk/psipred/
Jpred	http://jura.ebi.ac.uk:8888/
PredictProtein	http://cubic.bioc.columbia.edu/predictprotein/
Fold recognition:	
3D-pssm	http://www.bmm.icnet.uk/~3dpssm/
Threader	http://insulin.brunel.ac.uk/threader/threader.html
General tools:	
ExPASy	http://www.expasy.ch/
Entrez	http://www.ncbi.nlm.nih.gov/Entrez/
SWISS-PROT	http://www.expasy.ch/sprot/
Guide to structure:	
Prediction	http://www.bmm.icnet.uk/people/rob/CCP11BBS/

[a]Useful web sites for biological sequence analysis come and go. It is often possible to use general search tools, such as Google (http://www.google.com/) to locate resources of interest.

3. Peptide Cleavage Sites and Other Features Relevant to Proteomics

Biochemical methods such as two-dimensional (2D) gels, or mass spectrometry can provide information about proteins distinct from sequence, such as amino acid composition, pIs, masses, or peptide sequences. There are many tools designed to identify proteins given such data. A fairly extensive set can be found at the ExPASy site (http://www.expasy.ch/tools) or the Protein Prospector suite (http://prospector.ucsf.edu). It is possible to identify candidate proteins given a variety of data. It is also possible to predict sites of post-translational modifications that will affect details such as peptide masses.

4. Domain Analysis

It has long been recognized that many proteins are "modular," meaning that they contain multiple compact, units called domains. Domains are usually associated with discrete aspects of molecular function. For example, many signalling molecules contain a catalytic domain (e.g., a kinase or a phosphatase) that is connected to other domains that do things like recognize phosphorylation sites, or other protein surfaces. Interactions with other proteins serve to regulate the function of the catalytic domain.

As domains have distinct evolutionary histories searching a database with a multi-domain protein can be confusing, since "transitivity" can suggest that a protein performs a molecular function that is impossible given its domain structure. For example, one protein that contains a kinase catalytic domain and a *src* homology 2 (SH2; phosphotyrosine binding) domain can produce siginificant database hits to any protein containing an SH2 domain. To infer that all proteins containing an SH2 domain are kinases is erroneous, since many do not contain the catalytic domain. More generally: if a protein contains domains A & B, and is found to be similar to a protein containing domains B & C, one cannot infer that the first protein has the function conferred in domain C. For this reason, it is helpful to perform database searches both with a full sequence and with the protein split into domains.

The interpretation of database searches for multi-domain proteins can be difficult. In general, it is simpler to use a tool specifically designed to identify domains. Rather than finding which proteins a particular query sequence is homologous to, such tools seek to identify which domains it contains. This is done using a library of known domains; the query protein is searched against all members of the library, and a summary is presented.

Different domain databases represent domains in different ways. The oldest databases (for instance PROSITE) *(18)* simply used a characteristic sequence signature that was found in all members of a particular family. If a sequence contained this signature, it was classified into that domain family. Other databases such as PRINTS and BLOCKS

More modern databases tend to use an abstract representation of the domain, known as a Hidden Markov Model (HMM). The HMM is a statistical representation of the preferences for particular residue types to occur at specific points in the sequence alignment. The model also includes probabilities for insertion or deletion of residues in the domain. An alignment of representative members of a domain family is constructed, using normal sequence alignment software (although often optimized by hand editing). This is then converted into an

HMM by specialized software. (Such software is freely available from http://hmmer.wustl.edu/). Further programs can be used to search the HMM against a sequence database, or search a single sequence against a library of HMMs.

Two of the major domain databases adopting such an approach are Pfam (*[19]*; http://www.sanger.ac.uk/Pfam/) and SMART (*[20]*; http://smart.embl-heidelberg.de). Pfam offers unparalleled coverage of protein sequences – in the release current at 2001, (Pfam 5.5) 2/3 of SWISS-PROT & SP-TrEMBL entries match at least one Pfam entry. With respect to domain analysis, one minor caveat applies: in some cases, entries correspond to protein families (rather than domains *per se*), and thus may not represent distinct evolutionary units. The SMART database of domains focuses on some of the most mobile protein domains known, namely those found in intra- and extra-cellular signaling domains, as well as those found in nuclear proteins such as transcription factors. It currently contains around 500 domain models (the majority of which occur in different protein contexts, and so satisfy the more stringent definition of a protein domain). In some (though by no means all) cases, SMART may show greater sensitivity than Pfam. Both SMART and Pfam provide a facility to align your query sequence to any domains found within it. This can be extremely useful for the identification of functionally important residues.

The differences and strengths of the many domain databases available can be confusing to the uninitiated. The InterPro site (http://www.ebi.ac.uk/interpro/) is a 'meta-site' that enables multiple domain databases (including Pfam, PROSITE, & SMART) to be searched from a single web page. InterPro has been used in the *Drosophila* and Human genome annotation efforts. Apart from convenience, InterPro provides useful cross-referencing between member databases, and as it pools and extends annotation from contributing databases provides some of the best available reference material on domain function. Another recently developed 'meta-site' is the Conserved Domain Database from the NCBI (http://www.ncbi.nlm.nih.gov/Structure/cdd/cdd.shtml). This provides the ability to search SMART & Pfam with a modified 'reverse' version of the BLAST programs, and so is both fast and immediately familiar.

5. Multiple Sequence Alignment

A database search will usually reveal proteins that are homologous to a query sequence, and it is very useful to have a multiple sequence alignment of such a set. More sophisticated database search techniques require multiple sequence alignments. Moreover, they are essential for highlighting evolutionary conserved residues which are likely to be important for understanding function, or for attempts at reconstructing the phylogeny of a protein family. If one is dealing with a single domain, then a useful first step is to make use of databases providing pre-aligned sequences, such as SMART or Pfam. Failing this, an alignment must be constructed using one of a variety of programs. Before beginning it is important to remember that, despite the conceptual simplicity of the problem, methods may not align sequences accurately, and alignments should be inspected and edited by hand if necessary.

CLUSTAL *(21)* is perhaps the most widely used program for building multiple sequence alignments. It first aligns all sequences to each other, to see which pairs are most similar and then uses dynamic programming and a guide tree to align sequences hierarchically (most similar first) in order to produce the multiple sequence alignment. This strategy usually works reasonably well, although when sequence identities are very low, or sequences are of very different lengths, problems may arise.

In situations where one has an initial alignment and a set of other sequences that one wishes to add to it, Hidden Markov Models (HMMs, *see* above; e.g., *[22]*) can be very useful. Programs can be used to align new sequences to the set used to build the HMM. This will almost always give a more accurate alignment of a protein to a family than any pairwise alignment. The HMMer *(23)* and SAM *(24)* packages are the most widely used HMM programs.

As automatic alignment methods are prone to errors, it is often necessary to check and edit alignments. For instance, if the structure of one of the sequences in the alignment is known, it is useful to check that gap placement within the alignment is consistent with the known secondary structure. Thus, gaps are less likely to be found in α-helices or β-strands, and more likely to be found in the loops connecting them. Another useful check is that alignment positions corresponding to the buried hydrophobic core of the structure show conserved hydrophobic nature over all the sequences in the alignment. There are several programs that can help doing this, including Jalview (http://circinus. ebi.ac.uk:6543/jalview), Mview *(25)* and Seaview *(26)* (http://pbil.univ-lyon1. fr/software/seaview.html).

6. The Relationship Between Homology and Function

Protein domains are often members of large homologous families. Finding that a protein belongs to such a large family will generally give insights into the possible function, but there are several aspects of protein evolution and function that must be considered. Two fundamental processes can give rise to large protein families: "speciation" or intra-genomic "duplication." Proteins related by speciation only are referred to as "orthologs." Orthology suggests that the proteins will have the same function within two distinct species. Proteins related by duplications are referred to as paralogues. Successive rounds of speciation and intra-genomic duplication lead to situations where it is difficult to apply these terms in a simple manner.

Paralogous proteins can arise from whole genome duplication, or by more restricted duplication of part of a genome. To be maintained in a genome over time, paralogous proteins are likely to evolve different functions (or have a dominant negative phenotype, and so resist decay by point mutation). (27) Differences in function can range from subtle differences in substrate (e.g., malate versus lactate dehydrogenases), to only weak similarities in molecular function (e.g., hydrolases) to complete differences in cellular location and function (e.g., an intracellular enzyme homologous to a secreted cytokine). At the other extreme, the molecular function may be identical, but the cellular function may be altered, as in the case of enzymes with differing tissue specificities.

Generally, the degree of similarity in molecular function correlates with sequence identity. For example, mouse and human proteins with sequence identities in excess of 85% are likely to be orthologues, provided there are no other proteins with higher sequence identity in either organism (to be sure of this, the full proteomes of both species need to be available). Orthology between more distantly related species (e.g., human and *Drosophila*) is harder to assess, since the evolutionary distance between organisms can make it virtually impossible to distinguish orthologues form paralogues using simple measures of sequence similarity. In such cases, (and particularly when complete proteome information is available), operational definitions of orthology are often used. This is typically that the two proteins are each others best match in their respective proteomes. However, to confidently assign orthology, there is no substitute for constructing a phylogenetic tree of the protein family in question, to identify which sequences are related by speciation events.

Assignment of orthology and paralogy is perhaps the best way of determining likely equivalences of function. Unfortunately, complete genomes are unavailable for most organisms. Some rough rules of thumb can be used: function is often conserved down to 40% protein sequence identity, with the broad functional class being conserved to 25% identity (28). Obviously these numbers represent generalizations, and may not be true in any specific case.

7. Protein Three-Dimensional Structure

Database searches may reveal that a protein sequence has a homolog of known three-dimensional structure. In such instances, the best possible prediction of three-dimensional structure will be obtained by modelling the structure based on such a homology. When sequence similarity is high (e.g., $\geq 50\%$), then models providing a good approximation to the real three dimensional structure can generally be built, with deviations usually restricted to mutated residues, and short loops. However, when sequence identity is lower, modeling becomes more problematic.

Obtaining an accurate alignment of protein sequences is a key issue for homology modeling. Errors in the alignment will produce erroneous structures, since no modeling technique is currently able to correct such drastic deviations. When constructing an alignment for homology modeling, one should keep several things in mind:

1. That parts of obviously homologous proteins may not resemble each other at all. It is common for homologous proteins to have different embellishments, particularly at the N- or C- termini, where a helix in one protein may well be replaced by a strand in another. It is difficult to detect such things, but secondary structure prediction can provide a useful aid.
2. That gaps are least likely to occur within *core* secondary structure elements. Automated alignment methods may put insertions or deletions in a region that one knows to be a core α-helix or β-strand. It is usually best to correct such anomalies prior to building any model.
3. How residues are conserved across all homologues. In other words, do not just consider a pairwise alignment of a query protein to one of known three-dimensional structure. The more conserved a position, the less likely it should be to be missing/mutated in a member of the family. Particular care should be taken to maintain conserved hydrophobic residues.

Given an alignment of one protein to another of known three-dimensional structure, it is easy to construct an initial model via the WWW (e.g., SWISS-MODEL) *(29)* or via several free or commercial software packages (*see* **Table 1**).

7.1. Secondary Structure Prediction

In the absence of a homologue of known three-dimensional structure, structural insights can still be provided by methods of structure prediction, and those that are devoted to secondary structure prediction are the most accurate.

Methods of protein secondary structure prediction are generally derived from an analysis of a large number of proteins of known three-dimensional structure. They are thus somewhat biased towards those proteins where structures can be determined by nuclear magnetic resonance (NMR) or X-ray crystallography, namely soluble proteins that typically comprise α helices and β sheets. If these methods are run on proteins that are insoluble or atypical of those of known three-dimensional structure, then the predictions can be misleading. It is therefore important to identify trans-membrane segments, coiled coils and regions of low-complexity and treat them separately. While secondary structure prediction methods such as PHD *(30)* will sometimes predict coiled-coil or TM segments correctly as helices, it is preferable to use methods that are explicitly designed to predict these regions, since they will be more accurate.

There are too many methods for secondary structure prediction to permit any review of the subject in this short paper. For many proteins, it is now possible to predict secondary structures with near perfect accuracy (meaning that the agreement between the prediction and the 'true' secondary structure will be as good as the agreement between different methods of defining secondary structure from known structures). The improvement in accuracy has mostly been due to the availability of many sequences of proteins homologous to any particular protein of interest, but also due to sophisticated techniques such as neural networks, and to methods that derive a consensus from many methods. PHD *(30)*, PSI-pred *(31)* and Jpred2 *(32)* are probably the most widely used methods; all are available on the Internet (*see* **Table 1**).

It is also possible to augment predictions or remove ambiguity by looking at patterns of residue conservation that are indicative of particularly secondary structures. It is essentially these patterns that account for why homologous protein sequence information can improve prediction accuracy. Several patterns are shown in **Fig. 1**. Probably the best known example is that found within an *amphipathic* α-helix, where one face of the helix packs against the hydrophobic core and the other is exposed to solvent. Owing to the periodicity of the α-helix, the hydrophobic face of such helices has a characteristic pattern of hydrophobic residue conservation of i, i+3, i+4, i+7. Other patterns are also well known, including the i, i+2, i+4, etc. pattern characteristic of amphipathic β-strands.

In practice, it is best to consider a selection of predicted secondary structures together with an alignment of a protein or domain with its homologues, and arrive at a consensus prediction.

7.2. Fold Recognition

Accurate methods which predict protein structures *ab initio* with any degree of accuracy are still not a realistic possibility. In instances where sequence searches fail to identify a homologue of known three-dimensional structure, it is sometimes possible to detect a similarity by way of "fold recognition" or "threading." The unifying theme is that a protein sequence is compared to a database of protein structures to assess how well the sequence fits on to the fold. Hopefully, the best score (or lowest energy or smallest P-value, etc.) is given to a protein that has a similar structure to the query sequence.

There are now dozens of different fold recognition methods in existence. Many of them (e.g., THREADER *[33]*; TOPITS *[34]*; 3D-pssm *[35]*, etc.) are available on line. As with other methods of structure prediction, it is best to run several methods and try and build a consensus. Often fold recognition methods are most useful when they confirm or dismiss a marginal similarity identified by sequence comparison (i.e., a similarity with a poor score). An obvious, but

Amphipathic α–helix

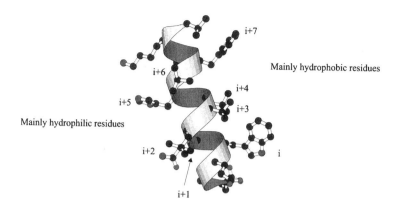

Mainly hydrophobic residues

Mainly hydrophilic residues

Amphipathic β–strand

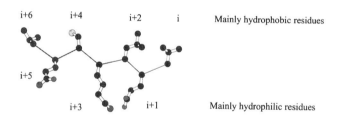

Mainly hydrophobic residues

Mainly hydrophilic residues

Buried β–strand

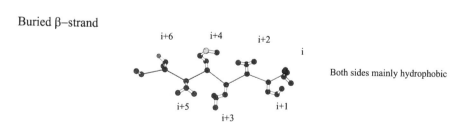

Both sides mainly hydrophobic

Fig. 1. Molscript *(39)* diagrams showing the periodicity of residue types for (top) an amphipathic α-helix, (middle) an amphipathic β-strand, and (bottom) a buried β-strand. When such periodicity is found conserved in a multiple sequence alignment, it is a strong indication that that type of secondary structure is present.

frequently forgotten caveat is that a protein may well adopt a fold that has not been seen before, so treat all results with caution. If no prediction stands out, then it is often best to abandon such methods.

8. Examples

8.1. Materials

Any modern PC or Macintosh with web browsing software and a connection to the internet can be used to perform small scale bioinformatics analyses, of the type described here, over the WWW. A vast number of programs and resources are available on the WWW, *see* **Table 1** for a selection of some of the most useful. Sequences are usually pasted directly into the appropriate box in the web page. It is then usually necessary to click on a "submit," or similarly named button.

8.2. Example 1: GABA$_B$ Receptor 1A

The analysis presented here uses a selection of the programs discussed above, in the order in which we feel they are most usefully applied. The specific programs are: Signalp, Coils, Seg, Blast, Pfam, and SMART. Of course, other programs will also provide useful information. Readers are encouraged to submit other sequences to other web servers, to find the tools most useful for their needs (*see* **Notes 1** and **2**).

Human GABA$_B$ receptor 1 A (GABA$_B$R1A) comprises 823 amino acids (NCBI accession number: 3776094). The sequence can be copied and pasted into the submission web pages of the relevant program. Most programs do not mind whether lower or upper case characters are used. In general, unless the web pages specify a particular format, just the sequence of amino acids is needed.

8.2.1. Before the Database Searches: SignalP, Coils, Seg

SignalP strongly predicts a secreted protein signal peptide ending at position 29 in the sequence, which is not surprising as such membrane proteins often have these signals. Considering the rest of the sequence, COILS finds a stretch of coiled coil between positions 761 and 807 (i.e., at the C-terminus), which is known to be involved in interactions with the GABA$_B$R2 subunit *(36)*. The sequence also contains several regions of low-complexity as identified with SEG, and contains several transmembrane segments as predicted by a number of methods. All of these regions are shown graphically in **Fig. 2**.

8.2.2. Database Searches: Blast

A simple BLAST search comparing the protein sequence to the NCBI nonredundant database (15 December 2000) finds 102 hits with a significance E-value smaller than 10^{-4}. These hits are from 17 species, and all of the longer hits (i.e., those that align to most of the query sequence) are from eukaryotes, including the major model organisms (Human, Mouse, *C. elegans*, *A. thaliana*, *D. melanogaster*, etc.). For bacterial proteins, bor-

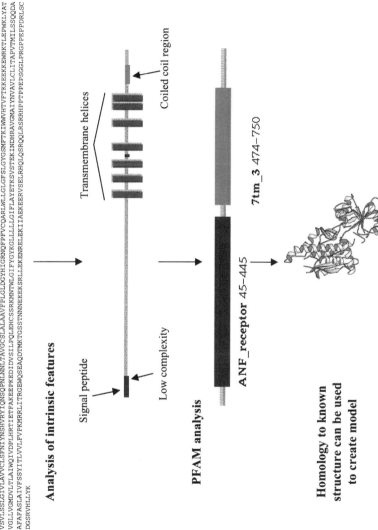

Fig. 2. Schematic detailing how the steps described in the text characterize the sequence of GABA$_B$ Receptor 1A.

derline hits to the N-terminal ligand-binding domain were found. The highest scoring hits from rat and mouse are clearly orthologues, and high scoring hits from *C. elegans* and *D. melanogaster* are likely to share considerable functional similarity with the human protein. The search also shows the considerable redundancy, since many nearly identical sequences are found (i.e., from human, mouse, etc.).

8.2.3. Domain Analysis: Pfam and SMART

A Pfam analysis of the sequence of $GABA_B$ reveals two distinct regions: an ANF-receptor family ligand-binding region, from residues 45-445 and a metabotropic glutamate family 7 transmembrane region from residues 474 to 750. Clicking on either of these regions leads to linked information describing the function of the region in question, and further links to other family members, including a multiple sequence alignment for the family. SMART also finds transmembrane segments, the coiled-coil region and the signal peptide.

8.2.4. A Region in the N-terminus is Homologous to a Protein of Known Three-Dimensional Structure

Receptors bind to their ligands in many different ways. Like many transmembrane receptors, $GABA_BR1$ contains a long N-terminal region that is known to interact with the ligand. $GABA_B$ is a small molecule, which binds to a cleft within a region that is homologous to a large family of small molecule binding proteins. BLAST detects weak similarities between the N-terminal region of $GABA_B$, and several proteins for which a structure is known: the hormone binding domain of rat atrial natriuretic peptide receptor; the ligand binding region of metabotropic glutamate receptor and Leucine binding protein (LBP). These all belong to a structural family of periplasmic binding proteins. The homology with Leucine binding protein (whose structure was available several years before the other two proteins) was used to argue that the receptor binds to GABA in a fashion that is similar to LBPs interaction with Leucine *(37)*. Although it is comparatively straightforward to use simple sequence analysis tools to discover this information, the homology between the receptor domain and LBPs is also recorded within the Pfam documentation of the ANF-receptor family.

8.2.5. Implications of the Analysis

$GABA_B$ (γ-aminobutyric acid type B) is a neurotransmitter that plays important roles in the brain. It elicits an effect by binding to a membrane bound receptor molecule, which in turn interacts with molecules inside the cell to cascade a signal in response. The sequence analysis of $GABA_BR1A$ adds further insight to its role as a membrane bound receptor, and, via the predicted

Fig. 3. The domain structure of the human proto-oncogene *vav*. Care must be taken when performing database searches with multi-domain proteins. Not all hits will share all domains; those which do not are likely to have different functions.

three dimensional structure, provides further lines of enquiry, suggesting, for instance, which residues are critical for function *(37)*.

8.3. Example 2: Proto-oncogene vav

Another example is provided by the proto-oncogene *vav*, which is probably an exchange factor for a small GTPase like protein. The $GABA_B$ protein can be divided into two separate regions, and so is not too difficult to analyze using simple sequence database search methods. However some proteins are made from many more distinct evolutionary units. Some of the most mobile biological elements are found in proteins involved in eukaryotic signalling processes. The domain structure of such proteins can easily be revealed by use of web servers such as SMART or Pfam (*see* **Table 1**), by simply pasting the sequence of interest into the web form.

Searches of the SMART or Pfam databases initiated with the proto-oncogene *vav* (NCBI protein accession number 586213) show that it contains 7 distinct domains. Using SMART, it is possible to perform follow up analyses ('architecture analysis') to identify other proteins that contain similar identical and non-indentical arrangements of protein domains in different organisms (e.g., *see* **ref.** *[38]*). This feature can be useful when trying to identify functional units within proteins, as simple BLAST searches show hits to many proteins only having two or three domains in common with *vav* (*see* **Fig. 3**).

9. Notes

1. It should be borne in mind that the World Wide Web is insecure and that third parties can potentially see anything submitted to a web server. If security is a major concern, it is necessary to obtain copies of the desired software and run searches on local machines, without recourse to the WWW. Having said that, we are not aware of any cases where sequences have fallen into the wrong hands.

2. Programs which search for intrinsic features of a protein, such as coiled coils, transmembrane helices, low complexity sequences or signal peptides, are in general very fast, taking, at most, a few seconds each to run. Sequence database searches are much slower. The time taken will obviously depend on the database searched. Comparing a protein a full Non-Redundant protein database can take a

few minutes. Obviously, in both cases, the exact amount of time taken when running these programs over the web will vary with other factors, such as the number of other people trying to use the programs at the same time.

References

1. Altschul, S. F., Gish, W., Miller, W., Myers, E. W., and Lipman, D. J. (1990) Basic local alignment search tool. *J. Mol. Biol.* **215**, 403–410.
2. Altschul, S. F., Madden, T. L., Schaffer, A. A., Zhang, J., Zhang, Z., Miller, W., and Lipman, D. J. (1997) Gapped BLAST and PSI-BLAST: a new generation of protein database search programs. *Nucleic Acids Res.* **25**, 3389–3402.
3. Barton G. J. (1993) An efficient algorithm to locate all locally optimal alignments between two sequences allowing for gaps. *Comput. Appl. Biosci.* **9**, 729–734.
4. Perutz M. F. (1999) Glutamine repeats and neurodegenerative diseases: molecular aspects. *Trends Biochem. Sci.* **24**, 58–63.
5. Wootton J. C. (1994) Non-globular domains in protein sequences: automated segmentation using complexity measures. *Comput. Chem.* **18**, 269–285.
6. Lupas, A., Van Dyke, M., and Stock, J. (1991) Predicting coiled coils from protein sequences. *Science* **252**, 1162–1164.
7. Berger, B., Wilson, D. B., Wolf, E., Tonchev, T., Milla, M., and Kim, P. S. (1995) Predicting coiled coils by use of pairwise residue correlations. *Proc. Natl. Acad. Sci. USA* **92**, 8259–8263.
8. Wolf, E., Kim, P. S., and Berger, B. (1997) MultiCoil: a program for predicting two- and three-stranded coiled coils. **Protein Sci. 6**, 1179–1189.
9. Hoffmann, K. and Stoffel, W. (1993) TMbase - A database of membrane spanning proteins segments *Biol. Chem. Hoppe-Seyler* **347**, 166
10. Sonnhammer, E. L. L., von Heijne, G., and Krogh A., (1998) A hidden Markov model for predicting transmembrane helices in protein sequences, in *Proceedings of Sixth International Conference on Intelligent Systems for Molecular Biology* (Glasgow, J., Littlejohn, T., Major, F., Lathrop, R., Sankoff, D., and Sensen, C., eds.) Menlo Park, CA, AAAI Press, pp. 175 –182.
11. Krogh, A., Larsson, B., von Heijne, G., and Sonnhammer, E. L. L. (2001) Predicting transmembrane protein topology with a hidden markov model: application to complete genomes. *J. Mol. Biol.* **305**, 560–580
12. Nielsen, H., Engelbrecht, J., Brunak, S., and von Heijne, G. (1997) Identification of prokaryotic and eukaryotic signal peptides and prediction of their cleavage sites. *Protein Eng.* **10**, 1–6.
13. Ladunga, I. (1999) PHYSEAN: PHYsical SEquence ANalysis for the identification of protein domains on the basis of physical and chemical properties of amino acids. *Bioinformatics* **15**, 1028–1038.
14. Nakai, K. (2000) Protein sorting signals and prediction of subcellular localization. *Adv. Prot. Chem.* **54**, 277–344.
15. Eisenhaber, B., Bork, P., and Eisenhaber, F. (1999) Prediction of potential GPI-modification sites in proprotein sequences. *J. Mol. Biol.* **292**, 741–758.

16. Emanuelsson, O., Nielsen, H., and von Heijne, G. (1999) ChloroP, a neural network-based method for predicting chloroplast transit peptides and their cleavage sites. *Protein Sci.* **8**, 978–984.

17. Emanuelsson, O., Nielsen, H., Brunak, S., and von Heijne, G. (2000) Predicting subcellular localization of proteins based on their N-terminal amino acid sequence. *J. Mol. Biol.* **300**, 1005–1016.

18. Hofmann, K., Bucher, P., Falquet, L., and Bairoch, A. (1999) The PROSITE database, its status in 1999. *Nucleic Acids Res.* **27**, 215–221.

19. Bateman, A., Birney, E., Durbin, R., Eddy, S. R., Finn, R. D., and Sonnhammer E. L. L. (1999) Pfam 3.1: 1313 multiple alignments match the majority of proteins. *Nucleic Acids Res.* **27**, 260–262.

20. Schultz, J., Copley, R. R., Doerks, T., Ponting C. P., and Bork, P. (2000) SMART: a web-based tool for the study of genetically mobile domains. *Nucleic Acids Res.* **28**, 231–234.

21. Thompson, J. D., Higgins, D. G., and Gibson, T. J. (1994) CLUSTAL W: improving the sensitivity of progressive multiple sequence alignment through sequence weighting, position-specific gap penalties and weight matrix choice. *Nucleic Acids Res.* **22**, 4673–4680.

22. Eddy, S. R. (1996) Hidden Markov models. *Curr. Opin. Struct. Biol.* **6**, 361–365.

23. Eddy, S. R. (1998) Profile hidden Markov models. *Bioinformatics* **14**, 755–763.

24. Karplus, K., Barrett, C., and Hughey, R. (1998) Hidden Markov models for detecting remote protein homologies. *Bioinformatics* **14**, 846–856.

25. Brown, N. P., Leroy, C., and Sander, C. (1998) MView: a web-compatible database search or multiple alignment viewer. *Bioinformatics* **14**, 380–381.

26. Galtier, N., Gouy, M., and Gautier, C. (1996) SeaView and Phylo_win, two graphic tools for sequence alignment and molecular phylogeny. *Comput. Appl. Biosci.* **12**, 543–548.

27. Gibson, T. J. and Spring, J. (1998) Genetic redundancy in vertebrates: polyploidy and persistence of genes encoding multidomain proteins. *Trends Genet.* **14**, 46–49.

28. Wilson, C. A., Kreychman, J., and Gerstein, M. (2000) Assessing annotation transfer for genomics: quantifying the relations between protein sequence, structure and function through traditional and probabilistic scores. *J. Mol. Biol.* **297**, 233–249.

29. Peitsch, M. C. (1996) ProMod and Swiss-Model: Internet-based tools for automated comparative protein modelling. *Biochem. Soc. Trans.* **24**, 274–279.

30. Rost, B. and Sander, C. (1993) Prediction of protein secondary structure at better than 70% accuracy. *J. Mol. Biol.* **232**, 584–599.

31. Jones, D. T. (1999) Protein secondary structure prediction based on position-specific scoring matrices. *J. Mol. Biol.* **292**, 195–202.

32. Cuff, J. A. and Barton, G. J. (2000) Application of multiple sequence alignment profiles to improve protein secondary structure prediction. *Proteins* **40**, 502–511.

33. Jones, D. T., Taylor, W. R., and Thornton, J. M. (1992) A new approach to protein fold recognition. *Nature* **358**, 86–89.

34. Rost, B. (1995) TOPITS: threading one-dimensional predictions into three-dimensional structures. *Ismb* **3**, 314–321.

35. Kelley, L. A., MacCallum, R. M., and Sternberg, M. J. (2000) Enhanced genome annotation using structural profiles in the program 3D-PSSM. *J. Mol. Biol.* **299**, 499–520.

36. White, J. H., Wise, A., Main, M. J., Green, A., Fraser, N. J., Disney, G. H., et al. (1998) Heterodimerization is required for the formation of a functional GABA(B) receptor. *Nature* **396**, 679–682.

37. Galvez, T., Parmentier, M. L., Joly, C., Malitschek, B., Kaupmann, K., Kuhn, R., et al. (1999) Mutagenesis and modeling of the GABAB receptor extracellular domain support a venus flytrap mechanism for ligand binding. *J. Biol. Chem.* **274**, 13362–13369.

38. Dekel, I., Russek, N., Jones, T., Mortin, M. A., and Katzav, S. (2000) Identification of the *Drosophila melanogaster* homologue of the mammalian signal transducer protein, Vav. *FEBS Lett* **472**, 99–104

39. Kraulis P. J. (1991) Molscript: a program to produce both detailed and schematic plots of protein structures. *J. Appl. Crystallog.* **24**, 964–950.

APPENDIX 1 _____

Letter Codes, Structures, Masses, and Derivatives of Amino Acids

James P. Turner

1. Introduction

This appendix cannot claim to be an exhaustive list of all known amino acids and their derivatives, because this seems to be ever-extending. It is intended, however, that it does include most of them, or at least the most common ones (with some concious exceptions - *see* below). To assist in calculation of masses of residues not included here, **Table 1** gives the atomic weights used in the calculations for this **Appendix**.

Table 2 describes genetically-encoded amino acids, including the 20 commonly recognized ones plus selenocysteine, which is coded by UGA (otherwise used as a stop codon). Insertion of selenocysteine during translation, instead of termination at UGA is dictated by the presence of particular secondary structure to the 3' side in the mRNA, and possibly other factors *(1)*. Selenocysteine is found at the active site of various prokaryotic and eukaryotic enzymes.

Table 3 lists modifications of these amino acids; some are natural (i.e., posttranslational), others are artificial. Some modifications give rise to residues that have been given trivial names, such as desmosine (and its isomer, isodesmosine), which derives from lysine and which is found in elastin. Other modifications are combinations of genetically encoded amino acids, arising by condensation or oxidative coupling between two or more amino acids. Cystine is the most common of these. Such residues are important in that they form crosslinks in or between polypeptides.

To calculate the mass of a given polypeptide, sum the masses of the constituent amino acid residues (*see* **Table 2**) and add the mass of one molecule of water (monoisotopic mass 18.0106; average mass 18.0153 Da or atomic mass

From: *Methods in Molecular Biology, vol. 211: Protein Sequencing Protocols, 2nd ed.*
Edited by: B. J. Smith © Humana Press Inc., Totowa, NJ

Table 1
Atomic Weights

Element	Monoisotopic atomic mass	Average atomic mass
Br	78.9183361	79.903527
	80.916289	
C	12.0	12.011037
Cl	34.968852721	35.452737
	36.96590262	
F	18.99840322	18.99840322
H	1.007825035	1.0079759
I	126.904473	126.904473
N	14.003074002	14.006723
O	15.99491463	15.999304
P	30.973762	30.973762
S	31.9720707	32.064387
Se	79.9165196	78.959586

Data from **ref.** *(8)*.

Monoisotopic = atomic weight of isotope which occurs most commonly in nature. Both isotopes of Cl and of Br are listed.

Average = where more than one isotope occurs naturally, the mean of the atomic weights of all isotopes, each weighted according to its abundance in nature.

Note that specific (e.g., radioactive) isotopes will have different atomic weights.

units). Exceptions to this would be cyclic peptides such as gramicidin-S, which has its N-terminus peptide-bonded to its C-terminus. (This peptide is also noteworthy for its inclusion of the arginine derivative, ornithine, and **D**-phenylalanine.) Where there are modifications to the polypeptide, make the relevant change in mass, as listed in **Table 3**. To assist in identification of modification(s) by observation of change in mass, modifications are listed in order of mass difference in **Table 4**.

Various biologically significant modifications are omitted from the present Appendix because of the complication of their heterogeneity. One such is the large number of derivatives of lysine (and similar residues such as hydroxylysine, allysine and hydroxyallysine) that form interchain crosslinks in proteins. Some representative examples have been included in **Table 3**, but for more detail of this type of modification *see* **ref.** *(2)*.

Another example is the GPI anchor, a modification whereby the C-terminal α-carboxyl of a protein becomes linked via an amide bond to a glycosylphosphatidyl "tail" which, being lipophilic, associates with the cell membrane and anchors the protein there. The modifying group has a core structure of: ethanolamine-phospho-6Manα1-2Manα1-6Manα1-4GlcNα1-6 myo inositol-1- phospholipid. This core is modified by addition of various groups, differently in different species. The mass of the GPI anchor is therefore in excess of 1900 Da, according to the detail of the groups added to the core structure. For further detail of GPI-anchor structures, *see* the review by Englund *(3)*.

Glycosylation is also a complex modification, typically of the side chains of serine and asparagine residues, but possibly others such as threonine or hydroxytyrosine. It is the subject of many reviews (e.g., **ref.** *[4]*). The masses of sugar residues are included in **Table 3** to assist in calculation of oligosaccharide mass. Note that detailed knowledge of the oligosaccharide's structure (branching; sialylation) is required for accurate calculation. Note also that nonenzymatic glycosylation (or glycation) may also occur, for instance of hemoglobin of diabetics, in whose serum glucose concentrations may be higher than normal. Glycation occurs by addition of glucose to an amine (N-terminus or lysine side chain) via an aldimine linkage, which rearranges to form a ketoamine product (reviewed in **ref.** *[5]*).

Ubiquitination is important in regulation of protein turnover and possibly other processes *(6)*. Ubiquitin is a 76-amino acid protein whose sequence has remained highly conserved throughout evolution, there being only three differences between animals and plants or yeast (as shown in **ref.** *[7]*). Animal ubiquitin itself has the following mass: monoisostopic, 8559.6161 Da; average 8566.0109 Da. The glycine at its C-terminus becomes linked via an isopeptide bond to the side chain (ε-amino) of lysine. A variable number of ubiquitin molecules may become linked to a protein, so modification entails mass increase in multiples of the ubiquitin residue (monoisotopic mass, 8541.6055 Da; average mass, 8546.9875 Da). Note that ubiquitin will contribute its own N-terminal sequence and peptides to an analysis of any protein modified in this way.

Table 2
Genetically Encoded Amino Acids: Structures and Other Characteristics

Amino Acid	3 Letter Code	1 Letter Code[1]	1 Letter Code Mnemonic	Composition	Structure	Monoisotopic Residue Mass[2]	Average Residue Mass[2]	%Relative Abundance[3]
Alanine	Ala	A	Alanine	C_3H_5NO		71.0371	71.0800	7.49
Arginine	Arg	R	aRginine	$C_6H_{12}N_4O$		156.1007	156.1901	5.22
Asparagine	Asn	N	asN	$C_4H_6N_2O_2$		114.0429	114.1054	4.53
Aspartic acid	Asp	D	asparDic	$C_4H_5NO_3$		115.0269	115.0900	5.22

Table 2 *(continued)*
Genetically Encoded Amino Acids: Structures and Other Characteristics

Cysteine	Cys	C	**C**ysteine	C_3H_5NOS	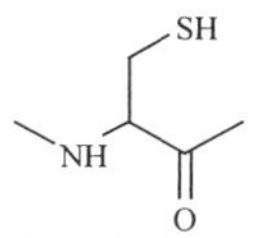	103.0092	103.1444	1.82
Glutamic acid	Glu	E	glu**E**tamic	$C_5H_7NO_3$	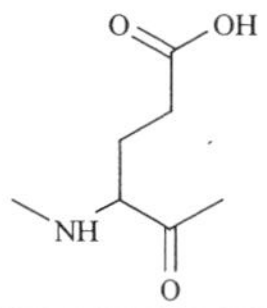	129.0426	129.1173	6.26
Glutamine	Gln	Q	**Q**utamine	$C_5H_8N_2O_2$	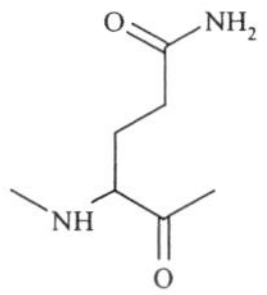	128.0586	128.1327	4.11
Glycine	Gly	G	**G**lycine	C_2H_3NO	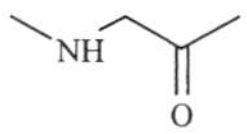	57.0215	57.0527	7.10
Histidine	His	H	**H**istidine	$C_6H_7N_3O$	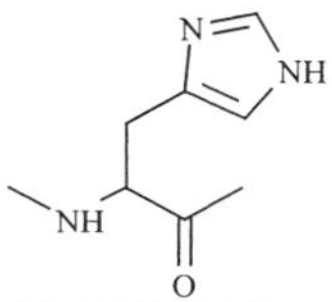	137.0589	137.1435	2.23

Table 2 (continued)
Genetically Encoded Amino Acids: Structures and Other Characteristics

					Structure			
Isoleucine	Ile	I	Isoleucine	$C_6H_{11}NO$		113.0841	113.1620	5.45
Leucine	Leu	L	Leucine	$C_6H_{11}NO$		113.0841	113.1620	9.06
Lysine	Lys	K	-	$C_6H_{12}N_2O$		128.0950	128.1767	5.82
Methionine	Met	M	Methionine	C_5H_9NOS		131.0405	131.1991	2.27
Phenylalanine	Phe	F	Fenylalanine	C_9H_9NO		147.0684	147.1801	3.91

Table 2 *(continued)*
Genetically Encoded Amino Acids: Structures and Other Characteristics

					Structure			
Proline	Pro	P	Proline	C_5H_7NO		97.0528	97.1187	5.12
Selenocysteine[^]	Sec	-	-	C_3H_5NOSe		150.9536	150.0396	-
Serine	Ser	S	Serine	$C_3H_5NO_2$		87.0320	87.0793	7.34
Threonine	Thr	T	Threonine	$C_4H_7NO_2$		101.0477	101.1066	5.96
Tryptophan	Trp	W	**Double** ring	$C_{11}H_{10}N_2O$		186.0793	186.2176	1.32

Table 2 (continued)
Genetically Encoded Amino Acids: Structures and Other Characteristics

Tyrosine	Tyr	Y	$C_9H_9NO_2$		163.0633	163.1794	3.25
Valine	Val	V	C_5H_9NO		99.0684	99.1347	6.48

[1] Deamidation (for instance during acid hydrolysis) converts asparagine to aspartic acid and glutamine to glutamic acid. In such cases where N and D or Q and E cannot be distinguished from one another the following single and 3-letter codes are used: N or D = B or Asx; Q or E = Z or Glx.

[2] Residue mass = amino acid mass minus one molecule of water.

[3] Data on relative abundance adapted with permission from **ref. (9)**. Data for the rare amino acid selenocysteine not included.

[4] Selenocysteine is coded by codon UGA (*see* text).

Table 3
Modifications to Amino Acids Found in Proteins and Peptides[a]

Name	Transformation	Monoisotopic Mol. Wt. Change	Mean Mol. Wt. Change	Site of Modification
Acetamidomethyl (ACM)		71.0371	71.0790	Cys-SH [2]
Acetyl		42.0106	42.0373	N-Term.[1]; Lys-N$^\varepsilon$ 1,2; Ser-OH [1]
N-Acetylgalactosaminyl		203.0794	203.1953	Ser-OH[1]; Thr-OH[1]
N-Acetylglucosaminyl-1-phosphoryl		283.0457	283.1749	Ser-OH [1]

[a]*See* Heading Key for Table 3 on page 459.

Table 3 (continued)
Modifications to Amino Acids Found in Proteins and Peptides

Adenylyl	329.0525	329.2093	Tyr-OH[1]
ADP-ribosyl	541.0611	541.3051	Arg-Nω[1]; Cys-SH[1]; Glu-Oγ[1]; Diphthamide-Nτ (qv.)[1]
Alaninohistidine[3]	-18.0106	-18.0153	Ser + His Nτ[1]; Ser + His Nπ[1]
Allysine	-1.0316	-1.0313	Lys[1]
Amide	-0.9840	-0.9846	C-Term.[1]
4-Anisyl	90.0470	90.1251	Glu[2]
α/β-Aspartylhydroxamate	17.9868 Minus residues C-Term. of Asn	17.9919	Asn[2]

Table 3 (continued)
Modifications to Amino Acids Found in Proteins and Peptides

Nε-(β-Aspartyl)lysine [3]		-17.0265	-17.0307	Asn + Lys [1]
Benzoyl		104.0262	104.1085	N-Term [2]
Benzyl		90.0470	90.1251	Asp-Oβ [2]; Cys-SH [2];Glu-Oγ [2]; Met-S [2]; Ser-OH [2]; Thr-OH [2]; Trp-N1 [2]; Trp-CArom. 2; Tyr-OH [2];Tyr-CArom. 2; Phosphate-OH (qv.)
Benzyloxycarbonyl (CBZ)		134.0368	134.1348	Lys-Nε [2]
Benzyloxymethyl (BOM)		120.0575	120.1514	His-Nπ [2]
Biotinyl		226.0776	226.2985	Lys-Nε [1]

Table 3 (continued)
Modifications to Amino Acids Found in Proteins and Peptides

Modification	Structure			Sites
3,3'-Bityrosine [3]		-2.0157	-2.0160	(Tyr)$_2$ [1]
Bromo		1: 77.9105(0.51) 79.9085(0.49) 2: 155.8210(0.26) 157.8190(0.50) 159.8169(0.24)	78.8957 157.8190	Tyr-C^3 [1]; Phe-C^2 [1]; Phe-C^3 [1]; Phe-C^4 [1]; Tyr-C^3,C^5 [1]
t-Butyl		56.0626	56.1080	Asp-O$^\beta$ 2; Cys-SH 2; Glu-O$^\gamma$ 2; Met-S 2; Ser-OH 2; Thr-OH 2; Trp-N1 2; Trp-C$^{Arom.}$ 2; Tyr-OH 2; Tyr-C$^{Arom.}$ 2
t-Butyloxycarbonyl (t-BOC)		100.0524	100.1176	N-Term. 2; His-N$^\tau$ 2; Lys-N$^\varepsilon$ 2; Trp-N^1 2
Carbamoylethyl		71.3711	71.0790	Cys-SH 2
Carboxy		43.9898	44.0096	Asp-c$^\beta$ 1; Glu-C$^\gamma$ 1; Trp-N1 2

Table 3 (continued)
Modifications to Amino Acids Found in Proteins and Peptides

Modification	Structure			Site
Carboxyethyl		72.0211	72.0636	Cys-S [2]
Carboxymethyl		58.0055	58.0366	Cys-SH [2]
Chloro		1: 33.9610(0.76) 35.9581(0.24) 2: 67.9221(0.58) 69.9191(0.36) 71.9162(0.06)	34.4448 68.8896	Tyr-C[3] [1] Tyr-C[3],C[5] [1]
2-Chlorobenzyloxycarbonyl		167.9978(0.76) 169.9949(0.24)	168.5795	Lys-Nε [2]
Citrulline		−0.9840	−0.9846	Arg [1]
Coenzyme-A		751.0839	751.4976	Cys-SH [1]
Cysteic acid		47.9847	47.9979	Cys [2]

Table 3 (continued)
Modifications to Amino Acids Found in Proteins and Peptides

Name	Structure		
Cystine [3]	(chemical structure)	-2.0157	-2.0160	$(Cys)_2$ [1,2]
Dansyl	(chemical structure)	233.0511	233.2899	N-Term. [2]
Dehydroalanine	(chemical structure)	-18.0106	-18.0153	Ser [1]
Dehydrobutyrine	(chemical structure)	-18.0106	-18.0153	Thr [1]
Desmosine [3]	(chemical structure)	-58.1344	-58.1478	$(Lys)_4$ [1]
2,6-Diaminopimeloyl	(chemical structure)	172.0848	172.1843	Lys-Nα[1]; Lys-Nε[1]

Table 3 (continued)
Modifications to Amino Acids Found in Proteins and Peptides

Name	Structure			Residue
4,4'-Dimethoxybenzhydryl		226.0994	226.2758	Asn-Nβ [2]; Gln-Nγ [2]
2,4-Dinitrophenyl		166.0015	166.0928	His-Nτ [2]; His-Nπ [2]
Dipalmitoylglyceryl		550.4961	550.9099	Cys-SH [1]
Diphthamide		143.1184	143.2097	Lys [1]
Farnesyl		204.1878	204.3570	Cys-SH [1]

Table 3 (continued)
Modifications to Amino Acids Found in Proteins and Peptides

Name	Structure	Monoisotopic	Average	Sites
8α-Flavin Adenine Dinucleotidyl (FAD)		783.1415	783.5428	Cys-SH[1]; His-Nτ[1]; His-Nπ[1]; Tyr-O4[1]
6-Flavin Mononucleotidyl (FMN)		455.0968	455.3415	Cys-SH[1]
9-Fluorenylmethoxycarbonyl (FMOC)		222.0681	222.2439	N-Term.[2]; Lys-Nε[2]; HMB-OH (qv.)[2]
Formyl		27.9949	28.0103	Gly-Nα[1]; Met-Nα[1]; Trp-N1[1,2]; Amino[2]; Kynurenin-N[2] (qv.)
Furanosyl (Arabinosyl, Xylosyl)		132.0423	132.1162	Ser-OH[1]; 4-HydroxyPro-O4[1]

Table 3 (continued)
Modifications to Amino Acids Found in Proteins and Peptides

	Structure			
Geranylgeranyl	(structure)	272.2504	272.4760	Cys-SH [1]
Glucuronyl	(structure)	176.0321	176.1259	N-Term. [1]
S-(γ-Glutamyl)cysteine [3]	(structure)	-18.0106	-18.0153	Cys + Glu [1]
Nε-(γ-Glutamyl)lysine [3]	(structure)	-17.0265	-17.0307	Gln + Lys [1]
γ-Glutamylsemialdehyde	(structure)	-42.0456	-42.0638	Arg [1]
Oβ-(γ-Glutamyl)serine	(structure)	-18.0106	-18.0153	Glu + Ser [1]
1-Glyceryl	(structure)	74.0368	74.0796	Cys-SH [1]; Glu-O$^{\delta}$ [2]

Table 3 (continued)
Modifications to Amino Acids Found in Proteins and Peptides

Name	Structure			Notes
Heme (Cytochrome)		616.1917	616.5058	2 x Cys-SH [1]
S-(2-Histidyl)cysteine [3]		-2.0157	-2.0160	His + Cys [1]
Homoserine		-29.9928	-30.0921	Met [2]
Homoserine lactone		-18.0106	-18.0153	C-Term. Homoserine (qv.) [2]
Hydroxy		1: 15.9949	15.9993	Allysine-C^δ (qv.) [1]; Asp-$C\beta$ [1]; Cys-SH (Sulfenic acid) [1]; Lys-C^δ [1]; Phe-$C\beta$ [1]; Pro-C^3 [1]; Pro-C^4 [1]; Trp-$C\beta$ [1]; Tyr-C^3 [1]; Tyr-$C\beta$ [1]; Lysinonorleucine-C^δ (qv.) [1]
		2: 31.9898	31.9986	Leu-C^γ, C^δ [1]; Pro-C^3, C^4 [1]; Tyr-C^3, C^6 [1] Lysinonorleucine-C^δ, $C^{\delta'}$ (qv.) (Syndesine qv.) [1]
		3: 47.9847	47.9979	Leu-C^γ, C^δ, $C^{\delta 1}$ [1]

Table 3 (continued)
Modifications to Amino Acids Found in Proteins and Peptides

2-Hydroxyethylthio		60.0040	60.1184	Cys-SH [2]
2-Hydroxy-4-methoxybenzyl (HMB)		136.0524	136.1507	Ala-N$^{\alpha}$ [2]; Gly-N$^{\alpha}$ [2]; Leu-N$^{\alpha}$ [2]; Val-N$^{\alpha}$ [2]; See also FMOC
N-Hydroxysuccinimido		97.0164	97.0734	C-Term. [2]
Hypusine		87.0684	87.1220	Lys [1]
Iminolysine		-2.0157	-2.0160	Lys [2]
Iminothiazolidinyl		-1.9918 Minus residues N-term of Cys	-1.9926	Cys [2]
Iodo		1: 125.8966 2: 251.7933	125.8965 251.7930	His-C^{4} [1]; Tyr-C^{3} [1] Tyr-C^{3},C^{5} [1]
α-Ketobutyryl		84.0211	84.0747	N-Term. [1]

Table 3 (continued)
Modifications to Amino Acids Found in Proteins and Peptides

Name	Structure			Location
Kynurenin		3.9949	3.9883	Trp [2]
Lanthionine [3]		-18.0106	-18.0153	Cys + Ser [1]
Lauroyl		182.1671	182.3072	N-Term. [1]
Lipoyl		188.0330	188.3121	Lys-Nε [1]
Lysinoalanine [3]		-18.0106	-18.0153	Ser + Lys [1]
Lysinonorleucine [3]		-17.0265	-17.0307	(Lys)$_2$ [1]
Mesitylenesulfonyl		182.0402	182.2421	Arg-Nω [2]; Trp-N1 [2]

Table 3 *(continued)*
Modifications to Amino Acids Found in Proteins and Peptides

Methionine sulfone		31.9898	31.9986	Met-S [2]
Methionine sulfoxide		15.9949	15.9993	Met-S [1,2]
4-Methoxybenzyl		120.0575	120.1514	Cys-SH [2]
4-Methoxy-2,3,6-trimethyl benzenesulfonyl (MTR)		212.0507	212.2684	Arg-N$^\omega$ [2]
4-Methoxytrityl (MMT)		272.1201	272.3477	Arg-N$^\omega$ [2]

Table 3 (continued)
Modifications to Amino Acids Found in Proteins and Peptides

	Structure			Sites
Methyl		1: 14.0157	14.0270	N-Term. [1]; Arg-Nδ [1]; Asn-Nβ [1]; Asp-Cβ [1]; Asp-Oγ [1]; Cys-SH [2]; Glu-Oδ [1]; Glu-Nα [1]; Gly-Nα (Sarcosine) [1]; His-Nπ [1]; Lys-Nε [1]; Ser-OH [1,2]; Thr-OH [1]; Tyr-OH [2]; C-Term. [1]; Phosphate-OH (qv.) [2]
		2: 28.0313	28.0540	Arg-Nω,Nω [1]; Arg-Nω,Nω [1]; Lys-Nε,Nε [1]
		3: 43.0548	43.0889	Lys-Nε,Nε,Nε [1]
Methyl Lanthionine [3]		-18.0106	-18.0153	Cys + Thr [1]
Myristoyl		210.1984	210.3612	N-Term. [1]
4-Nitrophenyl		121.0164	121.0955	C-Term. [2]
Ornithine		-42.0218	-42.0404	Arg [1,2]

Appendix 1 453

Table 3 (continued)
Modifications to Amino Acids Found in Proteins and Peptides

Name	Structure			Site
2-Oxohistidine		15.9949	15.9993	His-C[2 1]
Palmitoyl		238.2297	238.4152	Cys-SH[1]
Pantetheinephosphoryl		340.0858	340.3360	Ser-OH[1]
Pentafluorophenyl		165.9842	166.0503	C-Term.[2]
2,2,5,7,8-Pentamethylchroman-6-sulfonyl (PMC)		266.0977	266.3604	Arg-N[ω 2]
Phenylthiocarbamyl		135.0143	135.1882	C-Term. N[2], Arg-N[ω 2]

Table 3 (continued)
Modifications to Amino Acids Found in Proteins and Peptides

Phosphoropyridoxalyl		229.0140	229.1291	Lys-Nε [1]
Phosphoryl		79.9663	79.9797	Arg-Nω [1]; Asp-Oβ [1]; Cys-SH [1]; His-Nτ [1]; His-Nπ [1]; Lys-Nε [1]; Ser-OH [1]; Thr-OH [1]; Tyr-OH [1];
β-Propionamide		71.0371	71.0790	Cys-SH [2]
Pyranosyl (Glucosyl, Galactosyl, Mannosyl)		162.0528	162.1425	N-Term [1]; Asn-Nβ [1]; Cys-SH [1]; Ser-OH [1]; Thr-OH [1]; δ-HydroxyLys-Oδ [1]; β-HydroxyPhe-Oβ [1]; 4-HydroxyPro-O4 [1]; β-HydroxyTyr-Oβ [1]
4-Pyridylethyl		105.0579	105.1398	Cys-SH [2]

Table 3 (continued)
Modifications to Amino Acids Found in Proteins and Peptides

Name	Structure			Site
Pyroglutamate		-18.0106	-18.0153	Glu (N-Term.) [1]
Pyruvoyl		70.0055	70.0477	N-Term. [1]
Retinalyl		266.2035	266.4281	Lys-Nε [1]
Succinyl		100.0160	100.0740	Lys-Nε [2]
Sulfonyl		79.9568	80.0623	Tyr-OH [1]
Syndesine		14.9633	14.9679	(Lys)$_2$ [1]

Table 3 (continued)
Modifications to Amino Acids Found in Proteins and Peptides

Name	Structure			Amino acid site
Tetraiodothyronine		595.6128	595.6834	Tyr[1]
4-Toluenesulfonyl		154.0089	154.1881	Arg-N$^\omega$ 2; His-N$^\tau$ 2; His-N$^\pi$ 2
Trifluoroacetyl		95.9823	96.0086	Amino[2]
Triiodothyronine		469.7162	469.7869	Tyr[1]
Triphenylmethyl (Trityl)		242.1096	242.3214	Cys-SH[2]

Table 3 (continued)
Modifications to Amino Acids Found in Proteins and Peptides

3,3':3",5'-Tertyrosine		-4.0313	-4.0319	$(Tyr)_3$ [1]
Tryptophan-6,7-dione		29.9742	29.9827	Trp C^6,C^7 [1]
S-(3-Tyrosyl)cysteine [3]		-2.0157	-2.0160	Tyr + Cys [1]
2,4'-bis-Tryptophan-6',7'-dione [3]		27.9585	27.9667	$(Trp)_2$ [1]

Table 3 (continued)
Modifications to Amino Acids Found in Proteins and Peptides

Uridylyl		275.0515	275.1950	Tyr-OH [1]
Xanthyl		180.0575	180.2066	Asn-Nβ [2]; Gln-Nγ [2]

Key

Name gives the name of the modification, either as a prefix to the amino-acid name, or in some cases as the full name of the post-translationally generated amino-acid.

Transformation shows examples of the named modification that gives rise to the molecular weight changes quoted. It is not intended to represent a reaction scheme, which may be a complex series of modifications. For brevity, in most cases only the side-chain of the amino-acid is shown, the alpha-carbon being represented by "α."

Monoisotopic Molecular Weight Change gives the change in molecular weight for the transformation as represented, calculated from the atomic weights of the most naturally abundant isotope. In the case of Bromine and Chlorine, each of which has two naturally abundant isotopes, molecular weights are quoted incorporating each, with their characteristic relative abundance quoted in parentheses. Where multiple derivatizations are listed, multiples of the molecular weights are given for convenience.

Mean Molecular Weight Change gives the change in molecular weight based on the mean values of atomic weight.

Site of Modification shows the amino-acid residue involved and, in the case of substituent groups, the point of attachment. In the case of post-translational modifications, these are often unique and very specific, however in the case of synthetic protecting groups, much more variation is likely to be encountered and the information given with respect to these should be seen as an indication of the most probable sites of attachment. The history of a particular synthetic sample will give an indication as to which modifications are likely to be present, and their locations. However the possibility must also be considered of protecting group migration, which can occur as a result of postsynthetic handling. Possible sites of migration are also quoted.

Footnotes

1. A natural, post-translational modification that is found in proteins. Many of these are unique to a single known protein, often prosthetic groups involved with a catalytic or other function; although some are more widely seen.

2. A synthetic modification such as a protecting group for chemical synthesis, or a modification arising as an artefact of sample handling. The latter may be found in synthetic and naturally derived samples. Examples are quoted of adducts arising both in peptide synthesis and protein sample manipulation.

3. A post-translational modification involving two or more amino-acid residues that results in the formation of a cross-link.

Table 4
Modifications to Amino Acids Listed by Molecular Weight

Name	Monoisotopic mol. wt. change	Mean mol. wt. change
8a-Flavin Adenine Dinucleotidyl (FAD)	783.1415	783.5428
Coenzyme-A	751.0839	751.4976
Heme (Cytochrome)	616.1917	616.5058
Tetraiodothyronine	595.6128	595.6834
Dipalmitoylglyceryl	550.4961	550.9099
ADP-ribosyl	541.0611	541.3051
Triiodothyronine	469.7162	469.7869
6-Flavin Mononucleotidyl (FMN)	455.0968	455.3415
Pantetheinephosphoryl	340.0858	340.3360
Adenylyl	329.0525	329.2093
N-Acetylglucosaminyl-1-phosphoryl	283.0457	283.1749
Uridylyl	275.0515	275.1950
Geranylgeranyl	272.2504	272.4760
4-Methoxytrityl (MMT)	272.1201	272.3477
Retinalyl	266.2035	266.4281
2,2,5,7,8-Pentamethylchroman-6-sulphonyl (PMC)	266.0977	266.3604
Diiodo	251.7933	251.7930
Triphenylmethyl (Trityl)	242.1096	242.3214
Palmitoyl	238.2297	238.4152
Dansyl	233.0511	233.2899
Phosphoropyridoxalyl	229.0140	229.1291
4,4'-Dimethoxybenzhydryl	226.0994	226.2758
Biotinyl	226.0776	226.2985
9-Fluorenylmethoxycarbonyl (FMOC)	222.0681	222.2439
4-Methoxy-2,3,6-trimethyl benzenesulphonyl (MTR)	212.0507	212.2684
Myristoyl	210.1984	210.3612
Farnesyl	204.1878	204.3570
N-Acetylgalactosaminyl	203.0794	203.1953
Lipoyl	188.0330	188.3121
Lauroyl	182.1671	182.3072
Mesitylenesulfonyl	182.0402	182.2421
Xanthyl	180.0575	180.2066
Glucuronyl	176.0321	176.1259
2,6-Diaminopimeloyl	172.0848	172.1843
2-Chlorobenzyloxycarbonyl	167.9978	168.5795

Table 4 *(continued)*
Modifications to Amino Acids Listed by Molecular Weight

Name	Monoisotopic mol. wt. change	Mean mol. wt. change
2,4-Dinitrophenyl	166.0015	166.0928
Pentafluorophenyl	165.9842	166.0503
Pyranosyl (Glucosyl, Galactosyl, Mannosyl)	162.0528	162.1425
Dibromo	157.8190	157.8190
4-Toluenesulfonyl	154.0089	154.1881
Diphthamide	143.1184	143.2097
2-Hydroxy-4-methoxybenzyl (HMB)	136.0524	136.1507
Phenylthiocarbamyl	135.0143	135.1882
Benzyloxycarbonyl (CBZ)	134.0368	134.1348
Furanosyl (Arabinosyl, Xylosyl)	132.0423	132.1162
Iodo	125.8966	125.8965
4-Nitrophenyl	121.0164	121.0955
Benzyloxymethyl (BOM)	120.0575	120.1514
4-Methoxybenzyl	120.0575	120.1514
4-Pyridylethyl	105.0579	105.1398
Benzoyl	104.0262	104.1085
t-Butyloxycarbonyl (t-BOC)	100.0524	100.1176
Succinyl	100.0160	100.0740
N-Hydroxysuccinimido	97.0164	97.0734
Trifluoroacetyl	95.9823	96.0086
4-Anisyl	90.0470	90.1251
Benzyl	90.0470	90.1251
Hypusine	87.0684	87.1220
α-Ketobutyryl	84.0211	84.0747
Phosphoryl	79.9663	79.9797
Sulfonyl	79.9568	80.0623
Bromo	77.9105	78.8957
1-Glyceryl	74.0368	74.0796
Carboxyethyl	72.0211	72.0636
Carbamoylethyl	71.3711	71.0790
Acetamidomethyl (ACM)	71.0371	71.0790
β-Propionamide	71.0371	71.0790
Pyruvoyl	70.0055	70.0477
Dichloro	67.9221	68.8896
2-Hydroxyethylthio	60.0040	60.1184
Carboxymethyl	58.0055	58.0366
t-Butyl	56.0626	56.1080

Table 4 (continued)
Modifications to Amino Acids Listed by Molecular Weight

Name	Monoisotopic mol. wt. change	Mean mol. wt. change
Cysteic acid	47.9847	47.9979
Trihydroxy	47.9847	47.9979
Carboxy	43.9898	44.0096
Trimethyl	43.0548	43.0889
Acetyl	42.0106	42.0373
Chloro	33.9610	34.4448
Dihydroxy	31.9898	31.9986
Methionine sulfone	31.9898	31.9986
Tryptophan-6,7-dione	29.9742	29.9827
Dimethyl	28.0313	28.0540
Formyl	27.9949	28.0103
2,4'-bis-Tryptophan-6',7'-dione	27.9585	27.9667
α/β-Aspartylhydroxamate	17.9868	17.9919
Hydroxy	15.9949	15.9993
Methionine sulfoxide	15.9949	15.9993
2-Oxohistidine	15.9949	15.9993
Syndesine	14.9633	14.9679
Methyl	14.0157	14.0270
Kynurenin	3.9949	3.9883
Amide	− 0.9840	− 0.9846
Citrulline	− 0.9840	− 0.9846
Allysine	− 1.0316	− 1.0313
Iminothiazolidinyl	− 1.9918	− 1.9926
3,3'-Bityrosine	− 2.0157	− 2.0160
Cystine	− 2.0157	− 2.0160
S-(2-Histidyl)cysteine	− 2.0157	− 2.0160
Iminolysine	− 2.0157	− 2.0160
S-(3-Tyrosyl)cysteine	− 2.0157	− 2.0160
3,3';3'',5'-Tertyrosine	− 4.0313	− 4.0319
N^ϵ-(β-Aspartyl)lysine	− 17.0265	− 17.0307
N^ϵ-(γ-Glutamyl)lysine	− 17.0265	− 17.0307
Lysinonorleucine	− 17.0265	− 17.0307
Alaninohistidine	− 18.0106	− 18.0153
Dehydroalanine	− 18.0106	− 18.0153
Dehydrobutyrine	− 18.0106	− 18.0153
Homoserine lactone	− 18.0106	− 18.0153
Lanthionine	− 18.0106	− 18.0153
Lysinoalanine	− 18.0106	− 18.0153

Table 4 *(continued)*
Modifications to Amino Acids Listed by Molecular Weight

Name	Monoisotopic mol. wt. change	Mean mol. wt. change
Methyl Lanthionine	− 18.0106	− 18.0153
O$^\beta$-(γ-Glutamyl)serine	− 18.0106	− 18.0153
Pyroglutamate	− 18.0106	− 18.0153
S-(γ-Glutamyl)cysteine	− 18.0106	− 18.0153
Homoserine	− 29.9928	− 30.0921
Ornithine	− 42.0218	− 42.0404
γ-Glutamylsemialdehyde	− 42.0456	− 42.0638
Desmosine	− 58.1344	− 58.1478

References

1. Berry, M. J., Harney, J. W., Ohama, T., and Hatfield, D. L. (1994) Selenocysteine insertion or termination: factors affecting UGA codon fate and complimentary anticodon:codon mutations. *Nucleic Acids Res.* **22**, 3753–3759.
2. Gallop, P. M., Blumenfeld, O. O., and Sieftner, S. (1972) Structure and metabolism of connective tissue proteins. *Ann. Rev. Biochem.* **41**, 617–672.
3. Englund, P. (1993) The structure and synthesis of glycosyl phosphatidyl protein anchors. *Ann. Rev. Biochem.* **62**, 121–138.
4. Hart, G. (1992) Glycosylation. *Curr. Opin. Cell Biol.* **4**, 1017–1023.
5. Kennedy, I. and Lyons, T. J. (1989) Non-enzymatic glycosylation. *Br. Med. Bull.* **45**, 174–190.
6. Hershko, A. and Ciechanover, A. (1992) The ubiquitin system for protein degradation. *Ann. Rev. Biochem.* **61**, 761–807.
7. Rechsteiner, M. (1987) Ubiquitin-mediated pathways for intercellular proteolysis. *Ann. Rev. Cell Biol.* **3**, 1–30.
8. Lide, D. R. and Frederikse, H. P. R. (eds.), (1993) *The CRC Handbook of Chemistry and Physics*, 74th ed. CRC Press, Boca Raton, FL.
9. Correia, J. J., Lipscomb, L. D., and Lobert, S. (1993) Nondisulfide crosslinking and chemical cleavage of Tubulin subunits: pH and temperature dependence. *Arch. Biochem. Biophys.* **1**, 105–114.
10. Volkin, D. B., Mach, H., and Middaugh, C. R. (1997) Degradative covalent reactions important to protein stability. *Mol. Biotech.* **8**, 105–122.
11. Lewisch, S. A. and Levine, R. L. (1995) Determination of 2-oxohistidine by amino acid analysis. *Anal. Biochem.* **231**, 440–446.
12. Yoshino, K., Takao, T., Suhara, M., Kitai, T., Hori, H., Naura, K., et al. (1991) Identification of novel amino acid o-bromo-L-phenylalanine in egg-associated peptides that activate spermatozoa. *Biochemistry* **30**, 6203–6209.

13. Climent, I. and Levine, R. L. (1991) Oxidation of the active site of glutamine synthetase: conversion of Arginine-344 to gamma-glutamyl semialdehyde. *Arch. Biochem. Biophys.* **289**, 371–375.
14. Kim, J-S. and Raines, R. T. (1994) A misfolded but active dimer of bovine seminal ribonuclease. *Eur. J. Biochem.* **224**, 109–114.
15. Budavari, S. (ed.) (1996) *The Merck Index*, 12th ed. Merck & Co., Inc., Whitehouse Station, NJ.
16. Stryer, L. (1988) *Biochemistry*, 3rd ed. W.H. Freeman & Co., New York, NY.
17. Mitchelhill, K. (Ed.), (2000) *Delta Mass*, http://www.abrf.org/ABRF/Research Comittees/deltamass/deltamass.html.
18. Siegers, K., Heinzmann, S., and Entian, K-D. (1996) Biosynthesis of Lantibiotic Nisin. *J. Biol Chem.* **271**, 12294–12301.
19. Skaugen, M., Nissen-Meyer, J., Jung, G., Stevanovic, S., Sletten, K., Abildgaard, C. I. M., and Nes, I. F. (1994) In vivo conversion of L-Serine to D-Alanine in a ribosomally synthesized polypeptide. *J. Biol. Chem.* **269**, 27183–27185.

APPENDIX 2

Protein Consensus Sequence Motifs

Alastair Aitken

This article lists the many consensus sequences for short stretches of primary structure that may be recognized without necessity for computer analysis. These include motifs that are shared between distinct proteins, particularly those with distinct biological function and from diverse phyla. Covalent binding sites are also considered.

Predictions of higher-order structure are not considered when they are mainly secondary-structure motifs such as helix-turn-helix structures or β-sheet nucleotide binding domains. Protein "signatures"—which are sequence profiles that are confined to one particular family—are not included either. Cell-wall peptides and peptide antibiotics that are not synthesized on ribosomes are also outside the scope of this list. A full recent list of post-translational modifications on proteins is in **ref.** *(1)*.

Computer analysis of protein structure and consensus sequences may be carried out on the internet. The most useful of these programs, PROSITE, is accessed from the user-friendly ExPASy WWW server (from the University of Geneva), which also contains the SWISS-PROT, ENZYME, SWISS-2DPAGE, and SWISS-3DIMAGE databases. PROSITE lists consensus sequences as well as an exhaustive treatment of protein "signatures" *(2)* and can enable one to quickly recognize that a sequence from a new protein belongs to a particular family through its database of biologically significant sites, patterns, and profiles. The recent database issue of Nucleic Acids Research summarizes the current status of the general nucleotide and protein databanks and includes many databases available for gene and protein sequences from specific organisms. For the present purpose BLOCKS, PRINTS *(3)*, Pfam, and ProDom may be particularly useful.

From: *Methods in Molecular Biology, vol. 211: Protein Sequencing Protocols, 2nd ed.*
Edited by: B. J. Smith © Humana Press Inc., Totowa, NJ

Table 1
Abbreviations for the Nonstandard Amino Acids

Amino acid	Abbreviation 3 letter	single letter
Aspartic acid or Asparagine	Asx	B
Glutamic acid or Glutamine	Glx	Z
γ-carboxy glutamic acid	Gla	
Pyrollidone carboxylic acid	Glp	
Any amino acid	Xaa (or Yaa to distinguish specific sites)	X
Aromatic residues	Aro	
Aliphatic	Ali	
Hydrophobic amino acid	Hyd	
Hydrophilic amino acid (or a neutral residue)	Naa	
Oxygen-containing amino acid	Oaa	
Negatively charged residues (Asp, Glu, or phospho-)	Neg	
Acidic residues	Aaa	
Basic residues	Baa	
hydroxyproline	Hyp	
Hydroxylysine	Hyl	
If either Hyp or Hyl	Hyx	

An arrow indicates peptide bonds cleaved.

X^3 and X^4 indicates the third and fourth residues of that type in the sequence.

X_2 indicates two unspecified residues; X_{40-80} indicates between 40–80 unspecified residues.

X(Y) indicates that residue X is commonly at this position but that Y may also occur.

X^{-1} and X^{+3} refer to the residues one position N-terminal and 3 positions C-terminal (respectively) to the active site amino acid.

-A/B/C- indicates that any of these 3 residues may occur at this position, e.g., –L/I/V/M- indicates Leu or Ile or Val or Met may occur at that position.

$-(A/B/C)_2-$ indicates that any of these 3 residues occur in any combination at these 2 consecutive positions.

$-(Baa/X_2)-$ Baa may be at position 1, 2, or 3.

-(X)- indicates a residue that is usually present and forms part of the consensus sequence, but is not invariant.

1. Covalent Modifications

The covalently modified, liganded or active site residues are underlined. Throughout the chapter, apart from the non-standard residues, listed in **Table 1**, the standard IUPAC one-letter codes for the amino acids are used. Further details of the notation are in the footnote to **Table 1**.

1.1. Phosphorylation-Site Consensus Sequences

In many instances pseudosubstrate sequences have been identified where the phosphorylatable Ser or Thr is replaced by Ala. The optimal phosphorylation by kinases C is notably very similar to the pseudosubstrate sequences of the individual isoforms.

1.1.1. Serine/Threonine Kinases

Cyclic-AMP dependent protein kinase -R-X_{1-2}-S/T- or -R-R*-X-S/T-(Hyd)-

Lys can replace Arg* but with lower efficiency. Hyd is I, L or V mainly.

Cyclic-GMP dependent Protein kinase -R/K_{1-3}-X_{1-3}-S/T-R/K_{0-1}-

Akt (Protein kinase B or RAC) -R-X-R-X_2-S/T-Hyd-

Protein kinase C -R/K_{1-3}-X-S/T-X^{+1}-(R/K)$_{1-3}$-
where X is uncharged. X^{+1} mainly hydrophobic. This is a large family of kinases. C-terminal basic residues are the strongest determinants for type I (α, β, γ) and η but the other more recently described members δ, ϵ, ζ, and μ (PKD) prefer hydrophobic residues at these positions. All isoforms prefer basic residues at −3 and (with the exception of PKCμ/PKD) also prefer basic residues at −6, −4 and −2. Isoforms α, β, and δ prefer Arg at −5 while the others select hydrophobics.

Ca^{2+}-calmodulin-dependent Protein kinase II -R-X-X-S/T-Hyd- (Hyd especially Val)

Ribosomal Protein kinases p70S6 and p90S6(MAPKAP kinase-1) -(R/K)-X-R-X-X-S-(Hyd)-

MAPKAP kinase-2 -Hyd-Xaa-Arg-Xaa_2-Ser-

Phosphorylase kinase -R/K-X-X-S-Hyd-(Baa)- (Hyd is generally Val or Ile) (may also phosphorylate Tyr, in the presence of Mn^{2+})

Myosin light chain kinase(s) -K/$R_{\sim 3}$-X_2-R-X_2-S-
In contrast to most kinases, both myosin light chain and phosphorylase kinases phosphorylate peptide substrates with low efficiency and may only recognize these motifs in proteins where the sequence is in a structural feature that is recognized.

AMP-activated kinase (SNF1) -Hyd-(Baa/X_2)-X-S/T-X_3-Hyd-
Hyd is M, V, L, or I; Baa is K, R, or H (this may be at position −2, −3 or −4).

Proline-directed kinases

Cell-cycle kinases (K)-S/T-P-X-K/R- or (K)-S/T-P-(K/R)$_{1-2}$-

MAP kinase family -(P)-(L)-P-X-S/T-P-(P)-

Ceramide activated kinase -P-L $_{0-1}$-T-L-P- or -P-L_{1-2}-T-P-

Glycogen synthase kinase-3 -S/T-X^1-X-X- *Sp* -

Sp is already phosphorylated by the same or another kinase. The residue X^1 is frequently Pro.

Casein kinase I - Sp /Tp -X-X-S̲/T̲-

(the *Sp* or *Tp* that is already phosphorylated, can be substituted by D or E but this is a poorer substrate).

Casein kinase II -S̲/T̲-Aaa-Aaa-*Aaa³*-Aaa-Aaa-

where Aaa are acidic residues, Asp, Glu, phosphoserine, or phosphotyrosine, particularly at +3 (*Aaa³*).

Golgi casein kinase -S̲-X-E/Sp-

2-oxoacid dehydrogenase kinases -S̲-X_{1-2}-D/E-

Light harvesting complex II kinase -R-K-Naa-Naa-T̲-Naa/K-K/X-(K)- (Naa is S,T, or A)

NIMA kinase family

(originally named for n̲ever-i̲n-m̲itosis in *A̲spergillus nidulans*, essential for G2 –>M progression in the cell cycle).

-R-F-$R_{1,2}$-X-S̲/T̲- **R(Hyd)₂-Hyd₂-**

***Autophosphorylation-dependent kinase* (p21-activated kinase-2, PAK2)** -R-X_{1-2}-S̲/T̲-X_3-S/T-

1.1.2. Tyrosine-Protein Kinases

-Naa_{1-3}-X-Y̲-X_2-Hyd- where Naa is hydrophilic.

The consensus includes one or more acidic residues, E is preferred to D at -2 to -4. (The best substrate is also the optimal phosphorylated sequence for binding to the appropriate SH2 domain, *see* **Subheading 8.**)

Receptor tyrosine kinases -Naa_{1-3}-E-Y̲-Hyd-Hyd-Hyd- (e.g., insulin receptor -Naa_{1-3}-E-Y̲-M-Hyd-M-)

(preferentially phosphorylate peptides recognized by group III SH2 domains)

Cytosolic/intracellular tyrosine kinases - Naa_{1-3}-I/V-Y̲-E/G-E-Hyd- (in c-Abl: -Hyd-Y̲-A-A-P-)

(preferentially phosphorylate peptides recognized by their own or group I SH2 domains)

1.1.3. Dual-Specificity Kinase MAP Kinase Kinase (Mek Family)

-T̲-E-Y̲- both Thr and Tyr are phosphorylated in this motif.

1.2. Cofactors and Prosthetic Group Attachment Sites

1.2.1. Glycosylation

1.2.1.1. N-Glycosylation -N̲-X-S/T-

X is any amino acid but is rarely P or D (P, C-terminal to S/T is also rare). The motif -N̲-X-C- is also known but is rare.

1.2.1.2. O-Glycosylation

The consensus is not as clear-cut as N-glycosylation. There is frequent occurrence of S, T, A and P at positions -4 to +3. The glycosylated residue is often within 1 or 2 residues of Pro, e.g., <u>S</u>-X-X-P- i.e., Pro at +3.

Within EGF modules (*see* **Subheading 13.**) -C-X-<u>S</u>-X-P-C- and -C-X-X-G-G-<u>T/S</u>-C-

The latter is the attachment site for fucosyl residues.

Mucin-type O-glycosylation -X^{-1}-<u>T</u>-P-X^2-P-

Uncharged residues are preferred at position −1; small or basic residues preferred at +2.

Glucosaminoglycans -(two acidic residues at -2 to -4)-<u>S</u>-G-X-G-

1.2.1.3. Tryptophan-Glycosylation -<u>W</u>-X$_2$-W-

Recently, mannosylation of tryptophan residues (the 1st in the above motif) gives rise to a 3rd type of glycosylation. Mutagenesis to -<u>W</u>-X$_2$-F- reduces reaction 3 fold.

1.3. Tyrosine Sulphation -Neg$_{(1-5)}$-<u>Y</u>-Neg$_{(1-5)}$-

At least three negatively charged acidics, D, E or TyrSO$_4$, at positions ±1 to 5. The residue at position -1 (mostly D) is the strongest determinant. E at -2 to -5 or Tyr SO$_4$ at ±2 is frequently found.

1.4. Phosphopantetheine Binding Site

-L-G-X-D-<u>S</u>-L/I/T- (or -Hyd-G-Hyd-D/K-S-Hyd-)

1.5. Biotinyl-Lysine

-A/I/V-M-<u>*K*</u>-M*-

There is one exception where the 2nd methionine, M*, is replaced by an alanine

1.6. Lipoyl-Lysine Binding Sites

-Hyd^1-E-T/S-D-<u>*K*</u>-A-X- Hyd^1-Neg- Hyd^2-

where Hyd^1 is a non-aromatic hydrophobic residue (I, L, M, or V); Hyd^2 is any hydrophobic; Neg is D, E or occasionally G.

1.7. Bilin Attachment

-A-X-<u>*C*</u>-Hyd-R-D-

1.8. Dipyrromethane Cofactor

-G-S/A/G-<u>*C*</u>-X-V-P-

1.9. FAD Binding Site

-R-S-**_H_**-S/T-X$_2$-A-X-G-G-

1.10. Hypusine Attachment (in eIF-5A)

-T-G-**_K_**-H-G-X-A-K-

1.11. Retinal Binding

-L-D-Hyd-X-A-**_K_**-X$_2$-Aro- (Hyd is L,I,V or M and Aro is W, Y or F)

1.12. Carbamoyl-Phosphate Binding

- F-X-E/K-<u>S</u>-G/T-<u>R</u>-<u>T</u>-

1.13. Phosphohistidine Active Sites

In both types, Hyd is L, I, V, or M

1.13.1. ATP-citrate Lyase and Succinyl CoA Ligases

-Hyd-G-**_H_**-A-G-A-

1.13.2. Phosphoglycerate Mutase

-Hyd-X-R-**_H_**-G-X$_3$-N-

1.14. Acyl-Phosphate in Phosphomutases and Phosphatases*

The 1st Asp in the motif is the acyl-phosphate -<u>D</u>-V-D-X-T/V-
(*including L-3-phosphoserine phosphatase but not protein phosphatases).

2. N- and C-Terminal Consensus Sequences

2.1. N-Terminal Myristoylation

The initiator methionine, M, is removed in the consensus M-**_G_**-X^2-X$_2$-Naa-X^{6-} resulting in the modified N-terminal sequence, myr-**_G_**-X^2-X$_2$-Naa-X^{6-}

Positions 2 and 6 are not P; Naa is S, T, A, G, C, N; i.e., mainly small, uncharged, hydrophilic.

2.2. N-Terminal Palmitoylation

Proteins that are myristoylated in the above consensus are also palmitoylated if X^2 is cysteine, i.e., in the initial transcript sequence:- M-<u>G</u>-<u>C</u>----

2.3. N-Terminal Methylation

2.3.1. In Eukaryotes

-(Me)$_3$-A-P-K- and (Me)$_2$-P-P-K-

2.3.2. In Bacterial Pilins

-G-M/F*-S/T-L/T-Hyd-E-

The precursor is cleaved after G and M/F are methylated.
* starred residue is generally M or F but L, I, V, and Y can also occur.

2.3.3. Di-Methylation of Internal Arginine Residues

Occurs in RNA binding proteins with an N(G)-arginine methylase recognition sequence: -F/G-G-G-$\underline{R}$-G-G-F/G-

2.4. Bacterial Lipoprotein Glyceride Cysteine Thioethers

-L-A^2-G$^3_{\nearrow}$-$\underline{C}$-S^5-S^6-N- $\nearrow$ denotes cleavage site

The lipid is attached to the cysteine residue (underlined) which becomes the amino terminus. Other neutral residues occur at 2, 3, 5 or 6, mostly S, A or N.

2.5. C-Terminal Consensus Sequences

2.5.1. Amidation (Generally)

-X-X-G-R/K-R/K- becomes -X-X **CONH$_2$**
The precursor is cleaved between G and R/K then G is processed to an amide.
In **thyrothropin releasing hormone** and related peptide precursors -
-X-Q-X-P-G-X- Q is cyclised to pyroglutamate and mature hormone structure is **pyro**Glu-Xaa-Pro**NH$_2$**

2.5.2. C-A-A-X Boxes

At the C-terminus of Ras family GTP-binding proteins the motif -**Cys**-**Ali**^{1}-**Ali**^{2}-**Xaa**-**CO$_2$H** is processed by prenylation of the cysteine with a farnesyl or geranyl-geranyl moiety followed by proteolytic processing (axeing) of the –AAX sequence then carboxyl-methylation of the carboxyl group of the new C-terminal cysteine residue.

Ali^1 is generally Asp, Glu, Asn or Gln; Ali^2 is Leu, Ile, Met, or Val.
The C-terminus is **farnesylated** if X is S, C, M, A, or Q.
The protein is **geranylated** if X is L or F.
In the related YPT/rab proteins that end in -**Cys**-**Xaa**-**Cys** **CO$_2$H** then the latter Cys is isoprenylated (geranylated) and **carboxymethylated**; If the sequence ends in –**Cys**-**Cys** **CO$_2$H** then the C-terminal Cys isoprenylated but not carboxymethylated.

3. Metal Binding Motifs

3.1. Calcium Binding

3.1.1. EF Hand Consensus Sequence (as in Calmodulin)

X	*Y*	*Z*	*-Y*	*-X*	*-Z*

-Asp-Xaa-Asx-Xaa*-Asx-Gly-Oaa-I/V-Oaa-Xaa-Neg-Glu-F/L-

The most common residues are shown but even the glycine previously thought to be invariant can alter. The 4th residue, Xaa* , is not hydrophobic. Oxygen-containing residues (D, N, S, T, E, Q, or a cysteine thiol) or Gly (where a water molecule substitutes for the absence of a side chain) occur at all the coordination sites to Ca^{2+} (indicated *X,Y....-Z* above) except for -Y, where any residue can occur since the oxygen comes from the main chain.

General pattern of the helix-loop-helix in the EF hand:-

-E-Hyd-X_2-Hyd_2-X_2-Hyd-Oaa-X-Oaa-X-Oaa-G-X-Hyd-Oaa-X_2-Neg-Hyd-X_2-Hyd_2-X_2-Hyd-

<————————helix————————><————————loop————————><——helix——>

3.1.2. Calmodulin-Binding IQ Motif -I-Q-X_3-R-G-X_3-R-X_2-Y(W)-

Ile may be another hydrophobic residue. There are a lot of variations, and some calmodulin-binding proteins only have the motif, **-I-Q-X_3-R-G-X_3-R-** or even half the domain, often the conserved N-terminal part **-I-Q-X_3-R-**. Frequently a PEST region (rich in Pro, Glu, Ser, and Thr) may be adjacent and may be the site of cleavage by calpain (Ca^{2+} -activated protease).

3.1.3. Annexins (Ca^{2+}-Dependent Phospholipid Binding)

-K-G-Hyd-G-T-D-E-X-S/A/T/C-L/I-I/L/V/T-X-I/L/V-I/L/M-C/A/T/V-X-R-S/T-X_{26}-D/E-

The most common residues in the high-affinity Ca^{2+}-binding regions of repeats I, II and IV are shown (part of the approx 70 residue repeat). There is similarity to the phospholipase A_2 binding site. The consensus for Ca^{2+}-binding is underlined.

3.1.4. γ-Carboxyglutamic Acid (Gla)-Containing Proteins

-Gla-X_3-Gla-X-C-

3.2. Zinc Fingers

These are classically DNA-binding motifs but now are known to bind RNA and mediate protein-protein interactions. The underlined residues are zinc ligands.

$C_2 H_2$ class	-F/Y-X_1-<u>C</u>-X_{2-5}-<u>C</u>-X_3-F(Hyd)-X_5-L-X_2-<u>H</u>-X_{2-5}-<u>H</u>-
GATA /C_4 type	overall consensus/cysteine pattern -<u>C</u>-X_2-<u>C</u>-X_{17}-<u>C</u>-X_2-<u>C</u>-
GATA type	-<u>C</u>-X-N-<u>C</u>-X_4-T-X-L-W-R-R/K-X_3-G-X_3-<u>C</u>-N-A-<u>C</u>-
C_4 type -<u>C</u>-D/E/S-X-	<u>C</u>-X_3-I-X_3-R-X_4-P-X_4-<u>C</u>-X_2-<u>C</u>-
C_4 steroid finger	-<u>C</u>-X_2-<u>C</u>-X-D/E-X_5-H-F/Y-X_4-<u>C</u>-X_2-<u>C</u>-X_{15-17}-<u>C</u>-X_5-<u>C</u>-X_9-<u>C</u>-X_2-<u>C</u>-X_4-<u>C</u>-

LIM domains	(in homeobox and other proteins) -C-X$_2$-C-X$_{16-22}$-Hyd-H-X$_2$-C(H)-X$_2$-C-X$_2$-C-X$_{16-21}$-C-X$_{2-3}$-C/H/D-
Ring finger (C$_3$H$_4$ type)	-C-X-Hyd-C-X$_{9-27}$-C-X-H-X$_2$-C-X$_2$-C-X$_{6-17}$-C-P-X-C-
GAL4/fungal Zn$_2$Cys$_6$ binuclear cluster	-C-X$_2$-C-X$_3$-Baa-X$_2$-C-X$_{5-9}$-C-X$_2$-C-X$_{6-8}$-C-
TFIIS zinc ribbon	-C-X$_2$-C-X$_9$-Hyd-Q-T-R-S/T/A-X-D-E-P-X$_6$-C-X$_2$-C- (Hyd is L, I, V, or M)
Glo family	-C-X$_2$-C-X$_{16}$-C-X$_2$-C-
B box	-C-X$_2$-H-X$_7$-C-X$_7$-C-X$_2$-C-X$_5$-H-X$_2$-H-
BIR	(apoptosis inhibitory proteins) –C-X$_2$-C-X$_{16}$-H-X$_6$-C-
FYVE	(endosomal localization) -C-X$_2$-C-X$_{12}$-C-X$_2$-C-X$_4$-C-X$_2$-C-X$_{16}$-C-X$_2$-C-

3.2.1. Diacylglycerol/Phorbol Ester Binding

-His-X-Hyd-X$_{10-11}$-*Cys* -X$_2$-*Cys* -X$_3$-Hyd-X$_{(2-7)}$-A/G-Hyd-X-Cys-X$_2$-Cys-X$_4$-*His*-X$_2$-*Cys* -X$_{(6-7)}$-Cys-

(The residues underlined and italicized are the first and second zinc coordination sites, respectively)

Hyd is mainly Phe.

3.2.2. Thermolysin ("Metzincin"- Type)

- H-E-X$_2$-H-X$_2$-G-X$_2$-H-

3.2.3. Histidine Triad (HIT) family

The name derives from the general pattern -His-Hyd-His-Hyd-His-Hyd-N-X-G(E)-X$_2$-G/A-X-Q-T/S/E-V(I)-X-H-L/V(S/T)-H-Hyd-H-L/V(I)-L/I(F)- (Residues that are known to occur in unique cases are shown in parentheses). Only the 1st two histidines may in fact bind Zn.

3.3. Iron Binding Motifs

In the following three motifs, the cysteines are the iron ligands.

3.3.1. Ferredoxins and Related Iron-Sulphur Proteins

[2Fe-2S] cluster -C-X$_4$-C-X$_2$-C-X$_n$-C-
[4Fe-4S] cluster -C-X$_2$-C-X$_2$-C-X$_3$-C-P*- * there are two exceptions, with G or E at this position.

3.3.2. Reiske Iron-Sulphur Proteins

-$\underline{C}$-T-H-L-G-$\underline{C}$-L/I/V-X_n-$\underline{C}$-P-$\underline{C}$-H-G-S

3.3.3. Rubredoxin Turn (Knuckle)

-Hyd-X_3-W-X-$\underline{C}$-P-X-$\underline{C}$-G/A/D- Hyd is Leu, Ile, Met, or Val.

3.3.4. Covalent Haem-Binding Site (in C-Type Cytochromes)

-$\underline{C}$-X-X-$\underline{C}$-$\underline{H}$-X_n-$\underline{M}$-(P)-
The haem (covalently attached to the cysteines) supplies four ligands to the iron. H and M are the 5th and 6th. The 6th iron ligand is Y in cytochrome f; 5th and 6th are H in cytochrome b_5; 5th and 6th are M in bacterioferritin.

3.4. Copper Binding

The copper ligands are underlined.

3.4.1. Type I Copper Binding -$\underline{H}$-N-X_{4-39}-Y-X-Y/F-Y/F-$\underline{C}$-X-P-$\underline{H}$-X_{2-6}-$\underline{M}$-

3.4.2. Tyrosinase -W-$\underline{H}$-R-

3.5. Other Metal Binding Sites

Dehydrogenases	(Fe^{2+} or other M^{2+} with two oxidation states)
	-$\underline{H}$-X-L/I/M-X-$\underline{H}$-X_9-$\underline{H}$-G-
Nickel-dependent hydrogenases	-$\underline{C}$-X_2-$\underline{C}$-
Metallothionein	-$\underline{C}$-X-$\underline{C}$-X_3-$\underline{C}$-X-$\underline{C}$- or -$\underline{C}$-X_3-$\underline{C}$-X-$\underline{C}$-X_2-$\underline{C}$-X-$\underline{C}$-X_2-$\underline{C}$-
Phospholipase	A_2 -$\underline{C}$-$\underline{C}$-X_2-H-X_2-$\underline{C}$-
Aminopeptidase	-N-T-$\underline{D}$-A-$\underline{E}$-G-R-L- (D and E are Zn or Mn ligands)
Pyruvate kinase	-Hyd_2-X-$\underline{K}$-Hyd-$\underline{E}$-N/R-X-E/Q-G/A-

Hyd is L,I,V or M; K is active site residue and $\underline{E}$ is a Mg^{2+} ligand.

3.6. Redox Proteins

Thioredoxins	-W-$\underline{C}$-G-P-$\underline{C}$- (other residues than W, G, and P occur).
Glutaredoxin	-$\underline{C}$-P-Aro-$\underline{C}$-X_2-T/A- Aro is F, Y, or W

4. Proteases, Esterases, and Serine Active Sites

This includes lipases and acetylcholine esterases. -G-X-$\underline{S}$-X-G-

4.1. Proteases

The catalytic triads are underlined

4.1.1. Serine Proteinases

-T/S-Ala-Ala-<u>H</u>-C$_{\sim40}$-N-N-<u>D</u>-I-T/M/A-L-L-K—X$_{\sim85}$-G-X-<u>S</u>-X-G-

In the **subtilisin subclass**, the linear arrangement of the underlined active site residues is different:-

-D-....-H-....-S-

4.1.2. ATP-Dependent Serine Proteases (lon Family) -D-G-P/D-<u>S</u>-A-G/S-Hyd-T/A-L/I/V/M-

Hyd is small, non-aromatic hydrophobic.
The family also contains a P-loop; the ATP-binding motif (*see* **Subheading 6.**).

4.1.3. Cysteine (thiol) Proteinases

-G*-X-<u>C</u>-W/Y-.....-<u>H</u>-G/S/T/A/C/E-L/I/V/M-.....-<u>N</u>-S/T-W-
* One exception is E in papaya protease.

4.1.4. Aspartate Proteinases -<u>D</u>-T-G- (occurs twice)

The aspartates are generally found in the following consensus sequence:-
-F-<u>D</u>-T/S-G-S-........-I/L/V-V-<u>D</u>-T-G-

Active site aspartates in retroviruses show slight variation -L-V-<u>D</u>-T(S)-G-A-........-I/L/V-G-R-<u>D</u>-

S (in parentheses) is found very rarely in place of T.

4.1.5. Thioesterase Active Site Serine -G-X-<u>S</u>-X-G-........-G-B-<u>H</u>-X-X-L-

4.1.6. Cysteine Switch

In mammalian extracellular metalloproteinases (matrixins). The Cys in the propeptide chelates the active site Zn ion. This is the **cysteine switch** or autoinhibitor region.

-P-R-<u>C</u>-G*-X-P-D*-

Other residues, N and R respectively, are known at the positions asterisked.

5. Proteolytic Processing Motifs

5.1. Signal peptidase Cleavage Sites

-Xaa-X-Yaa-

Mainly small neutral residues, A, G, and S occur at the Yaa position - also common at position Xaa

In addition, L, I, and V occur at Xaa.

5.2. Mitochondrial Processing Peptidase

-R-X-Yaa-

5.3. Prohormone Processing

-K-R-X-; X-R-K-; or -R-R-X-

5.4. Viral Processing by Endogenous Proteinases

In picornaviruses
The recognition sequence is commonly: -T-X-G- where X is often Y, F or Q.
-G-G-A/G- (the G-A/G-G/A- motif is found in **alphaviruses**)
-G-G-Hyd- where Hyd is a hydrophobic residue, I, V, F, M, or A.

5.4.1. Aspartate Proteinases from Retroviruses

$-Hyd^1-Hyd^2-Pro-Hyd^3$
where the hydrophobic residue Hyd^1 is Leu or aromatic, Hyd^3 is a small hydrophobic residue (A, V, I, L) and Hyd^2 is more variable.

5.5. Thiol Ester

$$-P/G-X-G/S-C-G/A-E-Z-X-M/L/I/V-$$

$$\underset{S \; - \; C \, = \, O}{\lfloor}$$

Z (Glx) is the γ-thioglutamyl residue, specified in the gene by glutamine codons.

5.6. Protein Splicing (Intein) Motif

This is a post-transcriptional modification resulting from removal of an *in*ternal pro*tein* segment from a precursor and ligation of the *ex*ternal pro*tein* segments (exteins) to form a new native peptide bond.

There are conserved splice-junction residues, Cys, Ser, or Ala at the intein N-terminus; Asn at the intein C-terminus and Ser, Thr, or Cys following the downstream splice site (i.e., at the extein N-terminus).

The motif around the C-splice site including the nucleophilic Asn, is
-I/V(L,A)-L/Y/V(M)-V/A-H-N-C/S(T)-
The residues in parentheses have only one known occurence.

The general motif is -Hyd-Hyd-V/A-H*-N-C/S(T)- *A few spliced proteins have G, F, S, K, or A and the histidine (which assists in the C-terminal cleavage reaction), is elsewhere in the sequence but adjacent in the tertiary structure.

6. Nucleotide-Binding Proteins

general feature (**the P-loop**) $-G-X-X^3-X^4-X-G-(*K/Hyd)-$
There is normally a third glycine residue at positions X^3 or X^4, depending on the individual type of nucleotide binding protein. *lysine or a hydrophobic residue depending on sub-type.

6.1. Guanine Nucleotide Binding

-G-X^4-G-K-X_{40-80}-D-X_2-G-X_{40-80}-N-K-X-D-

The last four residues comprise the nucleotide specificity region. If the sequence -N-K-X-W- sequence is present, ITP as well as GTP may be utilized.

6.2. G-Protein Consensus Sequence

-K-Hyd_4-*G*-(A)-*G*-G/E-V/S-*G*-K-S-......-D-(T)-X-G-(Q)-........-N-K-X-D-

Hyd is L, I, or V. The A in the first motif may be replaced by G or N. Similarly, the other residues in parentheses are not totally invariant.

6.3. Protein Synthesis Initiation and Elongation Factors

-G-H-I/V-D-X-G-K-T/S-......-D-C/A/S/T-P-G-H.....-N-K-(M)-D-

X is frequently H, S, or A. C, V, or E may replace M.

6.4. Dinucleotide Binding Proteins

-*G*-X-*G*-X-X-*G*-Hyd- e.g., FAD-binding

6.5. Mononucleotide Binding Proteins

-*G*-X-X-*G*-X-*G*-K-S/T-

6.6. Protein Kinase Catalytic Domain Consensus Sequences

<——nucleotide binding——>   <——catalytic domain——>

-L/V/I-G-X-G-X-Y/F-G-X-V-X_{9-26}-A-X-K-X-Hyd-X_n-D-F-G-$X_{\sim20}$-*A-P-E*-

(The catalytic Lys and ATP-binding residues are underlined)

Consensus for Serine/threonine kinase specificity

-D-L-K-P-E-N- and -G-T/S-X-X-T/F-X-*A-P-E*-

Consensus for Protein-tyrosine kinase specificity

-D-L-A-A-R-N- and -P-I/V-L/R-W-T/M-*A-P-E*-

There is a variation in the Src viral tyrosine kinases first subdomain: -D-L-R-A-A-N-

6.7. DEAD- and DEAH- Box ATP-dependent Helicases

-Hyd_2-D-E-A-D-K/R/E/N- and -Hyd_3-D-E-A/L/I/V-H-C/R/E/N-

7. Protein Phosphates Active Sites and Interaction Motifs

7.1. Tyrosine-Specific Phosphatase Active Site Cysteine

The consensus is -I/V-H-C̲-X-A-G-X_2-R-S/T-G-

Residues known to occur -Hyd-H(V)-C̲-X_2-G-X_2-R-S/T-S/T/A/G-X-Hyd-

7.2. Protein Phosphatase -2C

-Hyd-Hyd-G/S/A/C-L/I/M/V-<u>D</u>-G-H-G/A/V-where D is divalent cation binding residue.

7.3. Binding Domain of Protein Phosphatase 1 Catalytic Subunit (PP1-c) Inhibitory and Targeting Proteins

-R/K-R/K-R/K-I/V-Q/S-F-

7.4. Protein Phosphatase-1 "Inhibitor 2" Type Binding Motif

-F-E-X$_2$-R-K-

8. Protein-Protein Binding

8.1. Binding Domains and Motifs in Signal Transduction Proteins

This section lists the domain itself and the short peptide motif (or phospholipid) to which it binds. The latter sequences are more easily recognized by eye. Some domains are long and variable although the identification in any protein can be easily verified by computer pattern prediction programs.

8.1.1. SH2 Domain

W-Y-F-G-X-I/L-G/S-X$_{0-5}$-R-K-D/E-A-E-X-L-L-X$_{3-11}$-G-S/T-F-L-V-R-E-S-X$_{5-7}$-Y-S-L-S-V-X$_{4-22}$-V-K-H-Y-K-I-X$_{3-25}$-Y-Y-I-X$_{4-6}$-F-X-S-L-Q-E-L-V-X-H-Y-
(This is the consensus sequence including some common alternatives only.)
SH2 domain binding general motif is -Yp-X-X-Hyd- Yp is phosphotyrosine
PI-3-kinase binding - Yp-X-X-M-; *Src kinase*, - Yp-E-E-I; *Grb2*, - Yp-X-N-X;
Crk, - Yp-X-X-P; *T-cell receptors*, - Yp-X-X-L/I-(repeated); *SYp/PTP2*, -
Yp-Hyd$_5$-.
(*See* **Subheading 1.1.2.** for some examples).

8.1.2. SH3 Domains

A-L-Y-D-Y-X-A-A-X$_{5-10}$-D/E-L-T-F-X$_{1-3}$-<u>K/R-G-D/E</u>-X-Hyd-Hyd-X-Hyd-Hyd-X$_{3-11}$-G-X- -W-W-X-A-X$_{3-9}$-G-X$_2$-G-Hyd-Hyd-P-S-N-Y-N-Y-V-
This is a consensus sequence only; some members contain an RGD, KGE, or KGD motif, underlined, that may be involved in interactions at the cell membrane (*see* **Subheading 11**).
SH3 domain binding
class I -R-X/P-L-P-(P)-X-P (more generally -P-(P)-X-P)
e.g., *PI3K*, R-X-L-P-P-R-P ; *Src*, R-(A)-L-P-P-L-P-.
class II -P-P-L-P-X-R- (bind in opposite orientation to class I)

examples of *SH3 domain-binding Abl*- -P-X-X-P-P-P-Hyd-X-P-; *Src*, -P-P-L-P-X-R- ;

amphiphysin -P-X-R-P-X-R(H)-R(H)-.

"Homer"-related synaptic proteins binding to glutamate and IP3 receptors -P-P-X$_2$-F-R- This may be related to SH3 binding motifs, e.g., X$_2$ is –S-P- in glutamate receptor.

8.1.3. WD-40 Repeats (β-Transducin Repeats)

-X$_{\sim 8}$-L/F-X-G-H-X$_3$-I/L/V-X$_2$-Hyd-X-Naa-X$_{\sim 6}$-Hyd-Hyd-S/T-G/A-G/A-X-D/N-X$_2$-Hyd-X-I/L/V-<u>W</u>(F/Y)-<u>D</u>(N)-

This is the approximately 42–43 residue repeat with the relatively conserved W and D underlined, hence the name "WD-40". Naa is uncharged.

8.1.4. PH (Pleckstrin) Homology Domain

This domain of ~100 residues binds to phosphatidylinositols, generally in the decreasing order of selectivity PtdIns-3,4,5-P$_3$ > PtdIns-4,5-P$_2$ > PtdIns-3,4-P$_2$ although the PH domains on some proteins may show similar affinity. Inositol phosphates such as Ins-1,3,4,5-P$_4$ can also bind. This is the consensus sequence plus some common alternatives only—many differences are known but the structure is highly conserved.

-V-I/V-K-E-G-Y-L-K-K-K-G-S-X$_n$-K-S-W-K-R-R-Y/W-F-V-L-R/T-D/E-X$_n$-L-S-Y-Y-K-D-S-X$_n$-P-K-G-L/S-I-D/P-L-E-N/G-I/C-Q-I/V-V-E-V-E-D-X$_n$-K-H-C-F-E-I-V-T-K/P-D-G-X$_n$-L-I/L-L-Q-A-E/S-S-E-E-E-R-E/Q-E-W-V/I-A/K-A-L/I-R/Q-R-A-I-

8.1.4.1. THE PHOSPHOTYROSINE-BINDING (PTB OR PI) DOMAIN

is a sub-class of PH domain. In Shc and IRS-1 the domain binds phosphotyrosine in the motif, -Hyd-X-N-P-X-Yp- where Yp is phosphotyrosine.

In Cbl proto-oncogene, PTB binding motif is -D-N/D-X-Yp-

There is also a 3rd type that interacts with -G-P-Yp-

8.1.5. PDZ Domain Binding

1st group-E-S/T-D-V-*CO$_2$H* ; 2nd group* -F/Y-Y-V/I/A- *CO$_2$H*

Other similar residues can occur at any of these positions. The 2nd type of PDZ domain* has a 2nd hydrophobic binding pocket, giving a generalized carboxylate binding motif, *-Hyd-Hyd-Hyd* CO$_2$H.

Examples, Neuronal nitric oxide synthase binds to –G-D/E-X-V-CO$_2$H;

Syntrophin PDZ domain binds a motif in voltage gated sodium channels -R/K/Q-E-S/T-X-V CO$_2$H; The modified motif in the GTPase-activating protein, "*RGS-GAIP*" is –S-E-A- CO$_2$H

8.1.6. WW Domain

Is composed of 38–40 residues. The most highly conserved region is that between the two tryptophan residues.

-X_2-(L)-(P)-X-(G)-**W**-(E)-X_{6-7}-(G)-X_2-F/Y-F/Y-Hyd-(N)-(H)-X-(T)-X-(T)-(T)-Ali-**W**-X_2-P-X_6-

Ali is mainly T or Q but S, C, and R also occur. Residues in brackets are not invariant but another amino acid (usually with similar properties) may occasionally occur.

WW domain binding

1st type	-(A/P)-P-P-X-Y- or -(A/P)-P-P-(A/P)-Y-
2nd type	-P-P-L-P-
3rd type	-$(Hyd)_x$-Tp/Sp-P-(R/Hyd)-where Tp or Sp are phosphorylated by a proline-directed kinase (*see* **Subheading 1.1.1.**).

8.1.7. 14-3-3 Protein Interaction Motif

-R-S-X-**S^p**-X-P- and -R-X-Y/F-X-**S^p**-X-P- where the serine, **S^p,** is phosphorylated.

8.1.8. Caveolin Binding Motif

-Aro-Xaa_4-Aro-Xaa_2-Aro- where Aro is Trp, Tyr, or Phe.

8.1.9. AKAPs (Cyclic AMP Dependent-Kinase Anchor Proteins)

-L/I-E-T(E)-A/K-S/A-K/R-L/I-V-Q/D/K-N/A-I/A/V-I-Q(E)-

The motif forms an amphipathic helix.

The binding determinants on RII regulatory subunit of cyclic-AMP-dependent kinase are in the 1st five residues.

S-H-I-Q-I-

An AKAP that is targeted to peroxisomes contains a -C-R-LCO_2H motif at its C-terminus (*see* **Subheading 12.**).

8.1.10. SOCS Box (Suppressor of Cytokines Signaling)

–Hyd/P-X-T/S/P-L-Q-H-Hyd-L-C-R-X_2-Hyd*-X_3-Hyd-X_{2-10}-Hyd-X_2-L-P-L*-P-X_2-Hyd-X-D*-Y-L-X_{1-3}-Y-

Hyd* is mainly I, L, or V; no residues are totally invariant, particularly those asterixed.

8.2. Cell Death/Apoptosis Motifs

There are a number of motifs present in the Bcl-related apoptotic regulating family of proteins.

These include a POZ motif, also known as the ***BTB domain***, of approx 120 residues.

BH3 motif

-L-R(A)-X-I*-G-D-E(D)-Hyd-D/E*-

I* is mainly I but other hydrophobics may occur; D/E* may be other aliphatic residues (N).

BH1 domain (critical for repression of apoptosis in Bcl-2) -N-W-G-R-

This is also present in galectin-3 (*see* also **Subheading 10.**).

8.2.1. TRAF Binding Motif

in a large number of proteins from diverse organisms that interact with the family of *t*umor necrosis factor *r*eceptor-*a*ssociated *f*actors.

-P-X-Q-X-T/S-

8.3. Other Protein-Protein Interaction Motifs

8.3.1. Actin-binding

-L-X_2-I-G-X_3-I/L-V-D-D-A/S/N-I-K-K-K-Hyd*-L-G-L-I-W-(T/N/Q)-(I)-I-L-

Hyd* is L, M, or F. There is some variation in hydrophobic residues.

8.3.2. Methionine Bristles in the Signal Recognition Particle

The SRP is a ribonucleoprotein that targets proteins to the mammalian ER. The 54kDa protein subunit in this complex contains a methionine-rich C-terminal M-domain in an amphipathic helix.

The consensus is -

-F-T-L-X_2-Hyd-R-X-Q-M-X_2-M-$(R/K)_2$-M-G-P-M-X_2-Hyd_2-X-M-L/I-P-G-M-G-$X_{1,2}$-M-P-.

Other hydrophobic residues (Leu or Phe) may replace Met and variations in most of the other residues are known.

8.3.3. Collagen

-G-P/A-Hyx- where Hyx is hydroxyproline or hydroxylysine.

8.3.4. Ankyrin Repeat

-G-X-T-P/A-L-H-A-A-X_7-V/A- X_2 -L-L-X_2-G-A-X_{2-6}-D/N-

8.3.5. G-Protein Coupled Receptors of the Rhodopsin-Family

-N/D-P-X_2-Y-

The motif is in the 7th transmembrane domain of members of this family and receptors with Asn (but not Asp) form complexes with the small G proteins ARF and RhoA.

9. DNA-Protein Binding

9.1. Binding to DNA Rich in AT Sequences

The binding proteins contain tandem repeats of -S-P-K-K-
(variations include -S-P-R-R-; -T-P-S-R-; -S-P-R-K-; -G-R-P-; -K-P-K- or -R-P-R-; -A-K-; -A-K-P-)

9.2. A+T Hook

Originally described in high-mobility group nonhistone chromosomal protein. Now observed in other DNA-binding proteins form a range of organisms. There are 3 types. The overall consensus is

-(K)-R(K)-X^1-R-G-R-P-X^2-K/G- where X^1 is P, K, G or R; X^2 is R, P or K.

9.3. Hox, Homeo Domain

Most proteins with this domain are transcription factors. This is a 60 residue domain that binds DNA through a helix-turn-helix structure. This would not be possible to identify by eye. Recently however, a short conserved (hexa)peptide motif as been identified upstream of the homeo domain

-I-Y-P-W-M-K-

9.4. Eukaryotic Transcription Regulation Motif

Leucine zipper

K-R-X-R-N-X^{*}-X-A_2-X-K-C-R-X-R-K-X_6-L-X_6-L-X_6-L-X_6-L-X_6-L-
<————basic motif————>   <——leucine repeat——>
* frequently K or R. The above is the particular basic motif for the fos/jun family.

9.5. Coactivator Motif

for nuclear receptor transcriptional activation
Two repeats of a motif with a core of -L-X_2-L-L-

10. Protein-Carbohydrate Binding

10.1. Sugar-Binding in Galectins

(galactoside binding lectins)
-W-G-X-E-X-R/K-   The motif in galectin-8 is -W-G-X-E-X-I-

10.2. Heparin-Binding Motif -K-K-T-R-

10.3. Glucose Transporters

GLUT1, GLUT3, and GLUT4 (high affinity glucose transporters) contain a -Q-L-S- motif that interacts with glucose at the C-1 position.

In GLUT2 (a glucose/fructose transporter) the motif is -H-V-A-

11. Cell Targeting, Adhesios, and Cell-Division Motifs

11.1. C-Terminal Endoplasmic Reticulum-Retention Signal

Vertebrates	-K-D-E-L.CO_2H
	-R-D-E-L maintains efficient retention
	-K-N-E-L, -D-K-E-L and -K-E-E-L
	maintain at least partial retention.
Drosophila *and* **C. elegans.**	-K-D-E-L
Viruses	-K-T-E-L
Budding yeast	-H-D-E-L
Fission yeast	-A-D-E-L
Plants	-K-D-E-L and -H-D-E-L.CO_2H.

11.2. Nuclear Localization Signal

-K/R-K/R-X_{10}-K/$R_{(3/4 \text{ out of } 5)}$-

11.3. Nuclear Export Signal

-L-X_3-L/I-X_2-L-X-L/I-

11.4. Nuclear Accumulation GR Motif

-R-G-R-A-P-$(X/A)_n$-(G)-G-R-G-R-G-R-(G-R-G)-A-P-$(X/A)_n$-R-G-(R-G)-.

11.5. EH (Eps Homology) Domain Binding

The EH domain itself is an 100 residue module associated with regulation of protein transport, sorting membrane trafficking and actin cytoskeleton.

There are 2 main classes of binding motif. **1st type** -N-P-F-(R)- **2nd type** -(N-P)-F-W-

11.6. Membrane Protein Sorting Signals

These are involved in targeting of transmembrane proteins to lysosomes, endosomal compartments and the trans-Golgi network. They bind to distinct sites on the clathrin endocytotic adaptor protein complexes.

11.6.1. Tyrosine Motif -Y-X-X-Hyd- where Hyd is L, I, M, or F

11.6.2. C-Terminal Di-Leucine Motif -D-X$_3$-L-L

11.7. C-Terminal Peroxisome Targeting Signal (PTS1)

-Ser-Lys-Leu.**CO$_2$H**. at the carboxy-terminus of the protein.

The original peroxisomal targeting signal was defined as -S-K-L but recently the consensus has broadened to

-**S**/**A**/C/K/N-**K**/**R**/**H**/Q/N/S/-L.**CO$_2$H**.

Residues in bold are the most common found at these positions.

The C-terminal microbody or "glycosomal import targeting signal" consensus is

-**S**/A/C/G/H/N/P-**K**/R/S/H/M/N-**L**/I/M/Y.**CO$_2$H**.

The **TPR (tetratricopeptide)** repeats of some proteins may interact with the peroxisomal targeting signal. TPR repeats are degenerate 34 residue tandem repeats and are found in a wide variety of proteins. They have been implicated in various processes including transcription control, cell-cycle regulation, mitochondrial (as well as peroxisomal) transport, protein kinase inhibition, protein folding and neurogenesis. TPR repeat consensus is

-Hyd-X$_2$-Hyd-X$_3$-A-X$_2$-Hyd$_2$-X$_4$-Hyd-X$_2$-A-Hyd-X$_2$-Hyd-X$_2$-A-Hyd-X-Hyd-X-(P)-X$_2$-

Hyd indicates a large hydrophobic residue. A is alanine or other small hydrophobic. Each repeat forms 2 antiparallel α-helices.

11.8. Nucleoporin Anchoring -X-F-X-F-G-

11.9. Ca^{2+}-Binding Parallel β-Roll Motif

(in proteins secreted by Gram negative bacteria) -G-G-X-G-X-D-

11.10. Cadherin (Cell–Cell Adhesion) Motif

-L-D-R-E-X$_4$-Y-X-L-

11.11. Cell-Adhesion Motif -R-G-D-

The motif -K-G-D- occurs in platelets and megakaryocytes, specific for the integrin GPIIb-IIIa.

11.12. Cell Division Motif

-B-X$_2$-C-X-T/E/S-X$_{1-8}$-D/E-E/D/T/S-D/E-
<——β-turn———> <——-α helix——>

12. Disulfide Bond Patterns

12.1. EGF Domain

$-C^1-X_{4-7}-C^2-X_{2-3}-G-X-C^3-X_{1-3}-D/N-X_4-F/Y-X-C^4-X-C^5-X_2-G-Aro-X_{0-20}-G-X_2-C^6-$

D/N may be β-hydroxy-Asp or Asn. Cysteines at 1,3 2,4, and 5,6 are in disulphide bonds.

12.2. Pancreatic Trypsin Protease Inhibitor (Kunitz Domain) Superfamily

$-C^1-X_8-C^2-\underline{X}^*-X_5-F/Y/W-F/Y-Y/F-X_6-C^3-X_2-F-X-Y/W-X-G-C^4-X_4-N-X-F-X-S/T-X_3-C^5-X_3-C^6-$

The six cysteines 1,6 2,4 and 3,5 are involved in disulphide bonds. *This residue forms an active site bond. For example R-X, K-X or F-X as would be expected for the proteolytic preference of trypsin or other serine proteases.

12.3. Four-Disulfide Core, WAP-Type

$-C^1-P-X_{10}-C^2-X_4-C^3-X_2-D/N-X_2-C^4- X_5-C^5-C^6-X_3-C^7-X_3-C^8-$

Cysteines at 1,6 2,7 3,5 and 4,8 are in disulphide bonds.

12.4. P-Domain (Trefoil Motif)

$-C^1-X_{3-4}-P-X_2-R-X-N/D-C^2-G-Y/F-P-G-I-T-X_2-Q/E-C^3-X_2-K/R-G-C^4-C^5-F-D-X-T/S-V/I-X_2-V/T-P/K-W-C^6-F-X-P-$

The consensus pattern is generally $-C-X_6-R-X_2-C-G-Hyd-X_{3,4}-S/T-X_3-C-X_4-C-C-Hyd-X_8-W-C-Hyd/H-$

Hyd is large hydrophobic. Cysteines at 1,5 2,4 and 3,6 are in disulphide bonds.

References

1. Aitken, A. (1995) Chap.23, Protein Chemistry Methods, Post-translational Modification, Consensus Sequences, in: *Proteins Labfax* (Price, N.C., ed.). Academic Press, San Diego, pp. 253–285.
2. Falquet, L., Pagni, M., Bucher, P., Hulo, N., Sigrist, C.J., Hofmann, K., and Bairoch, A. (2002) The PROSITE database, its status in 2002. *Nucleic Acids Res.* **30,** 235–238.
3. Attwood, T.K., Blythe, M.J., Flower, D.R., Gaulton, A., Mabey, J.E., Maudling, N., McGregor, L., Mitchell, A.L., Moulton, G., Pain, K., and Scordis, P. (2002) PRINTS and PRINTS-S shed light on protein ancestry. *Nucleic Acids Res.* **30,** 239–241.

Index